Advances in Fluvial Dynamics and Stratigraphy

Advances in Fluvial Dynamics and Stratigraphy

Edited by
PAUL A. CARLING
University of Lancaster, UK
and
MARTIN R. DAWSON
PGS Tigress Ltd, Aberdeen, UK

JOHN WILEY & SONS
Chichester · New York · Brisbane · Toronto · Singapore

Copyright © 1996 by John Wiley & Sons Ltd,
Baffins Lane, Chichester,
West Sussex PO19 1UD, England

National 01243 779777
International (+44) 1243 779777

Other Wiley Editorial Offices

John Wiley & Sons, Inc., 605 Third Avenue,
New York, NY 10158–0012, USA

Jacaranda Wiley Ltd, 33 Park Road, Milton,
Queensland 4064, Australia

John Wiley & Sons (Canada) Ltd, 22 Worcester Road,
Rexdale, Ontario M9W 1L1, Canada

John Wiley & Sons (Asia) Pte Ltd, 2 Clementi Loop #02–01,
Jin Xing Distripark, Singapore 0512

Library of Congress Cataloging-in-Publication Data

Advances in fluvial dynamics and stratigraphy / edited by Paul A.
    Carling and Martin R. Dawson.
        p.   cm.
    Includes bibliographical references and index.
    ISBN 0-471-95330-X
    1. Sedimentation and deposition. 2. Sediment transport.
I. Carling, Paul. II. Dawson, Martin.
QE571.A265 1996
551.3′53—-dc20   95–49524
                                                      CIP

British Library Cataloguing in Publication Data

A catalogue record for this book is available from the British Library

ISBN-471-95330-X

Typeset in 10/12pt Times by Mathematical Composition Setters, Salisbury, Wiltshire
Printed and bound in Great Britain by Bookcraft (Bath) Ltd
This book is printed on acid-free paper responsibly manufactured from sustainable
forestation, for which at least two trees are planted for each one used for paper production.

Contents

vi

List of contributors

James L. Best
Department of Earth Sciences, University of Leeds, Leeds LS2 9JT, UK

Andrew C. Brayshaw
BP Exploration (Alaska) Incorporated, 900 East Benson Boulevard, Anchorage, Alaska 99508–4254, USA

Gary Brierley
School of Earth Sciences, Macquarie University, Australia

Charles Bristow
Department of Geology, Birkbeck College, University of London, Malet Street, London WC1E 7HX, UK

Paul A. Carling
hysed (Hydrodynamics & Sedimentology Laboratory), Department of Geography, University of Lancaster, Lancaster LA1 4YB, UK

Patrick W. M. Corbett
Department of Petroleum Engineering, Research Park, Heriot-Watt University, Riccarton, Edinburgh EH14 4AS, UK

Glenn W. Davis
BP Exploration Operating Company Limited, BPX Technology Provision, Chertsey Road, Sunbury-on-Thames TW16 7LN, UK

Martin R. Dawson
PGS Tigress Ltd, The PSTI Technology Centre, Offshore Technology Park, Exploration Drive, Bridge of Don, Aberdeen AB23 8GX, UK

Richard D. Hey
School of Environmental Sciences, University of East Anglia, Norwich NR4 7TJ, UK

Adrian Kelsey
hysed (Hydrodynamics & Sedimentology Laboratory), Department of Geography, Lancaster University, Lancaster LA1 4YB, UK

Paul D. Komar
College of Oceanic & Atmospheric Sciences, Oregon State University, Corvallis, Oregon 97331–550, USA

Roger Kuhnle
United States Dept. of Agriculture, Agricultural Research Service, Mid South Area
National Sedimentation Laboratory, Airport Road, P.O. Box 115, Oxford,
Mississippi 38655–1157, USA

Colin P. North
Department of Geology and Petroleum Geology, University of Aberdeen, Kings
College, Aberdeen AB9 UE, UK

Peter C. B. Rainbird
School of Environmental Sciences, University of East Anglia, Norwich NR4 7TJ,
UK

Simon P. Todd
BP Exploration Operating Company Ltd, Farburn Industrial Estate, Dyce, Aberdeen
AB2 OPB, UK

Peter Whiting
Department of Geological Sciences, Case Western Reserve University, 10900
Euclid Avenue, Cleveland, Ohio 44106–7216, USA

Jon J. Williams
Proudman Oceanographic Laboratory, Bidston Observatory, Birkenhead L43 7RA,
UK

Preface

Modelling sediment transport in rivers and the interactions with the bed that give rise to bedforms has progressed remarkably within the last two decades. In similar vein the techniques of understanding the product of fluvial processes have also advanced such that ever more refined interpretations of stratigraphic sequences are now possible. There has been a significant increase in the understanding of the nature of turbulence, and an improvement in the ability to replicate mathematically simple flow–particle interactions observed experimentally. Better interpretations of fluvial sequences have developed from a growing understanding of the causative processes, often from such experimental work, and improved knowledge of the relationship between the sedimentary structures produced.

As always, additional knowledge has raised even more intriguing questions and the quest for explanation of phenomena has led to increased specialization such that few fluid dynamicists or sedimentologists can successfully span the two disciplines. As a consequence there has been a selective sharing of knowledge between related specialisms. The interest of hydrodynamicists in bedforms revolves around how geometry influences flow fields and fluvial geomorphologists have been concerned particularly with channel form; both these latter groups pay little attention to the detail of sedimentary complexes which result from a combination of geomorphic processes. This is despite the clues to past hydrodynamic behaviour which may be extracted from insightful interpretation of fluvial deposits. In contrast, whilst sedimentologists have been quick to pick up on advances in hydrodynamics which may be used, for example, to explain how bedforms are generated, some stratigraphers and engineers persist in making interpretations of fluvial sequences, based on an incomplete understanding of geomorphic processes and their relationships.

The purpose of this book is therefore two-fold. Firstly, it is not easy to keep abreast of all developments in such a broad field of interest and so we asked a number of individuals to review recent advances in their specialities in a manner which we hope is both accessible to the non-specialist and pertinent to those with more focused interests. The second objective is less tangible. We hope that the ideas contained within this compendium will promote an increased and fruitful dialogue between sedimentary geologists, geomorphologists and hydrodynamicists, indeed all those who have a common interest in fluvial systems. Consequently we anticipate that the volume will be of value to Earth scientists and engineers who work in the general area of fluvial systems as well as to graduate students requiring a source of information on the state-of-the-art. In general, although we presented each author with a brief, we have not demanded strict editorial control nor sought conformity of written style. Through this approach authors have been able to express their unfettered individual views. Cross-referencing and indexing, however, should help the reader find relevant sections. The majority of chapters in this volume were prepared in 1993. Consequently, it is inevitable that some important recent advances since this date are not acknowledged.

ACKNOWLEDGEMENTS

Primarily we must thank the many referees who provided thoughtful and thorough consideration of individual chapters. In addition, the Institute of Freshwater Ecology, Lancaster University, the Petroleum Science and Technology Institute and PGS Tigress Ltd made resources and time available to complete the task.

1 Turbulent Flow in Rivers

JON J. WILLIAMS
Bidston Observatory, Birkenhead, UK

FLUIDS AND TURBULENCE

Introduction

Turbulence in rivers, whether arising through shear with the bed or through flow separation at bluff bodies projecting into the flow, is a natural phenomenon occurring in almost all fluvial situations. Characterized by highly disordered and chaotic fluid motion over a wide range of length scales and frequencies, turbulence is linked closely with sediment transport mechanics and with mixing processes. Owing to the complexity of fluid motions involved, numerical descriptions of turbulent flow apply to only the most simplified cases and remain one of the major unsolved problems of classical physics.

However, through the use of novel experimental techniques in the past 25 years there has been a growing appreciation that turbulent fluid motion has an ordered structure and is not as random as it first appears. For example, in a series of laboratory experiments using hydrogen bubble flow visualization techniques, Kline *et al.* (1967) identified coherent structures in turbulent flow close to a boundary. These consisted principally of streaks and horseshoe vortices which were found to scale with flow properties and to maintain their physical characteristics for a considerable distance downstream. Although similar features had been noted by other researchers in the past, their role in turbulence production in the near-bed region had been overlooked. The term "fluid bursting" entered the literature to describe the sequence of violent events which give rise to turbulent kinetic energy production and dissipation in the region close to a solid surface. Whilst subsequent studies of these phenomena have led to significant advances in this difficult branch of fluid mechanics, in common with other aspects of turbulent flow, this line of experimental inquiry has turned up as many questions as it has answered.

Sedimentologists were quick to apply the new understanding of near-bed flow structure to explain such features as the parting lineations widely observed in sedimentary rocks (Bridge, 1978; Allen, 1982), in fine river sediments and beach sands in the swash zone. Further, the spatial occurrence of ripples and dunes

Advances in Fluvial Dynamics and Stratigraphy. Edited by P.A. Carling and M. R. Dawson.
© 1996 John Wiley & Sons Ltd.

observed in rivers has been related to scaled-up versions of the flow structures observed in laboratory flows (Jackson, 1976). In other work, concerned with the transport of sediment as bedload and in suspension, links between turbulent bursting processes and mass transport dynamics have been identified by Sutherland (1967), Müller *et al.* (1971), Grass (1971), Sumer & Oguz (1978), Sumer & Deigaard (1981) and Heathershaw & Thorne (1985). However, as the mathematical description of these newly discovered phenomena remains largely intractable, advances in sediment entrainment and transport models have been very limited. In most cases, turbulent structures have only provided a new conceptual framework to aid qualitative descriptions of fluid and sediment interactions.

In the first section of this chapter, through presentation of the continuity and Navier–Stokes equations, the basic principles underlying the physics of fluid motion and turbulence are introduced. Where possible, comments are restricted to situations pertaining to open channel flows. The second section concerns itself with coherent structures, fluid bursting processes and interactions between fluid and sediments. Owing to the limited range of field experiments examining the consequences of turbulence phenomena in rivers, examples from marine studies are included to illustrate aspects of this important and developing branch of fluvial hydraulics and sediment dynamics.

The Newtonian viscous fluid

All fluids are composed of molecules which are discretely spaced and in random colliding motion. In most applications, it is unnecessary to consider the motion of individual molecules, and thus to aid the study of fluid behaviour the continuum concept is invoked, in which the properties of a fluid are considered to vary continuously across the flow domain. Continuum fluid properties of special interest in fluvial studies include velocity, shear stresses, pressure, density and viscosity.

Under identical external forcing conditions different fluids are observed to flow at different rates. The differences in behaviour between, for example, water and syrup, stem from a difference in resistance to internal shearing forces and, through analogy with solid mechanics, may be thought of as internal fluid friction. In the case of Newtonian fluids, the rate of internal fluid shear is found to be proportional to the shearing force (τ). Thus,

$$\tau = \mu \, \frac{\mathrm{d}u}{\mathrm{d}x} \tag{1.1}$$

where $\mathrm{d}u/\mathrm{d}x$ is the rate of change of velocity in direction x and the constant of proportionality (μ) is the dynamic viscosity of the fluid. In rivers, where flow is unsteady, acceleration resulting from shear stresses is proportional to the dynamic viscosity and inversely proportional to the density (ρ). Thus, fluid acceleration is proportional to the ratio μ/ρ, (kinematic viscosity). In Newtonian fluids, where viscosity is independent of the applied shear stress, changes in μ are brought about principally by changes in fluid temperature and pressure.

In the case of non-Newtonian fluids studied in rheology, fluid viscosity changes in response to either increasing or decreasing rates of shear and in some cases may be time-dependent as well. The response to shearing forces by non-Newtonian geophysical fluids such as magma, high density turbidity currents, mud flows and avalanches is highly complex and relatively little is known about their behaviour. Fortunately, whilst being subject to complex motions, most fluvial situations exhibit Newtonian behaviour and thus are more readily amenable to measurement and analysis.

Shear flow instability

In a series of experiments in 1883, Osborne Reynolds examined the flow of liquid in a pipe. By injecting dye into pipe flow at low velocities, the resulting dye streak was observed to persist along the length of the test section showing that the fluid moved in parallel streamlines (or laminae). The term "laminar flow" is used to describe this flow state. Increasing the flow velocity resulted in the development of small wave-like disturbances in the dye streak which tended to break down to chaotic, irregularly convoluted patterns. At higher velocities, all the water in the pipe was observed to be coloured owing to vigorous mixing processes. The break-down of laminar flow to a turbulent flow state, where flow conditions are intermittently turbulent before returning to laminar conditions, is termed "transition" (see Robinson, 1983). Water running from a tap provides a familiar example of these different flow states. At low flows, the surface of the water is smooth and glassy in appearance, (laminar flow), whilst at higher speeds it becomes rough and irregular, (turbulent flow). In contrast, syrup remains laminar regardless of how quickly it is poured as any flow irregularities are quickly damped by strong viscous action. In virtually all geophysical flow situations, in which the processes of entrainment, erosion and transport operate, flow is turbulent.

Through dimensional analysis, Reynolds determined that for a given fluid and for a given flow situation, transition should occur at a fixed value of the ratio between inertial and friction (viscous) forces. This ratio, called the Reynolds number (Re), defines flow state in terms of fluid velocity, density and viscosity and a characteristic length scale. Thus

$$Re = \frac{\text{inertial force}}{\text{friction force}} = \frac{\rho U L}{\mu} \qquad (1.2)$$

where U and L are the characteristic velocity and length scales, respectively, of the flow being considered. In general, turbulent flow is characterized by Re values greater than 10^3. An interesting historical review of the Reynolds number is given by Rott (1990) and other aspects of shear flow instability are explored by Maslowe (1981).

In situations where gravity plays a significant role in determining flow characteristics a further important dimensionless parameter applicable to both laminar and turbulent flow with a free surface (or interface) was introduced by the naval architect William Froude. The Froude number (Fr) is defined as the ratio of inertia force to the gravity force in the form

$$Fr = \frac{\text{inertial force}}{\text{gravity force}} = \left[\frac{U}{\sqrt{lg}}\right] \tag{1.3}$$

where g is the acceleration due to gravity. When values of the term $\sqrt{lg}$, describing the velocity (celerity) of a gravity wave in shallow water, are less than unity, then gravity waves travel faster than the flow and the flow is said to be subcritical or tranquil. At values greater than unity, flow is said to be supercritical or shooting. At values equal to unity in decelerating flow, a hydraulic jump occurs and flow depth increases to restore tranquil flow. The Froude number is therefore a useful parameter to define the conditions leading to a significant change in the character of open channel flow.

Average flow conditions in channel flow

In open channel flows, such as a short reach of a river, a balance exists between the gravitational force and frictional resistance in the form

$$\tau = \rho g h \sin \beta \tag{1.4}$$

where h is flow depth and β is the water slope. Based upon the assumption that flow per unit area of a river bed is proportional to flow velocity and that the effective component of the gravity force is equal to the channel friction, Chezy proposed that

$$U'' = C(r\beta)^{1/2} \tag{1.5}$$

where U'' is the cross-sectional mean flow, C is the Chezy resistance factor, r is the hydraulic radius $= A/(2d + w)$, A, w and d are the cross-sectional area, width and depth of the channel respectively. In a similar equation proposed by Manning

$$U'' = \frac{1}{n} r^{2/3} s^{1/2} \tag{1.6}$$

where factors controlling channel resistance (e.g. sediment, vegetation etc.) are combined in a single empirical parameter n with dimensions $L^{-1/3}T$ and assuming values in the range 0.01 to 0.05. The Chezy coefficient is related to Mannings' n by $C = 1/n \ (r^{1/6})$. Whilst only strictly applicable to steady, uniform flows, these and other empirical equations are frequently still used to estimate average flow in unsteady, turbulent rivers.

The nature of turbulence

Grass *et al.* (1991) speculate on the likely conditions arising in a world where flow can only be laminar. In such a world, large rivers would attain surface flow velocities greater than $100 \ \text{km h}^{-1}$ ($28 \ \text{m s}^{-1}$) and sediment transport in suspension would not exist. Further, the smaller tractive force exerted on the bed by laminar flow would greatly reduce the population of mobile bed material, and erosion and accretion processes would be modified significantly. It is paradoxical, therefore, that through the action of chaotic, random fluid motion, river, atmospheric and tidal flow turbulence reduces flow velocities and thereby actually promotes order in natural systems.

Observations of turbulent flows in a range of different situations show that fluid motion is characterized by a wide range of velocity fluctuations and scales (e.g. McLean & Smith, 1979). This apparently random, stochastic phenomenon is not readily amenable to analytical approaches (Bradshaw, 1971; Sreenivasan, 1990). Owing to the complexity of turbulent motion, full description of flow is impossible at present and interpretation of flow visualization and detailed flow measurements present formidable difficulties.

Turbulent flow is populated by fluid eddies which have irregular, near-random motions and highly erratic paths. This results in rapid internal mixing lateral to, and in the direction of the free stream. The rather vague term "eddy" is used here to characterize a typical flow pattern. The eddies range in size between the smallest, governed by viscous dissipation, and the largest, limited by the boundary conditions. Numerous eddies exist within a volume of fluid and possess varying rotational motion or vorticity. In the case of rivers, the boundary limits may be set and scaled by a combination of the depth and width of the flow. Interactions between different sizes of eddies lead to rapid and effective mixing of fluid elements and ultimately, through a cascade of eddy sizes, to molecular diffusion. This results in enhanced rates of momentum, heat and mass transfer and provides a mechanism to oppose gravity to allow solid particles to remain in suspension in the fluid.

In common with studies of other aspects of fluid flow, turbulence research has benefited enormously from the vast increase in computational power over the last two decades (see Launder & Spalding, 1972; Rogers & Moin, 1987; Kline & Robinson, 1989; Robinson et al., 1989; Lee et al., 1990). Despite many advances, however, Grass et al. (1991) conclude that it may well be several decades yet before it is possible to simulate numerically even the simplest geophysical flow situation. These problems are compounded in sediment-laden river flows where characterization of multiphase flow and kinematic descriptions are difficult and where the relatively low concentrations of suspended particles raise questions regarding the suitability of the continuum concept. Further, non-linear coupling and feedback between sediments and turbulent structure must be accounted for in any physically realistic hydrodynamic and sediment transport models (Lyn, 1986).

Equations of motion and Reynolds stresses

When considering any fluid flow problem, the analytical solution involves the determination of the three components of velocity, u, v and w, and the static pressure in terms of a spatial coordinate system (e.g. Cartesian x, y and z, respectively) at a given time t. Using the distribution of static pressure over the boundary enables evaluation of pressure forces whilst shearing forces may be determined from the velocity gradients at the surface. Together, these forces constitute the resultant force acting at a boundary or on an obstacle in the flow. In the simplest terms, the equations governing fluid motion may be derived by equating the net rate of outward transfer of mass and momentum, respectively, through the faces of an infinitesimal, fixed volume of fluid to their net sources in the control volume. Whilst the resulting equations for mass conservation (the continuity equation) and for momentum conservation (the Navier–Stokes equations) may appear to be mathematically complicated, they are actually based on relatively simple principles concerned with the conservation of mass and momentum stated in Newton's second law.

The equation of continuity expresses the fact that for unit volume of a fluid, a balance exists between masses entering and leaving per unit time and the changes in density. For incompressible fluid, where density (ρ) is constant, the equation of continuity is expressed in Cartesian component notation as

$$\frac{\partial u}{\partial x} + \frac{\partial v}{\partial y} + \frac{\partial w}{\partial z} = 0 \qquad (1.7)$$

where u, v and w are velocity components in directions x, y and z, respectively. The Navier–Stokes equations of fluid motion below are exact mathematical statements of dynamic conditions in a viscous fluid with constant density. With acceleration terms written out fully, the Navier–Stokes equations take the following form

$$\rho\left(\frac{\partial u}{\partial t} + u\frac{\partial u}{\partial x} + v\frac{\partial u}{\partial y} + w\frac{\partial u}{\partial z}\right) = X - \frac{\partial p}{\partial x} + \mu\left(\frac{\partial^2 u}{\partial x^2} + \frac{\partial^2 u}{\partial y^2} + \frac{\partial^2 u}{\partial z^2}\right) \qquad (1.8a)$$

$$\rho\left(\frac{\partial v}{\partial t} + u\frac{\partial v}{\partial x} + v\frac{\partial v}{\partial y} + w\frac{\partial v}{\partial z}\right) = Y - \frac{\partial p}{\partial y} + \mu\left(\frac{\partial^2 v}{\partial x^2} + \frac{\partial^2 v}{\partial y^2} + \frac{\partial^2 v}{\partial z^2}\right) \qquad (1.8b)$$

$$\rho\left(\frac{\partial w}{\partial t} + u\frac{\partial w}{\partial x} + v\frac{\partial w}{\partial y} + w\frac{\partial w}{\partial z}\right) = Z - \frac{\partial p}{\partial z} + \mu\left(\frac{\partial^2 w}{\partial x^2} + \frac{\partial^2 w}{\partial y^2} + \frac{\partial^2 w}{\partial z^2}\right) \qquad (1.8c)$$

where p is pressure. Thus with known body forces X, Y and Z, there are four unknown terms, u, v, w and p. Whilst analytical solutions to 1.8 are possible for some flow situations (e.g. laminar pipe flow and some boundary layers), verification of the Navier–Stokes equations is generally only possible through experimental work. Full derivation of these equations is beyond the scope of this chapter and readers wishing to explore this subject in greater detail should refer to the full derivation given by Batchelor (1970), Cebeci & Bradshaw (1977) and Schlichting (1979). In the brief summary below, only approximate solutions for turbulent flow are described.

When considering the instantaneous properties of turbulent flow, principles applied when describing laminar flow (with constant viscosity and density) are still valid since the smallest eddies are still several orders of magnitude greater than the mean free path of fluid molecules. However, whilst possible in theory, the large range of eddy sizes present in turbulent flows at even moderate Reynolds numbers makes calculations of instantaneous values impossible with existing computer facilities. Practical solution to this problem is possible, however, by use of time-averaged versions of the continuity and Navier–Stokes equations. By substitution of instantaneous values by the sum of the average and fluctuating velocity components the continuity equation (1.7) may be written in Cartesian coordinates as

$$\frac{\partial U}{\partial x} + \frac{\partial V}{\partial y} + \frac{\partial W}{\partial z} = \frac{\partial \bar{U}}{\partial x} + \frac{\partial \bar{V}}{\partial y} + \frac{\partial \bar{W}}{\partial z} + \frac{\partial u'}{\partial x} + \frac{\partial v'}{\partial y} + \frac{\partial w'}{\partial z} = 0 \qquad (1.9)$$

where U, V and W are the three components of velocity and $\bar{U}$, $\bar{V}$, $\bar{W}$ and u', v', w' are the time-averaged and fluctuating velocity components, respectively. Averaging terms in the continuity equation over a time period gives

$$\frac{\partial \bar{u}'}{\partial x} + \frac{\partial \bar{v}'}{\partial y} + \frac{\partial \bar{w}'}{\partial z} = 0 \qquad (1.10)$$

hence

$$\frac{\partial \bar{u}}{\partial x} + \frac{\partial \bar{v}}{\partial y} + \frac{\partial \bar{w}}{\partial z} = 0 \qquad (1.11)$$

and

$$\frac{\partial u'}{\partial x} + \frac{\partial v'}{\partial y} + \frac{\partial w'}{\partial z} = 0 \qquad (1.12)$$

A similar procedure can be applied to the Navier–Stokes equations by substitution of the instantaneous velocity components by the sum of the average and fluctuating values, expansion of the differentials and by taking the mean over some time interval (the mean of the fluctuating component must be zero). This yields

$$\rho\left(\bar{u}\frac{\partial \bar{u}}{\partial x} + \bar{v}\frac{\partial \bar{u}}{\partial y} + \bar{w}\frac{\partial \bar{u}}{\partial z}\right) = -\frac{\partial \bar{p}}{\partial x} + \mu\nabla^2\bar{u} - \rho\left(\frac{\overline{\partial u'^2}}{\partial x} + \frac{\overline{\partial u'v'}}{\partial y} + \frac{\overline{\partial u'w'}}{\partial z}\right) + X \qquad (1,13a)$$

$$\rho\left(\bar{u}\frac{\partial \bar{v}}{\partial x} + \bar{v}\frac{\partial \bar{v}}{\partial y} + \bar{w}\frac{\partial \bar{v}}{\partial z}\right) = -\frac{\partial \bar{p}}{\partial y} + \mu\nabla^2\bar{v} - \rho\left(\frac{\overline{\partial u'v'}}{\partial x} + \frac{\overline{\partial v'^2}}{\partial y} + \frac{\overline{\partial v'w'}}{\partial z}\right) + Y \qquad (1,13b)$$

$$\rho\left(\bar{u}\frac{\partial \bar{w}}{\partial x} + \bar{v}\frac{\partial \bar{w}}{\partial y} + \bar{w}\frac{\partial \bar{w}}{\partial z}\right) = -\frac{\partial \bar{p}}{\partial z} + \mu\nabla^2\bar{w} - \rho\left(\frac{\overline{\partial u'w'}}{\partial x} + \frac{\overline{\partial v'w'}}{\partial y} + \frac{\overline{\partial w'^2}}{\partial z}\right) + Z \qquad (1,13c)$$

where ∇^2 is Laplace's operator. The left-hand terms in eqns 1.13a–1.13c are identical to the steady-state Navier–Stokes equations above (eqns 1.8a–1.8c) if the velocity components u, v, w are replaced with time average values. In contrast, however, eqns 1.13a–1.13c also contains terms u', v' and w' which describe the turbulent fluctuations in velocity about a mean value.

Eqns 1.13a–1.13c are often called Reynolds equations as they contain the turbulence terms generated from convective accelerations which give rise to apparent or Reynolds stresses (e.g. $-\rho\overline{u'w'}$), collectively termed the Reynolds stress tensor. Reynolds stresses have opposite signs to their respective velocity gradients and in most circumstances are several orders of magnitude greater than their laminar counterparts, reflecting the enhancement of momentum transfer (see Bernard & Handler, 1990).

In order to solve the time-averaged Reynolds equations for turbulent boundary layer flow it is necessary to model the behaviour of the Reynolds stress tensor. Fortunately, in shear layer flows, the principal Reynolds stress resulting from a velocity gradient normal to the flow is $-\rho \overline{u'v'}$ and other terms such as $-\rho \overline{u'^2}$ and $-\rho \overline{v'^2}$ in eqns 1.13a and 1.13b can be ignored. Using simplifications based on Prandtl's boundary layer approximation, the Navier–Stokes equations for thin, two-dimensional turbulent flow reduce to (Cebeci & Bradshaw, 1977; Tritton, 1977)

$$\rho \left(u \frac{\partial \bar{u}}{\partial x} + v \frac{\partial \bar{u}}{\partial y} \right) = -\frac{\partial \bar{p}}{\partial x} + \frac{\partial}{\partial y} \left(\mu \frac{\partial \bar{u}}{\partial y} - \rho \overline{u'v'} \right) \tag{1.14}$$

Equation 1.14 provides a convenient method to compute boundary shear stress and is applicable to many situations encountered in geophysical flows.

Eddy viscosity and mixing length concepts

In an analogy with the viscosity coefficient in Stokes' law for laminar flow (eqn 1.1) an eddy viscosity coefficient (ε) is frequently used for turbulent flows to equate the average velocity gradient with shear stress (Boussinesq hypothesis), i.e.

$$\tau = -\rho \overline{u'w'} = \varepsilon \frac{\mathrm{d}\bar{U}}{\mathrm{d}z} \tag{1.15}$$

However, unlike μ which is determined by the physical properties cf the fluid, ε is dependent principally upon the flow Reynolds number. An analogy between the relatively large-scale, near-random motion of turbulent eddies and the motion of fluid molecules led Prandtl to suggest that the mean distance travelled by typical eddies before mixing with the surrounding flow can be characterized by a mixing length dimension similar in principle to the mean free path of fluid molecules. Hence

$$\tau = -\rho \overline{u'w'} = \rho L^2 \left| \partial \bar{U}/\partial z \right| \overline{\partial U/\partial z} \tag{1.16}$$

where L is the mixing length which is determined experimentally. Whilst not being strictly valid in physical terms, mixing length theory provides empirical equations that can be applied in a wide range of flow situations. More elaborate formulations are given by Launder & Spalding (1972).

The boundary layer

In order to simplify mathematical flow solutions in classical hydrodynamics, fluids are assumed to be ideal. Such fluids are assumed to be incompressible and to offer no resistance to flow. Whilst the first assumption is broadly acceptable when considering river flows, the second is not and thus inviscid theory must be abandoned when considering problems involving fluvial flows and sediment interactions.

At the boundary between a flowing fluid and a solid surface, fluid elements do not slip, irrespective of the external flow conditions (within the range of temperatures and pressures for which the fluid may be treated as a continuum). A velocity gradient exists, therefore, between the boundary and the unimpeded free stream flow at some distance away from the surface. The relatively narrow region occupied by retarded flow is called the boundary layer and extends a distance δ above a surface. Flow within this layer may be laminar, transitional or fully turbulent and, in certain circumstances may be amenable to analysis using the Navier–Stokes equations of fluid motion. In conditions giving rise to the entrainment of particles in rivers, in the atmosphere and in marine situations, flow in the boundary layer is nearly always turbulent.

In addition to the flow within the boundary layer it is also necessary to consider flow around individual roughness elements (i.e. sand grains, pebbles, boulders, etc. in a river). Using a term expressing the intensity of flow turbulence (U_*) the character of the flow acting upon grains is determined by the grain Reynolds number (Re^*) defined as

$$Re^* = U_* d_g / v \qquad (1.17)$$

where U_* is shear velocity $= (\tau/\rho)^{0.5}$, d_g is a representative measure of grain dimensions and v is kinematic viscosity $= \mu/\rho$. At Re^* values greater than 3.5, grains shed eddies from their lee face and contribute directly to turbulence generation. Such flows, typical of most geophysical situations, are described as being hydraulically rough and can lead to a situation where single grains are exposed to much higher bed shear stresses than their fellows (Bagnold, 1941). At Re^* less than 3.5, a semiviscous, non-turbulent layer of fluid clings to each grain and flow becomes smooth. Drag is distributed more evenly over the bed and thus grains are less easily disturbed. However, these conditions are rarely encountered in most fluvial situations.

The turbulent boundary layer is characterized by highly distorted streamlines where individual fluid elements are subjected to chaotic random motion. The average velocity profile for steady flow within the boundary region is plotted on logarithmic axis in Figure 1.1a. Here, average velocity is defined by

$$\bar{U} = 1/\Delta t \int_0^{\Delta t} U \, dt \qquad (1.18)$$

where $\bar{U}$ is the time-average velocity, Δt is the averaging time and z is the distance from the boundary. In Figure 1.1a both average velocity and distance from the boundary are normalized using friction velocity (U_*) and kinematic viscosity (v). Three regions of flow within the turbulent boundary layer are shown in Figure 1.1a and include (a) the viscous sublayer, (b) the log-law region, and (c) the outer region. In reality, however, in the smooth transition of velocity from the surface to the free stream, there is no strong demarcation between these regions of the flow.

The viscous (or linear) sublayer, as the name implies, is a region of flow, extending less than 50 $v/\bar{U}_*$ from the boundary, in which flow is essentially laminar

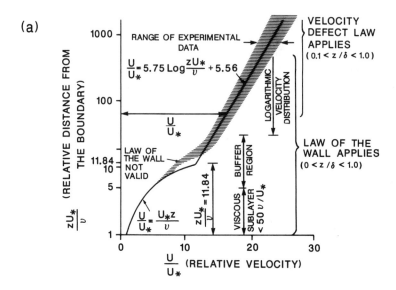

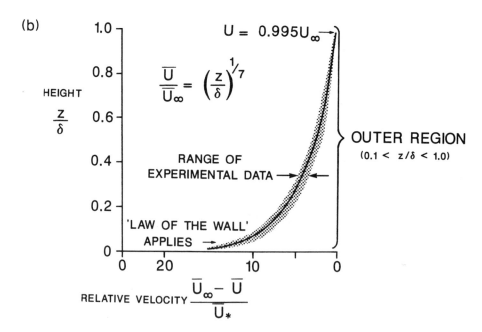

Figure 1.1   The average velocity profile in a turbulent boundary layer

and shear stresses are transmitted to the surface entirely through the action of molecular viscosity (Coles, 1978). The velocity distribution in this region is linear and conforms to the relationship

$$\bar{U}/\bar{U}_* = \bar{U}_* z/v \qquad (1.19)$$

Above this thin layer is a region where velocity increases linearly with the logarithm of height so that

$$\bar{U}(z)/\bar{U}_* = A_k \ln(z/z_0) + B_k \qquad (1.20)$$

where z_0 is a parameter describing the surface roughness (generally called the roughness length), and A_k and B_k are experimentally determined constants. Great care must be taken when attempting to fit a logarithmic profile to observed fluvial data owing to the disturbing effects of positive and negative flow accelerations, variability in upstream roughness, stratification and zero datum errors associated with the siting of current measurement devices. These factors are considered further by Dyer (1986).

At $z = 0$, the Reynolds shear stress is zero owing to the no-slip boundary condition (i.e. $-\rho u' v' = 0$). Away from the boundary, turbulent fluid motion generates Reynolds stresses which are large compared to viscous forces. The total shear stress

$$\tau = \mu \frac{\partial u}{\partial z} - \overline{\rho u' v'} \qquad (1.21)$$

close to the boundary cannot vary significantly without there being a large mean acceleration which can only be sustained by an unlikely flow distribution. Thus, both viscous stresses and Reynolds stresses vary rapidly away from the wall and, through their combined effect, maintain a region of constant stress.

The velocity distributions described by eqns 1.19 and 1.20 are often called the law of the wall and meet at a $z\bar{U}_*/v$ value of 11.8. Figure 1.1a shows, however, that in a zone $5 < z\bar{U}_*/v < 30$, called the buffer region, neither equation defines correctly the velocity distribution. Together, the viscous sublayer, buffer layer and log-law region comprise the inner layer of a turbulent boundary layer flow.

Above the log-law region, the velocity distribution is described by the velocity defect law which extends from $z/\delta > 0.1$ to the free stream (i.e. 80–90% of the boundary layer thickness). Here the velocity distribution follows the relationship

$$(\bar{U}_\infty - \bar{U})/\bar{U}_* = f_\psi(z/\delta) \qquad (1.22)$$

where U_∞ is the mean free stream velocity (Figure 1.1b). The function f_ψ is affected greatly by the streamwise pressure gradient $(\partial \bar{p}/\partial x)$. In most geophysical flows, however, the velocity profile in this region may be approximated by the 1/7 power law in the form

$$\bar{U}/\bar{U}_\infty = (z/\delta)^{1/7} \qquad (1.23)$$

Figure 1.2, showing normalized turbulence energy production as a function of distance from the wall (z), indicates that the inner region of the boundary layer accounts for at least 80% of total turbulence production. This dominance of the

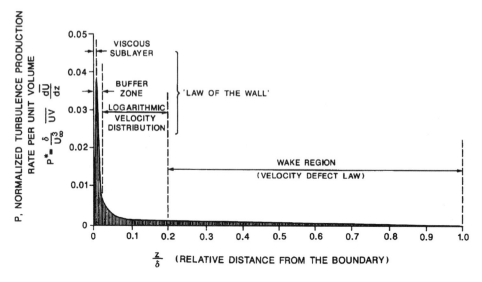

Figure 1.2   Production of turbulent kinetic energy (after Klebanoff, 1956)

inner region in the production of turbulent energy within a wall-bounded shear flow is demonstrated elegantly by Favre *et al.* (1966) who showed that a turbulent boundary layer could be relaminarized by removing the inner region by suction. The mechanisms giving rise to turbulent energy production within this region are considered below.

Analysis of turbulence

It is clear from the sections above that direct analysis of flow turbulence is limited to all but the simplest flow situations and limited by computational power. A useful alternative approach is to consider the statistical properties of turbulence. The value of the mean velocity and the spatial and temporal resolution of turbulent motion for a given flow situation depends on a number of interacting and, in some circumstances, conflicting factors which include the record length, stationarity, sampling variability, sampling interval and sensor response. These factors are discussed in detail by Soulsby (1980). For steady flow (i.e. the mean does not vary through time) the mean is defined by eqn 1.18. When turbulent flow is not stationary (e.g. in the tidal reaches of a river), ensemble averaging, filtering and trend removal techniques are used to estimate mean flow conditions.

Whilst $\bar{U}$ values give a reference velocity, they do not provide information on the amplitude of turbulent velocity component fluctuations about the mean. This statistic, often referred to as turbulence intensity, is obtained by squaring all velocity values to eliminate negative signs, calculating the mean and then taking the square root to give the root mean square (r.m.s.) values of turbulent fluctuations, $\hat{u}$. Thus

$$\hat{u} = \left(1/\Delta t \int_0^{\Delta t} (u)^2 \right)^{1/2} \tag{1.24}$$

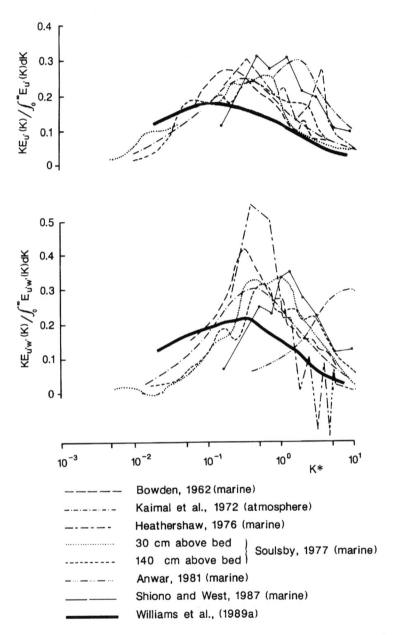

Figure 1.3 Normalized, wavenumber-weighted turbulence spectra and cospectra for u' and $u'w'$ from marine, estuarine, fluvial and atmospheric boundary layers

In many applications it is useful to estimate the scale of eddies in a given turbulent flow regime. Using an extension of the Eulerian version of Taylor's (1935) technique, the duration of the macroscale of turbulence (T_a) can be estimated by integration of the normalized autocorrelation curve so that

$$T_a = \int_0^\infty R(\gamma)\mathrm{d}\gamma \qquad (1.25)$$

where $R(\gamma)$ is the normalized autocorrelation function for a given flow component and γ is the lag. The product of T_a and mean turbulent flow velocity gives an estimated eddy length scale.

An alternative technique, widely used in studies of turbulence, utilizes the fast fourier transform (FFT) to obtain frequency spectra from Eulerian flow measurements. Large eddies (the macroscale) attain a maximum size determined by the boundary layer thickness. In contrast, the minimum dimensions of small eddies (the microscale) are determined principally by viscosity. At a fixed point in the flow the velocity fluctuations corresponding to small eddies will be very rapid owing to their small physical size. Large eddies, on the other hand, will give rise to lower frequency velocity modulation. Therefore, a sensitive velocity probe in turbulent flow will record a wide range of frequencies through time. However, since the apparent frequency of turbulence measured at a fixed point will vary as a function of flow velocity, use of the frequency spectrum does not allow direct comparison between different flow regimes. To achieve this, wavenumber spectra are obtained by multiplying the reciprocal of frequency (i.e. period) by $U/2\pi$. As kinetic energy is a function of flow velocity, analysis of turbulence records in the frequency domain using an FFT to give energy spectra provides useful information on the distribution of kinetic energy according to eddy size. The product of average flow velocity and the reciprocal of spectral wavenumber indicates a typical eddy length scale.

The wavenumber spectrum has many useful applications in the study of turbulence. These include the calculation of bed shear stress through consideration of turbulent energy dissipation rates (e.g. Huntley, 1988), calculation of typical velocity fluctuations at a given length scale and determination of the most appropriate frequency response for instruments measuring turbulence. In Figure 1.3 energy spectra and cospectra for u' and $u'w'$ respectively, obtained in the near-bed region of boundary layer flow (i.e. $<0.1\delta$) are shown in the conventional normalized wavenumber weighted form. Close similarities between atmospheric, marine and fluvial boundary layer flows are clearly evident. Turbulent spectra from rivers are considered in detail by Ishihara & Yokosi (1967).

COHERENT STRUCTURES AND FLUID BURSTING

Introduction

In general, near-bed flow phenomena described below have been studied in controlled laboratory conditions where flow is steady and generally less than one metre deep and the solid boundary is hydraulically smooth. Further, flow stratification induced by suspended sedimentary particles and density effects is not considered. Clearly, in natural situations, whether in air, in rivers or the sea, variations in density brought about by temperature and salinity modify fluid behaviour significantly and high concentrations of suspended solids may induce significant deviation from ideal Newtonian fluid behaviour. Care must be exercised, therefore, when drawing analogies between laboratory results and natural systems. Further, attempts to explain natural phenomena using conceptual models for simple fluid flow must be approached with caution.

Coherent structures

Whilst most theories attempting to explain turbulent motion concern themselves with the consequences of turbulence and not with its nature, the recent advances in understanding of coherent structures in turbulent flow go some way in redressing this imbalance (cf. Clifford et al., 1993). In the simplest terms a coherent structure in turbulent boundary layer flow may be thought of as an eddy (or similar flow configuration) that maintains its characteristics within the main body of the flow for a considerable distance downstream (Laufer, 1975; Cantwell, 1981; Wygnaski & Peterson, 1987; Liu, 1989; Robinson, 1991; Wark & Nagib, 1991). More specifically, coherence infers vorticity and thus observations of intermittency in a given flow through the measurement of velocity alone is not necessarily indicative of coherent structures. Similarly, analyses of linear momenta in a range of different flow situations often give statistically significant correlations across the flow width even though no coherent structure is present in the flow (Hussain, 1983). Eddy shedding phenomenon such as the familiar Kármán vortex sheets serve as good illustrations of how apparent coherence may be brought about.

Well documented large coherent structures identified by many workers include vortex rings, loops, spirals, etc. These structures, in turn, are probably composed of complex arrangements of smaller scale features referred to collectively as coherent substructures. These include hairpin vortices, typical eddies, pockets and streaks. In all cases they are characterized by a high degree of coherent vorticity, Reynolds stress, heat and mass transport and turbulence production. As the basis for a conceptual model, these structures can be visualized as being embedded in a non-coherent (i.e. phase-random) flow. A comprehensive summary of length scales, convection velocities, persistence distances and duration of coherent structures, including streamwise vortices, energetic near-wall eddies, typical eddies and large-scale eddies, is given by Cantwell (1981) and Gyr (1983).

Whilst being implicit in mixing length theory, the presence of large organized motions in turbulent flow was first suggested by Townsend (1956) and investigated in detail by Grant (1958), and Keffer (1965). Through the pioneering work of Corrsin (1957), Hama & Nutant (1963) and Kline et al. (1967), quasi-deterministic and quasi-periodic structures were identified in turbulent flow using novel flow visualization based on the hydrogen bubble technique. These and other subsequent studies aided in the advancement of theories to explain the production and dissipation of turbulent kinetic energy close to the boundary through turbulent bursting models (see below).

At the macroscale, coherent structures are responsible for the transport of significant mass, heat and momentum. They may achieve this without necessarily being energetic themselves. Coherent structures cannot overlap spatially and thus interactions between adjacent structures are inherently non-linear (Hussain, 1983) and involve tearing and pairing to form new structures which retain the characteristics of both parents (Hussain & Clark, 1981).

Coherent structures in turbulent boundary layers

Although progress in the understanding of coherent motion in free shear flows (e.g. jets, wakes, etc.) has advanced significantly over the last decade, wall-bounded shear flows still pose many unsolved problems. These problems are further exacerbated when considering turbulent boundary layers developed over hydraulically rough, irregular surfaces such a bedforms and large clasts in rivers. Even in the laboratory situation, where sophisticated flow visualization and measurement techniques are available, progress in understanding near-bed flow has been limited owing to the thinness of the active flow zone. As described above, the existence of the inner and outer regions and the viscous dominated region adjacent to the surface in turbulent boundary layers gives rise to three distinct length and time scales. This inherent complexity gives rise to many kinds of coherent motions which may or may not be genetically related to common parents.

Following a series of innovative wind tunnel studies and stimulated by the ideas of Theodorsen (1952) and Black (1966, 1968), Head (1978, 1979) and Head & Bandyopadhyay (1981) consider vortex loops to be the fundamental structures in turbulent wall-bounded shear flows. These structures are illustrated schematically in Figure 1.4. In a conceptual model the boundary layer is visualized as being composed of many stretched vortex loops which are essentially straight over a substantial portion of their length, limited in extent only by δ and inclined at approximately 45° to the bed. It is postulated that distinct vortex features such as loops, elongated loops, and elongated hairpins identified in turbulent flow constitute the fundamental elements of eddies, and using the elegant analogy of Kovasznay (1970) are considered to be the "sinews and muscles of turbulence". On average, when combined in complex arrays, these features are considered to give rise to the coherent features such as horseshoe vortices defined by Cantwell (1981). Excellent reviews of vortical structures and concepts are given by Wallace (1985), Kim & Moin (1986), Acarlar & Smith (1987), Gulliver & Halverson (1987) and Robinson (1989).

Observations within the viscous sublayer of bounded shear flows by Kline et al. (1967), Grass (1971, 1982) and many other researchers has revealed a streaky structure composed of spanwise alternating regions of relatively high and low speed fluid arranged apparently at random (e.g. Grass, 1971); Streaks exhibit significant streamwise vorticity and under normal circumstances are found to be unstable. Various laboratory experiments conducted in a limited range of Reynolds numbers have determined average streak spacing (λ_s) to be

$$\lambda_s U_*/\nu \equiv y_s^+ \approx 100 \tag{1.26}$$

where y_s^+ is a dimensionless number in a range 60–200 (Hussain, 1983), derived from inner flow variables (Kline et al., 1967).

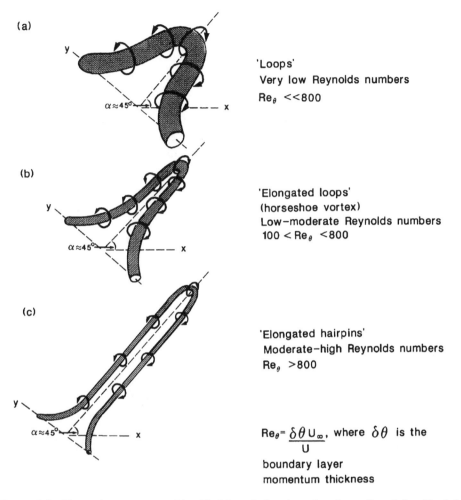

(a)

'Loops'
Very low Reynolds numbers
$Re_\theta \ll 800$

(b)

'Elongated loops'
(horseshoe vortex)
Low-moderate Reynolds numbers
$100 < Re_\theta < 800$

(c)

'Elongated hairpins'
Moderate-high Reynolds numbers
$Re_\theta > 800$

$Re_\theta = \dfrac{\delta\theta U_\infty}{U}$, where $\delta\theta$ is the boundary layer momentum thickness

Figure 1.4   Vortex loop structures identified in turbulent boundary layer flow (after Head &
Bandyopadhyay, 1981)

Various analytical techniques suggest that the streaky pattern results from
containment of stagnant fluid between pairs of counter-rotating longitudinal
boundary vortices but their precise origin and their role in the transport of
momentum, heat and mass remains elusive. Brown & Thomas (1977) and Coles
(1978) have proposed that streaks result from a Taylor–Götler instability
mechanism which give rise to random vortex pairs. Head & Bandyopadhyay (1981)
conceive streaks to be the trailing leg of loops which diffuse into each other
periodically in a complex process which cancels their respective vorticity. In such
cases, the vortex tips will persist and migrate in the outer layer as no such
mechanism exists for the destruction of transverse vorticity. Indeed Grass (1982)
suggests that the slowly rotating motion associated with the large-scale eddy
structures in the outer region may be a manifestation of enclosed fluid circulation
around surviving vortex tips.

Turbulent bursting processes

In studies of streaks, Kline *et al.* (1967) also observed the violent breakdown of the semiorganized near-bed flow during violent events they termed bursting. Subsequent studies by Kim *et al.* (1971), Nychas *et al.* (1973), Brodkey *et al.* (1974), Offen & Kline (1975), Blackwelder & Eckelmann (1979) and Lu & Smith (1991) have provided evidence with which to construct a conceptual model of the bursting process. In this model, the occurrence of bursts is equated with the lifting and stretching of vortex loops from the surface. Rapid elongation of the loops occurs in the steep velocity gradient and violent mixing ensues. During this stage of the

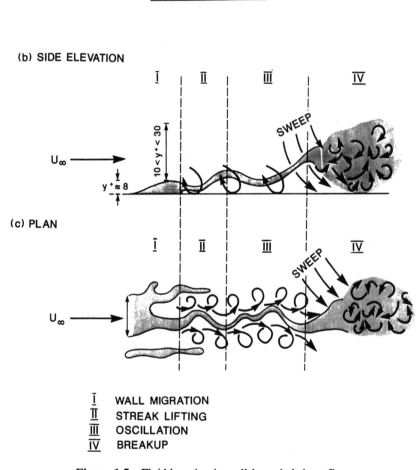

Figure 1.5   Fluid bursting in wall-bounded shear flow

bursting cycle, additional vortices are generated which extend to occupy the entire boundary layer region (Bogard & Tiederman, 1987). Following the burst (or ejection), fluid from adjacent regions rushes in to replace the displaced fluid in a poorly understood process called a sweep. To an observer travelling with the flow, the sweep appears as a local recirculation cell beneath the burst. At a fixed point, however, the sweep appears as a velocity deficit close to the bed with enhanced velocity over a small streamwise distance just away from the boundary. The sweep is therefore a mechanism by which near-boundary vorticity is intensified by lateral spanwise stretching and by which new vorticity may be generated. The turbulent bursting concept is illustrated schematically in Figure 1.5.

Fluid bursting is typically identified and analysed using conditional sampling techniques and a classification scheme for Eulerian flow measurements away from the boundary developed by Lu & Willmarth (1973) called quadrant analysis (Figure 1.6). Here quadrants one and two define fluid moving away from the boundary and quadrants three and four define fluid moving towards the boundary. Quadrants one and three give a negative contribution to kinematic stress ($u'w'$) and hence a positive Reynolds stress. Conversely, quadrants two and four give rise to a positive kinematic stress and thus result in a negative Reynolds stress close to the boundary (i.e. bed shear stress). All available observations confirm that as expected quadrants one, corresponding to bursts, and three, corresponding to sweeps, provide the most significant contributions to total Reynolds stress in the near-bed region of the flow. Studies by Corino & Brodkey (1969), Willmarth & Lu (1972) and Williams *et al.* (1989a) show that whilst being infrequent and brief (occurring for less than 20% of the total time), bursts and sweeps contribute the bulk of Reynolds stress (more than 80% of the total stress). This intermittency is a principal characteristic of fluid bursting and is reported from a range of observations in geophysical flows by Gordon & Witting (1971), Heathershaw (1974, 1979), Anwar (1981), Anwar & Atkins (1980), West & Oduyemi (1989) and Shiono & Knight (1990). The distribution of stress through time is considered by Williams & Tawn (1991) using ARIMA and GEV modelling approaches.

EVENT		U'	W'	U'W'	Fluid motion relative to the mean flow
EJECTION (BURST)	✳	<0	>0	<0	⬆+W' ⎿ -U'
INRUSH (SWEEP)	✳ ✳	>0	<0	<0	+U' ⬇ -W'
INWARD INTERACTION		<0	<0	>0	-U' ⬇ -W'
OUTWARD INTERACTION		>0	>0	>0	⬆+W' +U'

Principal events associated with sediment transport: ✳ suspended load; ✳✳ bedload

Figure 1.6 Quadrant analysis of turbulent motions contributing to total Reynolds stress

Mechanisms giving rise to the periodic breakdown of low speed streaks have remained largely speculative despite many attempts to understand the process. Brown & Thomas (1977) and Guezennec *et al.* (1989) suggest that momentary increases in bed shear stress brought about by the downstream passage of horseshoe vortices in the outer region of the boundary layer may be sufficient to trigger bursting. In this case the leading edge of the vortex would give rise to a sweep and the trailing edge would initiate an ejection. Further, despite many efforts it is still unclear whether the loops in the viscous sublayer result from a concentration of streamwise vorticity as suggested by Lighthill (1963) or a warping of the transverse vorticity. It could conceivably result from the joint action of both.

Visualization and theoretical modelling studies conducted by Smith *et al.* (1989) and Walker (1990) demonstrate the formation of low speed streaks in the wall region and the subsequent violent flow perturbations characterizing bursts. Such events are found to be triggered under the single limb of a horseshoe-like vortex and thus vortex symmetry, implied in the model of Kline *et al.* (1967) and other researchers, is not necessary for the production of turbulent kinetic energy in the boundary layer by this mechanism. Further, studies reported by Grass *et al.* (1991) demonstrate clearly that similar mechanisms operated irrespective of the surface roughness.

In contrast with scaling parameters for streaks, the spatial occurrence of bursts in controlled laboratory conditions has been found to scale with outer flow variable in the form (Rao *et al.*, 1971)

$$U_\infty T_b / \delta \equiv T^+ \simeq 5 \qquad (1.27)$$

where T_b is the interval between observed bursts at a point and T^+ is the dimension-less burst frequency. Over a limited range of experimental conditions T^+ is found to lie in a range 3–7. However, as the temporal occurrence of bursts must be related to the uplift frequency of streaks, the apparent outer layer scaling of burst frequency presents one of the many conflicting results associated with bursting theory (Luchik & Tiederman, 1987). Similar conflicting issues arise when considering burst frequency in geophysical flows (Jackson, 1976; Clifford, 1990; French & Clifford, 1992).

The occurrence of bursts is apparently not confined to the inner region of the boundary layer. Fiedler & Head (1966), Imaki (1968) and Kaplan & Laufer (1969) also observe burst-like phenomena in the outer region of the flow and find high correlation between observed bursts in the outer region and large-scale motion in the inner region. Kovasznay (1970) visualizes a reverse cascade of eddies in the boundary layer beginning with frequent small-scale events near the wall which either grow or coalesce to produce the large-scale eddies which compose the outer region of the flow.

Geophysical turbulent bursting and sediment dynamics

Evidence of quasi-periodic burst-like processes in geophysical flows are numerous and often overlooked. The twisting, sinuous streams of aeolian sand blowing along a beach characterized by periodic uplift and dispersion has been cited by many observers as being indicative of large-scale bursting processes. Similar patterns of fluid motion can be seen when observing the sporadic and sometimes energetic oscillations of aquatic plants in shallow streams or seaweed fronds in tidal channels. Further, during moderate or strong winds the motion of cereal crops can reveal the spatial and temporal distribution of turbulence (Kaimal & Businger, 1970). In these and other examples, one is struck both by the complexity of the fluid motion and the similarity between one situation and the next.

In a detailed discussion of turbulent boundary layers developed over rough surfaces Grass (1971) and Grass *et al.* (1991) cite poor understanding of the mechanism leading to the growth of structures in turbulent boundary layers as being the principal weakness in all existing models of near-bed flows. This limitation is especially relevant when considering large-scale geophysical flows such as rivers characterized by large Reynolds numbers and rough boundaries. In such cases the ratio between the largest eddy size ($\cong \delta$) to the near-wall streak dimensions (y_s^+) is very large and mechanisms linking the small-scale flow of the inner region of the boundary layer to the large-scale motions of the outer layer are unclear. In a qualitative sense, the surface boils in river and tidal channels indicate strongly that burst-like events are responsible for vertical momentum transport, but at a scale far larger than that associated with the near-bed events observed in laboratory flows (Jackson, 1976). Further, observations of sand and dust transport and smoke blown along the ground provide convincing evidence of large-scale burst-like processes.

In a conceptual model proposed by Grass *et al.* (1991) and supported in part by the findings of Rogers & Moin (1987) and Lee *et al.* (1990), a mechanism by which large-scale coherent structures could evolve is proposed. In this model the thin layer of high vorticity close to a rough boundary (the so-called carpet layer) is considered to be susceptible to the same regenerative mode of instability as that identified in the viscous sublayer. Thus, through a mechanism similar to streak oscillation and break-up in the near-bed region, the carpet layer can also give rise to additional horseshoe vortex structures which scale with eddy viscosity and constitute a further layer of turbulent structure. Extension of this model allows assembly of a hierarchical scale of structures which conforms broadly with the Townsend (1976) and Perry & Chong (1982) models and structures are limited in size only by the depth of the boundary layer.

In a paper linking laboratory-scale bursts with boils observed on the surface of the Wabash River, USA, Jackson (1976) found that, in common with results obtained by Rao *et al.* (1971), boil frequency scaled with the flow variables δ and U. Based on this indirect evidence Jackson speculated that boils and their associated kolks in rivers are the geophysical manifestation of fluid bursting phenomena. However, in work reported by Levi (1991) in which the universal Strouhal law is applied to observed boil periodicity (Levi, 1983) it is apparent that boils can also be explained by eddy shedding processes on the river bed. Without further data, these two conflicting interpretations of the same physical phenomenon remain equally plausible.

The discussion of turbulent bursting in the sections above and field observations indicate strongly that the ejection phase of fluid bursting is a principal vehicle for the upward diffusion of momentum and its role in maintaining particles in suspension is generally considered to be pivotal (Matthes, 1947; Coleman, 1969;

Rood & Hickien, 1989; Soulsby *et al.*, 1984, 1985, 1989). In laboratory experiments Grass (1974, 1982) reports observations which clearly implicate the ejection phase of turbulent bursting in the suspension of fine particles. Similarly, studies by Weedman & Slingerland (1985) and Gyr *et al.* (1989) show dynamic connections between suspended particulate matter (SPM) and intermittent turbulent events. In a study of oscillatory flow Inman *et al.* (1986): demonstrated that links between SPM and bursting are not confined to unidirectional flow situations.

Progress in understanding links between geophysical turbulent bursting events and suspension dynamics, however, has recently been made possible through the development of fast response sensors to measure SPM. In studies of tidal resuspension of sand using electromagnetic current meters and impact probes in the Taw Estuary, UK, Soulsby *et al.* (1985, 1989) show that SPM is characterized by alternate and irregular high and low values. Through cross-correlation analyses of SPM data obtained at different heights above the boundary, Soulsby *et al.* (1989) demonstrate that definite large-scale structures exist within the main body of the flow which contain high SPM values and are similar in character to flow features associated with the downstream passage of horseshoe vortices. Limited by the resolution of vertical SPM measurements, however, the apparent similarities have remained speculative until the advent of acoustic backscatter sensors (ABS), which provide a non-invasive and rapid measurement of SPM profiles (Thorne *et al.*, 1991). The high spatial and temporal resolution measurements obtained using this instrument now permit unambiguous correlation between flow structure and SPM profiles (Thorne *et al.*, 1993).

In a study of near-bed turbulence and suspended sediments using an electromagnetic current meter and optical backscatter device in the Fraser River, British Columbia, Lapointe (1992) examined the relationship between SPM and fluid motion (Figure 1.7). Derived from measurements one metre above the bed, this figure shows high frequency and smoothed modulation of SPM time series by the vertical flow component (w'). In common with other turbulent mixing processes Lapointe (1992) obtains a weak correlation (+0.35) between high frequency w' and SPM values. In further analyses of the flow records, Lapointe examined the instantaneous turbulence series and reported a stronger association between large-scale ejection events and SPM. In this way local upwelling events associated with ejections are shown to be the principal agents for the promotion and maintenance of sediment suspension. However, the scale and frequency of burst-like events in this study were found to be highly variable with no simple scaling for the intensity of recurrence. On the basis of these field measurements Lapointe (1992) concludes that attempts to model alluvial suspension processes in terms of bursts with a simple recurrence period scaled using outer flow variables (e.g. Jackson, 1976) are likely to be premature. Thus, whilst geophysical bursting phenomena provide a useful conceptual framework to aid description of sediment dynamics, the present paucity of experimental data restricts progress in the advancement of quantitative transport models of the type suggested by Grass (1970) and driven by realistic simulations of turbulent flow.

Owing to their role in generating locally high bed shear stresses, sweeps are considered by many to be the principal agent when considering bedload transport and

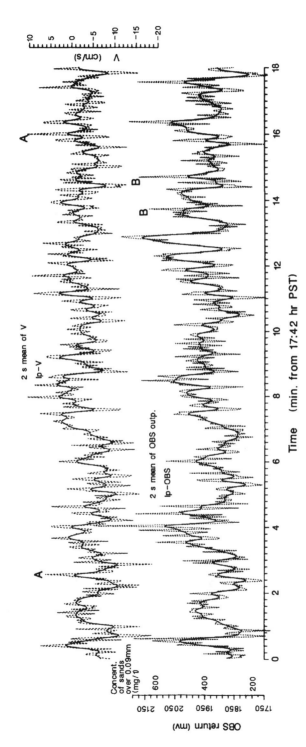

Figure 1.7 High frequency and smoothed modulation of suspended sediment concentrations by vertical velocity fluctuations in the Fraser River, Canada (after Lapointe, 1992)

the mobilization of bed grains (e.g. Sutherland, 1967; Grass, 1971; Nakagawa *et al.*, 1980; Drake *et al.* 1988; Thorne *et al.*, 1989; Williams *et al.*, 1989b; Williams, 1990). In common with studies reported above, the significance of sweeps has only become apparent with the advent of novel experimental techniques. In a study of bedload in Duck Creek, Wyoming, Drake *et al.* (1988), utilizing motion-picture photography to obtain quantitative relationships between sweeps and grain motion on the bed, found that 70% of total transport occurring in less than 9% of the total time was sweep-derived.

Findings similar to those reported by Drake *et al.* (1988) are given by Thorne *et al.* (1989) for marine gravel transport measured using a passive acoustic device (Figure 1.8). In this figure, showing the instantaneous flow and bedload transport (Q_b) time series, the four turbulent flow events identified using quadrant analysis are shown. It is clear that whilst outward interactions (OI) and sweeps (S) ($u' > 0$) give rise to significant peaks in the Q_b record, the inward interactions (II) and ejections (E) ($u' < 0$) cause minimal bed disturbance even though kinematic stress may be roughly equivalent in both cases. It is also instructive to note from Figure 1.8 that although making a negative contribution to total stress, OI events give rise to significant bedload transport. However, owing to their low return period and magnitude, their overall effect in determining total Q_b is relatively small.

The response of bed sediment to the passage of a typical sweep event is illustrated in Figure 1.9 (Williams, 1990). This figure shows (A) direction and intensity of grain motion recorded using a video camera (schematic), (B) the maximum mobile grain diameter, (C) the instantaneous horizontal flow component and (D) the

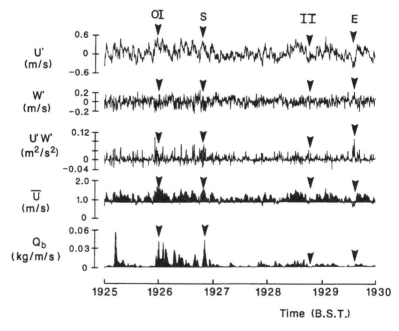

Figure 1.8 Bursting events giving rise to bedload transport of marine gravel (after Thorne *et al.*, 1989). E = ejections; S = sweeps; OI = outward interactions; II = inward interactions

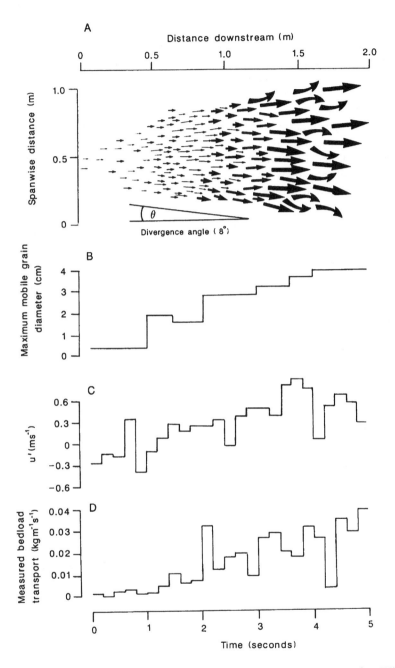

Figure 1.9 Response of marine gravel to the passage of a sweep event (after Williams, 1990)

instantaneous bedload transport rate. In a qualitative sense such events are identical in all respects to the sweeps visualized in the laboratory using fine sand (Grass, 1971). Whilst differing in size by nearly an order of magnitude it would appear that the same flow structures are responsible for mobilization of bed material in both cases. The magnitude and frequency of such events over periods of less than one hour are found to have unusual probability density functions (cf. Williams & Tawn, 1991) and are effective in mobilizing bed material at mean flow velocities less than the threshold (Williams *et al.*, 1989b). These observations are broadly in accord with sediment dynamics reported in the fluvial situation by Drake *et al.* (1988).

The studies described above clearly implicate geophysical turbulent bursting as being the fundamental forcing agent in sediment transport dynamics in fluvial situations. What remains less clear, however, is the method by which this knowledge can usefully be applied to the practical problem of estimating sediment flux. In particular, whether the formulation of new sediment transport models based on kinematic description of multiphase flow is possible remains an open question. However, it seems likely that with recent progress in experimental techniques and instrumentation in fields closely related to fluvial geomorphology, the response of sediment to turbulent bursting events, downstream sorting processes and bedform growth and decay are some of the unresolved issues in fluvial sediment dynamics that may be addressed through studies of turbulence structure in rivers.

ACKNOWLEDGEMENTS

This work was undertaken as part of the MAST II "Circulation And Sediment Transport Around Banks, (CSTAB)" programme. It was funded jointly by the NERC and by the Commission of the European Communities Directorate General for Science and Education, Research and Development under contract number MAS2-CT92–0024C.

NOMENCLATURE

A_k	constant	l	characteristic length scale
B_k	constant	n	Mannings' coefficient
C	Chezy resistance factor	p	dynamic/static pressure
d	channel depth	R	normalized autocorrelation
d_g	average grain diameter		function
E	spectral energy density	Re	Reynolds number
Fr	Froude number	Re_*	grain Reynolds number
f_ψ	stream-wise pressure gradient	Re_θ	momentum thickness Reynolds
	function		number
g	acceleration due to gravity	r	hydraulic radius
k	wavenumber	T_b	"burst" interval
k^*	scaled wavenumber (kz)	T^+	scaled "burst" frequency
L	mixing length	T_α	duration of macroscale turbulence

t	time	w	channel width
U	velocity	X, Y, Z	orthogonal body forces
U''	cross-sectional mean flow	x, y, z	orthogonal coordinates
U_∞	free stream velocity	y_s^+	scaled streak spacing
U_*	shear velocity	z_0	roughness length
$\bar{U}$	mean streamwise flow	β	channel slope
$\hat{u}$	r.m.s turbulence intensity	γ	autocorrelation lag
u	stream-wise flow component	δ	boundary layer thickness
u'	fluctuating stream-wise flow component	δ_θ	boundary layer momentum thickness
$\bar{V}$	mean cross-stream flow	ε	eddy viscosity
v	cross-stream flow component	λ_s	average low speed streak spacing
v'	fluctuating cross-stream flow component	μ	dynamic viscosity
$\bar{W}$	mean vertical flow	ν	kinematic viscosity
w	vertical flow component	ρ	fluid density
w'	fluctuating vertical	τ	shear stress
		∇	Laplace's operator

REFERENCES

Acarlar, M.S. & Smith, C.R. (1987) A study of hairpin vortices in a laminar boundary layer. Part II: hairpin vortices generated by fluid injection. *Journal of Fluid Mechanics*, **175**, 43–83.

Allen, J.R.L. (1982) *Sedimentary Structures*, Volume 1, Elsevier, Amsterdam, 593 pp.

Anwar, H.O. (1981) A study of the turbulent structure in a tidal flow. *Estuarine, Coastal and Shelf Science*, **13**, 373–387.

Anwar, H.O. & Atkins, R. (1980) Turbulence measurements in simulated tidal flow. *Journal of Hydraulic Engineering*, **106**, 1273–1289.

Bagnold, R.A. (1941) *The Physics of Blown Sand and Desert Dunes*, Methuen, London, 265 pp.

Batchelor, G.K. (1970) *The Theory of Homogeneous Turbulence*, 2nd edn, Cambridge University Press, London.

Bernard, P.S. & Handler, R.A. (1990) Reynolds stress and the physics of turbulent momentum transport. *Journal of Fluid Mechanics*, **220**, 99–124.

Black, T.J. (1966) Some practical applications of a new theory of wall turbulence. *Proceedings of the Heat Transfer and Fluid Mechanics Institute*, Stanford University Press, 57 pp.

Black, T.J. (1968) *An Analytical Study of the Measured Wall Pressure Field under Supersonic Turbulent Boundary Layers*, NASA Report **CR–888**, 56 pp.

Blackwelder, R.F. & Eckelmann, H. (1979) Streamwise vortices associated with the bursting phenomenon. *Journal of Fluid Mechanics*, **94**, 577–594.

Bogard, D.G. & Tiederman, W.G. (1987) Characteristics of ejections in turbulent channel flow. *Journal of Fluid Mechanics*, **179**, 1–19.

Bowden, K.F. (1962) Measurement of turbulence near the sea bed in a tidal flow. *Journal of Geophysical Research*, **67**, 3181–3186.

Bradshaw, P. (1971) *An Introduction to Turbulence and its Measurement*, Pergamon, Oxford, 218 pp.

Bridge, J.S. (1978) Origin of horizontal laminations under turbulent boundary layers. *Sedimentary Geology*, **20**, 1–16.

Brodkey, R.S., Wallace, J.M. & Eckelman, H. (1974) Some properties of truncated turbulence signals in bounded shear flows. *Journal of Fluid Mechanics*, **63**, 209–224.

Brown, G.L. & Thomas, A.S.W. (1977) Large scale motions in a turbulent boundary layer. *Physics of Fluids Supplement*, **20**, 243–252.

Cantwell, B.J. (1981) Organised motion in turbulent flow. *Annual Review of Fluid Mechanics*, **13**, 457–515.

Cebeci, T. & Bradshaw, P. (1977) *Momentum Transfer in Boundary Layers*, Hemisphere, Washington, USA.

Clifford, N.J., (1990) *The Formation, Nature and Maintenance of Riffle–Pool Sequences in Gravel-bedded Rivers*, unpublished PhD thesis, University of Cambridge, UK.

Clifford, N.J., French, J.R. & Hardisty, J. (eds) (1993) *Turbulence: Perspectives on flow and sediment transport*, John Wiley & Sons, Chichester, 360 pp.

Coleman, J.M. (1969) Brahmaputra river-channel processes and sedimentation. *Sedimentary Geology*, **3**, 129–239.

Coles, D.E. (1978) A model for flow in the viscous sublayer. *In* Smith, C.R. & Abbott, D.E. (eds), *Lehigh Workshop in Coherent Structures in Turbulent Boundary Layers*, Lehigh University, PA, USA, 98–109.

Corino, E.R. & Brodkey, R.S. (1969) A visual investigation of the wall region in turbulent flow. *Journal of Fluid Mechanics*, **37**, 1–30.

Corrsin, S. (1957) *Proceedings of the First Naval Hydro. Symposium*, National Research Council, Washington.

Drake, T.G., Shreve, R.L., Dietrich, W.E., Whiting, P.J. & Leopold, L.B. (1988) Bedload transport of fine gravel observed by motion picture photography. *Journal of Fluid Mechanics*, **192**, 193–217.

Dyer, K.R. (1986) *Coastal and Estuarine Sediment Dynamics*, John Wiley, Chichester, 342 pp.

Favre, A., Dumas, R., Verollet, E. (1961) Couche limite sur paroi plane poreuse avec aspiration. *Publication Scientifiques et Techiques du Ministère de l'air*, No. 277, 27 pp.

Fiedler, H. & Head, M.R. (1966) Intermittency measurements in a turbulent boundary layer. *Journal of Fluid Mechanics*, **25**, 719–735.

French, J.R. & Clifford, N.J. (1992) Characteristics and 'event-structure' of near-bed turbulence in a macrotidal saltmarsh channel. *Estuarine, Coastal and Shelf Science*, **34**, 49–69.

Gordon, C.M. & Witting, J. (1977) Turbulent structure in a benthic boundary layer. In: Nihoul, J. (ed.), *Bottom Turbulence, Proceedings of the 8th International Liege Colloquium on Ocean Hydrodynamics*, Elsevier Oceanographic Series, **19**, 59–81.

Grant, H.L. (1958) The large eddies of turbulent motion. *Journal of Fluid Mechanics*, **4**, 149–190.

Grass, A.J. (1970) Initial instability of fine bed sand. Journal of Hydraulic Division, *ASCE*, **HY3**, 619–632.

Grass, A.J. (1971) Structural features of turbulent flow over smooth and rough boundaries. *Journal of Fluid Mechanics*, **50**, 233–255.

Grass, A.J. (1974) Transport of fine sand on a flat bed. *Proceedings of EUROMECH 48, Turbulence and Suspension Mechanics*, Technical University of Denmark, Copenhagen.

Grass, A.J. (1982) The influence of boundary layer turbulence on the mechanics of sediment transport. *Proceedings of Euromech 156: Mechanics of Sediment Transport*, Balkema, Istanbul, 3–17.

Grass, A.J., Stuart, R.J. & Mansour-teharni, M. (1991) Vortical structures and coherent motion in turbulent flow over smooth and rough boundaries. *Philosophical Transactions of the Royal Society of London*, A, **336**, 35–65.

Guezennec, Y.G., Piomelli, U. & Kim, J. (1989) On the shape and dynamics of wall structures in turbulent channel flow. *Physics of Fluids*, **A1**, 764–766.

Gulliver, J.S. & Halverson, M.J. (1987) Measurement of large streamwise vortices in an open channel flow. *Water Resources Research*, **23**, 115–123.

Gyr, A. (1983) Towards a better definition of the three types of sediment transport. *Journal of Hydraulic Research*, **21**(1), 1–15.

Gyr, A., Muller, A. & Schmid, A. (1989) Observation of self stabilization processes in sediment transport linked to scales of coherent structures. In: *Proceedings of Technical Session A*, Turbulence in Hydraulics, *IAHR XXIII Congress*, 21–25 August 1989, Ottawa, Canada, 31–38.

Hama, F.R. & Nutant, J. (1963) *Proceedings of the Heat Transfer and Fluid Mechanics Institute*, Stanford University.

Head, M.R. (1978) Combined flow visualization and hot-wire measurements in turbulent boundary layers. In: Smith, C.R. & Abbott, D.E. (eds), *Lehigh Workshop in Coherent Structures in Turbulent Boundary Layers*, Lehigh University, PA, USA, 98–109.

Head, M.R. (1979) Flow visualization of turbulent boundary layer structure. *AGARD Conference Proceedings*, **271**, 25–39.

Head, M.R. & Bandyopadhyay, P. (1981) New aspects of turbulent boundary layer structure. *Journal of Fluid Mechanics*, **107**, 297–338.

Heathershaw, A.D. (1974) "Bursting" phenomenon in the sea. *Nature*, **248**, 394–395.

Heathershaw, A.D. (1976) Measurements of turbulence in the Irish Sea benthic boundary layer. In: McCare, I.N. (ed.), *The Benthic Boundary Layer*, Plenum Press, New York, 11–31.

Heathershaw, A.D. (1979) The turbulent structure of the bottom boundary layer in tidal currents. *Geophysical Journal of the Royal Astronomical Society*, **58**, 395–430.

Heathershaw, A.D. & Thorne, P.D. (1985) Sea-bed noises reveal role of turbulent bursting phenomenon in sediment transport by tidal currents. *Nature*, **316**, 339–342.

Huntley, D.A. (1988) A modified inertial dissipation method for estimating seabed stresses at low Reynolds number, with applications to wave/current boundary layer measurements. *Journal of Physical Oceanography*, **18**, 339–346.

Hussain, A.K.M.F. (1983) Coherent structures – reality and myth. *Physics of Fluids*, **26**, 2816–2849.

Hussain, A.K.M.F. & Clark, A.R. (1981) On the coherent structure of the axisymmetric mixing layer: a flow-visualization study. Journal of Fluid Mechanics, **104**, 263–294.

Imaki, I. (1968) The structure of the turbulent boundary layer's "superlayer". *Bulletin of the Institute of Space and Aeronautical Science*, **4**, University of Tokyo, Japan, 448–536.

Inman, D.L., Jenkins, S.A., Hicks, D.M. & Kim, H.K. (1986) *Oscillatory Bursting over Beds of Fine Sand*. SIO Reference Series **86–13**, University of California, Scripps Institution of Oceanography, 46 pp.

Ishihara, Y. & Yokosi, S. (1967) The spectra of turbulence in a river flow. *Proceedings of the 12th Congress, International Association for Hydraulic Research*, Colorado, 290–297.

Jackson, R.G. (1976) Sedimentological and fluid-dynamic implications of the turbulent bursting phenomenon in geophysical flows. *Journal of Fluid Mechanics*, **77**, 531–560.

Kaimal, J.C. & Businger, J.A. (1970) Case studies of a convective plume and a dust devil. *Journal Applied Meteorology*, **9**, 612–620.

Kaimal, J.C., Wyngarrd, J.C., Izumi, Y. & Cote, O.R. (1972) Spectral characteristics of surface layer turbulence. *Quarterly Journal of the Royal Meteorological Society*, **98**, 563–589.

Kaplan, R.E. & Laufer, J. (1969) Further measurements in the intermittent region of a turbulent boundary layer. *Boeing Symposium on Turbulence*, Seattle, Washington, USA.

Keffer, J.F. (1965) The uniform distortion of a turbulent wake. *Journal of Fluid Mechanics*, **22**, 135–159.

Kim, H.T., Kline, S.J. & Reynolds, W.C. (1971) The production of turbulence near a smooth wall in a turbulent boundary layer. *Journal of Fluid Mechanics*, 50, 133–160.

Kim, J. & Moin, P. (1986) The structure of the vorticity field in turbulent channel flow. 2. Study of ensembled-averaged fields. *Journal of Fluid Mechanics*, **162**, 363–389.

Klebanoff, P.S. (1954) *Characteristics of Turbulence in a Boundary Layer with Zero Pressure Gradient*. NACA **TN–3178**, 68 pp.

Kline, S.J. & Robinson, S.K. (1989) Turbulent boundary layer structure: progress, status and challenges. *Second IUTAM Meeting on Structure of Turbulence and Drag Reduction*, Zurich.

Kline, S.J., Reynolds, W.C., Schraub, F.A., & Rundstadler, P.W. (1967) The structure of turbulent boundary layers. *Journal of Fluid Mechanics*, **30**, 741–773.

Kovasnay, L.S.G. (1970) The turbulent boundary layer. *Annual Review of Fluid Mechanics*, **2**, 95–112.

Lapointe, M. (1992) Burst-like sediment suspension events in a sand bed river. *Earth Surface Processes and Landforms*, **17**, 253–270.

Laufer, J. (1975) New trends in experimental turbulence research. *Annual Review of Fluid Mechanics*, **7**, 307–326.

Launder, B.E. & Spalding, D.B. (1972) *Lectures in Mathematical Models of Turbulence*, Academic Press, London, 169 pp.

Lee, M.J., Kim, J. & Moin, P. (1990) Structure of turbulence at high shear rate. *Journal of Fluid Mechanics*, **216**, 561–583.

Levi, E. (1983) A universal Strouhal law. *Journal of Engineering Mechanics*, ASCE, **109**(3), 718–727.

Levi, E. (1991) Vortices in hydraulics. *Journal of Hydraulic Engineering*, **117**(4), 399–413.

Lighthill, M.J. (1963) In: Rosenhead, L. (ed.) *Introduction*; *Boundary Layer Theory Laminar Boundary Layers*, Clarendon Press, Oxford, 46–113.

Liu, J.T.C. (1989) Coherent structures in transitional and turbulent shear flows. *Annual Review of Fluid Mechanics*, **21**, 285–315.

Lu, L.J. & Smith, C.R. (1991) Use of flow visualization data to examine spatial–temporal velocity and burst-type characteristics in a turbulent boundary layer. *Journal of Fluid Mechanics*, **232**, 303–340.

Lu, S.S. & Willmarth, W.W. (1973) Measurements of the Reynolds stress in a turbulent boundary layer. *Journal of Fluid Mechanics*, **60**, 481–511.

Luchik, T.S. & Tiederman, W.G. (1987) Timescale and structure of ejections and bursts in turbulent channel flows. *Journal of Fluid Mechanics*, **174**, 529–552.

Lyn, D.A. (1986) *Turbulence and Turbulent Transport in Sediment-laden Open-channel Flows*, report no. **KR–R–49**, W.M. Keck Laboratory of Hydraulics and Water Resources, California Institute of Technology, Pasadena, USA, 244 pp.

Maslowe, S.A. (1981) Shear flow instabilities and transition. In Swinney, H.L. & Gollub, J.P. (eds), *Hydrodynamic Instabilities and Transition Turbulence*, Springer-Verlag, Berlin, 121–228.

Matthes, G.H. (1947) Macroturbulence in natural stream flow. *Transactions of the American Geophysical Union*, **28**, 255–265.

Mclean, S.R. & Smith, J.D. (1979) Turbulence measurements in the boundary layer over a sand wave field. *Journal of Geophysical Research*, **84**(C12), 7791–7808.

Müller, A., Gyr, A. & Dracos, T. (1971) Interaction of rotating elements of the boundary layer with grains of the bed: a contribution to the problem of the threshold of sediment transport. *Journal of Hydraulic Research*, **9**, 373–411.

Nakagawa, H., Tsjimoto, T. & Hosokawa, Y. (1980) Statistical mechanics of bed-load transportation with 16 mm film analysis of behaviors of individual sediment particles on a flat bed. *Third International Symposium on Stochastic Hydraulics*, 5–7 August 1980, Tokyo, ,Japan, 12 pp.

Nychas, S.G., Hershey, H.C. & Brodkey, R.S. (1973) A visual study of turbulent shear flow. *Journal of Fluid Mechanics*, **61**, 513–540.

Offen, G.R. & Kline, S.J. (1975) A proposed model of the bursting process in turbulent boundary layers. *Journal of Fluid Mechanics*, **70**, 209–228.

Perry, A.E. & Chong, M.S. (1982) On the mechanism of wall turbulence. *Journal of Fluid Mechanics*, **119**, 173–217.

Rao, K.N., Narasimha, R. & Narayanan, M.A.B. (1971) Bursting in a turbulent boundary layer. *Journal of Fluid Mechanics*, **48**, 339–352.

Robinson, A.L. (1983) How does fluid flow become turbulent? *Science*, **221**, 140–143.

Robinson, S.K. (1989) A review of vortex structures and associated coherent motions in turbulent boundary layers. *Second IUTAM Meeting on Structure of Turbulence and Drag Reduction*, Zurich.

Robinson, S.K. (1991) Coherent motions in the turbulent boundary layer. *Annual Review of Fluid Mechanics*, **23**, 601–639.

Robinson, S.K., Kline, S.J. & Spalart, P.R. (1989) *A Review of Quasi-coherent Structures in a Numerically Simulated Turbulent Boundary Layer*. NASA Technical Memorandum **102191**.

Rogers, M.M. & Moin, P. (1987) The structure of the vorticity field in homogeneous turbulent flows. *Journal of Fluid Mechanics*, **176**, 33–66.

Rood, K.M. & Hickien, (1989) Suspended sediment concentration in relation to surface flow structure in Squamish River estuary, southwestern British Columbia, *Canadian Journal of Earth Sciences*, **26**, 2172–2176.

Rott, N. (1990) Note on the history of the Reynolds number. *Annual Review of Fluid Mechanics*, **22**, 1–11.

Schlichting, H. (1979) *Boundary Layer Theory* 7th edn, McGraw-Hill, New York, 817 pp.

Shiono, K. & West, J.R. (1987) Turbulent perturbations of velocity in the Conwy estuary. *Estuarine, Coastal & Shelf Science*, **25**, 533–553.

Shiono, K. & Knight, D.W. (1990) Turbulent open-channel flows with variable depth across the channel. *Journal of Fluid Mechanics*, **222**, 617–646.

Smith, C.R., Walker, J.D.A., Haidari, A.H. & Taylor, B.K. (1989) Hairpin vortices in turbulent boundary layers: the implications for reduced surface drag. In: Gyr, A. (ed.), *Proceedings of the Second IUTAM Symposium on Structure of Turbulence and Drag Reduction*, Springer-Verlag, New York.

Soulsby, R.L. (1977) Similarity scaling of turbulence spectra in marine and atmospheric boundary layers. *Journal of Physical Oceanography*, **7**, 934–937.

Soulsby, R.L. (1980) Selecting record length and digitization rate for near-bed turbulence measurements. *Journal of Physical Oceanography*, **10**(2), 208–219.

Soulsby, R.L., Salkield, A.P. & Le Good, G.P. (1984) Measurements of the turbulence characteristics of sand suspended by a tidal current. *Continental Shelf Research*, **3**(4), 439–454.

Soulsby, R.L., Salkield, A.P., Haine, R.A. & Wainwright, B. (1985) Observations of the turbulent fluxes of suspended sand near the seabed. *Proceedings of EUROMECH 192, Transport of Suspended Solids in Open Channels*, Munich, 186–193.

Soulsby, R.L., Atkins, R. & Salkield, A.P. (1989) Observation of the turbulent structure of a suspension of sand in a tidal current. *Proceedings of EUROMECH 215, Mechanics of Sediment transport in Fluvial and Marine Environments*.

Sreenivasan, K.R. (1990) Turbulence and the tube. *Nature*, **334**, 192–193.

Sumer, B.M. & Deigaard, R. (1981) Particle motions near the bottom in turbulent flow in an open channel. *Journal of Fluid Mechanics*, **109**, 311–337.

Sumer, B.M. & Oguz, B. (1978) Particle motions near the bottom in turbulent flow in an open channel. *Journal of Fluid Mechanics*, **86**, 109–127.

Sutherland, A.J. (1967) Proposed mechanism for sediment entrainment by turbulent flows. *Journal of Geophysical Research*, **72**, 6183–6194.

Taylor, G.I. (1935) Statistical theory of turbulence. *Proceedings of the Royal Society of London, A*, **151**, 421–454.

Theodorsen, T. (1952) Mechanisms of turbulence. *Proceedings of the Second Midwestern Conference on Fluid Mechanics*, Ohio State University.

Thorne, P.D., Hardcastle, P.J. & Soulsby, R.L. (1993) Analysis of acoustic measurements of suspended sediments. *Journal of Geophysical Research*, **98**, 899–910.

Thorne, P.D., Williams, J.J. & Heathershaw, A.D. (1989) *In situ* acoustic measurements of marine gravel threshold and transport. *Sedimentology*, **36**, 61–74.

Thorne, P.D., Vincent, C.E., Hardcastle, P.J., Rehmans, S. & Pearson, N. (1991) Measuring suspended sediment concentrations using acoustic backscatter devices. *Marine Geology*, **98**, 7–16.

Townsend, A.A. (1956) *The Structure of Turbulent Shear Flow*, Cambridge University Press, Cambridge, 315 pp.

Townsend, A.A. (1976) *The Structure of Turbulent Shear Flow*, 2nd edn, Cambridge University Press, Cambridge, 429 pp.

Tritton, D.J. (1977) *Physical Fluid Dynamics*, Von Nostrand Reinhold, London, 346 pp.

Walker, J.D.A. (1990) Models based on dynamical features of the wall layer. *Applied Mechanics Review*, **43**, S232–S239.

Wallace, J.M. (1985) *The Vortical Structure in Bounded Shear Flow*, Lecture Notes in Physics **235**, Springer-Verlag, 253–268.

Wark, C.E. & Nagib, H.M. (1991) Experimental investigation of coherent structures in turbulent boundary layers. *Journal of Fluid Mechanics*, **230**, 183–208.

Weedman, S.D. & Slingerland, R. (1985) Experimental study of sand streaks formed in turbulent boundary layers. *Sedimentology*, **32**, 133–145.

West, J.R. & Oduyemi, K.O.K. (1989) Turbulence measurements of suspended solids concentration in estuaries. *Journal of Hydraulic Engineering*, ASCE, **115**, 457–474.

Williams, J.J. (1990) Video observations of marine gravel transport. *Geo-Marine Letters*, **10**, 157–164.

Williams, J.J. & Tawn, J.A. (1991) Simulation of bedload transport of marine gravel. *Coastal Sediments '91 Proceedings*, ASCE, 25–27 June 1991, Seattle, USA, 703–716.

Williams, J.J., Thorne, P.D. & Heathershaw, A.D. (1989a) Measurements of turbulence in the benthic boundary layer over a gravel bed. *Sedimentology*, **36**, 959–971.

Williams, J.J., Thorne, P.D. & Heathershaw, A.D. (1989b) Comparison between acoustic measurements and predictions of the bedload transport of marine gravels. *Sedimentology*, **36**, 973–979.

Wygnaski, I. & Peterson, R.A. (1987) Coherent motion in excited free stream flows. *AIAA Journal*, **25**, 201–213.

2 Three-dimensional Flow in Straight and Curved Reaches

R. D. HEY AND P. C. B. RAINBIRD

School of Environmental Sciences, University of East Anglia, UK

INTRODUCTION

Observations of rivers, particularly during high in-bank flows, indicate that there are zones where the flow is downwelling, as evidenced by lines of helical vortices and floating debris, and where upwelling occurs and the water surface is smoother and appears to be gently boiling. These features are indicative of major helical structures in the flow as downwelling results from the convergence of surface flow while upwelling produces surface flow divergence.

The first scientific investigation of helical flows, often referred to as secondary currents, was carried out by Thompson (1876). In laboratory experiments he established that helical flows in meander bends resulted from the imbalance in radial forces in the bend. Subsequent work by Möller (1883) and Stearns (1883) also indicated that the maximum velocity filament occurred below the free surface as a result of secondary circulation. Further support for Thompson's results was provided by Eakin (1935), from investigations of the transport of pea-coal and limestone chat around bends in the Mississippi River and by Friedkin (1945), Chacinski (1954) and Rozovskii (1961) from model and flume experiments.

This chapter reviews the processes controlling velocity distributions in turbulent flow, outlines procedures for determining three-dimensional flows, identifies how channel form controls the generation of secondary flows and establishes their effect on velocity and shear stress distributions in both straight and meandering rivers. The principles underlying the mathematical modelling of velocity and shear stress distributions in meander bends are introduced, and form the basis for more advanced morphological and sedimentological models.

VELOCITY DISTRIBUTION IN NATURAL CHANNELS

Introduction

Flow in rivers is principally retarded by the drag imparted on the flow by the surface roughness of the boundary. A very thin layer near the bed and banks is stationary,

Advances in Fluvial Dynamics and Stratigraphy. Edited by P.A. Carling and M.R. Dawson.
© 1996 John Wiley & Sons Ltd.

but further from the boundary, the velocity progressively increases as the resistance to shear between adjacent fluid layers is much less effective than the surface roughness in retarding flows. The boundary layer defines a region in which a velocity gradient occurs as a result of boundary drag. In rivers the boundary layer can be quantitatively defined as extending to a distance away from the boundary at which the velocity is $0.99 u_0$, where u_0 is the free stream velocity. However, this is more usually approximated to extending to the water surface (sometimes referred to as the free surface).

The state of flow in rivers is essentially turbulent with a Reynolds number above the critical value of 2000; typical values are of the order of 250 000 at bankfull stage. This indicates that inertial forces are more significant than viscous ones (see Chapter 1). The velocity profile for an infinitely wide channel (avoiding complication due to the presence of the banks) has been shown to vary with distance from the boundary (Prandtl, 1952; Figure 2.1).

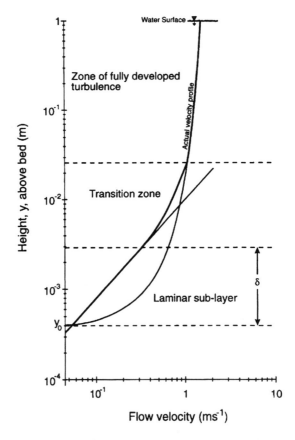

Figure 2.1 Vertical velocity profile in turbulent uniform flow over a rough surface. Actual (thicker line) and computed logarithmic velocity profiles plotted to show deviation within laminar sublayer and transition zone. Flow depth 1.0 m, roughness height of bed material 0.01 m, and channel slope 0.001. Logarithmic scales used on both axes to facilitate comparison

The vertical velocity gradient, dv/dy, is defined by:

$$\frac{dv}{dy} = \frac{\tau}{(\mu + \varepsilon)} \tag{2.1}$$

where τ is the temporally averaged shear stress at any point in the flow, μ is the molecular viscosity and ε is the eddy viscosity. Momentum transfer within the flow results from molecular exchange (defined by μ) and from the dispersive action of turbulent eddies (defined by ε). The velocity profile is controlled by the eddy viscosity as the molecular exchange is negligible in comparison. High momentum fluid is transferred by eddies from near the water surface, towards the channel bed, where the flow is slower and *vice versa*, producing a steeper near-bed velocity gradient. The flow has a cross-stream axis, spanwise vorticity and is essentially two-dimensional.

The eddy viscosity is not simply a property of the fluid, but also depends on the characteristics of the flow and the distance from the boundary. Prandtl (1952), identified that:

$$\varepsilon = \rho l^2 \frac{dv}{dy} \tag{2.2}$$

where ρ is the density of water and l the mixing length characterizing the eddy size. To determine the velocity gradient, Prandtl (1952) assumed that the mixing length is proportional to distance, y, from the bed, giving,

$$l = \kappa y \tag{2.3}$$

where κ is a constant of proportionality, and that the shear stress is constant throughout the flow and equal to the boundary shear stress τ_0. Combining eqns (2.1), (2.2) and (2.3) and disregarding the molecular viscosity, μ, gives:

$$\tau_0 = \left(\rho \kappa^2 y^2 \frac{dv}{dy}\right) \frac{dv}{dy} \tag{2.4}$$

Solving for the velocity, v, at distance y from the boundary gives:

$$v = \frac{1}{\kappa} \left(\frac{\tau_0}{\rho}\right)^{1/2} \ln y + c \tag{2.5}$$

where $(\tau_0/\rho)^{1/2}$ defines the shear velocity, v_*. This indicates that the velocity distribution in turbulent flow is logarithmic.

The constant of integration, c is equal to $-\ln y_0$, where y_0 is the distance above the bed where the flow velocity appears to be zero. Hence:

$$v = \frac{v_*}{\kappa} \ln\left(\frac{y}{y_0}\right) \tag{2.6}$$

For the more mathematically inclined, the derivation of the logarithmic velocity profile can be more rigorously done using asymptotic analysis (see Raupach *et al.*, 1991).

In reality, due to the no-slip property of water, the actual velocity at the bed is zero. Near the bed, velocities and Reynolds numbers are low enough for the flow to be laminar. This zone is referred to as the laminar sublayer, which has a thickness δ (Figure 2.1). Experiments have shown that its average thickness is given by:

$$\delta = \frac{4\nu}{v_*}$$
(2.7)

where ν is the kinematic viscosity, indicating that the sublayer is thinner under conditions of high bed shear stress when turbulence penetrates closer to the bed. It should be noted that there is no distinct cut-off point between the laminar sublayer and the zone of fully developed turbulence, but there is a transitional zone where the Reynolds number is considered to be such that the flow is neither laminar nor turbulent. The thickness of the transition zone will depend on both the properties of the fluid and the bed roughness.

Close to a smooth channel bed (i.e. when the grain size is less than $\delta/3$), in the laminar sublayer, a regularity of flow structure is evident (Leeder, 1982), with longstream elongate streaks spaced regularly across the flow. These streaks are composed of lower-than-average fluid velocity (Grass, 1970) and oscillate within the sublayer, occasionally jumping outward into the main flow. Outward migrations of low momentum fluid are termed bursts and are associated with lower-than-average longstream velocities and upward vertical velocities. These low-velocity streaks are alternated with high-speed streaks associated with downward velocities and are termed sweeps. With increasing distance from the bed, the streaked structure becomes less distinct, with the bursts dissipating as they rise through the boundary layer. Higher in the flow, turbulence is more random and in the uppermost zone (often termed the wake region) turbulence is intermittent and of a large scale with turbulent blobs reaching the surface as boils elevating the water surface.

In natural channels the flow is hydraulically rough. This is because the roughness elements, grains which make up the boundary, prevent the formation of a laminar sublayer. For uniform granular material, the effective roughness height, k_s, approximates the diameter of the grains comprising the surface layer of the bed. In the case of non-uniform material, the value of k_s is a function of the grain size frequency distribution and particle shape and can be characterized by an above-average grain diameter. The effective roughness height was found to be $2D_{90}$ (Kamphuis, 1974) for sand-bedded channels and $3.5D_{84}$ (Hey, 1979) for gravel-bed rivers, where 90 and 84% of the surface bed material is less then or equal to D_{90} and D_{84}, respectively. Provided k_s exceeds five times the thickness of the laminar sublayer then the flow is hydrodynamically rough. When turbulent flow exists throughout the water column Nikuradse (1930) showed that y_0 is related to k_s by:

$$y_0 = \frac{k_s}{30}$$
(2.8)

Equally, the constant of proportionality, κ, referred to as Von Kármán's constant of turbulent exchange for channel flow, has a value of 0.4. Substituting for κ, y_0 and converting Naperian logarithms to logarithms to the base ten in eqn 2.6 gives:

$$v = 5.75v_* \log\left(\frac{30y}{k_s}\right) \qquad (2.9)$$

In rivers, it has been observed (e.g. Stearns, 1883; Chow, 1959; Bathurst *et al.*, 1977; Thorne & Hey, 1979; Nezu *et al.*, 1993) that the maximum velocity often lies below the water surface, especially in channels with small width to depth ratios and in meander bends, and that the velocity distribution in the cross-section does not exactly correspond to the shape of the channel (Figure 2.2). This indicates that helical, or secondary flow, is occurring and that flow is three-dimensional and exhibits a streamwise vorticity. This results from: (i) the spatial variation in turbulence due to vortices resulting from differences in boundary roughness (Reynolds stresses) between bed and banks, or changes in channel planform geometry; (ii) density differences due to variations in suspended sediment load (Vanoni, 1946); (iii) the interaction of the flow with stagnation points in the corners of rectangular or trapezoidal sections (Nikuradse, 1930) which essentially produces regions of reduced bed roughness as the flow is moving over stagnant water.

Measurement of velocity distributions

As flow in rivers is generally three-dimensional, the actual velocity at any one point can be determined by simultaneously measuring the three orthogonal components of velocity (two in the horizontal plane and one vertical).

Electromagnetic (EM) flow meters can be used to determine secondary flows in the field. The meter consists of a spherical or ellipsoid shaped head (*c*. 2 to 5 cm diameter) which contains a doughnut shaped wire coil. By passing a small direct current through the coil, a magnetic field is generated which, when placed in flowing water (a conducting medium), generates a back electromotive force in the coil which is proportional to the local flow velocity. They enable an accuracy of plus or minus 10 mm s^{-1} to be achieved (Bathurst *et al.*, 1979). The original meters

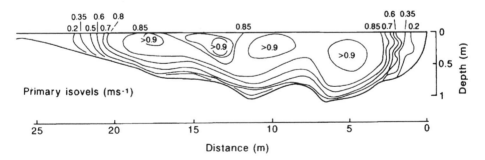

Figure 2.2 Primary velocity distribution in a natural channel indicating non-conformance with channel cross-section due to secondary flows. Penstrowed, River Severn, UK, downstream from apex of meander bend with a discharge of $13 \text{ m}^3 \text{s}^{-1}$. From Thorne & Hey (1979)

allowed only two orthogonal components of velocity to be measured simultaneously and, to determine the third component, the head of the instrument had to be rotated through 90° and the measurements repeated. However, meters are currently being developed for field use which enable all three components to be measured simultaneously.

The meter has to be deployed from a wading rod as it is vital to maintain its orientation in a fixed, and known, plane for the duration of the measurement. Readings are conventionally taken for 60 s to average out any turbulent fluctuations in the flow. It is customary to align the sensors parallel and perpendicular to the river bank at the measurement section to determine the long- and cross-stream flow velocities. The small size of the head enables velocities to be measured to within 5 cm of the bed. However, it should be noted that if the EM flow meter head was placed 5 cm from a rough bed with roughness elements approximately 5 cm in diameter, then the readings may be inaccurate.

The basic data, paired long- and cross-stream velocities, measured at several points in a vertical and at a number of verticals in the cross-section, have to be processed in order to determine the primary and secondary velocities. The direction of the primary velocity is defined to be in the plane of maximum discharge, with secondary flow occurring in a plane normal to the primary flow.

In rivers, the primary velocity is unlikely to be parallel to the banks, particularly in meander bends, and its mean orientation has to be established. Three methods have been devised to determine the orientation of the mean primary flow and thereby obtain local primary and secondary flow components.

The method developed by Bathurst *et al.*, (1977, 1979), based on that of Rozovskii (1961), is the simplest one and requires a minimum of data. The mean primary flow at each vertical is defined by establishing the orientation where the net secondary discharge in the vertical is zero. This can vary between each vertical in a cross-section as each is assessed independently of any other.

At each vertical, the long- and cross-stream velocities, defined as u and w, respectively, produce a resultant velocity q at any depth y making a horizontal angle δ to the long-stream direction (Figure 2.3). Hence:

$$u = q \cos \delta \qquad (2.10)$$

$$w = q \sin \delta \qquad (2.11)$$

The mean primary velocity for the vertical section is at a fixed angle α to the long-stream direction. The angle α defines the vector of primary flow, so the primary velocity at any depth, U_p, is found by resolving velocities in that direction:

$$U_p = q \cos(\delta - \alpha)$$
$$= u \cos \alpha + w \sin \alpha \qquad (2.12)$$

The secondary velocity U_s is found by resolving in a plane normal to the plane of primary velocity:

$$U_s = q \sin(\delta - \alpha)$$
$$= u \cos \alpha - w \sin \alpha \qquad (2.13)$$

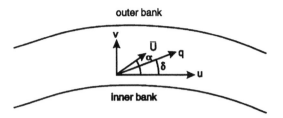

Figure 2.3 Resolution of primary and secondary flows based on maintenance of continuity at a vertical: method of Bathurst *et al.* (1977, 1979)

The angle α is unknown, but can be determined by integrating the secondary velocities over the flow depth, d_m, as the mean secondary discharge is assumed to be zero at each section.

$$\int_0^{d_m} q \sin(\delta - \alpha)\,dy = 0 \tag{2.14}$$

$$\tan \alpha = \frac{\int_0^{d_m} q \sin \delta\,dy}{\int_0^{d_m} q \cos \delta\,dy} \tag{2.15}$$

substituting from eqns 2.10 and 2.11 gives:

$$\tan \alpha = \frac{\int_0^{d_m} w\,dy}{\int_0^{d_m} u\,dy} \tag{2.16}$$

As u and w are known from the original data, α can be determined by calculating the area between the curve and the ordinate axis on a graph of velocity against height above the bed. This corresponds to the integration of eqn 2.16. By substituting α into eqns 2.12 and 2.13, the primary and secondary velocities can be resolved.

Although this method has the virtue of simplicity and it can be applied when full cross-sectional data are not available, it does not accurately portray secondary flows near the inner bank in meander bends. Here, the flow may be radially outward throughout the vertical, yet this model is prescribed to generate balancing secondary flows.

Where a full set of cross-sectional velocity data is available, orientation of the mean primary flow can be achieved by establishing the direction in which the net secondary discharge in the whole cross-section is zero (Paice, 1990). This method enables unidirectional secondary flow to occur at individual verticals. Effectively, reorientation produces the maximum downstream primary discharge for the whole section. The analysis proceeds as follows. (i) The discharge is computed using the mid-section method. This assumes that the velocity profile measured at a vertical is representative of the area bounded by the mid-points between adjacent profiles. (ii) For a range of reorientation angles, between 0° and 360°, the velocities and associated discharges are calculated. The direction that produces the maximum discharge through the cross-section defines the mean primary flow. The method assumes that the correct angle of reorientation is achieved when the secondary discharge is zero. When the mean primary discharge is at a maximum, the mean secondary discharge will be zero. (iii) The velocity data are then rotated by the angle between the downstream direction (raw data) and the primary flow velocity (calculated data) to generate the local primary and secondary components of flow. This method is physically more realistic than that of Bathurst *et al.* (1977, 1979) as it enables radial flows on the point bar to be modelled.

Provided velocity data are available at several closely spaced cross-sections, the procedure developed by Dietrich & Smith (1983) can be applied. This represents the most accurate method for establishing secondary flows as the direction of mean primary flow is obtained by determining the orientation where the net secondary discharge between adjacent cross-sections is zero. The observed long-stream velocity distribution between successive sections is used to compute the nett cross-stream velocity to maintain continuity. Each cross-section is then rotated until the observed secondary discharge matches the computed value. Dietrich & Smith's cross-sections were approximately 5 m apart for a channel up to 6 m wide, suggesting that a width spacing ratio of, at the most, unity is required.

Channel-fitted curvilinear coordinates are used (Figure 2.11) and the continuity equation is given by:

$$\langle u_n \rangle h = \frac{-1}{1-N} \int_{-w/2}^{n} \frac{\partial \langle u_s \rangle h}{\partial s} \, dn \qquad (2.17)$$

where u_n is the cross-stream velocity, u_s is the long-stream velocity, h is the local flow depth, s is the long-stream curvilinear coordinate, N is the ratio n/R in curvilinear coordinates, n is the cross-stream curvilinear coordinate, R is the hydraulic radius and w is the channel width. Angled brackets indicate that the enclosed quantity has been vertically averaged. Vertical averaging of the flow field assumes that the pressure distribution is hydrostatic, the velocity distribution is uniform over the flow depth and the channel bed slope is small.

The use of a curvilinear coordinate system allows the channel boundaries to be described along a single axis which makes their determination much easier than with orthogonal Cartesian coordinates. Also the direction of long-stream flow is a function of one vector and not a combination of two orthogonal vectors which vary with respect to distance around the bend.

In order to compute the total secondary discharge of water, Q_{nw}, eqn 2.17 has to be integrated from bank to bank:

$$Q_{nw} = \int_{-w/2}^{w/2} \langle u_n \rangle h \, dn = \int_{-w/2}^{w/2} \frac{-1}{1-N} \int_{-w/2}^{n} \frac{\partial \langle u_s \rangle h}{\partial s} \, dn \, dn \qquad (2.18)$$

The average direction of flow, θ_w, between successive sections with an average downstream discharge Q_{sw} is:

$$\theta_w = \arctan \frac{Q_{nw}}{Q_{sw}} \qquad (2.19)$$

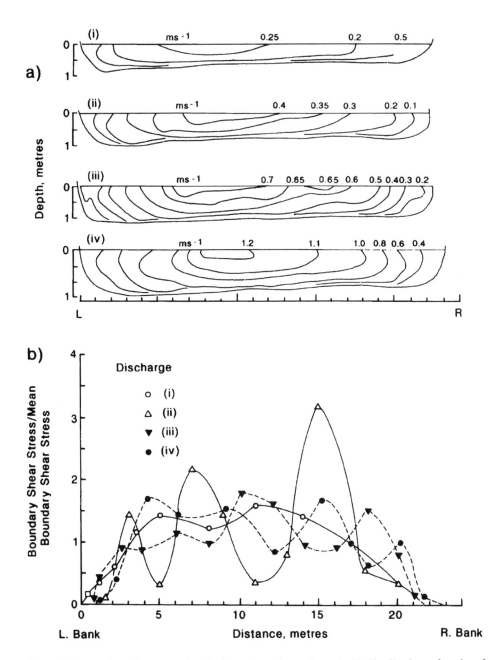

Figure 2.4   (a) Isovel pattern (vertical exaggeration ×2) and (b) distribution of ratio of boundary shear to mean boundary shear stress in a pool section on a straight channel; Llandinam, River Severn, UK, at discharges (i) 2.62, (ii) 4.74, (iii) 10.60 and (iv) 21.68 m³s⁻¹ (from Bathurst, 1979; reproduced by permission of Garnett Williams)

The down- and cross-stream components of flow can then be rotated by the angle θ_w to give the primary and secondary flow directions. The magnitudes of these "new" flow components can then be adjusted using the Pythagoras rule to give the primary and secondary flow velocities as the resultant velocity must remain constant throughout reorientation.

Straight channels

Spatial variations in the turbulence (variations in Reynolds stresses) generated at the channel boundary and the free surface are responsible for the development of secondary currents in straight channels (Prandtl, 1952; Perkins, 1970; Bradshaw, 1971; Nezu & Nakagawa, 1984). Three-dimensional flow in straight channels was probably first recorded by Stearns (1883), following experiments in the Sudbury conduit, Boston, USA, where flow was observed to well up at the sides of the conduit with surface currents directed towards mid-channel. Stearns concluded that secondary flow was composed of a pair of longitudinal helices or vortices.

Measured velocity distributions in straight natural channels confirm these observations. For pool sections on straight reaches, Bathurst (1979) showed that the velocity distribution was relatively uniform (Figure 2.4). Although secondary currents were not measured directly, they influence the long-stream velocity distribution. Where faster surface flows converge, the associated downwelling increases the near-bed velocity gradient and hence shear stress. The reverse will be true in regions where upwelling is occurring. As there was little cross-sectional variation in the ratio of peak to mean shear stress, nor any discernible change with discharge, it was concluded that this reflected a multicellular system of weak stress-induced secondary circulation, with adjacent cells rotating in opposite directions, generating alternate zones of upwelling and downwelling with associated troughs and peaks in the shear stress distribution.

At the approach to riffles, peak shear stress was observed in mid-channel in association with the zone of maximum velocity (Figure 2.5). Bathurst (1979) concluded that this was in response to flow acceleration as the relatively shallow riffle section was approached. In order to obey the laws of continuity, the flow must either accelerate, or, in the case of a widening channel, diverge. Acceleration would tend to cause compression of the isovels in the zone of maximum velocity thus increasing the shear stress peak. In the case of flow divergence, vertical velocities would be generated to maintain continuity, which would modify the shear stress distribution.

Mean secondary currents, V (vertical), W (cross-stream) in straight channels are governed by the vorticity equation of the streamwise component of vorticity, Ω (Einstein & Li, 1958; Nezu et al., 1993):

$$\left[V\frac{\partial \Omega}{\partial y} + W\frac{\partial \Omega}{\partial z} \right] = \left[\frac{\partial^2}{\partial y \partial z}\,(\overline{w^2} - \overline{v^2}) \right] + \left[\left\{ \frac{\partial^2}{\partial y^2} - \frac{\partial^2}{\partial z^2} \right\}\overline{vw} \right] + \quad [w\nabla^2\Omega] \qquad (2.20)$$

[advection terms] = [generation term] + [Reynolds stresses] + [viscous term]

$$\Omega \equiv \frac{\partial V}{\partial z} - \frac{\partial W}{\partial y} \qquad (2.21)$$

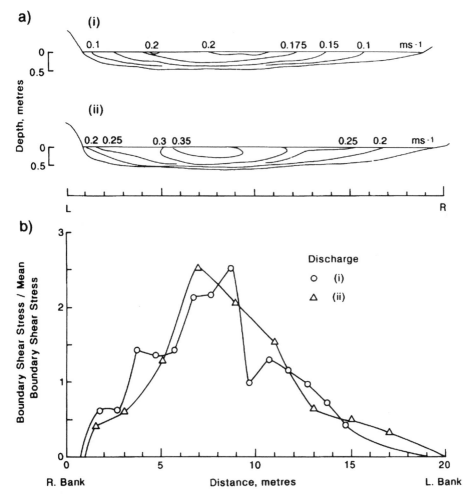

Figure 2.5 (a) Isovel pattern (vertical exaggeration ×2) and (b) distribution of ratio of boundary shear to mean boundary shear stress on a straight reach upstream of a riffle; Llanidloes, River Severn, UK, at discharges (i) 0.76 and (ii) 2.02 m³s⁻¹ (from Bathurst, 1979; reproduced by permission of Garnett Williams)

where v, w are the point velocity fluctuations in the z (vertical) and y (cross-stream) planes, respectively, the overbars indicating that the value has been averaged. The vortex advection term $V\partial\Omega/\partial y + W\partial\Omega/\partial z$ defines the strength of the secondary circulation, but the terms are difficult to measure and their calculation involves finding the solution to a second-order partial differential equation. The first and second terms on the right-hand side of eqn 2.20 relate to the processes generating and suppressing secondary currents. The viscous term is negligible except for regions near the wall where it becomes dominant. Nezu & Nakagawa (1984) verified experimentally and Demuren & Rodi (1984) explained numerically, that the generation term $\partial^2(\overline{w}^2 - \overline{v}^2)/\partial y\partial z$ and the Reynolds stress term $(\partial^2/\partial y^2 - \partial^2/\partial z^2)\overline{vw}$

are the dominant ones in eqn 2.20. Hence the advection term can be approximated from the first two terms in eqn 2.20. The mechanisms controlling the generation of secondary currents in straight channels are illustrated in Figure 2.6.

Secondary currents influence the primary flow distribution which, in turn, affects the near-bed velocity gradient and the boundary shear stress, τ_0. As the gradient of the primary velocity influences the primary Reynolds stress ($-\bar{u}\bar{w}$ and $-\bar{u}\bar{v}$), secondary currents will transfer momentum, mass and energy in a channel and the distribution of secondary currents will vary in all three dimensions.

Stress-induced secondary flows can occur in narrow non-circular ducts (Nikuradse, 1930) due to the influence of the banks. These currents are often termed corner flows as they flow from the core to the corners of the duct. Similar flow may be generated in natural channels if the width to depth ratio is less than five (Nezu et al., 1993), as the sidewall generates a turbulent flow comparable to that in ducts. For wider channels (width/depth ratio greater than five), wall effects do not influence the central zone of the channel and two-dimensional flow is exhibited in the long-term averaged turbulent structures.

Thus, the velocity distribution in a straight channel is not a simple two-dimensional phenomenon as characterized by the log velocity law and consequently it does not reflect the cross-sectional shape of the channel. Weak stress-induced circulation depresses the zone of maximum velocity below the surface due to the displacement of high-momentum fluid following the transportation of low-momentum fluid in

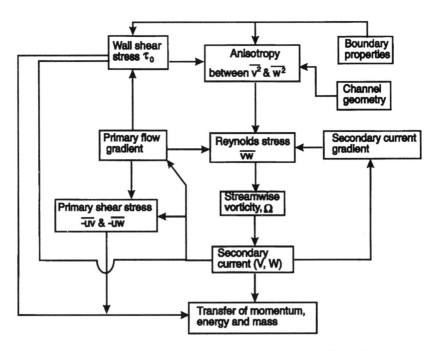

Figure 2.6 Generation of secondary currents in straight channels (from Nezu et al., 1993); reproduced from *Journal of Hydraulic Engineering*, vol. 119, no. 5, May, 1993. © ASCE, by permission of the American Society of Civil Engineers)

the cross-channel direction by secondary currents (Sarma *et al.*, 1983). The more intense the secondary circulation, the greater the downward displacement of the maximum velocity filament below the free surface.

Meandering channels

Stronger secondary currents are observed in meander bends due to the skewing of cross-stream vorticity in a streamwise direction as the flow is curved around the bend. Flow curvature generates a centrifugal force which is proportional to the square of the point velocity and inversely related to the radius of curvature of the channel centre-line. The faster surface flow will experience the greatest radial force driving it towards the outer bank in the bend and causing super-elevation of the free surface. In turn, this generates a cross-stream pressure gradient force (given by pgΔh, where ρ is the density of water), which results in a return flow of near-bed water from the outer to inner bank. As a result, a large secondary, or helical, flow cell is generated which is characteristic of flow in meander bends (Figure 2.7). This is generally confined to the thalweg section of the channel. Effectively this transfers high-momentum surface water towards the outer bank which locally accelerates the flow, while low-momentum near-bed water is transferred to the inner bank decelerating the flow. This advective transport of primary flow momentum by the secondary flow has a significant effect on the distribution of primary flow velocities. The intensity of this process is inversely related to bend curvature.

Whilst secondary flows in meander bends are principally generated by curvature effects, they are also influenced by changes in the cross-sectional shape of the channel. River bends have an asymmetric cross-section with a deep pool adjacent to

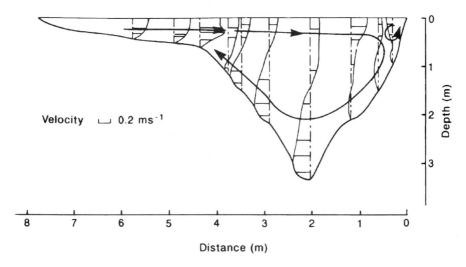

Figure 2.7 Secondary flow pattern at the apex of a meander bend showing main secondary cell, outer bank cell and radial flow over the point bar; Fall River, Colorado, USA (from Thorne & Rais, 1983; reproduced from *River Meandering, Proceedings of the Conference on Rivers 1993* (Ed. C. M. Elliot) by permission of the American Society of Civil Engineers)

the outer bank at, or immediately downstream of, the bend apex, and a shallow point bar on the inner bank, these features both reflecting and influencing secondary flows. Flow approaching the point bar experiences a rapid reduction in the flow depth which, to satisfy continuity, has either to accelerate or diverge. Acceleration increases the radial forces acting on the fluid promoting flow towards the outer bank (Figure 2.7), and this would be reinforced by flow divergence (Dietrich et al., 1979; Thorne et al., 1985).

Laboratory and field data indicate that there can also be a small cell, with a reverse rotation to the main cell, adjacent to the outer bank. This was first noted in laboratory studies (Einstein & Harder, 1954; Rozovskii, 1954) and was later detected in natural channels (Hey & Thorne, 1975; Bathurst et al., 1979; Dietrich et al., 1979). It is caused by the interaction of the main cell with the outer bank and appears where the bank is steep (Figure 2.7). The combination of a strong secondary velocity and steep outer bank will produce a stagnation point at the bank face and generate flow up and down the bank. Upwelling of faster flowing water creates the outer bank cell. Although small compared with the main cell, extending out from the bank to a distance of half to one and a half times the bank height, it significantly influences bank erosion and planform development. On shelving banks an outer bank cell can not develop as bank roughness progressively reduces lateral velocities and vertical flow is constrained by the shallow depth.

The strength of the secondary circulation in a bend varies with discharge. It is weakest at low discharges, even though the radius of curvature is small. At intermediate discharges, velocities increase and this can offset reduced curvature to produce maximum centrifugal forces and hence secondary flows. During bankfull discharges, if the river straightens its course by cutting across the point bars then secondary flows can be reduced in strength (Bathurst et al., 1979).

Although it is often assumed that secondary flow is locally adapted to the channel curvature, Gottlieb (1976), Kitanidis & Kennedy (1984) and Ikeda & Nishimura (1986) have indicated that within an irregular bend, there may be a phase lag between secondary current strength and channel curvature. This, according to Ikeda & Nishimura (1986), is independent of flow depth and simply depends on the local radius of curvature of the channel centre-line. Thus, the magnitude of the secondary flow in an irregular bend varies from that for uniformly curved channels. This phase lag is induced by inertia in the downstream advective acceleration of the secondary flow.

Between bends, the secondary circulation of the upstream bend generally decays to be replaced by the new skew-induced circulation in the downstream bend. Flume studies with rectangular fixed bed channels indicate that the replacement is lateral, with cells side by side when width/depth ratios are low (Toebes & Sooky, 1967), but stacked vertically where width/depth ratios are high (Chacinski & Francis, 1952). Field results have shown a two-cell stacked pattern in a river where the width/depth ratio is high (Thorne & Hey, 1979; Figure 2.8). This probably resulted from local variations in bed topography as the thalweg is more sinuous than the bankline in meandering channels. At the inflection point, the thalweg crosses the channel and this produces curvature of the stream lines

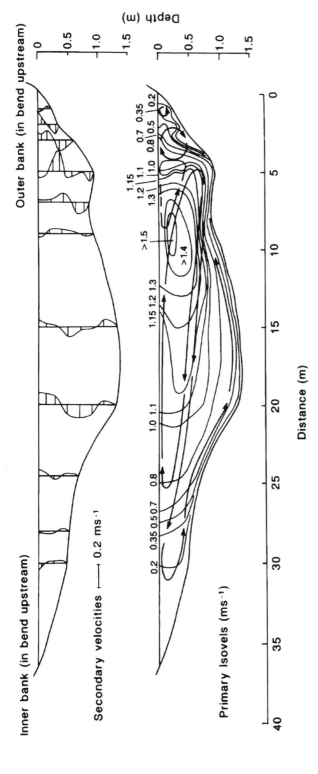

Figure 2.8 Secondary flow pattern at inflection point between meander bends, showing vertical stacking with new flow cell developing at the bed and the relict cell being displaced at the surface; Penstrowed, River Severn, UK (from Thorne & Hey, 1979; reprinted with permission from *Nature*, vol. 280, p. 227. Copyright 1979 Macmillan Magazines Limited)

near the bed, and skewing of the shear stress field thereby generating the new secondary circulation.

The core of maximum primary velocity in meanders is also influenced by bend geometry. During low flows, the core of maximum velocity at the entrance to a single bend is located between centre channel and the inner bank due to the favourable pressure gradient close to the inner bank at the bend entrance. In the bend, the skew-induced secondary circulation develops and eventually breaks down the free vortex carrying the core of maximum velocity towards the outer bank (Ippen & Drinker, 1962; Dietrich et al., 1979). Consequently the position of the core of maximum velocity depends on the strength of the secondary circulation. At medium discharges the secondary circulation is relatively strong and so the point at which the core of maximum velocity crosses over towards the outer bank is located early in the bend. However, as discharge increases, this point drifts downstream towards the bend exit (Bhowmik & Stall, 1978).

In a series of meander bends Bridge & Jarvis (1976) suggest that if relatively high velocities occur near the convex bank of river bends, then they are likely to be the remnant of the velocity field from the adjacent upstream bend. This implies a "memory effect", yet on the Squamish River, Hickin (1978) observed high velocities over the point bar when the velocity distribution at the entrance to the bend was sensibly symmetrical. Consequently velocity distributions in these bends are likely to be the result of downstream advective acceleration modified by secondary circulation. These contrasting conclusions probably reflect the fact that the meanders on the South Esk are closely spaced, whereas those on the Squamish are separated by relatively long straight sections.

SHEAR STRESS DISTRIBUTIONS

The average shear stress at a cross-section can be obtained by equating the downstream gravity component against the frictional resistance opposing the flow which acts upstream. Under uniform flow conditions, with no local long-stream changes in average velocity or flow depth, the two forces balance and the average boundary shear stress, τ_0, can be determined. This is given by:

$$\tau_0 = \gamma dS \qquad\qquad (2.22)$$

where γ is the specific weight of water, d the average flow depth and S the average water surface slope.

This assumes that all the energy in the flow is dissipated in order to overcome surface drag at the channel boundary whereas, in reality, energy can also be lost due to secondary flows, especially in meandering channels, local acceleration and deceleration, bedforms, reverse flows, bed and bank vegetation and surface waves. Consequently, the actual shear stress at the bed (often referred to as the boundary shear stress due to skin friction) can be less than that calculated from eqn 2.22. Only where the channel is straight, the bed and bank are composed of similar material and are unvegetated and the flow is perfectly uniform will all the energy be dissipated at the boundary, overcoming skin friction.

The bed shear stress can best be determined from measured velocities as the gradient of the log velocity profile (eqn 2.9) enables the shear velocity v_* to be determined. This can be converted to the bed shear stress, τ'_0, since:

$$\tau'_0 = \rho v^2 \qquad (2.23)$$

However, care has to be exercised when determining the bed shear stress using this method as eqn 2.9 was derived on the assumptions that the flow was two-dimensional, that the shear stress in the fluid equalled the bed shear stress and that the eddy mixing length is proportional to the distance from the bed. In theory, where flows are three-dimensional, the above assumptions restrict its use to the lowest 20% of the flow depth. Within this "inner region" turbulence is fully developed, whilst in the "outer region" the mixing length becomes constant. Although in practice eqn 2.9 often fits throughout the full flow depth, it is still preferable to determine the shear stress from point velocity measurements taken within the lowest 15–20% of the flow depth. Peak values of the bed shear stress occur where the isovels are compressed near the bed, namely below the maximum velocity filament, and in zones of downwelling. Low values occur where near-bed primary velocities are low and in zones of upwelling.

Local values of bed shear stress vary considerably both at and between cross-sections. Straight reaches show contrasting patterns between pools and riffles (Bathurst, 1979). At a pool, the distribution shows peaks and troughs with no clear maximum (Figure 2.4). The ratios of peak to mean values are between 1.5 and 2. This pattern might result from a multicellular stress-induced secondary circulation as described by Perkins (1970). Such a circulation would produce alternate regions of upwelling and downwelling between cells. On the approach to a riffle, the distribution shows a central peak below the core of maximum velocity and smaller shoulder peaks (Figure 2.5). Ratios of peak to mean shear stress lie in the range 1.5 to 2.5. This pattern may result from acceleration of primary flow over the riffle and secondary circulation associated with flow divergence.

Channel bends show two peaks of bed shear stress, one associated with the core of maximum velocity and the other with the region of downwelling between the main and outer bank cells (Figure 2.9). Peak values below the core of maximum velocity were in the range 1.5 to 2.5 greater than the section mean, which agrees with the observations by Apmann (1972) and Hooke (1975). Downwelling produces peaks up to four times the section mean, depending on the strength of the outer bank cell. Where the outer bank cell is absent, downwelling is too diffuse to produce any peak. The relative size of the two peaks varies with discharge. At low flows both primary and secondary velocities are small and either peak may be the larger. At medium discharges, the effects of secondary circulation seem to be at their strongest and often the outer bank peak is the greater. However, at high discharges, primary flow effects dominate over secondary flow effects and the stress peak below the core of maximum velocity is the largest (Figure 2.9).

The average cross-sectional bed shear stress, based on the average of point values measured at 2 m intervals across the channel using eqn 2.23, has been shown to vary between pools and riffles with respect to discharge (Carling, 1991; Keller &

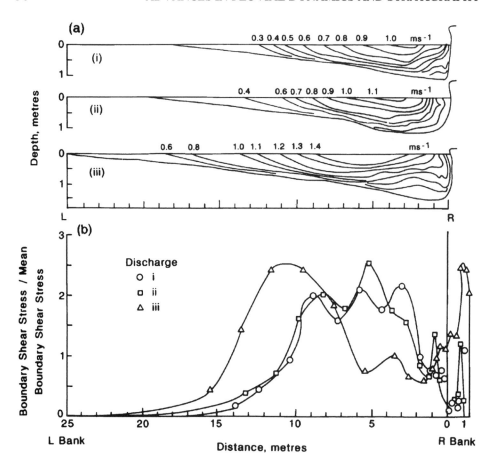

Figure 2.9 (a) Isovel pattern (vertical exaggeration ×2) and (b) distribution of ratio of boundary shear to mean boundary shear stress at the apex of a mender bend, Caersws, River Severn, UK, at discharges (i) 5.22, (ii) 6.84 and (iii) 15.48 $m^3 s^{-1}$ (from Bathurst, 1979; reproduced by permission of Garnett Williams)

Florsheim, 1993). At low to intermediate flows, the bed shear stress on the riffle exceeds that in the pool, due principally to higher water surface slopes offsetting lower flow depths. As flow increases, up to bankfull, the bed shear stress on the riffle may actually decline, as it is drowned out and the water surface gradient is reduced faster than the flow depth is increased. In contrast, the bed shear stress in the pool continues to increase in response to further increases in the water surface slope and flow depth (Figure 2.10, from Carling, 1991). Variations in bed shear stress between low and high discharges mean that pools scour during peak flows and some of the coarser material is deposited in the riffle. Scouring occurs on the riffle during flood recession and material is transported into the adjacent downstream pool. Consequently during floods, bed material is transported through pool–riffle sequences in a series of pulses.

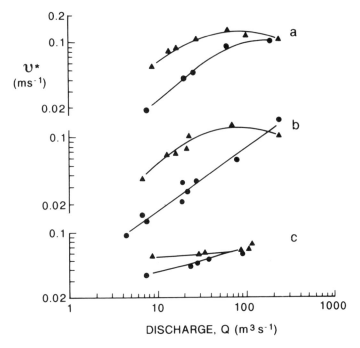

Figure 2.10   Variation of average shear velocity with discharge at three consecutive pool/riffle sections (a, b and c). Circles represent pools; triangles represent riffles (from Carling, 1991)

MODELLING VELOCITY AND SHEAR STRESS DISTRIBUTIONS IN BENDS

Velocity and shear stress distributions in natural channels are controlled by the morphology and sedimentary characteristics of the river and vary with discharge. These, in turn, affect sediment entrainment and transport which are responsible for bed scour, bank erosion and bar deposition and for the morphological evolution of the river. An understanding of the dynamic interaction between the flow, the channel boundary and the sediment is fundamental to the establishment of physically deterministic models of channel development and associated river and floodplain sedimentation. In this section the basic principles underlying the mathematical modelling of velocity and shear stress distributions in meander bends are presented for flow in fixed bed channels.

Modelling flow in bends

Equations of motion and coordinate system

The flow of water in an open channel with a free surface is fully described by the equation of mass continuity and the Navier–Stokes force–momentum equation (eqn 2.24) (see Paterson, 1983) which contains both hyperbolic and parabolic components:

$$\rho\,\frac{Dv_i}{Dt} = \rho F_1 - \frac{\partial p}{\partial x_i} + \frac{\partial}{\partial x_j}\left\{\mu\,\frac{\partial v_i}{\partial x_j} + \mu\,\frac{\partial v_j}{\partial x_i}\right\} + \frac{\partial}{\partial x_i}\left\{\left(K - \frac{2\mu}{3}\right)\frac{\partial v_k}{\partial x_k}\right\} \qquad (2.24)$$

where ρ is the fluid density, v is the fluid velocity in the i, j (horizontal) and k (vertical) directions, F is the frictional forces exerted by the fluid, p is the fluid pressure, x is a scale length, μ is the viscosity and K is a constant. The value of K is partly dependent on the fluid temperature and density but to determine its value it is necessary to solve the equation using local information for a particular flow condition. This can then be used to model different flows through the river reach.

For most practical applications, the full equations are more complex than generally required, or are impossible to solve owing to data limitations. Indeed there are no general solutions to the full set of equations. This problem can be avoided by the use of depth-averaged equations of motion, which removes the vertical velocity component from the equations.

In three dimensional flow, the vertical pressure distribution is controlled by the mass fluxes in all three directions. As this pressure distribution also alters the three dimensional velocity distribution, it is unclear whether it is the velocity distribution which controls the pressure distribution, or *vice versa*. This makes it difficult to determine the five variables required (three components of velocity, flow depth and fluid pressure). Depth-averaging follows from the assumption that the pressure at any depth in the flow can be approximated to the hydrostatic pressure. Thus the velocity distribution in the vertical will be uniform and, as its value can be calculated, terms containing the vertical velocity in the Navier–Stokes equations can be neglected. This means that cross-channel variations in depth-averaged velocity, as they vary with channel form and roughness, are incorporated into the model. Inevitably, any model based on vertical averaging will not predict the velocity distribution *per se* (Bernard & Schneider, 1992) as any secondary flows obtained from the calculations are in the form of an average net transfer in the long- and cross-stream directions. In a bend, only in regions of strong upwelling or downwelling will the convective effects of secondary flow be important and it should be borne in mind that the output from a model may become rather poor in these regions (De Vriend & Geldof, 1983). For the majority of the reach, where advective components dominate, model predictions will remain valid. By considering the orders of magnitude of the terms in the equations, they can be further simplified so that they apply to gradually varied flow (i.e. avoiding hydraulic jumps). In meandering rivers, the depth of flow, h, is small compared to the width, w, and thus $h/w < 1$. Also, the radius of curvature is generally larger than the width. Under such conditions, experiments have shown (Rozovskii, 1961; Yen, 1972; Odgaard & Bergs, 1988) that the cross-stream flow velocity, v_n, is of the order $v_s(h/w)$, where v_s is the long-stream velocity, and that the vertical velocity, v_z, is of the order $v_n(h/r)$, where r is the radius of bend curvature. Thus, all terms in the driving equations containing v_z can be eliminated.

The assumption of a hydrostatic pressure distribution will break down in the presence of a significant cross-channel flow as super-elevation of the water surface will develop adjacent to the outer bank. However, in the interests of the development of simple algorithms, this problem is disregarded on the assumption that these defects are local in nature. This can very often lead to large problems when these algorithms are included in a mathematical model as these local small-scale approximations may be significant enough to generate areas of mathematical instability in the model which are capable of perturbating through the whole system, making any data generated worthless. Instabilities such as this can have their greatest effect when they are located at or near the bank boundaries as, owing to the hyperbolic nature of the driving equations, errors at these boundaries are rapidly propagated into the rest of the numerical domain (Anderson *et al.*, 1984). Further simplifications can be made provided that the bed of the channel is rigid, bed roughness is dominant and all other roughnesses can be neglected (e.g. bank roughness), frictional losses are calculated assuming that steady-state conditions prevail, and the channel bed slope is small (less than 5°).

Generally the proviso that the bed roughness is dominant is a valid one. However, near the boundaries (at a distance from the bank equal to the local flow depth), bank roughness will have an increasing effect on the flow. Whereas bed roughness produces a deceleration acting in the vertical plane, bank friction acts in a plane perpendicular to the bank which, on shallow slopes, approaches that of the bed. However, on the outside of the bend, where the bank slopes are steep, bank roughness effects need to be considered.

Given the above restrictions and assumptions, the Navier–Stokes equations can be reduced to the two-dimensional St Venant (St Venant, 1871) shallow-water wave equations (eqns 2.25–2.27).

(i)   Continuity

$$\frac{\partial h}{\partial t} + \left[\frac{1}{\left(1 + \dfrac{n}{R}\right)} \frac{\partial}{\partial s} \langle v_s \rangle h + \frac{\partial}{\partial n} \langle v_n \rangle h + \frac{\langle v_n \rangle h}{R\left(1 + \dfrac{n}{R}\right)} \right] = 0 \qquad (2.25)$$

[temporal change in flow depth] + [fluid divergence] = 0

(ii)   Momentum equation in *s*-direction (long-stream)

$$\left[\frac{\partial}{\partial t} \langle v_s \rangle h \right] + \left[\frac{1}{\left(1 + \dfrac{n}{R}\right)} \frac{\partial}{\partial s} \langle v_s^2 \rangle h + \frac{\partial}{\partial n} \langle v_s \rangle\langle v_n \rangle h + \frac{2\langle v_s \rangle\langle v_n \rangle h}{R\left(1 + \dfrac{n}{R}\right)} \right] + \left[\frac{g}{\left(1 + \dfrac{n}{R}\right)} \frac{\partial}{\partial s} \left(\frac{h^2}{2} \right) \right]$$

$$= \left[\frac{gh}{\left(1 + \dfrac{n}{R}\right)} (S_{os} - S_{fs}) \right] \qquad (2.26)$$

[temporal change in long-stream depth-averaged velocity] + [main flow inertia]
+ [long-stream change in potential energy] = [long-stream bed friction]

(iii) Momentum equation in n-direction (cross-stream)

$$\left[\frac{\partial}{\partial t}\langle v_n\rangle h\right] + \left[\frac{1}{\left(1+\dfrac{n}{R}\right)}\frac{\partial}{\partial s}\langle v_s\rangle\langle v_n\rangle h + \frac{\partial}{\partial n}\langle v_n^2\rangle h + \frac{h\langle v_n^2\rangle - \langle v_s^2\rangle}{R\left(1+\dfrac{n}{R}\right)}\right] + \left[g\frac{\partial}{\partial n}\left(\frac{h^2}{2}\right)\right]$$

$$= [gh(S_{on} - S_{fn})] \quad (2.27)$$

[temporal change in cross-stream depth-averaged velocity] + [secondary flow inertia]
+ [cross-stream change in potential energy] = [cross-stream bed friction]

where $\langle v_s\rangle$ is the long-stream depth-averaged flow velocity, $\langle v_n\rangle$ is the cross-stream depth-averaged flow velocity, h is the flow depth, S_0 is the channel bottom slope, S_f is the friction slope, R is the hydraulic radius, n is the cross-channel curvilinear coordinate and the suffixes s and n refer to long-stream and cross-stream components, respectively.

Equations 2.25–2.27 are given in a curvilinear (or channel-fitted), rather than a Cartesian coordinate system. Curvilinear coordinates are defined (Figure 2.11) such that the origin, O, is at the centre of curvature of the channel bend, the long-stream, s, axis is given by concentric arcs centred at the origin, and the cross-stream, n, axis is defined to be orthogonal to the s axis at all points. The scaling factor $1/(1 + (n/R))$ is generated during the transformation from a Cartesian to a curvilinear coordinate system, and only applies to terms involving the long-stream axis as it is curved.

Channel-fitted coordinates are employed for three important reasons. First, the boundaries of the model can be imposed along one axis of the coordinate system and can be defined as a single function for the whole boundary. Second, the fact that the boundary of the channel is an axis of the coordinate system removes the "stair step" type boundary (Figure 2.11) that would be present if using a Cartesian coordinate system. This removes a source of error from the numerics of the scheme. Third, the use of such a coordinate system gives a very good geometric representation with a small number of grid points (Jovanovic & Officier, 1984).

The purpose of this change of coordinates is to establish a transformation from an arbitrary physical domain to a rectangular domain in which the difference equations can be solved. This introduces the coefficient $1/(1 + (n/R))$ in eqns 2.25–2.27. However, since the cross-stream axis is straight, its scale factor is unity. Although this complicates the driving equations, it removes a possible source of error. If the channel is natural, then it is unlikely that this form of circular curvilinear coordinate can be used to map the boundaries of the channel accurately, in which case natural curvilinear coordinates may be adopted (Jovanović & Officier, 1984). Thus, the cross-stream axis may also have to be curved to produce an accurate representation of the channel. This introduces a scaling factor for terms along the cross-stream axis which further complicates the equations.

Numerical grid

All numerical models require a grid in which to solve the driving equations. This basically consists of a network of data points (whose size and shape is defined by

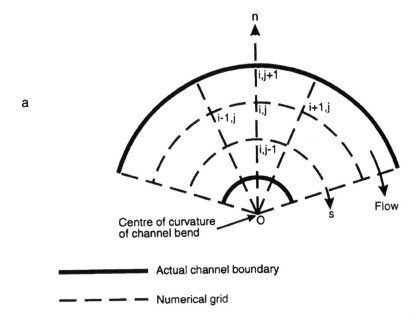

a

Centre of curvature
of channel bend

━━━━━━━━━ Actual channel boundary

─ ─ ─ ─ ─ Numerical grid

b

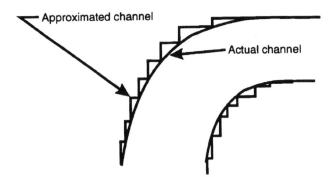

Figure 2.11 Fitting channel boundaries with (a) channel-fitted coordinates and (b) Cartesian coordinates

the geometry of the river and the coordinate system) to facilitate computation. A convenient grid would consist of the locations of each sampling point in the meander reach at which data were collected to form the initial conditions for the model. Equations 2.25–2.27 are solved at each intersection point of the grid. For numerical purposes, the grid point at which a calculation is being made is referred to as having the coordinate i, j (i and j are integers). Thus, for example, the adjacent point downstream in the s direction given by $i + 1, j$ and the adjacent point towards the inner bank is given by $i, j - 1$ (see Figure 2.11).

The spacing of the grid is important for two reasons: first, to resolve processes occurring in the flow, and second for numerical stability (this will be discussed below). In order to mathematically describe and resolve the turbulent eddies in the flow, the grid spacing has to be small enough so that a single eddy cannot exist within one grid box. Failure in this respect will prevent the effect of these "subgrid-scale" eddies on secondary flows from being modelled and parameterization will be necessary which can lead to errors in the numerical results. However, as the grid size becomes smaller, the computational requirement increases. Hence, the desire for accuracy has to be balanced with the computational power available.

Steady-state flow models

Steady flow implies that there is no change in the flow properties over time. Consequently, the temporal derivatives in eqns 2.25–2.27 can be removed, which further simplifies the model. If there is only a small change in the flow properties with respect to time then steady-state equations can still be employed on the assumption that the flow is able to equilibrate with the new flow conditions within the time interval (time-step).

Steady-state flow models are often used to verify and improve understanding of the physical processes controlling the velocity distribution in the bend. In addition to the limitations detailed above, such models (e.g. Kalkwijk & De Vriend, 1980; Bridge, 1982; De Vriend & Geldof, 1983; Odgaard, 1989a,b) are further restricted to a shallow and gently curving flow and to a Froude number that is so small that the rigid lid approximation can be applied to the water surface. This approximation leaves the water surface free in terms of the pressure gradient, but not in terms of the water depth.

The requirement for uniform shallow, gently curving flow in channels with "mildly" sloping banks inevitably restricts the wider application of such models. In rivers, the outer bank in a meander bend is generally steep, which is at variance with the requirements of the model. The profile of the banks and its variation around the bend has a strong effect on the flow. In regions with shallow sloping banks, the frictional resistance has a more dominant effect on the velocity distribution than in deeper sections with steep banks. At the shallow inner bank, flow depth progressively decreases towards the bend apex with the result that water is directed radially away from the bank. In contrast, surface water is converging at the outer bank producing downwelling and a near-bed return of water to the inner bank. Shallow banks at both sides of the channel will not allow this mechanism to develop properly.

In De Vriend & Geldof's model the strength of secondary circulation is proportional to the local properties of the primary flow whereas, in reality, it has its own inertia which retards its development. Even in sharp bends, the growth of the secondary flow to its full strength can take a large part of the bend. This means that in a natural channel the flow at a particular section reflects upstream conditions. By assuming that the secondary circulation is dependent only on local properties, there may be errors in the spatial distribution of the secondary circulation. Also, in sharp bends, the stronger secondary flows that are generated will tend to interact with the main flow. This leads to a considerable decrease in the maximum secondary flow intensity immediately downstream and, in the case of a steep outer bank, can give rise to the formation of a small, counter-rotating secondary cell (Bathurst *et al.*, 1979). Restriction to mildly sloping banks will prevent this phenomenon from being described in the model.

The model of De Vriend & Geldof (1983), and others that are based on similar approximations, are therefore only appropriate for shallow, curved and rather short bends which have a uniform symmetrical cross-section and mildly sloping banks. They should not be applied in other circumstances as significant numerical errors may be generated.

Unsteady-state flow models

The modelling of unsteady flow conditions requires that the time derivatives of the driving equations are considered. Previously known or calculated values for the fields of velocity and flow depth at time $t = t$ are used to predict new values at time $t = t + \Delta t$. These new values are then used to predict forward to further time levels $t = t + 2\Delta t ... t = t + n\Delta t$, where n is an integer. The value of Δt, the time-step, must be chosen so that no fluid that enters a grid box at the beginning of the time-step can travel across and out of the grid box during the time-step. The values at each grid point at the new time level are calculated using the surrounding "star" of grid points. This mechanism of time-stepping is illustrated in Figure 2.12, where two time levels are shown. Specific points in the grid represent the same spatial point at each time-step, but show temporal variations in the calculated field.

The full St Venant shallow-water wave equations are a set of non-linear hyperbolic partial differential equations whose analytical solutions are only available in idealized cases (Dressler, 1952). However, they can be solved in realistic situations numerically, using either finite difference or finite element methods. Finite difference techniques involve discretizing the differential equations to produce difference equations so that there is no need to define the limit of the differential coefficients in eqns 2.25–2.27. Thus, the system is approximated by the use of small spatial differences whose relationship can be used to predict the next time level. The finite element approach solves the equations in integral form by employing a grid which is divided into finite elements each of known area. This has the advantage over finite difference schemes in that it is a very simple matter to change the grid size without the need to alter the equations as is the case with finite difference models. Presently the finite difference approach is the most commonly used method, although finite element models are being developed, especially by engineers.

Two methods are available to solve the difference or integral equations at each time-step, implicit and explicit, which are equally applicable to either finite difference of finite element schemes. Implicit schemes solve for the whole numerical grid simultaneously using only data from the current time level. Explicit schemes solve for the grid using information from the previous time level and thus do not require the whole grid to be solved simultaneously. Explicit schemes have the advantage that they are generally easier to formulate as they do not require some

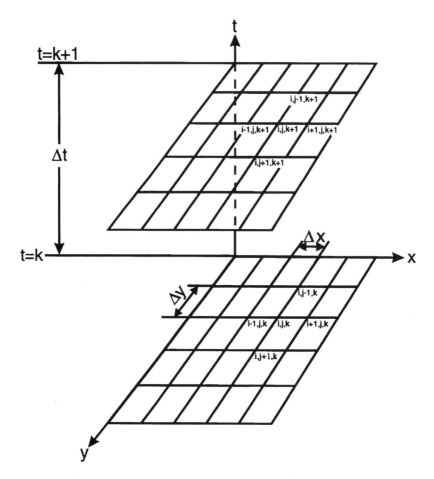

Figure 2.12   Finite difference grid showing two time levels

form of simultaneous matrix algorithm to be incorporated into their code. However, implicit schemes are not as restricted as explicit ones to the small time-steps and thus require less computational time. Also, extension to higher orders of spatial accuracy can be achieved with a minimal increase in computational effort with an implicit scheme (Beam & Warming, 1975). This benefit is reduced by the fact that second-order accurate schemes are usually sufficient for most applications, i.e. terms in the Taylor series expansion with a power higher than two are ignored. However, in some cases, it has been shown that second-order schemes can achieve a higher accuracy than third-order ones (Anderson, 1974).

A few models have been developed for unsteady flow conditions (e.g. Garcia & Kahawita, 1986; Dammuller *et al.*, 1989; Bernard & Schneider, 1992; Garcia-Navarro & Saviron, 1992. They all use two-dimensional shallow-water wave equations (St Venant, 1871) in channel-fitted coordinates (Figure 2.11) which are solved numerically using the second-order explicit MacCormack (1969) finite difference scheme.

As with steady-state models, the depth-averaging assumptions allow errors to be incorporated into these models at an early stage. Equally, the assumption of a rigid bed will not allow bedforms to develop with time. For a given time-step, the model may compute a depth of erosion/deposition at a grid point but, as a result of this assumption, it is not fed back into the model at the next time-step and the bed profile remains constant with time, which may be the case for dynamic equilibrium between flow and sediment transport, but is not necessarily so. Essentially they were developed to predict changes in water surface elevations during dam-break flows on the assumption that the bed remained fixed. For morphological modelling, however, this tends to defeat the object of solving unsteady-flow equations. Unsteady-flow models could enable the effects of changing sediment supply on channel development to be investigated. Sediment transport does not occur constantly with respect to time, and field evidence indicates that pulses of sediment are transported downstream. This could produce a bed that is stable for most of the time, but which adjusts to increased sediment supply over a short period of time. The models referred to above are not capable of reproducing this process. Also, as bed erosion and deposition due to local sediment imbalance only occur over relatively long time periods, much longer than for changes in flow hydraulics, mobile-bed models need to run for a long time in order to produce significant changes in channel topography. The overall error in any numerical scheme is a function of the number of time-steps through which the solution is advanced (Turkel, 1974). Thus, if the flow model has to be run for an extremely long time in order for appreciable sediment transport to occur, the overall numerical error could be significant.

Computation of frictional losses in these models may or may not cause errors, depending on the conditions being modelled. If the flow varies on a time-scale significantly larger than the time-step, then the model can be considered as being in steady state for each time-step and significant errors should not be introduced. However, if the flows are changing rapidly with respect to the time-step then the model is unsteady and frictional losses can be incorrectly computed causing mathematical instability in the model. The friction slopes used in the majority of these models are functions of the Manning roughness coefficient, which is taken as the average value of high- and low-flow values. This can be better represented by the slope of the total energy surface (Henderson, 1966), which has the advantage that it can be calculated directly from the flow field without any problems at the boundaries or stagnation points in the flow where the flow velocity could be set to zero. Also, the friction slope can be allowed to vary within the grid, which is more realistic than holding it constant with respect to space and time.

Errors also tend to be found at points close to the entrance to the bend, possibly as a result of the sharp decrease in the radius of curvature from infinity to some finite value. These errors may result from the boundary conditions as the uniform grid produces a coarse resolution at the boundary. Large velocity gradients will be generated near this boundary which will produce instabilities in the scheme. Use of a sinusoidal cross stream grid spacing would overcome this problem by overlaying a fine-resolution grid at the points where the highest gradients occur. Although this is complicated, it should not be impossible to implement in finite difference schemes provided more research is carried out on wall boundary conditions.

It has been observed that Dammuller *et al.*'s (1989) model becomes unstable as the flow approaches critical Froude numbers. This results from the use of an explicit scheme. With an implicit scheme, the new data for each time-step are calculated simultaneously and thus information for the whole flow can be transmitted to other points at the actual wave speed and any changes in the flow field will be apparent to all points in the grid. In an explicit scheme, this information transport is at the calculated wave speed, which can differ from the actual wave speed on the approach to critical Froude numbers. Explicit schemes use data from the present time level to calculate the new values at time $t + \Delta t$, whilst implicit schemes solve for all grid points simultaneously using only the new time level. On the approach to critical conditions, the computational wave speed was found to be slower than the actual wave speed and thus information on the flow could not be sent upstream at the required rate, with the result that the calculated flow is not a true reflection of the flow/bed interactions. When this occurs errors will appear in the grid. As the computed wave speed lags the actual speed on the approach to critical conditions, it implies that some of the assumptions in the model are unrealistic. There are two possible explanations. First, there may be an error in Dammuller *et al.*'s model producing this instability, or second, even though the MacCormack scheme has been shown to be stable in both sub- and supercritical conditions (MacCormack, 1971; Anderson, 1974; Moretti, 1978; Garcia & Kahawita, 1986; Garcia-Navarro & Saviron, 1992), it may not be stable when the flow is close to critical Froude numbers. There is insufficient evidence to identify what is actually causing the instability.

Modelling shear stress distributions

Once the flow field has been determined, the shear stress field can be calculated, which, if desired, can be used to compute sediment transport and channel evolution. This is not a simple process because the modelled flow field that is calculated is depth-averaged and, as such, there is no velocity gradient from which the shear velocity can be calculated.

Smith & McLean (1984) developed a steady-state model that overcomes this problem. The main differences between this model and earlier ones (e.g. Engelund, 1974; De Vriend, 1976) are the inclusion of some of the more complex convective acceleration terms and the recognition that the momentum flux terms in rivers with pool – riffle topography are of primary, not secondary, importance in the equations of motion.

By vertically averaging the flow, the following force balance equations are derived:

$$(\tau_{zs})_b = \frac{-\rho g h}{(1+N)} \frac{\partial E}{\partial s} - \frac{\rho}{(1+N)} \frac{\partial}{\partial s} \langle v_s^2 \rangle h - \rho \frac{\partial}{\partial n} \langle v_s v_n \rangle h + 2\rho \frac{\langle v_s v_n \rangle h}{(1+N)R} \qquad (2.28)$$

$$(\tau_{zn})_b = -\rho g h \frac{\partial E}{\partial n} - \rho \frac{\langle v_s^2 \rangle h}{(1+N)R} - \frac{\rho}{(1+N)} \frac{\partial}{\partial s} \langle v_s v_n \rangle h - \rho \frac{\partial}{\partial n} \langle v_n^2 \rangle h + \rho \frac{\langle v_n^2 \rangle h}{(1+N)R}$$

$$(2.29)$$

where $(\tau_{zs})_b$ and $(\tau_{zn})_b$ are the bed shear stresses in the long- and cross-stream directions, respectively, $\partial E/\partial n$ and $\partial E/\partial s$ are the cross- and long-stream water surface slopes, respectively, and N, R, v_s, v_n, s, n, h are as defined in eqns 2.25–2.27.

There are several problems with this scheme. First, the bed shear stress is indirectly dependent on the bed sediment size and form friction effects, being parameterized via the water surface slopes. Second, as the time derivatives are excluded, the model is restricted to steady flow. Consequently, it can only be applied to data sets that have been collected at an instant in time, or under steady-state conditions. If time derivatives of these expressions are not employed in a mathematical model, then the exclusion of energy losses as a result of the flow/bed interactions are effectively ignored. This will not produce an accurate model of a channel. Third, although in principle eqns 2.28 and 2.29 can be used to determine the shear stress from flow measurements, they represent a delicate balance of large terms, some of which nearly cancel each other out. In spite of the fact that this is a physically realistic model, it has been suggested (Dietrich, 1982) that it is impractical to apply as a result of measurement and sampling errors. However, Smith & McLean (1984) have shown that it realistically models free-surface elevations and boundary shear stress, which also agrees with the sediment distribution theory of Hooke (1975).

FUTURE PROSPECTS

At present, measurements of flow and channel geometry are carried out by what are effectively fairly primitive means, involving temporally spaced point measurements. Clearly, this does not produce an instantaneous or ideal picture of the flow processes that are occurring. With the increasing availability of portable computers, side-scan sonar techniques are now becoming available for surveying bed morphology in a natural channel. Side-scan sonar works on the same principle as an echo sounder, using a reflected sound wave to determine the reflector's distance from the source. The instruments include a pan and tilt mounting which enables the sound pulse to be directed in a wide arc to sample a large area of the channel bed. Given the position of the sonar (usually mounted on a moored boat) determined by standard surveying procedures, a detailed map of the bottom topography can be built up by processing the recorded information in a computer plotting package such as Surfer or Stanford Graphics. Use of this technique allows a much larger area of the bed to be surveyed, in a shorter time, than when employing more traditional methods, thereby giving a virtually instantaneous picture of the channel with a significantly higher resolution. The high frequency beam of modern sonar enables measurements to be obtained even with suspended sediment loads up to 5000 ppm. Using these techniques it is possible to produce an analogue rather than digital (point sampled) picture of the channel. In addition to sonar scanning, acoustic doppler current meters have also been developed for obtaining near point velocity measurements in shallow flow. Acoustic velocity meters can be deployed in tandem with sonar scanners, but several current sensors mounted in a vertical stack are required to give a vertical velocity profile. Their main advantage over an EM flow meter is that the sampling rate can be set much higher so that turbulent fluctuations can be measured, possibly even burst and sweep phenomena. In the case of both side-scan sonar and acoustic velocity meter, electronic loggers or, more usually, notebook computers can be coupled to their output ports to enable long-term remote sensing of the flow and channel morphology, especially in flood conditions, with sampling rates and scanning cycles preprogrammed into the computers. However, acoustic equipment is extremely expensive to purchase and it is likely that the more "traditional" methods will continue to be in use for some time to come.

The increased data resolution with acoustic sampling methods will provide an invaluable aid to numerical modellers who require data to define the initial flow and channel conditions for a numerical model. The ability to acquire multipoint data almost instantaneously will remove most doubts concerning the integrity of measuring instantaneous features. Secondary circulation, it should be noted, is a transient but persistent feature of a flow, and the patterns observed will fluctuate about some long-term mean condition. The ability to sample data at a higher rate removes some of the inaccuracies of attempting to sample this variation. Also, there is a spatial component to data integrity for a model. Sampling with EM and survey staffs, usually at equal intervals along the section, produces a digital representation of the flows and channel form. This may not be at the same density as the model grid and either interpolation up to the model grid or, more rarely, the removal of data to align the model grid will be required. This is a rather subjective process and may involve the smoothing or elimination of important features of the flow. Using acoustic methods to gain a full analogue description of the flow will remove any need for interpolation or elimination of data as it can be extracted from the sonar scans and profiles at exactly the model resolution.

The next major step as far as numerical models go is in the area of mobile-bed simulations. As sedimentary processes occur over a much greater time-scale than flow processes (several orders of magnitude greater), there are problems with interfacing hydraulic and sediment models. The overall error in any numerical scheme is a function of the number of time-steps through which the solution is advanced (Turkel, 1974); it is therefore impractical to attempt to simulate the sediment transport per time-step, as the potential for numerical error is great. Perhaps the simplest solution is to run the flow model for a certain number of time-steps, to allow for equilibrium to be reached from the initial conditions, and then apply a sediment transport model for a longer time using the equilibrium flow data from the model to calculate shear stresses and hence sediment discharges. Resultant changes in bed topography can then be fed back into the flow model, using the existing numerically calculated flow field before repeating the modelling process. Although this is physically unrealistic, it does allow the sediment transport solution to be advanced through a significant time interval with minimal numerical error in the flow field computation. Allowing the flow field to come to equilibrium with the adjusted topography should not produce significant errors overall.

Thus it can be seen that although significant progress has been made in understanding the processes occurring in natural channels, there are still problems with obtaining a full numerical representation of three-dimensional flow. However, advancements in data acquisition and numerical refinements will eventually allow full numerical simulations to be performed with the capability to investigate "what if?" scenarios. The only major step left is the extension of two-dimensional models to three dimensions in order to fully describe the flow in natural channels, but a general solution on this scale is still some way in the future.

REFERENCES

Anderson, D.A. 1974. A comparison of numerical solutions to the inviscid equations of fluid motion, *Journal of Computational Physics*, **15**, 1–20.

Anderson, D.A., Tannehill, J.D. & Pletcher, R.H. 1984. *Computational Fluid Mechanics and Heat Transfer*, McGraw-Hill, New York.

Apmann, R.P. 1972. Flow processes in open channels, *Journal of the Hydraulics Division*, American Society of Civil Engineers, **98**, 795–810.

Bathurst, J.C. 1979. Distribution of boundary shear stress in rivers, in Rhodes, D.D. & Williams, G.P. (eds), *Adjustments of the Fluvial System*, Proceedings of the 10th Annual Geomorphology Symposium, Binghampton, New York, 21–22 September 1979, 372 pp.

Bathurst, J.C., Thorne, C.R. & Hey, R. D. 1977. Direct measurements of secondary currents in river bends, *Nature*, **269**, 504–506.

Bathurst, J.C., Thorne, C.R. & Hey, R. D. 1979. Secondary flow and shear stress at river bends, *Journal of the Hydraulics Division, American Society of Civil Engineers*, **105**(HY10), 1277–1295.

Beam, R.M. & Warming, R.F., 1975. An implicit finite difference algorithm for hyperbolic systems in conservation-law form, *Journal of Computational Physics*, **22**, 87–110.

Bernard, R.S. & Schneider, M.L. 1992. *Depth Averaged Numerical Modelling for Curved Channels*, Technical Report **HL-92–9**, Waterways Experiment Station, US Army Corps of Engineers, Vicksburg, Mississippi, 44 pp.

Bhowmik, N.G. & Stall, J.B. 1978. Hydraulics of flow in the Kaskaskia River, *American Society of Civil Engineers Proceedings, Special Conference on Verification of Mathematical Models in Hydraulic Engineering*, University of Maryland, 9–11 August, 79–86.

Bradshaw, P. 1971. *An Introduction to Turbulence and its Measurement*, Pergamon Press, New York, 218 pp.

Bridge, J.S. 1982. A revised mathematical model and FORTRAN IV program to predict flow, bed topography, and grain size in open channel bends, *Computers and Geosciences*, **8**(1), 91–95.

Bridge, J.S. & Jarvis, J. 1976. Flow and sedimentary processes in the meandering river South Esk, Glen Clova, Scotland, *Earth Surface Processes and Landforms*, **1**, 303–336.

Carling, P.A. 1991. An appraisal of the velocity reversal hypothesis for stable pool riffle sequences in the River Severn, England, *Earth Surface Processes and Landforms*, **16**, 19–31.

Chacinski, T.M. 1954. Patterns of motion in open channel bends, *International Association of Scientific Hydrology*, **38**(3), 311–318.

Chacinski, T.M. & Francis, J.R.D. 1952. Discussion of 'Werner. P.W.. 1951. On the origin of river meanders. Transaction of the American Geophysical Union, **32**, 898–902, *Transaction of the American Geophysical Union*, **33**, 771–773.

Chow, V.T. 1959. *Open Channel Hydraulics*, McGraw Hill, London, International Edition, 680 pp.

Dammuller, D.C. Bhallamundi, S.M. & Chaudhry, M.H. 1989. Modelling of unsteady flow in curved channels. *Journal of Hydraulic Engineering, American Society of Civil Engineers*, **115**, 1479–1495.

Demuren, A.O. & Rodi, W. 1984. Calculation of turbulence driven secondary motion in non-circular ducts, *Journal of Fluid Mechanics*, **140**, 189–222.

De Vriend, H.J. 1976. A mathematical model of steady flow in curved shallow channels, *Journal of Hydraulic Research*, **15**, 37–54.

De Vriend, H.J. & Geldof, H.J. 1983. *Main Flow Velocity in Short and Sharply Curved River Bends*, Report No **83-6**, Communications on Hydraulics, Department of Civil Engineering, Delft University of Technology.

Dietrich, W.E. 1982. *Flow, Boundary Shear Stress, and Sediment Transport in a River Meander*, PhD dissertation, Department of Geological Sciences, University of Washington, Seattle.

Dietrich, W.E. & Smith, J.D. 1983. Influence of the point bar on flow through curved channels, *Water Resources Research*, **19**(5), 1173–1192.

Dietrich, W.E., Smith, J.D. & Dunne, T. 1979. Flow and sediment transport in a sand bedded meander, *Journal of Geology*, **87**, 305–315.

Dressler, R.F. 1952. Hydraulic resistance effect upon the dam-break functions, *Journal of Research, National Bureau of Standards*, **49**(3), 217–225.

Eakin, H.M. 1935. Diversity of current direction and load distribution on stream bends, *Transactions of the American Geophysical Union*, **2**, 467–472.

Einstein, H.A. & Harder, J.A. 1954. Velocity distribution and the boundary layer at channel bends, *Transaction of the American Geophysical Union*, **35**(1), 114–120.

Einstein, H.A. & Li, H., 1958. Secondary currents in straight channels, *Transaction of the American Geophysics Union*, **39**, 1085–1088.

Engelund, F. 1974. Flow and bend topography in channel bends, *Journal of the Hydraulics Division, American Society of Civil Engineers*, **100**, 1631–1648.

Friedkin, J.F. 1945. *A Laboratory Study of the Meandering of Alluvial Rivers*, US Army Waterways Experiment Station, Vicksburg, Mississippi.

Garcia, R. & Kahawita, R.A. 1986. Numerical solution of the St. Venant equations with the MacCormack finite difference scheme, *International Journal for Numerical Methods in Fluids*, **6**, 259–274.

Garcia-Navarro, P. & Saviron, J.M. 1992. MacCormack's method for the numerical simulation of one dimensional discontinuous unsteady open channel flow, *Journal of Hydraulic Research*, **30**(1), 95–105.

Gottlieb, L. 1976. *Three Dimensional Flow Pattern and Bed Topography in Meandering Channels*, Series Paper **11**, Institute of Hydrodynamics and Hydraulic Engineering, Technical University of Denmark, Copenhagen.

Grass, A.J., 1970. Initial instability of fine bed sand, *Journal of the Hydraulics Division, American Society of Civil Engineers*, **96**, 619–632.

Henderson, F.M. 1966. *Open Channel Flow*, Collier MacMillan Publishers, London, 522 pp.

Hey, R.D. 1979. Flow resistance in gravel bed rivers, *Journal of the Hydraulics Division, American Society of Civil Engineers*, **105**(HY4), 365–379.

Hey, R.D. & Thorne, C.R. 1975. Secondary flows in river channels, *Area*, **7**, 191–195.

Hickin, E.J. 1978. Mean flow structure in meanders of the Squamish River, British Columbia, *Canadian Journal of Earth Sciences*, **15**(11), 1833–1849.

Hooke, R. LeB. 1975. Distribution of sediment transport and shear stress in a meander bend, *Journal of Geology*, **83**, 543–565.

Ikeda, S. & Nishimura, T. 1986. Flow and bed profile in meandering sand-silt rivers, *Journal of Hydraulic Engineering ASCE*, **112**(7), 262–579.

Ippen, A.T. & Drinker, P.A. 1962. Boundary shear stress in curved trapezoidal channels, *Journal of the Hydraulics Division, American Society of Civil Engineers*, **88**(HY3), 143–179.

Jovanović, M.B. & Officier, M.J. 1984. *An Example of Application of Curvilinear Coordinates in Numerical Modelling of Complex Flow Patterns*, Hydrocomp 1984.

Kalkwijk, J.P. & De Vriend, H.J. 1980. *Computation of the flow in shallow river bends*, *Journal of Hydraulic Research*, **18**(4), 327–342.

Kamphuis, J.W. 1974. Determination of sand roughness for fixed beds, *Journal of Hydraulic Research*, **12**(2), 193–203.

Keller, E.A. & Florsheim, J.L. 1993. Velocity reversal hypothesis: a model approach, *Earth Surface Processes and Landforms*, **18**, 733–740.

Kitanidis, P.K. & Kennedy, J.F. 1984. Secondary current and river meander formation, *Journal of Fluid Mechanics*, **114**, 217–229.

Leeder, M.R. 1982. *Sedimentology*, Process and Product, Unwin Hyman, London, 344 pp.

MacCormack, R.W. 1969. *The Effect of Viscosity in Hypervelocity Impact Cratering*, Paper **69–354**, American Institute of Aeronautics, Astronautics, Cincinnati, Ohio.

MacCormack, R.W. 1971. Numerical solution of the interaction of a shock wave with a laminar boundary layer, *Lecture Notes in Physics*, **8**, 151–163.

Möller, M. 1883. Studien über die Beweauno das Wassers in Flüssen mit Bezugnahme auf die Ausbildung Flussprofiles, *Zeistsharift für Bauweses*, 201.

Moretti, G. 1978. The λ Scheme, *Computers and Fluids*, **7**, 191–205.

Nezu, I. & Nakagawa, H. 1984. Cellular secondary currents in straight conduit, *Journal of Hydraulic Engineering*, ASCE, **110**, 173–193.

Nezu, I., Tominaga, A. & Nakagawa, H. 1993. Field measurements of secondary currents in straight rivers, *Journal of Hydraulic Engineering*, ASCE, **119**(5), 598–614.

Nikuradse, J. 1930. Turbulente Strömunga in Nichkreisfömigen Röhren, *Ingenier-Archiv*, **1**.

Odgaard, A.J. 1989a. River meander model. I: development, *Journal of Hydraulic Engineering*, ASCE, **115**(11), 1433–1450.

Odgaard, A.J., 1989b. River meander model. II: applications, *Journal of Hydraulic Engineering*, ASCE, **115**(11), 1451–1464.

Odgaard, A.J. & Bergs, M.A. 1988. Flow processes in a curved alluvial channel, *Water Resources Research*, **24**(1), 45–56.

Paice, C. 1990. *Hydraulic Control of River Bank Erosion: An Environmental Approach*, Ph.D. dissertation, School of Environmental Sciences, University of East Anglia, Norwich, UK.

Paterson, A.R. 1983. *A First Course in Fluid Dynamics*, Cambridge University Press, Cambridge, 528 pp.

Perkins, H.J. 1970. The formation of streamwise vorticity in turbulent flow, *Journal of Fluid Mechanics*, **44**, 721–740.

Prandtl, L. 1952. *Essentials of Fluid Dynamics*, Blackie, London, 452 pp.

Raupach, M.R. Antonia, R.A. & Rajagopalan, S. 1991. Rough-wall turbulent boundary lavers, *Applied Mechanics Review*, **44**(1), 1–25.

Rozovskii, I.L. 1954. *Concerning the Question of Velocity Distribution in Stream Bends*, DAN URSR (Reports of the Academy of Sciences of the Ukr. SSR), No. **1**.

Rozovskii, I.L. 1961. *Flow of Water in Bend of Open Channels*, Israel Programme for Scientific Translations, Jerusalem.

Saint Venant, B. 1871. Theorie du movement non-permanent des eaux avec application aux crues des rivieres et a l'introduction des marees dans leur lit, *Acad. Sci. Comptes Rendus*, **73**, 148–154, 237–240.

Sarma, K.V.N. Lakshminarayana, P. & Rao, N.S.L,. 1983. Velocity distribution in rectangular open channels, *Journal of Hydraulic Engineering*, ASCE, **109**, 270–289.

Smith, J.D. & McLean, S.R. 1984. A model for flow in meandering streams, *Water Resources Research*, **20**, 1301–1315.

Stearns, F.P. 1883. On the current meter, together with a reason why the maximum velocity of water flowing in open channels is below the surface, *Transactions of the American Society of Civil Engineers*, **12**, 301–338.

Thompson, J. 1876. On the origin and winding of rivers in alluvial plains, with remarks on the flow around bends in pipes, *Proceedings of the Royal Society of London*, **25**(5), 5–8.

Thorne, C.R. & Hey, R.D. 1979. Direct measurements of secondary currents at a river inflexion point, *Nature*, **280**, 226–228.

Thorne, C.R. & Rais, S. 1983. Secondary current measurements in a meandering river, in Elliot, C.M. (ed.), *River Meandering, Proceedings of the Conference on Rivers 1983*, American Society of Civil Engineers, New York, 675–686.

Thorne, C.R. Zevenbergen, L.W. Pitlick, J.C. Rais, S. Bradley, J.B. & Julien, P.Y. 1985. Direct measurements of secondary currents in a meandering sand-bed river, *Nature*, **316**. 746–747.

Toebes, G.H. & Sooky, A.A. 1967. Hydraulics of meandering rivers with flood plains, *Journal of Waterways and Harbors Division, American Society of Civil Engineers*, **93**, 213–236.

Turkel, E. 1974. Phase error and stability of second order methods for hyperbolic problems, *Journal of Computational Physics*, **15**, 226–250.

Vanoni, V.A. 1946. Transportation of suspended sediment by water, *Transactions of the American Society of Civil Engineers*, **3**, 67–133.

Yen, B.C. 1972. Spiral motion of developed flow in wide curved open channels, in Shen, H.W. (ed.), *Sedimentation*, Fort Collins, Colorado, Chapter 22.

3 The Fluid Dynamics of Small-scale Alluvial Bedforms

JIM BEST
Department of Earth Sciences, University of Leeds, UK

INTRODUCTION

The development of small-scale bedforms in alluvial channels involves complex interactions between the turbulent flow structure, transport of sediment and morphology of the bed. Although the structure of the turbulent flow provides a template which controls the motion of sediment, this structure itself will be altered by both the grain and form roughness and the quantity and type of material being carried in suspension. Hence, our ideas on the type, rate of development and morphology of different small-scale bedforms under differing hydraulic regimes must take into account these complex interactions and feedbacks. Much early work upon small-scale alluvial bedforms concentrated on sand-grade material and only recently has research included examination of both gravel (e.g. Brayshaw, 1984; Dinehart, 1989, 1992; Robert, 1990) and mixed sand–gravel loads where the sorting of the sediment may be critical in influencing the nature and patterns of bedform generation (e.g. Chiew, 1991; Wilcock, 1992; Wilcock & Southard, 1989). Few attempts have been made to correlate the bedform fields between different size sediments, despite the fact that the turbulence structure, flow processes and sediment transport characterizing one particle size mix may have many parallels when interpreting bedforms generated in a quite different grain size.

This chapter reviews the fluid dynamics of small-scale bedforms (i.e. those that scale with the flow depth or are smaller and not barforms that scale with channel width) generated in alluvial channels over a wide range of grain sizes from silts to gravels, and presents a unified bedform phase diagram across this size range. Additionally, the review also considers the changing processes associated with the critical transitional areas between different bedforms. This synthesis allows comparison and extension of a genetic bedform phase diagram (Leeder, 1983a) which identifies the physical processes important in the generation of each bedform stability field. This permits an appreciation of the areas in which our understanding of bedform stability is most incomplete and suggests several areas of fruitful and urgently needed research.

Advances in Fluvial Dynamics and Stratigraphy. Edited by P.A. Carling and M.R. Dawson.
© 1996 John Wiley & Sons Ltd.

THE STRUCTURE OF TURBULENT BOUNDARY LAYERS

Before proceeding to examine the morphology and dynamics of small-scale bedforms, it is useful to present a brief synopsis of the structure of turbulent boundary layers and define terms that will be used later. Comprehensive, recent reviews of turbulent boundary layer structure can be found in Kline & Robinson (1989, 1990), Robinson (1990a,b), Robinson *et al.* (1990) and Nezu & Nakagawa (1993), whilst reviews with sedimentological and geomorphological applications can be found in Best (1993) and chapter 1.

Turbulent flow over a *smooth*-wall boundary can be characterized by four major categories of motion:

(1) *Low-speed streaks.* These are flow-parallel zones of less than average downstream velocity whose spanwise or lateral spacing, λ, scales with the inner-wall variables of shear velocity, u_*, and kinematic viscosity, ν, such that:

$$\lambda^+ = \lambda u_* / \nu \sim 100 \qquad (3.1)$$

Streak spacing has been found to increase with distance above the wall (Smith & Metzler, 1983) whilst individual streaks are detectable up to approximately $y^+ = 40$ away from the boundary ($y^+ = y u_* / \nu$) and may extend downstream for hundreds and perhaps thousands of wall units (Smith & Metzler, 1983; Kline & Robinson, 1990), where a wall unit $= n u_* / \nu$, where n is a characteristic length. The velocity of a streak is typically around $0.5U$ (where U is the mean flow velocity). Streaks may be envisaged as low-speed vortices with an elongation in the downflow direction.

(2) *High-speed sweeps.* These consist of wall-directed inrushes of higher than average downstream velocity (quadrant 4 events, Figure 3.1) which have been held responsible for large contributions to the Reynolds stress (e.g. Willmarth & Lu, 1972, 1974) and are considered to be important in the entrainment of sediment (Grass, 1970; Drake *et al.*, 1988; Williams *et al.*, 1989; Best, 1992). Sweeps commonly have a spanwise width of 50–70 wall units.

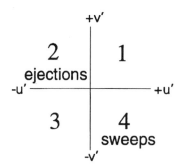

u & v = downstream & vertical components of flow
u' & v' = deviations from the mean

Figure 3.1   Quadrant plot of deviations of downstream (u') and vertical (v') velocities from their respective means. Quadrant 2 and 4 events represent bursts/ejections and sweeps, respectively

(3) *Ejections of fluid*. These motions are characterized by lower than average downstream momentum fluid being advected away from the wall (quadrant 2 events, Figure 3.1). Many studies have sought to investigate the structure of these ejections and detail their velocity characteristics (e.g. Grass, 1971; Rao *et al.*, 1971; Bogard & Tiederman, 1986; Talmon *et al.*, 1986; Luchik & Tiederman, 1987; Tiederman, 1990). One or several ejections may group together and form "bursts" of fluid that move away from the wall (Bogard & Tiederman, 1986). Quadrant 2 events have also been shown to be the dominant contributors to the Reynolds stress, with values of 70% being common (Willmarth & Lu, 1972, 1974).

Although initial research suggested a relationship between burst period and the outer-wall variables of mean flow velocity, U, and boundary layer thickness, Y, such that

$$T_b = TU/Y \sim 3-7 \tag{3.2}$$

subsequent work has suggested that this may not be the case and that an inner or mixed inner–outer wall variable scaling may be equally appropriate (Luchik & Tiederman, 1987). The relevance of burst scaling to bedform generation will be examined in more detail below and has been influential in speculation concerning the dynamics and origin of dune bedforms (e.g. Jackson, 1976, 1978; Levi, 1991; Lapointe, 1992; Best, 1993; Kostaschuk & Church, 1993; see also Chapter 1).

(4) *Large-scale vortices*. This category covers a wide range of structures including hairpin and horseshoe shaped vortices, larger coherent motions and outer-flow scale motions. The nature of these vortices is reviewed in the recent articles of Kline & Robinson (1989, 1990) and Robinson (1990a,b). A combination of both flow visualization and quantification has indicated that as low-speed streaks become uplifted they may form ejections and bursts and adopt a hairpin or horseshoe shape (Head & Bandyopadhyay, 1981; Smith, 1984; Smith & Walker, 1990; Smith *et al.*, 1991) although see Falco (1991) for an alternative view. Streaks which form the legs of such vortices become stretched at higher Reynolds numbers (Head & Bandyopadhyay, 1981). Smith & Walker (1990) show how an initial hairpin vortex may subsequently induce the growth of subsidiary hairpins to the side of the original vortex and secondary hairpins upstream to its rear (Figure 3.2a). Growth of such a patch of hairpin vortices within a boundary layer has many striking visual similarities to flow properties at the laminar–turbulent transition (Perry *et al.*, 1981) and may define a process capable of generating the larger coherent motions described by many authors (e.g. Falco, 1977, 1991; Brown & Thomas, 1977; Head & Bandyopadhyay, 1981) which have an interface which dips upflow by approximately 45° (Figure 3.2b).

These four categories define the basic elements of the structure of turbulent flow over a smooth wall. These structures may be greatly modified by both grain and form roughness (Blinco & Partheniades, 1971; Grass, 1971; Ligrani & Moffat, 1986; Kirkgoz, 1989; Grass *et al.*, 1991) as well as by changes to the flow (for instance, by the addition of suspended sediment). Thus, the critical questions that arise are: (1) how do these elements of turbulent boundary layer structure interact with bedform generation? and (2) what feedbacks are exerted by the bedforms upon the

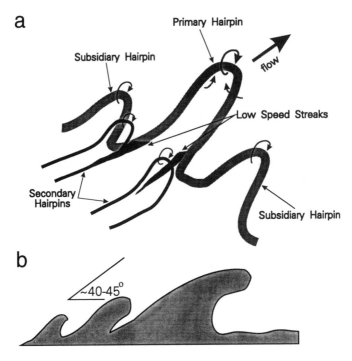

Figure 3.2 (a) Synthesized model of the structure of near-wall turbulence near a hydraulically smooth bed (after Smith & Walker, 1990). Hairpin vortices are generated from uplifted low-speed streaks and may generate secondary and subsidiary hairpins vortices as they propagate (from Best, 1993). (b) Section through a hairpin grouping illustrating the 40–45° interface with the mean flow. Note the similarity between this and the large outer scale coherent motions that have been described in many turbulent boundary layer studies (Falco, 1977, 1991; Brown & Thomas, 1977)

flow? In order to address these questions, the form and fluid dynamics of each category of principal bedform across the silt–gravel size range will be examined and used to construct a unified bedform phase diagram.

SMALL SCALE ALLUVIAL BEDFORMS

A range of small-scale alluvial bedforms may be defined and plotted as a bedform phase diagram across a wide range of grain sizes (Figure 3.3). This plot, using shear stress as the ordinate variable, defines regions of bedform stability that are discussed below and extends previous stability diagrams into coarse sands and gravels. Although this plot delineates broad areas of bedform stability, it suffers problems in areas of overlap and from the fact that bedform stability may not be solely a function of grain size and applied fluid stress (for instance, the influence of flow depth on the stability field position). Progress towards a more unified view of bedform generation must rely on the detailed study of the formative processes and turbulent flow field characteristics of bedforms. The following discussion analyses

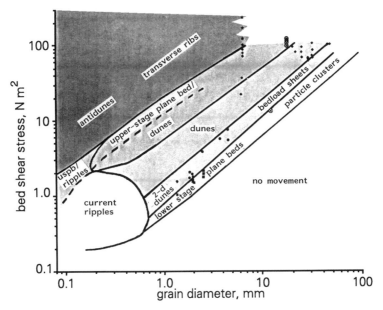

Figure 3.3 A unified bedform phase diagram across a range of sand and gravel sizes. This diagram is partly based upon the data and bedform stability fields presented in figure 11 of Southard & Boguchwal (1990). Additional data sources have been drawn from Ikeda (1983), Brayshaw (1983), Hubbell *et al.* (1987), Drake *et al.* (1988), Whiting *et al.* (1988), Wilcock (1992), Bennett (1992) and Dinehart (1992). Symbols representing data drawn from these studies are: ■, lower-stage plane beds; ⊗, particle clusters; •, two-dimensional dunes/bedload sheets; ♦, dunes; ⊙, upper-stage plane beds; ★, antidunes. No movement/movement threshold after Miller *et al.* (1977) and see Chapter 4. The dashed line between the ripple/dune and upper-stage plane bed stability fields denotes a boundary that has much overlap and possesses transitional bedforms between the two bedstates. Note that some transitions, such as between dunes – antidunes and particle clusters – transverse ribs, are possible that are not indicated by this bedform phase diagram – in these cases in very shallow, fast flows. See text for discussion

bedforms in terms of their associated turbulent flow structures and uses these to propose a modified genetic bedform phase diagram as advocated by Leeder (1983a).

Current ripples

Current ripples are generated and stable in sediment from silts to approximately 0.7 mm diameter (Allen, 1982). The height and wavelength of current ripples are independent of the flow depth, recent work showing that ripple forms will inevitably become three-dimensional and sinuous given a long enough development time (Baas, 1993; Baas *et al.*, 1993). Current ripples have been proposed to scale with the grain size, D, such that their wavelength, $\lambda_r \sim 1000D$. The control upon ripple wavelength, however, is the size of the flow separation zone that exists in its lee, this being dictated by the flow velocity and ripple height (Karahan & Peterson, 1980).

The initiation of current ripples upon a flat sediment bed has been explained through many mechanisms (see Best, 1992), the most widely held within the earth sciences involving flow separation and downstream propagation from a sweep-induced bed defect (Raudkivi, 1963, 1966; Williams & Kemp, 1971, 1972; Southard & Dingler, 1971). Williams & Kemp (1971) proposed that sweep impacts upon a flat bed, once exceeding the critical threshold bed shear stress of the sediment, entrain particles which are then transported downstream as flow-parallel ridges (Grass, 1970, 1983), decelerate and are eventually deposited. Once deposited, this sediment forms a small mound of grains (or bed "defect") which, upon reaching a critical height given by a defect Reynolds number, Re_d, of approximately 4.5 (where $Re_d = (u_* d)/v$, where d = defect height), triggers flow separation over the defect (Williams & Kemp, 1971). Gyr & Schmid (1989) describe three types of bed defect generated by either (1) deposition, (2) erosion of a flat bed at low transport stages, or (3) by an instability in sediment transport at high bed shear stresses. Once a sufficient size bed defect is generated, flow separation creates a zone of high instantaneous bed shear stress within the reattachment area (Raudkivi, 1963, 1966) which causes erosion and formation of a second bed defect downstream. Consequently, a rippled bedstate can evolve from flow separation and defect amplification downstream from an initial sweep-induced sediment mound (Figure 3.4). It is clear

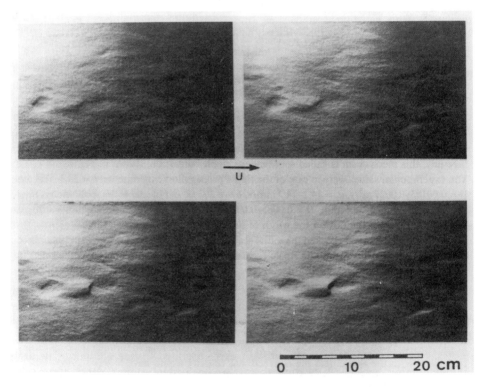

Figure 3.4 Ripple development from a bed defect generated through sediment transport by an intense sweep event ($u_* = 0.145$ cm/s; time between frames is 4 min) (from Gyr & Schmid, 1989)

that flow separation is a key process to ripple propagation but that these separation effects are restricted to the near-wall flow and do not influence the entire flow field as has been proposed for dune bedforms (see below). Once bed defects are formed they begin to modulate the sediment transport and form of the bed surface (Gyr & Schmid, 1989).

Best (1992) has further reasoned that the formation of ripples can be more completely explained through consideration of the multiple hairpin structure of the turbulent boundary layer. He argues that the multiple, nested hairpin vortices documented by Smith and co-workers (Smith & Walker, 1990; Smith et al., 1991) may give rise to either multiple sweep impacts upon a bed or sweep impacts larger than one sweep width. The consequence of this is to generate a bed defect that is wider than one sweep width and more in accord with the sizes of bed defects that appear upon sediment beds developing current ripples (Figure 3.4). This contention finds support in the observations of Gyr & Schmid (1989) and Gyr et al. (1989) who document "arrow"-shaped bed defects in planform that develop between the erosion threshold and current ripple bedstate. Gyr & Schmid (1989) and Gyr et al. (1989) suggest that these arrow-shaped defects are caused by the interactions between sweeps and larger coherent motions in the boundary layer and their scaling may be linked to the streak spacing. However, the lateral width of incipient ripples and the arrow-shaped defects noted by Gyr & Schmid (1989) are often larger than the streak spacing (see Figure 3.4); the "nested" hairpin vortex model of the turbulent boundary layer defined above may provide a more complete explanation for this phenomenon and the formation of current ripples. Furthermore, Best (1992) presents evidence that any initial flow-parallel ridges that form through sweep impacts may become stabilized in their position as they may concentrate the position of subsequent low-speed streaks over the ridges. The corollary of this streak concentration is that subsequent sweeps may be focused between the ridges and hence add to the amplification of the bed defect until it is large enough to create significant flow separation and downstream defect propagation. The influence of flow-parallel sediment ridges upon flow structure has also been demonstrated by Nezu & Nakagawa (1989) who show that higher flow velocities are present between rather than over the ridges; this also suggests subsequent increased sediment transport between the ridges as postulated by Best (1992). These ideas linking turbulent boundary layer structure and sediment transport may be used to define a more complete schematic model of ripple initiation upon a flat sediment bed (Figure 3.5).

Gyr & Schmid (1989) argue that flow around the initial bed defect may be characterized by longitudinal counter-rotating vortices (similar to flow around an obstacle) but that once a spanwise flow separation vortex begins to dominate the longitudinal vortices, then the ripples begin to align and become more two-dimensional with crestlines extending across the flow. Detailed experimental work by Baas (1994) has demonstrated that ripples generated in fine and very fine sands will inevitably become three-dimensional given a long enough development time (Figure 3.6). This is in contradiction to earlier work (e.g. Allen, 1968) which suggested that ripple sinuosity reflected the bed shear stress, with straight ripples forming at low flow powers and linguoid ripples at high shear stresses. These

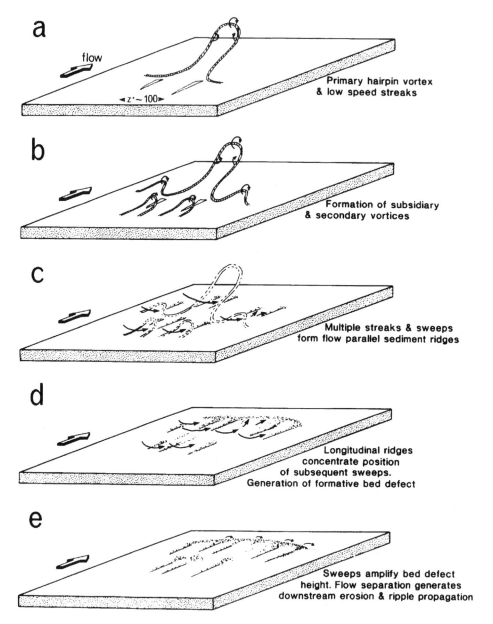

Figure 3.5 Model of bed defect and ripple development in sands. Downstream flow separation and defect propagation after stage (e) allow ripple development over the entire bed (from Best, 1992)

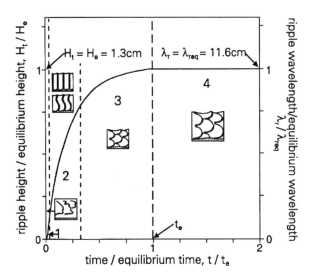

Figure 3.6 Development curve of ripples generated in very fine sands ($D_{50} = 0.095$ mm) demonstrating the development of ripple morphology with time (from Baas, 1993). Four stages of planform development are depicted until the bedform reaches an equilibrium height (H_e) and wavelength (λ_{req}) after a period of time designated the equilibrium time (t_e). Inset sketches illustrate the crestline planform

conclusions upon ripple development time have many important consequences for palaeohydrological interpretations of ancient fluvial sandstones.

Lower-stage plane beds and particle clusters

Once the entrainment threshold is exceeded, sediment greater than 0.7 mm in diameter remains as a flat, mobile bed and does not to deform into a rippled bedstate. Leeder (1980) reasoned that the absence of ripples in coarse sands could be explained through disruption of the viscous sublayer by the grain roughness and that, once sediment grains begin to protrude through the viscous sublayer, they enhance vertical mixing within the near-bed flow (Figure 3.7b). Increases in the vertical component of turbulence over larger grain roughnesses have been noted by Grass (1971) and may be related to fluid ejections from between the grains or flow separation (and shear layer development) behind the grains. This enhanced vertical mixing and separation effect downstream of large single particles is evident in the work of Kirkbride (1993) on larger roughness elements. Ligrani & Moffat (1986) demonstrate that larger particle Reynolds numbers produce progressively lower downstream component turbulence intensities but that the vertical component intensity is increased (Figure 3.7).

Whatever the source of this enhanced vertical mixing, the net effect appears to be to inhibit the formation of pressure gradients sufficient to generate flow separation or flow separation that causes erosion within the downstream reattachment area (Leeder, 1980; Gyr & Schmid, 1989). In the absence of flow separation, bed defects generated by sweep impacts remain stable and do not transform into current ripples

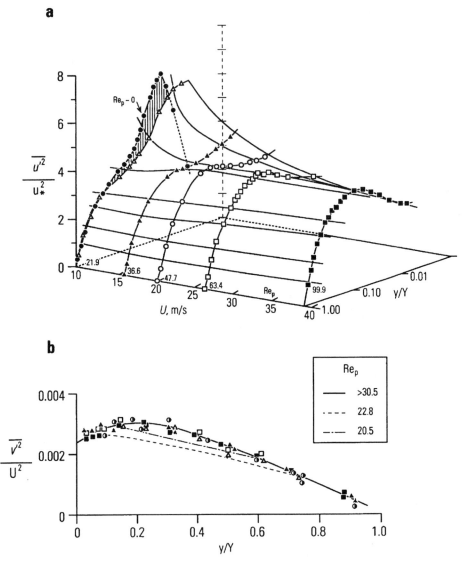

Figure 3.7 Variation in turbulence intensity as a function of height above the bed and grain roughness (after Ligrani & Moffat, 1986). (a) Variation in turbulence intensity of the downstream component of velocity, u, with height above bed and different grain roughnesses. The downstream turbulence intensity is made dimensionless through division by the shear velocity, u_*. Re_p represents the particle Reynolds number ($Re_p = (Du_*)/v$, where D is particle size and v is kinematic viscosity) and is varied by increasing the flow velocity (in this case in a wind tunnel) for a constant grain size. Height above the boundary, y, is shown relative to the total flow depth, Y. (b) Variation in turbulence intensity of the vertical component of velocity, v, with height above bed and different grain roughnesses. Turbulence intensity has been made dimensionless through division by the mean downstream flow velocity, U. The symbols and solid line illustrate that for particle Reynolds numbers >30.5 the vertical turbulence intensity profiles are similar whereas for the smaller particle Reynolds numbers of 20.8 (--- line) and 22.5 (--·-- line) the vertical turbulence intensities are less

(Figure 3.8). Leeder (1980) recorded small ridges of sediment upon a mobile coarse sand bed that did not amplify into bedforms. An additional factor in the stability of such ridges may be linked to the subsequent concentration of streaks and sweeps over and between such flow-parallel ridges, which may enhance their temporal persistence (Best, 1992 and see above). Small, flow-parallel ridges or flow transverse mounds may therefore remain as quasi-stable features upon a lower-stage plane bed; they may either form singly or anastomose together and are likely to be reworked by subsequent sweep impacts of sufficient magnitude.

The role of particle roughness and protrusion upon the flow becomes progressively more important in coarser grained sediments where the particles eventually create their own flow field capable of influencing the erosion, transportation and deposition of other grains. This is the case with particle clusters which have been observed within a range of gravel-bed streams (e.g. Dal Cin, 1968; Teissyre, 1977; Laronne & Carson, 1976; Brayshaw, 1984) and have been held as instrumental in the timing and rates of bedload transport by many authors (e.g. Reid et al., 1984; Reid & Frostick, 1986; de Jong, 1991; Hoey, 1992). Large single grains within a flow create two principal vortex systems (Figure 3.9). (a) A flank or standing vortex upstream of the particle is generated by "rolling down" of the boundary layer in front of the clast (Figure 3.9a,c). This vortex has a horseshoe shape in planform (but should not be confused with the genetically separate horseshoe vortices of turbulent boundary layers, see above). (b) Secondly, a three-dimensional leeside

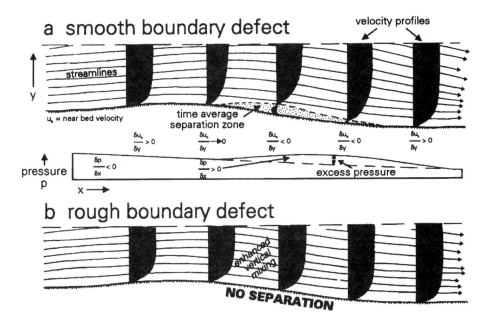

Figure 3.8 Model for the absence of ripples and stability of lower-stage plane beds (after Leeder, 1980). Enhanced vertical mixing of flow over a rough boundary defect (b) does not permit pressure gradients to be established that allow flow separation as over a smooth boundary defect (a). In the absence of significant flow separation the bed defect remains stable (b) and ripples do not propagate as they do on smooth boundaries (a)

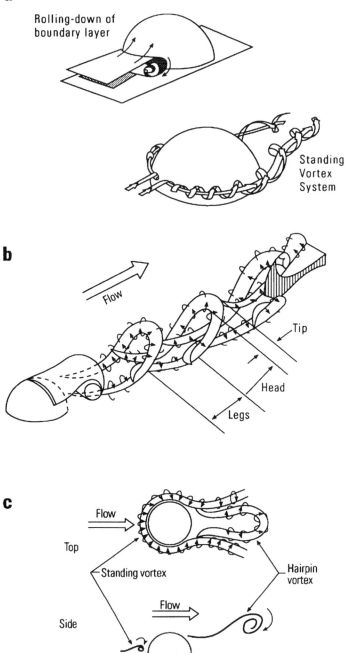

Figure 3.9 Vortex systems associated with an isolated particle. (a) Rolling down of the boundary layer in front of the clast and generation of a standing vortex system (after Best & Brayshaw, 1985). (b) Horseshoe vortices generated from three-dimensional flow separation downstream of a particle (after Acarlar & Smith, 1987). (c) Composite model for standing and hairpin vortices around an isolated clast (after Acarlar & Smith, 1987)

separation zone is created which has its own bounding free-shear layer along which Kelvin–Helmholtz instabilities are created and shed both outwards into the flow and downwards towards the boundary (Figure 3.9b,c). The leeside separation zone may produce a series of horseshoe-shaped or arch vortices that propagate into the outer flow (Figure 3.9b) which have similar characteristics to boundary layer horseshoe vortices (Acarlar & Smith, 1987). Kawanisi *et al.* (1993) also document primary and secondary hairpin vortices generated downstream of hemispheres at high Reynolds numbers. However, they found that the long upstream legs of the hairpin vortices, observed in the lower Reynolds number studies of Acarlar & Smith (1987; Figure 3.9b), were not present.

The frequency of eddy shedding from such obstacles may be expressed by the Strouhal number, *St*, given by

$$St = (fD)/U \qquad (3.3)$$

where f is frequency, D the particle size and U the mean flow velocity. Strouhal numbers of around 0.2 are characteristic of low Reynolds number flows but can become higher, up to ~0.40, at higher Reynolds numbers. Acarlar & Smith (1987) and Klebanoff *et al.* (1992) illustrate how the Strouhal number varies as a function of Reynolds number and roughness element shape. Given a turbulent flow with varying instantaneous velocities outside and within the leeside separation zone, it is likely that the shear layer will be characterized by a range of eddies with different periodicities and sizes. One flow will therefore be characterized by a distribution of eddy sizes and a spread of values around a mean Strouhal number. Much remains to be addressed concerning the details of these interactions along the free-shear layer, yet they are critical in the vertical mixing of fluid over rough beds (and therefore in the generation of velocity gradients and instantaneous shear stress) and in the suspension of sediment over bed roughness.

Isolated particles hence create their own bed pressure and velocity fields which are instrumental in influencing the movement of other clasts by producing an area of relative positive pressure upstream of the clast and a zone of relative low pressure in the downstream separation zone (Figure 3.10a). A region of relative positive pressure is also present at flow reattachment of the separation zone free-shear layer. Particle clusters are generated in response to the flow field created by the larger particles within the bedload, the particles larger than the 84th percentile of the grain size distribution perhaps being the most influential obstacles (Brayshaw, 1984). Brayshaw *et al.* (1983) illustrate how the lift and drag forces exerted upon one clast that is adjacent to another are a function of both proximity to the obstacle clast and the relative size of the clasts (Figure 3.10b). Particles in the lee of the obstacle suffer much reduced lift and drag forces than if they were exposed singly within the freestream, this clearly being related to the influence of the leeside separation zone. The proximity of clasts to one another will therefore determine the nature of the leeside separation zones: these zones may either act in an isolated manner, interact or, if the particles are very close together, no true separation zones may exist (these equating to the isolated roughness, wake interference and skimming flows of Nowell & Church (1979)). Small particles caught within the leeside separation zone also suffer drag with an upstream vector, reflecting reversing flows within this area

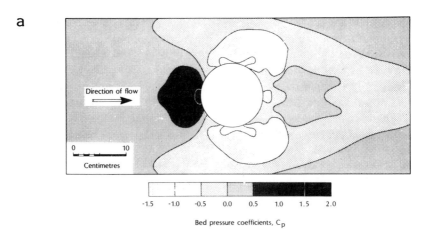

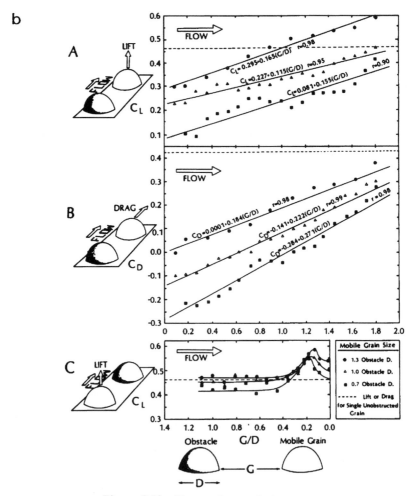

Figure 3.10 *For caption see facing page*

(Figure 3.10b). Larger particles which protrude more into the flow are able to counteract this reverse flow, even though their lift and drag forces are much reduced compared to the freestream values. Particles finer than the obstacle are therefore trapped within the leeside separation zone, where the magnitude of flow is typically around 20% of the mean freestream flow velocity, and this dictates the size and density of clasts transported into the leeside region. The lift forces exerted on particles upstream of the obstacle increase near to the clast itself and show the influence of flow acceleration around the flanks of the grain. Small particles may therefore be quickly transported around this area whilst larger particles may tend to become imbricated against the clast and become relatively immobile. The shear layer generated by the flank "standing" vortex may provide an area of high bed shear stresses in which particles can be rapidly transported, perhaps being subsequently delivered and entrained into the separation zone. Paola *et al.* (1986) discuss flow and skin friction patterns behind hemispheres and analyse flow in the far-field downstream from reattachment. This region may be dominated by the counter-rotating vortices formed from the standing vortex generated around the upstream clast flanks (Figure 3.11a). Increased skin friction and divergent flow along the centreline downstream from reattachment (Figure 3.11b) were found to confirm the model of counter-rotating vortices although this flow pattern is difficult to reconcile with the presence of tails of sediment downstream of the obstacle that may be many times longer than the leeside separation zone (Paola *et al.*, 1986). In this case, temporal variability in the direction of current flow or shallow flow depths over the clast, which may modify the separation vortex systems, may have to be invoked to explain these sedimentation zones (Paola *et al.*, 1986). Boyer & Roy (1991) examined the bed morphology and flow dynamics of isolated obstacles in shallow flows over sand beds. They found that the morphology of the downstream current crescent varied with (1) flow velocity and (2) the angle of the frontal wave generated on the upstream side of the particle. The angle of refraction of the wave front was found to control the growth and propagation of secondary streamwise vortices downstream of the obstacle which are responsible for producing different shaped current crescents (Boyer & Roy, 1991). At Froude numbers much greater than unity, the formation of secondary streamwise vortices in the lee of the obstacle

---

Figure 3.10 Bed pressure and lift and drag forces acting upon hemispheres (both after Brayshaw *et al.*, 1983). (a) Bed pressure coefficients around an isolated clast illustrating (i) the relative positive pressure upstream of the clast, (ii) lower pressures in the zones of flow acceleration around the clast flanks and flow separation downstream of the clast, and iii) higher pressures in the reattachment region. The bed pressure coefficient, $C_p = (p - p_0)/0.5\rho U^2$, where p is the local bed pressure, p_0 is the freestream static pressure, ρ is fluid density and U is the free stream velocity. (b) The influence of particle separation and size upon the lift and drag forces exerted on a mobile grain in proximity to an obstacle clast: (A) coefficient of lift for particle in the lee of another clast; (B) coefficient of drag for a particle in the lee of another clast; (C) coefficient of lift for a particle upstream of another clast. Each graph shows the relationships for different sizes of mobile particle in relation to the size of the obstacle clast, D (0.7, 1.0 and 1.3D). Note reversal of abscissa scale in (C). $C_L = F_L/0.5A\rho U^2$, $C_D = F_D/0.5A\rho U^2$, where F_L and F_D are the lift and drag forces, respectively, A is area of the particle and the other terms are defined above

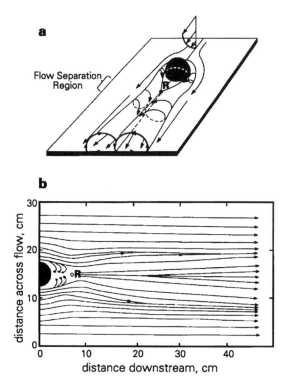

Figure 3.11   Flow dynamics around an isolated obstacle clast (after Paola *et al.*, 1986). (a) Vortex system around an obstacle on a flat bed. *R* denotes the reattachment point. Flow downstream of reattachment is characterized by counter-rotating streamwise vortices that originate around the flanks of the obstacle. (b) Skin-friction direction field behind an isolated obstacle. Note the divergence of flow downstream from reattachment

was inhibited and only a single sediment tail was found, this being in contrast to the multiple tails found at Froude numbers less than unity.

The deformation of the turbulent boundary layer by large particles, and generation of the vortex systems described above, therefore produces particle clusters which have a distinct, although varied, morphology and grain size segregation (Brayshaw, 1984; de Jong, 1991) (Figure 3.12). The origin of these bedforms in coarse-grained channels is intimately linked to the mutual interference and interactions between grains. Coarse sands display sweep-induced ridges and small flow-skewed mounds that are generated by particle interaction as the sweep decays downstream. The grain sizes in between these two sizes seem likely, therefore, to be characterized by the transition between these two bedforms that exist upon mobile lower-stage plane beds.

Two-dimensional dunes and bedload sheets

In sediments coarser than 0.7 mm, a plane bed transforms at higher applied bed shear stresses to a bedstate with irregular bedforms. The nature of these bedforms, termed

Figure 3.12 The planform (a) and side view (b) of a well developed particle cluster formed in poorly sorted fluvial gravels. Note the well imbricated clasts on the stoss side and the relatively finer grained sediment in the leeside flow-separation zone. Lens cap for scale. Flow is from left to right

"bars" by Costello (1974) and "2-D dunes" by Costello & Southard (1980), has been a matter of some debate with interpretations that they represent both distinct bedforms (Costello, 1974) or transitional dunes (Allen, 1983a, 1985; Ashley, 1990). These forms are termed "two-dimensional dunes" here in common with Costello & Southard

a

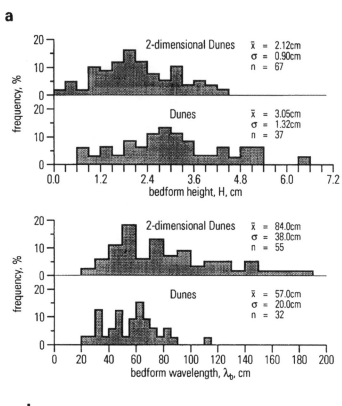

b

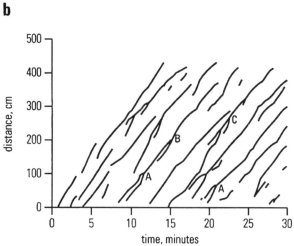

Figure 3.13 *For caption see facing page*

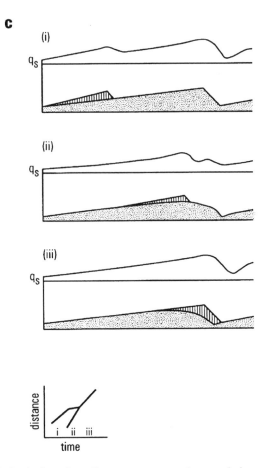

Figure 3.13 Morphological and sediment transport characteristics of two-dimensional dunes. (a) Histogram of bedform heights and wavelengths for two-dimensional and fully developed dunes (after Costello & Southard, 1980). (b) Plot of position of the fronts of two-dimensional dunes as a function of time. Bedform celerity is seen to decrease (A), remain constant (B) or increase (C) after bedform amalgamation (from Ditchfield & Best, 1992). (c) Schematic model of sediment transport rate, q_s, and bedform height as two, two-dimensional dunes amalgamate. Bottom inset shows schematic time–distance plot for this amalgamation. One form migrates up the stoss side of another (i) and at some stage sediment supply to the downstream bedform becomes starved as the smaller bedform approaches the crest of the downstream form (ii). Some erosion at the crest may occur at this point; bedform height and celerity increase once amalgamation has occurred (iii). (after Ditchfield & Best, 1992)

(1980), thus avoiding use of the term "bar" which is best restricted to macroforms that scale with the channel size (Bridge, 1985; Ashley, 1990). However, it should be noted that these bedforms are different in morphology and formative process than the dunes generated at higher shear stresses which possess significant erosion at flow reattachment and scale with flow depth (see below).

Costello (1974) and Costello & Southard (1980) have illustrated that the morphological characteristics of two-dimensional dunes are different from dunes formed in the same coarse sands: two-dimensional dunes tend to be lower, are more varied in wavelength and possess a lower height:wavelength ratio (Figure 3.13a). These bedforms may be no more than a few grain diameters in height but may reach up to several centimetres (Ditchfield & Best, 1992). Costello & Southard (1980) also argued that the movement of two-dimensional dunes may be modelled as kinematic waves, the height of the bedform dictating its celerity and bedform amalgamations, or kinematic shock waves, giving rise to a range of two-dimensional dune heights. Ditchfield & Best (1992) show that two-dimensional dune celerity may either increase, decrease or remain constant after bedform amalgamation (Figure 3.13b); further, they illustrate that local sediment supply to a bedform crest, and the influence of an upstream bedform catching up a slower-moving two-dimensional dune, is critical in influencing bedform amalgamation. Starving of sediment supply to a two-dimensional dune crest because of the presence of an upstream bedform can lead to a decrease in celerity of the downstream bedform and perhaps its degradation prior to any amalgamation with a greater-celerity, upstream form (Figure 3.13c).

Little data exist concerning flow over these bedforms. However, several important qualitative points are clear. First, flow separation does exist over their leesides but separation is relatively ineffective in causing downstream scour: two-dimensional dunes do not possess deep scour troughs and relatively little sediment is eroded at reattachment. Indeed, Costello & Southard (1980) note that sediment suspension in the troughs increases as these bedforms transform to full three-dimensional dunes at higher bed shear stresses. Second, the irregularity in wavelength of these bedforms (Figure 3.13a) suggests that flow separation may not be important in promoting regularly spaced bedforms as is apparent with ripples and perhaps dunes (see below). Third, particle interaction is still important in producing flow-parallel ridges over the stoss sides of these bedforms and smaller two-dimensional dunes may migrate rapidly up the back of slower-moving forms.

Very similar bedforms to two-dimensional dunes have been described in poorly sorted coarse-grained sands and fine gravels and have been termed "bedload sheets" (Whiting *et al.*, 1988). The similarity between two-dimensional dunes and bedload sheets is striking (Figure 3.14a): the sheets may be only several grains high, possess wavelength:height ratios of between 25 and 300 (Whiting *et al.*, 1988) or 56 to 338 with a modal value of approximately 130 (Bennett, 1992; Bennett & Bridge, 1995), generate flow separation over their leesides and undergo frequent amalgamations as sheets of different size and celerity interact. Observations of the behaviour of amalgamating sheets (Figure 3.14b) are similar to those of two-dimensional dune coalescence (Ditchfield & Best (1992) and see above). Whiting *et al.* (1988) ascribe the initiation of bedload sheets to the slowing and stopping of coarse particles (mutual particle interactions) which subsequently permits finer sediment to accumulate in the interstices. They argue that this causes the coarse particles to become mobile once again, possibly by decreasing the angle of friction of the larger grains.

From comparison with bedforms observed in coarse sands, it therefore seems unlikely that bedload sheets are a distinct and separate genetic bedform class. However, bedload sheets have a more distinct sorting of sediment over the bedform

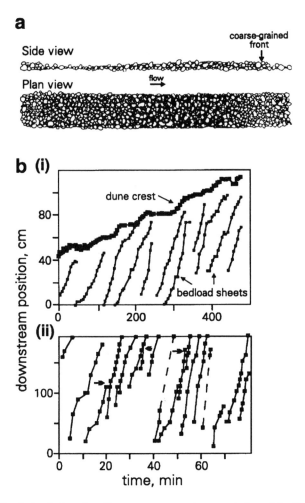

Figure 3.14 Morphological and sediment transport characteristics of bedload sheets (from Whiting *et al.*, 1988). (a) Schematic cross-sectional and planform morphology of bedload sheets formed in fine gravels. Flow arrow for scale is 5 cm long. (b) (i) Time–distance plot of the migration of bedload sheets up the stoss side of a dune illustrating the periodic supply of sediment to the dune crest; (ii) position of fronts of bedload sheets in time illustrating bedform amalgamations (arrowed)

because of the greater range of available grain sizes. It is noticeable that Wilcock (1992) describes bedload sheets in sediment mixtures that are poorly sorted but that very low, long, straight crested bedforms (equivalent to two-dimensional dunes?) are found in some sediment mixtures with a finer D_{50} or better sorting where the fine and coarse components do not behave independently. Particle interactions and a "catch and mobilize" process may be important as may the fact that bedload sheets will present areas of clear roughness transitions which undergo marked changes in turbulence structure (e.g. Antonia & Luxton, 1971, 1972, 1974; Robert *et al.*, 1992) and may strongly influence local sediment transport. Particle interactions and mutual particle interference may therefore dominate the generation of these forms; bedload sheets may only develop when sand is present with fine gravel (Whiting *et al.*, 1988) but this does not mean that the same genetic type of bedform ceases to exist in finer sediments that are better sorted. The bedforms that lie above the lower-stage plane bed/cluster field therefore encompass both two-dimensional dunes and bedload sheets and are possibly controlled by the same formative processes (particle interactions, bedform amalgamations and form-dependent shear stress fields). Neither of these bedforms, however, is characterized by the large macroturbulent flow structures that are associated with dune bedforms (see below).

Dunes

Dunes are generated from ripples, two-dimensional dunes and bedload sheets as the applied bed shear stress is raised over each of these bedforms. Dunes have several distinguishing characteristics: (1) they interact more strongly with the water surface than ripples in an out-of-phase manner; (2) their height has been found to scale with the flow depth (Jackson, 1976; Yalin, 1977); (3) dunes may possess a far more regular wavelength than two-dimensional dunes/bedload sheets (Figure 3.13a); and (4) dunes are a larger bedform although with similar asymmetric profiles to ripples (Figure 3.15; Yalin & Karahan, 1979; Haque & Mahmood, 1985, 1986; Allen, 1982). Dunes have been documented in both fine and medium sands (Allen, 1982), in coarse sands (Costello, 1974; Costello & Southard, 1980) and gravels (Figure 3.16; Dinehart, 1989, 1992).

The initiation of a dune bedstate appears to require a bedform of sufficient size that will significantly alter the entire flow and turbulence field to one that becomes

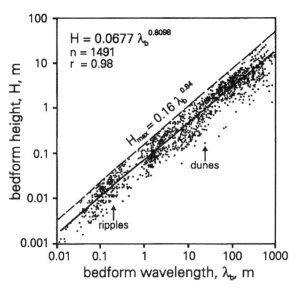

Figure 3.15 Plot of bedform height vs. wavelength for ripples and dunes. Both bedforms show a similar form but there is a gap between the two populations at approximately 0.8 m wavelength (from Ashley, 1990)

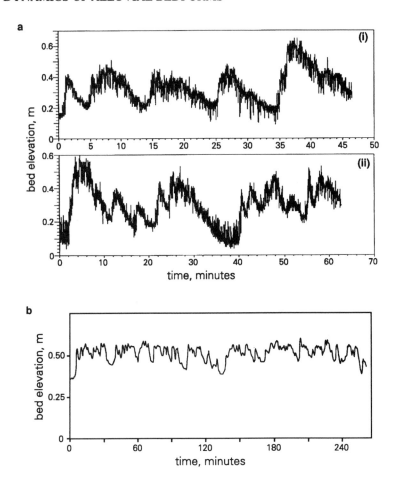

Figure 3.16   (a) Sonar records of dune bedforms generated in gravels ($D_{50} \sim 30$ mm). Plots show bed elevation at one point in time revealing the migration of dune forms (from Dinehart, 1992). (i) Regular wavelength dunes, (ii) plot illustrating a range of bedform heights and wavelengths with smaller bedforms on the stoss side of larger dunes. (b) Dune bedforms documented in laboratory experiments with bedload of $D_{50} = 23.5$ mm (from Hubbell *et al.*, 1987). Note the superimposition of smaller bedforms migrating upon the larger dunes

dominated throughout its depth by the bedform. The formation of this "larger than average" bedform may revolve around production of a "rogue ripple" (Leeder, 1983a) or amalgamation of bedforms to give a larger than average form. This transition is well illustrated by the two-dimensional dune – dune transition which can be triggered by amalgamation of smaller features (Costello, 1974). Costello & Southard (1980) note that the two-dimensional dune – dune transition is accompanied by an increase in the depth of scour at flow reattachment which must therefore lead to a downstream increase in the sediment transport rate. Little evidence exists to test the links between rogue bedform height and the effectiveness of flow separation in either enhancing erosion at reattachment or in triggering macroturbulence that scales with the flow depth. Whiting *et al.* (1988) also view bedload sheets as providing the "seeds" for dune development and show that the local supply of sediment to a dune crest in poorly sorted sediments may be through periodic input from bedload sheets as they reach the dune crest, this possibly controlling grading on the dune foreset. Furthermore, superimposition of bedload sheets may also be responsible for the growth of dunes that have been documented in gravels (Dinehart, 1992). It should also be noted that the superimposition of ripples upon fully developed dunes in sands is a common feature and the result of the modified shear stress field over the dune form (Bridge, 1981, 1982).

Gabel (1993) presents field data on dune geometry and kinematics in a sand-bed river and illustrates the complex three-dimensional form of dunes and their rapid change in morphology during migration. From detailed echo-sounder records, Gabel documented four types of dune creation and destruction: (1) dune splitting, perhaps through localized intense erosion in the reattachment region, where one larger dune form becomes modified into two (or more) forms; (2) dune combination where a faster-moving upstream dune amalgamated with a slower-migrating downstream bedform – in this case, the downstream dune can become starved of sediment through the influence of the upstream separation zone, a feature also noted in studies of coarse sand bedforms (see Figure 3.13c); (3) dune attenuation where the dune form became smaller until it could not be traced; and (4) "spontaneous" dune creation where a dune suddenly appeared on the sounder records, this type of creation possibly being due to the spanwise migration of a three-dimensional dune under the sounding line. Gabel (1993) linked dune creation and destruction to localized, unsteady sediment transport conditions and found that splitting and combination were the most common mechanisms of growth and destruction.

Flow over dunes is dominated by the presence of large-scale flow separation in the dune leeside and the presence of macroturbulent structures (Jackson, 1976; Müller & Gyr, 1982, 1986; Lapointe, 1992; Kostaschuk & Church, 1993; Nezu & Nakagawa, 1993; Bennett & Best, 1994, 1995) which may convect through the entire flow depth and reach the flow surface; these events have been linked to the suspension of sediment (Coleman, 1969; Jackson, 1976; 1978; Lapointe, 1992) in "kolks" and "boils". Several studies have detailed the characteristics of flow over dunes in both the laboratory and field, although to date there has been no comprehensive and detailed study of the three-dimensional flow and turbulence characteristics of dune bedforms. Several studies (e.g. Karahan & Peterson, 1980; Engel, 1981) have investigated the characteristics of the leeside flow separation zone, its length commonly being between four and seven times the dune height. Engel (1981) also demonstrated the length of the flow separation zone to be independent of the Froude number of the flow but to show some dependence on both the dune steepness and, for flat dunes, on the grain roughness. Raudkivi (1963, 1966), using a fixed bed moulded in metal from dune profiles, examined the bed pressure coefficients, mean velocity profiles and turbulence intensities at several locations over a dune profile (Figure 3.17). This seminal work realized several important points which are not only relevant to dune bedforms, but also guided Raudkivi's thoughts upon current ripple initiation (Raudkivi, 1963, 1966; see above). The principal points arising from this work were (see Figure 3.17):

(1) Flow downstream of the dune is dominated by the leeside separation zone which provides a wake region and a developing boundary layer downstream from reattachment. Subsequent numerical modelling of flow over dunes has begun to account for these factors (e.g. Nelson & Smith, 1989; Mendoza & Shen, 1990; van der Knaap et al., 1991; Lyn, 1993; Nelson et al., 1993).

(2) Bed pressure varies significantly over the bedform, reaching a maximum at or just downstream of reattachment and a minimum at the bedform crest where the velocity is greatest (Figure 3.17d). The bed pressure maximum is related to the impingement of shear layer eddies upon the bed (Mendoza & Shen, 1990).

(3) Turbulence intensities in both the downstream and vertical planes (u and v components, respectively) are highest in the free-shear layer associated with the leeside separation cell (Figure 3.17b,c). Peak turbulence intensities occur near reattachment and become both smaller and further from the bed as the boundary layer develops on the next dune back. A similar pattern was found in the Reynolds stress distribution.

(4) Mean shear stress increases towards the bedform crest although subsequent work has illustrated that the shear stress maximum is reached just upstream from the crest and hence deposition may take place at the dune crest as the shear stress declines (Nelson & Smith, 1989; see Figure 3.18e).

Subsequent investigations have improved our knowledge of dune dynamics. Rifai & Smith (1971) examined mean flow and turbulence profiles (u-component only) over fixed triangular elements representing dunes. Their findings were similar to those of Raudkivi (1963, 1966) confirming the maximum bed pressure at reattachment, boundary layer development downstream from this point and maximum velocities near the crest. However, Rifai & Smith (1971) were also able to examine the structure of turbulence by considering the energy spectra of the downstream component at several profiles. Rifai & Smith (1971) came to the significant conclusions that: (1) the leeside eddy is significantly more turbulent than the flow above it; (2) the structure of turbulence over dunes differed to that over a plane bed due to ejection of higher frequency turbulence from the leeside eddy into the outer flow; and (3) the scale of macroturbulence could be related to the dune height and not the flow depth. This argument will be further examined below. In a similar study, Nelson & Smith (1989) measured detailed velocity profiles over a fixed dune form and used these to verify a two-dimensional model of flow over dunes. Their analysis allows for the influence of flow separation and the effects of combined wakes downstream from the separation zone associated with each bedform (Figure 3.18a). Clearly, wake interaction and the interactions of vortices shed off the shear layer will produce a complex velocity field over the bedforms. McLean et al. (1994) illustrate how near the dune bed a positive skewness in the downstream velocity component and a negative skewness of the vertical component are indicative of wake turbulence encroaching on the boundary layer. Measurements of the wake downstream from dunes (Figure 3.18b) illustrate that wake height grows with distance from the dune crest. Velocity profiles over dunes show gradual development of the boundary layer downstream from reattachment (Figure 3.18c) and may be used to calculate the shear stress distribution over the dune back. This

92

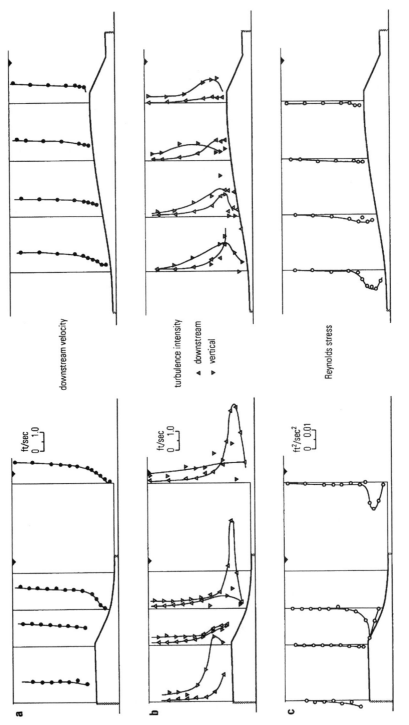

a

ft/sec
0   1.0

downstream velocity

b

ft/sec
0   1.0

turbulence intensity

◄ downstream
▶ vertical

c

ft²/sec²
0   0.01

Reynolds stress

Figure 3.17 *For caption see facing page*

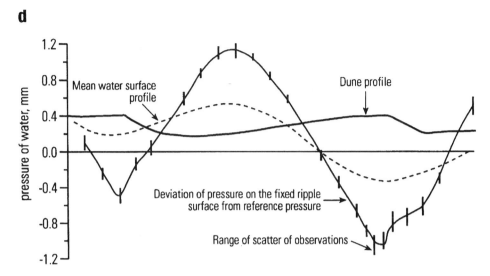

Figure 3.17 Velocity, turbulence intensity and bed pressure distributions at various points over a dune bedform (after Raudkivi, 1966). (a) Velocity profiles of downstream, v, component. (b) Turbulence intensity profiles for the downstream and vertical components of turbulence (Δ and ∇ symbols respectively). (c) Reynolds stress ($u'v'$) profiles. (d) Water surface profile and bed pressure (relative to a reference pressure upstream of bedform)

shear stress pattern (Figure 3.18d) reveals a peak just upstream of the crest, thus allowing deposition at the crest, and has been used to explain the growth of a perturbation from an initial defect (Smith, 1970; Nelson & Smith, 1989; Nelson *et al.*, 1993). Theoretical calculations suggest that the shear stress maximum lies upstream of the crest for wavelength:height ratios greater than 10 but is located at the crest for steeper features (Figure 3.18e; Nelson & Smith, 1989). The interaction of the wake region with the new internal boundary layer downstream from reattachment and its role in generating the turbulence structure over dunes are further discussed in the recent work of McLean *et al.* (1994).

Important data upon turbulence fields over dune bedforms have also come from field measurements (Smith & McLean, 1977; McLean & Smith, 1979). McLean & Smith (1979) show that the peak in the $-\rho u'u'$ Reynolds stress moves away from the boundary, being located near the crest at the bed but over the trough further up in the flow. This pattern was related to turbulence production being at a maximum in the zone of greatest shear near the bed which, as the turbulence diffuses away from the boundary, is convected downstream and hence rises in the flow. This time-averaged view clearly ignores the important non-uniform turbulence production associated with the leeside separation zone. Additionally, McLean & Smith (1979) found that the $-\rho u'v'$ stress increased in magnitude away from the boundary.

Despite these studies, the generation of dunes and their controlling fluid dynamic processes are poorly understood and thought upon this topic has been guided for many years by the untested hypothesis that dunes scale with bursts generated within a turbulent boundary layer. Early work upon boundary layer mechanics suggested

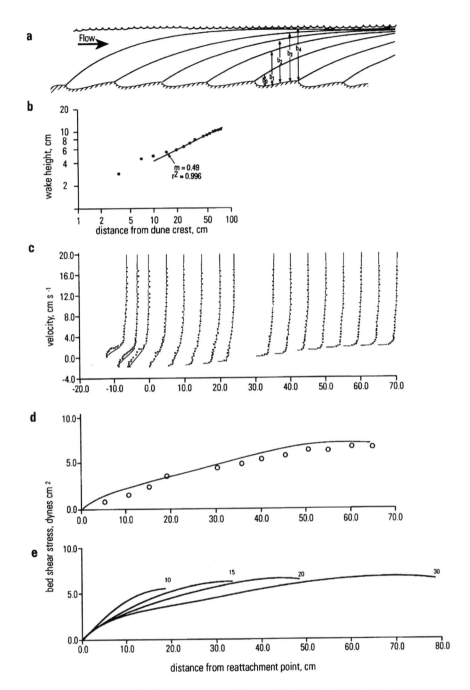

Figure 3.18 Fluid dynamic characteristics of dunes (after Nelson & Smith, 1989). (a) Wake formation and expansion over a dune bed; b_0–b_4 define wakes from different dune bedforms. (b) Wake height above a surface as a function of distance downstream from the dune crest. (c) Experimentally measured (circles) and theoretically predicted (lines) velocity profiles over a dune bedform. (d) Experimentally measured (circles) and theoretically predicted (lines) bed shear stresses over a dune bedform. (e) Calculated bed shear stress distribution over dune bedforms of wavelength 30, 45, 60 and 90 cm (form indices (wavelength:height) of 10, 15, 20 and 30). The maximum bed shear stress is located upstream of the crest for form indices greater than 10 but is located at the crest for steeper features

that bursts scale with the outer wall variables of mean flow velocity and depth (eqn 3.2), although subsequent research has illustrated that either an inner-wall variable or mixed inner/outer variable combination may equally be used to scale burst periodicity (Luchik & Tiederman, 1987, and see above). Yalin (1976, 1977, 1992) and Jackson (1976, 1978) used the outer-wall variable scaling of bursts to propose a link between the formation of dunes and boundary layer bursting within alluvial flows. The hypothesis evolved from two central points: (1) that a similar scaling can be found between upwellings or "boils" associated with dune fields and boundary layer bursting (Figure 3.19a and eqn 3.2); and (2) that the wavelength of dunes, λ_d, scales with the flow depth, Y, such that

$$\lambda_d \sim 5-7Y \qquad (3.4)$$

Hence, since $T \sim 3-7 \, Y/U$ and $\lambda \sim 5-7Y$, Yalin and Jackson used this similarity to propose that dunes are inherently linked to and controlled by boundary layer bursting. Further, Yalin (1976, 1977, 1992) assumed that the wavelengths of the growing dunes were set by the mean path length of macroturbulent eddies within the flow (Figure 3.19b). However, several problems are apparent with this hypothesis.

(i) Records of turbulence over dunes do not always show the intermittency in velocity and Reynolds stress signatures that would be expected from bursting, and the structures may be of shorter duration than those that would be expected through burst scaling (McLean & Smith, 1979).

(ii) It is probable that turbulent boundary layer bursts are not exclusively related to scaling with the outer-wall variables (see above), and these eddies associated with dunes are more likely due to Kelvin–Helmholtz instabilities and vortex shedding off the leeside free-shear layer (Rood & Hickin, 1989; Müller & Gyr, 1982, 1986; Levi, 1984, 1991; Best, 1993; Kostaschuk & Church, 1993; Nezu & Nakagawa, 1993; Williams, 1996; Bennett & Best, 1994, 1995). Research on mixing layers has also demonstrated the presence of secondary streamwise vortices associated with the braid region between the Kelvin–Helmholtz instabilities (see Metcalfe et al., 1987; Lasheras & Choi, 1988;. Silveira Neto et al., 1993). Müller & Gyr (1982, 1986) present visual evidence for the initiation, growth and shedding of eddies off the leeside shear layer which may reach the flow surface (Figure 3.19d). It is difficult to view the schematic model of Jackson (1976; Figure 3.19c) as being physically possible, where bursts of fluid originate from within the separation zone and advect through the free-shear layer to the flow surface. Levi (1984, 1991) considered that the periodicity of bursts associated with dunes could be explained through consideration of the Strouhal number (eqn 3.3), which is obtained by substituting burst frequency for burst period in eqn 3.2 and rearranging. This resulted in values of approximately 0.16 for dune-related bursts – the characteristic Strouhal number value. It should be noted here that Levi retains the flow depth as a length scale within the Strouhal number rather than using the dune height. Plotting the data of Jackson (1976) and Ikeda & Asaeda (1983) (Figure 3.19a) lends support to the contention that the origin of these macroturbulent events may lie in eddy shedding. It should also be noted that these periods are derived from visually apparent bursts and may not reflect the true occurrence of upward-directed coherent motions which may be

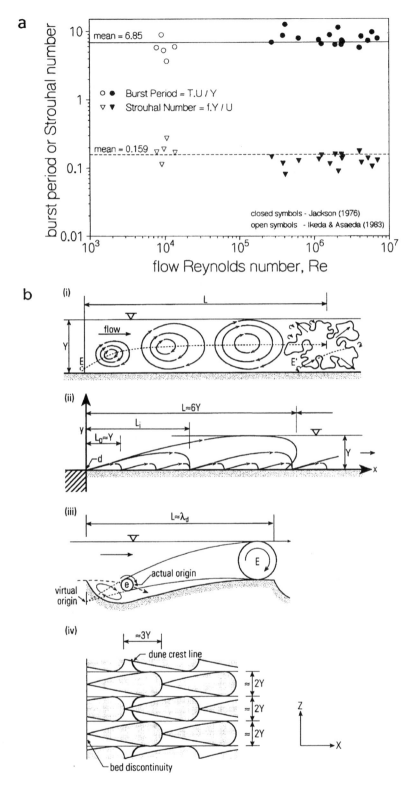

Figure 3.19   *For caption see facing page*

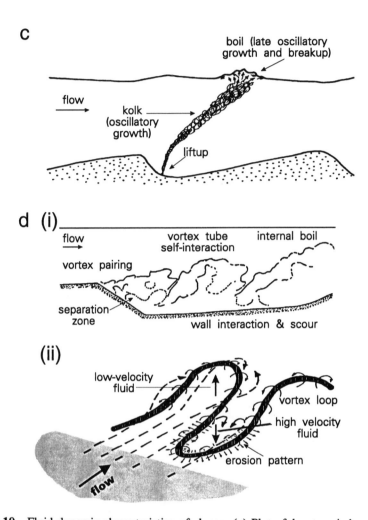

Figure 3.19 Fluid dynamic characteristics of dunes. (a) Plot of burst period and Strouhal number for the dune data sets reported by Jackson (1976) and Ikeda & Asaeda (1983). (b) Yalin's (1976, 1977, 1992) model of dune formation. (i) The lengthscale, L, of eddies within the flow, E, is set by the mean burst wavelength which scales with flow depth, Y. E' denotes burst breakup. (ii) Different scales of burst may coexist (L_o, L_i, L) with the dominant wavelength, L, being approximately equal to $6Y$. Bursts originate at a discontinuity in the bed, d (e.g. a negative step/dune crest brinkpoint). (iii) Origin of burst (e) may be associated with the free-shear layer and causes modulation of sediment transport downstream from a bed defect. The wavelength of the burst, E, determines the wavelength of the dune, λ_d. (iv) Bursts have a spanwise scale, z, of approximately $2Y$ which may determine the cross-stream dune width. Amalgamations of forms may then produce longer sinuous dune crestlines. (c) Turbulence generation associated with dunes and the origin of "kolks" and surface boils (after Jackson, 1976). (d) Vortex generation associated with the free-shear layer of the dune flow separation zone (after Müller & Gyr, 1986): (i) vortex development and boil generation downstream of a dune; (ii) "horseshoe" shaped vortex originating on free-shear layer of flow separation zone

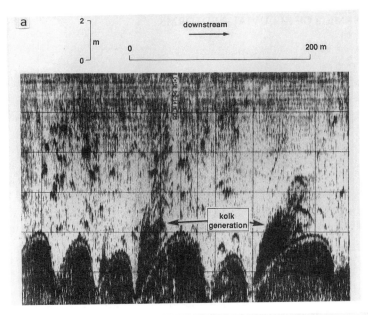

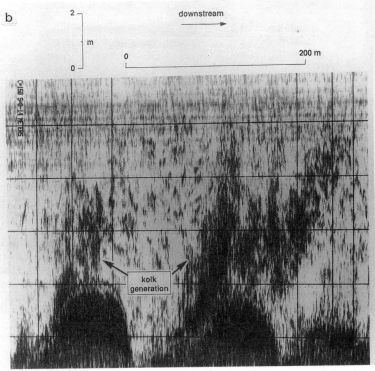

Figure 3.20 Suspension of sediment over dunes as visualized in the field using frequency acoustic profiling on the Fraser River, Canada (from Kostaschuk & Church, 1993). These suspension events have been linked with dune-associated macroturbulence. (a) Two large "kolks" originating on the lower stoss side of dunes. (b) "Kolks" generated over dunes with origin nearer the bedform crest. Downstream dune has two kolks inclined at 7.6° and 2.9° to the flow

of smaller magnitude and not recognized within the flow (especially if they do not transport appreciable quantities of suspended sediment). Itakura & Kishi (1980) also report Strouhal numbers of 0.14 associated with dunes which lends support to the idea of shear-layer-generated macroturbulence.

Kostaschuk & Church (1993) provide field evidence of macroturbulence associated with dunes through use of a frequency acoustic profiler. These images (Figure 3.20) illustrate the high sediment concentrations associated with ejections downstream of dunes; electromagnetic current meter records were also able to link these suspension plumes to quadrant 2 events (Kostaschuk & Church, 1993). The acoustic profiler images show that "bursts" or "ejections" may originate low down on the dune stoss side (Figure 3.20a). This is similar to the observations of Iseya & Ikeda (1986) and Nezu & Nakagawa (1993) who document large "kolk-boils" as originating from near the reattachment region in the dune lee (Figure 3.21). However, some eddies may also originate closer to the dune crest and upper part of the separation-zone shear layer (Figure 3.20b). Bennett & Best (1994, 1995) present experimental data of flow over fixed dunes, possessing angle-of-repose slipfaces, that illustrate the dominant role of the separation-zone shear layer in generating quadrant 2 events (Figure 3.22). Strong ejections may therefore occur all along the shear layer and not just from the reattachment region, although the potential for sediment suspension by these events may be greater at reattachment where vortices impact upon the bed and entrain sediment. Soulsby *et al.* (1991) reason that sediment may be ejected into suspension from the crest of dunes, whilst Onslow *et al.* (1993) point to the entrainment of sediment from within the separation-zone reverse flow and ejection of a plume of sediment, which is then entrained by the shear layer, into the outer flow. Kostaschuk & Church (1993) also argue that eddy shedding may occur from dunes with low-angle slipfaces that do not possess flow separation. In this case Kelvin–Helmholtz instabilities are generated from a shear layer formed in the region of flow expansion downstream from the bedform. It is clear that much

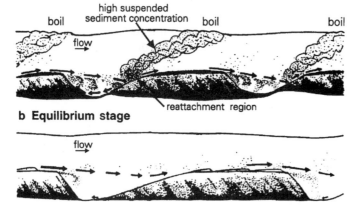

Figure 3.21 "Kolk-boils" developed downstream from the reattachment point of dunes (after Iseya & Ikeda, 1986). (a) Developing stage and high suspended-sediment concentration in boils. (b) Lessening of boil activity in equilibrium conditions

FLOW

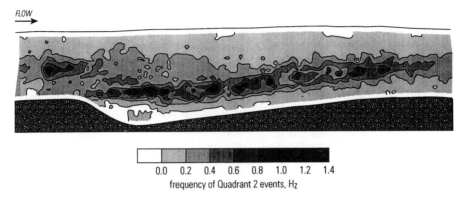

0.0   0.2   0.4   0.6   0.8   1.0   1.2   1.4
frequency of Quadrant 2 events, Hz

Figure 3.22 Frequency of quadrant 2 events over an equilibrium dune profile (after Bennett & Best, 1994). Contours show the frequency of quadrant 2 events exceeding a threshold, B, of 4 (where $B = uv/u''v''$, where u'' and v'' are the r.m.s. values of the downstream and vertical components of velocity respectively; see Willmarth & Lu (1972, 1974) for details of quadrant analysis). Note the strong association of quadrant 2 events with the free-shear layer associated with flow separation

work remains to be conducted on the nature of dune-related macroturbulence, for instance in establishing the influence of dune shape on the origin and nature of shear-layer eddy generation (Figure 3.21). Nezu & Nakagawa (1993) provide an outline of the feedback mechanisms between the bedform, "burst/boil" generation and the free-shear layer (Figure 3.23) although how this acts to control the generation of dunes is still largely a matter of speculation (see below).

(iii) No clear link has been made with *how* any large "bursts" dune control wavelength through the control of sediment transport. Several hypotheses may be outlined:

- Bursts control the suspension and therefore fallout distance of sediment that is uplifted by such events. The onset of dune-induced macroturbulence may therefore generate suspension by bursting and the subsequent downstream settling/accumulation of this load modulates the local deposition rate and therefore dune wavelength. This could reinforce existing dunes or may act as a trigger in dune creation where two smaller forms amalgamate into a larger than

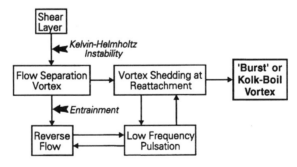

Figure 3.23 The mechanism of burst or kolk–boil vortices associated with dune bedforms (after Nezu & Nakagawa, 1993)

average bedform at the dune transition. Whilst it is evident that bursts, of whatever origin, associated with dunes are capable of suspending sediment (Iseya, 1984; Rood & Hickin, 1989; Lapointe, 1992; Kostaschuk & Church, 1993; Nezu & Nakagawa, 1993; Soulsby *et al.*, 1991; Onslow *et al.*, 1993), this argument cannot be applied to gravel beds in which dunes have also been found to form (Figure 3.16; Dinehart, 1989, 1992; Hubbell *et al.*, 1987). Furthermore, the settling of sediment from a burst will involve progressive fallout of different grain sizes in response to a decaying vertical component of turbulence – the deposited sediment load may therefore settle over a wide area downstream. The higher at-a-point deposition rates required to amplify a downstream ripple, two-dimensional dune or bedload sheet into a dune may therefore not be present. The settling distance of the suspended sediment also depends upon both the magnitude of the burst *and* the grain size distribution, inferring some grain size control upon dune wavelength which is not present in dunes.

- Bursts control the instantaneous rate of bedload transport and locally increase sediment transport downstream. This increased and unsteady bedload transport leads to deposition and results in downstream bedform amplification and subsequent dune growth. This process may also be intimately linked to the high stress producing quadrant 4 events that occur within the reattachment region.

The formation and maintenance of a dune field requires that the sediment transport rate across these bedforms is greater than those from which they have initiated (e.g. ripples in fine–medium sands, two-dimensional dunes or bedload sheets in coarse sands–gravels). The increased transport rate results in greater bedform heights and generation of larger flow separation cells. If we accept that an initial dune results from amalgamation of smaller bedforms, then it is the subsequent modification of the downstream bedload transport rate that appears critical. Suspended-sediment flux may also be increasingly important in finer sands and silts. However, the principal mechanism of dune growth must lie in a control on bedload transport which will produce dunes in sands or gravels. It also seems likely that this influence upon bedload transport occurs not through the influence of bursts advecting upwards within the flow, but upon high instantaneous Reynolds stresses (e.g. quadrant 4 events) acting upon the bed. One mechanism of providing these stresses is through higher bedforms, generated by amalgamation of smaller forms, generating larger eddies along the separation-zone free-shear layer. These Kelvin–Helmholtz instabilities may increase both erosion within the reattachment region and subsequent downstream deposition. This contention finds qualitative support in the observation that the two-dimensional dune – dune transition is marked by both larger bedform troughs and greater erosion at reattachment, but the characteristics of this region have not been detailed in the ripple and bedload sheet – dune transitions. Perturbations along a turbulent shear layer may have different strengths that are related to the height of the bed surface discontinuity with respect to the flow depth (Bradshaw & Wong, 1972). Bradshaw & Wong (1972) define three strengths of perturbation – weak (confined to the inner layer of the flow), strong, and overwhelming – where the latter two cases significantly influence the boundary layer outside the separated flow region. As the thickness of the upstream boundary layer becomes smaller with respect to the step height (or separation-zone

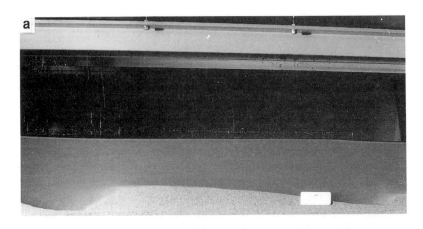

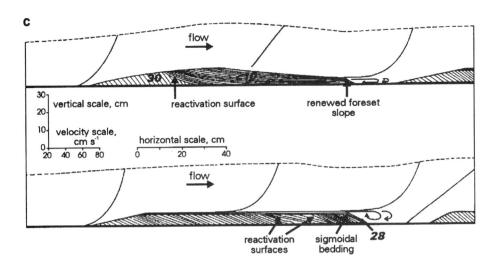

Figure 3.24   *For caption see facing page*

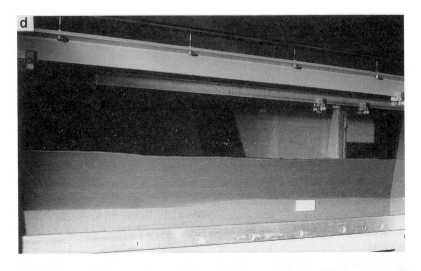

Figure 3.24 Bedforms across the dune – upper-stage plane bed transition. (a) Fully developed equilibrium dunes of wavelength ~0.75 m and height ~0.035 m. (b) Detail of flow separation zone shown in (a). Streak lines in flow are paths of neutrally buoyant particles. Note slower, recirculating flow in separation zone and paths of particles rising at ~45–75° to the bed above the upstream dune back which visualize uplifting eddies shed from the upstream separation zone shear layer. (c) Morphology and velocity profiles of humpback dunes (after Saunderson & Lockett, 1983). Bold italic numbers denote angle of foreset surfaces. (d) Washed-out dunes with low leesides. Note largely bed-parallel streak lines of neutrally buoyant particles in the flow. (e) Upper-stage plane beds with low amplitude bedwaves of height ~0.4 cm. Flow depth in (a), (b), (d) and (e) approximately 10 cm and from right to left (from Bridge & Best, 1988)

length), more of the shear layer is deflected upstream within the separation bubble which increases both the velocity gradient across the shear layer near reattachment and the turbulence intensity in the reattachment region (Bradshaw & Wong, 1972). It is also evident that large-scale turbulent eddies with length scales at least that of the step height are associated with the separation-zone free-shear layer and these eddies may pair and coalesce as they convect downstream (for a comprehensive review see Simpson, 1989). McLean *et al.* (1994) also highlight that events from other quadrants may be important in the sediment flux, for instance the higher occurrence of quadrant 1 events downstream from reattachment on the stoss side of the dune.

The nature of the dune flow separation-zone shear layer, its interaction with a mobile bed and role in the transition to, and stability of, dune bedforms remain critical research problems. However, the transition to dunes from either ripples or two-dimensional dunes/bedload sheets seems intimately connected with: (i) a sufficiently high bedform that can trigger significant flow separation and macroturbulence associated with the separation-zone shear layer; (ii) a flow depth that is often shallow enough to allow these shear-layer instabilities to reach the water surface and not dissipate through mixing in the flow above the bedform; and (iii) the increased influence of the separation-zone shear layer in downstream sediment transport either at/downstream of reattachment (bedload) or along the shear layer (suspension). The increased influence of the shear layer may be due to the greater absolute velocity differential across the shear layer over the dune transition (and therefore higher instantaneous velocities/shear stresses at reattachment and higher-magnitude fluid ejections from along the shear layer, see Bennett & Best, 1995). It is clear that there is a pressing requirement for research on both the flow fields over dunes and the nature of turbulence and sediment transport across the dune transition over a range of grain sizes.

Upper-stage plane beds

Recent work has illustrated that upper-stage plane beds are not completely flat but are characterized by a series of very low-relief bedwaves (Bridge & Best, 1988, 1990; Paola et al., 1989; Best & Bridge, 1992), although these waves have also been postulated to be antidunes (Cheel, 1990a,b). The transition from dunes to upper-stage plane beds has been interpreted as intimately associated with the suppression of near-bed turbulence and its effect on either (1) the generation of a bedform from a defect upon a plane bed (Allen & Leeder, 1980), or (2) the modification of dune morphology and sediment transport as shear stress is increased over the transition between these two bedstates (Bridge & Best, 1988). Work detailing the dune – upper-stage plane bed transition in sands has revealed the progression (Figure 3.24) from typical asymmetric dunes to humpback dunes to low-relief dunes and eventually to low-relief bedwaves on an apparently flat bed (Saunderson & Lockett, 1983; Bridge & Best, 1988; Best & Bridge, 1992). This series of transitional forms has also been recognized in ancient sandstones (Roe, 1987; Chakraborty & Bose, 1992).

Bridge & Best (1988) measured velocity and turbulence profiles over representative bedforms across this transition and their results indicate that the downstream component of turbulence intensity in the bedform trough decreases as the dunes become humpbacked (Figure 3.25). The lower 25% of flow over the dune backs, however, shows increasing turbulence intensity across the transition. Additionally, Bridge & Best (1988) observed a clear decrease in the number of upward-directed turbulent motions from the reattachment region. Thus, they reasoned that the dune – upper-stage plane bed transition is caused by the increasing influence of suspended sediment which acts to suppress turbulence, especially in the free-shear layer of the separation zone and at the reattachment region. These measurements support the earlier theoretical reasoning of Bagnold (1966) and Allen & Leeder (1980) who

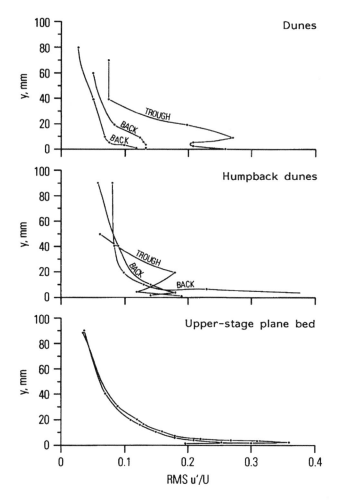

Figure 3.25 Downstream turbulence intensity profiles across the backs and troughs of fully developed and humpback dunes and over an upper-stage plane bed (after Bridge & Best, 1988)

proposed that a plane bed will remain stable if the bed is effectively occluded from turbulence within the flow by higher sediment concentrations. Allen & Leeder (1980) show that Bagnold's theoretical reasoning for upper-stage plane bed stability works fairly well for decelerating flows and propose a model for the stability and non-propagation of a bed defect upon an upper-stage plane bed (Figure 3.26). This model shows that a bed defect will not amplify upon an upper-stage plane bed if the fractional grain concentration in the bedload layer exceeds approximately 0.1.

These arguments concerning turbulence suppression across the dune – upper-stage plane bed transition are supported by empirical studies of dune stability which show that the dune stability field occupies a progressively smaller range of applied bed shear stresses as the sediment size becomes finer (Figure 3.3; see also Engelund &

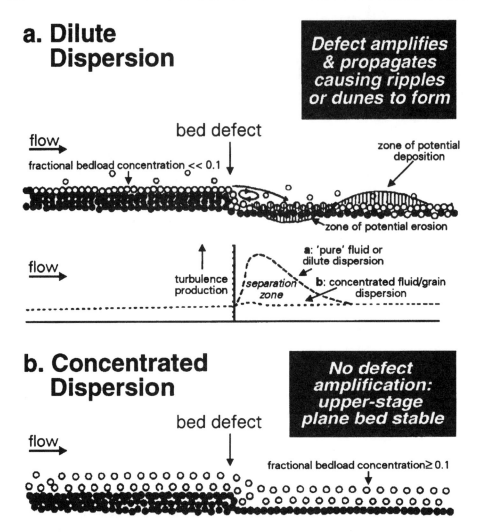

Figure 3.26 Model of the stability of upper-stage plane beds (after Allen & Leeder, 1980). In dilute dispersions (a) bed defects may propagate downstream due to flow separation; in concentrated dispersions (b) where turbulence is suppressed and occluded from the bed, defects will not amplify. Middle graph illustrates schematic turbulence production in each of these cases

Fredsøe, 1974; Watanabe & Hirano, 1988). This reduced stability range reflects the fact that in finer sediments the higher suspended-sediment concentrations required to influence dune stability are reached at lower bed shear stresses. This point is further supported by van den Berg & van Gelder (1993) who document the transition of ripples to an upper-stage plane bed in sediment of 33 μm diameter, and by Baas (1993) and Baas *et al.* (1993) who document a transition from ripples to washed-out (or flattened) ripples through to a plane bed in sediment of 95 μm mean diameter. Bedforms across the transition in this latter sediment become progressively smaller in height (Figure 3.27) and flatter in form, although bedform wavelength changes little across the transition, a point also noted by Bridge & Best (1988). Baas (1994) also links the upper-stage plane bed transition in silts with turbulence suppression due to suspended sediment. Upper-stage plane beds are dominated by very low-relief bedwaves (Bridge & Best, 1988; Best & Bridge, 1992) which, as they migrate and aggrade, are responsible for the production of planar laminae. Allen (1984) proposed from theoretical reasoning that these bedwaves could be produced by a slowly varying bed shear stress pattern generated by the advection of larger-scale coherent turbulent structures over the bed surface. However, the scaling of these eddies has not been found to match the kinematic characteristics of the bedwaves (Bridge & Best, 1988).

Several workers have also found that dunes are replaced at higher bed shear stresses in coarse sands by antidunes and not by an upper-stage plane bed (Williams, 1967; Southard & Boguchwal, 1990). The nature of the dune – antidune transition in coarse sands is complex, with bedforms that are both in- and out-of-phase with the water surface existing at one mean flow condition. However, in shallow flows over coarse sand beds, the Froude number required to generate antidunes (approximately >0.84 (Kennedy, 1961, 1963)) may be attained *before* the flow is able to suspend enough sediment required to suppress turbulence and generate an upper-stage plane bed. Hence, in this case the lack of sufficient vertical velocities required to suppress turbulence by the time a critical Froude number is reached leads to antidune formation after a dune bed state. The symmetric bedforms noted by Dinehart (1989) may depict antidunes that coexist with dunes in a gravel-bed river. A few observations suggest that a plane bed may be reached at much higher Froude numbers (~1.44) in coarse sands (Williams, 1970) but there is very little information

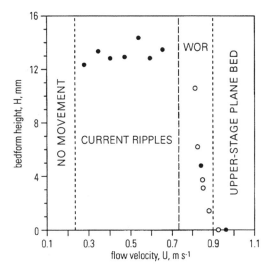

Figure 3.27   Plot of the height of ripples formed in fine sands ($D_{50} = 0.095$ mm) as a function of flow velocity illustrating the rapid decrease in height across the upper-stage plane bed transition. WOR = washed out ripples (from Baas *et al.*, 1993)

on these flow and bedform conditions. In deeper flows, where the Froude number can remain lower at the shear stresses required to obtain a dense bedload or traction layer, plane beds do exist in gravels (Dinehart, 1992). Dinehart (1992) argues that these plane beds may represent some kind of "rheologic bed stage" (*sensu* Moss, 1972) present at higher shear stresses where sediment concentrations near the bed are high and a carpet of saltating grains may be present. The nature of any bed topography on such a nominally flat bed in these conditions remains to be established.

A characteristic feature of upper-stage plane beds in sand is the ubiquitous occurrence of primary current lineations upon these surfaces, although such lineations are produced on beds at lower shear velocities. Such lineations have been attributed to the actions of sublayer sweeps and streaks in the lower part of the boundary layer (Allen, 1982, 1985; Mantz, 1978; Weedman & Slingerland, 1985). Primary current lineations have been found to scale with the streak spacing (eqn 3.1) at low shear velocities and tend to change in planform from straight – intersecting – wavy as shear velocity is increased (Weedman & Slingerland, 1985). This change in planform sinuosity has been interpreted as evidence for the uplift and wandering of low-speed streaks as they form ejections within the boundary layer (Figure 3.28). Additionally, Weedman & Slingerland (1985) suggest that the spacing of the lineations, and therefore the low-speed streaks, increases at higher shear velocities when the concentration of sediment within the bed layer is higher. It may be that this modification to the boundary layer structure mirrors those changes that have been documented within drag-reducing polymer flows in which streak spacing increases and burst periodicity decreases as the amount of polymer within the flow (and therefore the degree of drag reduction) increases (Tiederman *et al.*, 1985; Luchik & Tiederman, 1988; Best & Leeder, 1993). An upper grain size limit to the formation of primary current lineations between 0.29 and 1.38 mm has also been suggested by

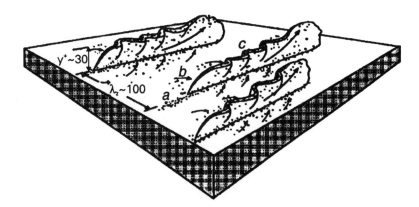

Figure 3.28 Model for the formation of primary current lineations and sand streaks (after Weedman & Slingerland, 1985). Lineations adopt a spanwise scaling reflecting the low-speed streaks within the boundary layer ($\lambda_z \sim 100 = \lambda u_*/v$) and may become wavy in their upper parts (labelled c) due to the sinuosity of uplifting streaks ($y^+ > 10$). Lineations may originate within (labelled a) and adjacent to (labelled b) lifting low-speed streaks

Weedman & Slingerland (1985), with no lineations being observed in the 1.38 mm sediment. Flow-parallel ridges have, however, been documented upon lower-stage plane beds (Leeder, 1980, and see above).

Antidunes and transverse ribs

As flow becomes supercritical, a plane bed or dune bedstate transforms into one dominated by antidunes, although the transition may not be abrupt, and complex antidune–dune bedstates may coexist under the same mean flow conditions (see above). Kennedy (1961, 1963) found that antidunes may form at Froude numbers greater than 0.84; the formation of antidunes relies upon the presence of a free-water surface and the in-phase coupling of the bed and water surfaces. Antidunes generated in sands have been found to possess a range of migration characteristics, with antidunes forming trains which migrate downstream, become stationary and then migrate upstream as bed shear stress is progressively increased (Kennedy, 1961, 1963; Barwis & Hayes, 1985; Langford & Bracken, 1987; Cheel, 1990a). Kennedy (1961, 1963) proposed that antidune wavelength, λ_a, may be determined from:

$$\lambda_a = 2\pi U^2/g \qquad (3.5)$$

whilst maximum antidune steepness could be given by:

$$2H/\lambda_a \sim 0.142 \qquad (3.6)$$

where U is the mean flow velocity, g is gravitational acceleration and H is the antidune height.

As antidunes grow in height, owing to deposition on the upstream side of the bedform, the steepness of the form increases. Breaking of the water surface wave and partial destruction of the antidune may be due to two processes (Kennedy, 1963): (1) creation of adverse pressure gradients causing the onset of flow separation and subsequent flow instability on the upstream side of the antidune; and (2) instability of the surface waves as they increase in height. Antidunes which migrate downstream and display quasi-periodic breaking have also been found to be associated with the formation of water bores in shallow, rapid streams (Foley & Vanoni, 1977) which give rise to pulsing flows in steep alluvial channels.

Antidunes within coarser gravel sediments have also been recorded in the form of transverse ribs (McDonald & Banerjee, 1971; McDonald & Day, 1978; Koster, 1978). Transverse ribs are low ridges of particles no more than a few grain diameters high which are oriented transverse to flow (Figure 3.29) and are especially common in bar-top settings where fast, shallow flows dominate. Transverse ribs show a clear relationship between their wavelength and particle size and have been associated with standing waves and antidunes (Koster, 1978). An antidune origin for transverse ribs seems likely although the relatively large roughness present in coarse-grained channels may offer many sites (i.e. large clasts) over which standing waves can initially form and propagate. Large particle clusters, often formed in lower regime conditions, may form favourable sites from which to generate antidunes at higher Froude number flows. A related form to the transverse rib is the transverse clast dam (Bluck, 1987). These bedforms are somewhat larger than transverse ribs

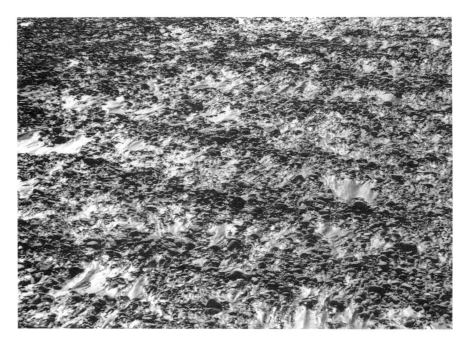

Figure 3.29   Transverse ribs formed in the gravelly Sunwapta River, Alberta, Canada. Rib wavelength is ~0.40 m and the width of the ribs is ~4 m. Note finer grained silt/fine sand infill between transverse ribs. Flow towards bottom left

(sometimes over 1.5 m high (Bluck, 1987)) and have a more stair-like appearance with finer clasts forming a "riser" upstream of the coarser clast dam (Richards & Clifford, 1991). Both bedforms, however, have similar particle size:wavelength ratios (Richards & Clifford, 1991) and ribs may also have an infill of fines between the transverse ridges (see Figure 3.29). Transverse ribs and clast dams can therefore be ascribed to a common origin as antidune/standing wave bedforms (Bluck, 1987).

It is interesting to note that the experiments of Iseya & Ikeda (1987), which documented distinct sorting patterns in sand – gravel mixtures, utilized flows that were largely supercritical and antidune waves were documented on the water surface. Indeed, Iseya & Ikeda (1987, p.19) attribute the formation of their "transitional" bedstate (between a gravel-rich congested bedstate and a smooth "sand-only" surface) to antidune waves forming above a gravel clast jam. The interaction between supercritical flow, antidune formation and the longitudinal sorting of sediment, which becomes more visible in poorly sorted sediments, may therefore be important in initiating some bedforms characteristic of poorly sorted mixtures.

THE INFLUENCE OF SEDIMENT SORTING UPON BEDFORM GENERATION

The sorting of the bed sediment is critical in determining bedform type and morphology (see sections above). Poorly sorted sediments, which produce a more marked segregation of grain sizes in response to variable bed shear stresses over a bedform, can produce apparently different bedforms to their counterparts in better sorted sediments even though the formative processes and genetic origin may be very similar or identical (see discussion of two-dimensional dunes and bedload sheets above). Recent studies (Wilcock, 1992; Wilcock & Southard, 1989; Chiew, 1991; Bennett, 1992) have demonstrated that the stability of bedforms may be significantly influenced by sediment sorting. In poorly sorted sediments the relatively finer-grained fractions may become mobilized so that bedforms, such as dunes, form in finer sediment and override the underlying coarser armour layer (Chiew, 1991; Wilcock, 1992; Bennett, 1992). Any relatively coarser-grained particles that become entrained due to their high relative protrusion are then rapidly transported over this fine sand bed as coarse particles overpass the finer sands (Allen, 1983b; Carling & Glaister, 1987; Carling, 1990). Wilcock (1992) found that bedform height (principally low dunes) decreases whilst bedform wavelength increases as the mixture sorting increases. However, Snishchencko *et al.* (1989) report that for dunes formed in coarse sediments ($D_{50} = 1.55-8.5$ mm) bedform height and wavelength become smaller in more poorly sorted sediments and that this causes a consequent increase in the bedform migration rate. Wilcock (1992) also found that dunes superseded a lower-stage plane bed at higher bed shear stresses in sediments with a log-normal grain size distribution, whilst bedload sheets formed after a lower-stage plane bed in bimodal mixtures. The dunes generated in the log-normal mix possess straight flow-perpendicular crests and may be similar to the two-dimensional dunes discussed above. The sorting of the sediment mixture in this case allows a different spatial grain size differentiation over what are apparently similar bedforms (i.e. two-dimensional dunes and bedload sheets).

Several authors have also documented flow-parallel longitudinal stripes of sediment within heterogeneous bed materials (Ferguson *et al.*, 1989; Tsujimoto, 1989a,b). The formation of narrow longitudinal stripes has been linked to the formation of cellular secondary currents (Tsujimoto, 1989a,b; Nezu & Nakagawa, 1989, 1993) which are generated by the *lateral* variation in bed shear stress between different grain roughnesses. These lateral roughness contrasts were found to generate secondary currents with a spanwise velocity near the wall of 2–5% that of the downstream component (Tsujimoto, 1989a,b). This transverse flow was then postulated to influence the movement of the finer sediment, generate subsequent lateral sediment sorting and further secondary flow cells and sediment ridges (Figure 3.30a). Additionally, studies of flow over and between these sediment ridges (Figure 3.30b) have shown that downstream velocities may be higher over the troughs than the ridges (see also earlier discussion upon current ripple initiation), and that the vertical velocities at these two sites confirm the presence of upwellings over the ridges and downwellings over the troughs (Figure 3.30c,d). Bed shear stress in the troughs is therefore higher than that on the ridges (Figure 3.30e). These velocity characteristics therefore serve to amplify and stabilize the sediment ridge. Nezu & Nakagawa (1989) suggest that such secondary flow cells may be initiated by the interaction of a flow with the corner eddy generated at the edge of a channel and that these secondary flows may propagate into the central region of the channel when the channel width:depth ratio is less than five. Additionally, Nezu & Nakagawa (1989)

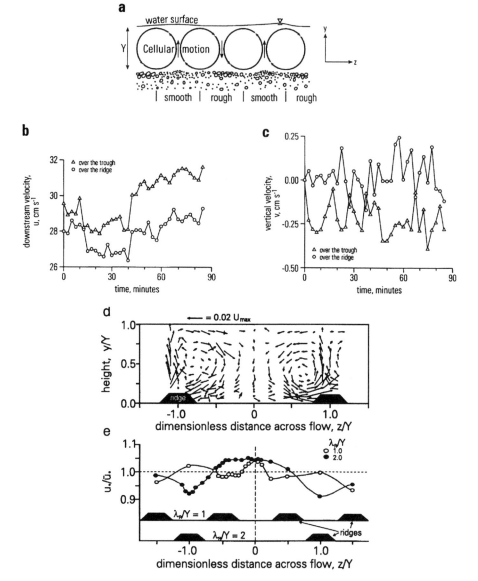

Figure 3.30 Characteristics of longitudinal sediment ridges. (a) Schematic model of cellular secondary currents over longitudinal smooth and rough stripes (after Tsujimoto, 1989a,b). Sketch is in the spanwise ($y-z$) plane. Secondary flows may laterally sort the finer sediment and lead to formation of longitudinal sediment stripes. (b) Downstream (u) and (c) vertical (v) velocities near the bed measured over and between sediment ridges (after Nezu & Nakagawa, 1989). Each velocity point represents a 30 s average. The plots illustrate the higher downstream velocities present between the ridges, the upwellings over the ridge and downwellings in the trough region (see Figure 3.30a). (d) Vector pattern of secondary flow cells over artificial ridges in open-channel flow (from Nezu & Nakagawa, 1993). Height in the flow, y, and spanwise distance across the flow, z, are made dimensionless through division by the flow depth, Y. Scale arrow at the top of the plot shows magnitude of secondary flows in relation to the maximum downstream velocity, U_{max}. (e) Spanwise distribution of shear velocity over longitudinal ridges and troughs (from Nezu & Nakagawa, 1993). Shear velocity u_* is made dimensionless though division by the mean shear velocity, $\bar{u}_*$. Shear velocity distribution is shown for two ratios of ridge spacing: flow depth, λ_R/Y

postulate that such currents may also be triggered around obstacles in the flow. Further work is clearly required to establish the occurrence, significance and fluid dynamic characteristics of these bedforms.

Chiew (1991) argues that in mixtures with a D_{50} of 0.6 mm but with a geometric standard deviation (σ_g) of greater than 2.3 (where $\sigma_g = D_{84}/D_{50}$), antidunes do not form even when Froude numbers exceed unity. This feature was attributed to the formation of a "dynamic armoured layer" or "pavement" in these poorly sorted mixtures which inhibited the formation of antidunes on the bed. The influence of grain size heterogeneity is less clear in lower-regime bedforms and requires attention in the feedbacks between grain roughness and fluid turbulence (for instance, over roughness transitions (Antonia & Luxton, 1971, 1972, 1974; Robert *et al.*, 1992)) and the resultant bed morphology.

THE INFLUENCE OF SUSPENDED SEDIMENT UPON BEDFORM MORPHOLOGY AND GENERATION

Most investigations of the morphology and stability of bedforms within mobile sediments have been conducted in the absence of appreciable fine-grained suspended sediment. However, it is clear that the role of fine, suspended sediment becomes increasingly important at higher concentrations and may cause significant changes to the morphology and stability fields of different bedforms. Previous sections have argued that the presence of suspended sediment in the near-bed layer becomes increasingly important at higher bed shear stresses and, indeed, is instrumental in causing the ripple/dune – upper-stage plane bed transition. Simons *et al.* (1963) and Wan (1982, 1983) illustrate how bedforms may become greatly

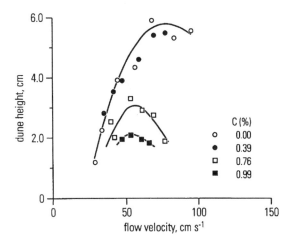

Figure 3.31 Dune height as a function of flow velocity for clear water and three concentrations of bentonite clay (after Wan, 1982). Note how bedform height is severely reduced by clay concentration and how only low-amplitude bedforms exist at any velocities for the maximum bentonite concentrations of 0.99%. Sediment concentration is expressed as a volume percentage

modified by the presence of high fine-sediment concentrations. Dune bedforms decrease in height as the suspended-sediment concentration increases (Figure 3.31). Although the exact processes by which these morphological changes occur are not known, it seems probable that they are connected with the dampening of the turbulence within the flow; these observations lend qualitative support to the arguments concerning the transition from dunes to upper-stage plane beds. The suspension of sediment has been linked to an anisotropy in the vertical component of turbulence such that fluid moving upwards and away from the boundary is dominated by high-magnitude, short-frequency events whilst flow down towards the boundary requires low-magnitude, high-frequency events (Bagnold, 1966; Leeder, 1983b; Wei & Willmarth, 1991). Such suspension has been linked to ejections/bursts of fluid away from the bed and larger coherent motions in the turbulent boundary layer. Suspension of mass therefore requires energy expenditure from the turbulent eddies and a reduction in the turbulent kinetic energy of the flow (Yalin, 1977; Adams & Weatherley, 1981). This energy expenditure, combined with the influence of increasing fluid viscosity at high sediment concentrations (and eventual non-Newtonian flow behaviour), may serve to either lessen the vertical distance over which ejections may rise from the bed or interact with either the turbulence or bed

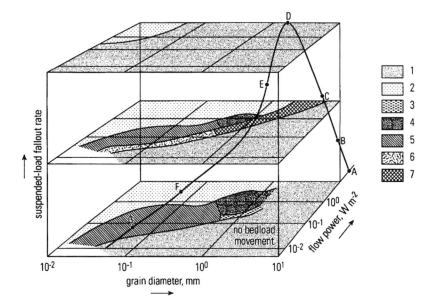

Figure 3.32 Lowe's (1988) schematic bedform phase diagram with respect to different conditions of suspended fallout rate as applied to turbidity currents. Suspended sediment may impart different stability fields for bedforms and must therefore be considered in palaeohydraulic reconstructions (see the path of a hypothetical flow from point A to G). The bedform stability fields defined here are with respect to the Bouma sequence (1 = Bouma A division of massive deposition; 2 = upper-stage plane beds; 3 = lower-stage plane beds; 4 = dunes; 5 = ripples; 6 = Bouma D and E divisions); 7 = inversely graded traction carpets (see Lowe, 1988). Similar modifications to the bedform stability diagram may be necessary for alluvial flows which transport large quantities of suspended sediment

pressure fields associated with flow separation, leading to modification of the flow field and hence bedform. The influence upon bedform stability of fine suspended sediment, or high concentrations of bedload at high bed shear stresses, is a poorly researched but important field that holds many key questions within a wide range of sedimentary environments. For example, Lowe (1988) qualitatively illustrated how suspended load fallout rate may influence the stability fields of bedforms generated in turbidity currents (Figure 3.32). Such modifications to the bedform stability diagram may be relevant in alluvial flows with high suspended-sediment concentrations such as the Yellow River (Wan, 1982; Wan & Niu, 1989; van den Berg & van Gelder, 1993).

CONCLUSIONS – A "UNIFIED" BEDFORM PHASE DIAGRAM

A synthesis of current information on the fluid dynamics and morphological characteristics of bedforms (Figure 3.3) suggests that several bedforms owe their origin to common processes across a range of grain sizes. This plot may be modified to express the principal flow processes that control the formation of these bedforms in a "genetic" bedform phase diagram (Figure 3.33, modified after Leeder, 1983a). The uses of this plot are: (1) to link common, formative processes of bedform generation and stability across a range of grain sizes, (2) to highlight the significant

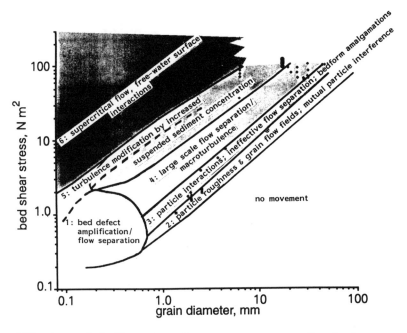

Figure 3.33 A genetic bedform phase diagram across a range of grain sizes summarizing the controlling fluid dynamic and sediment transport processes of different bedform fields. Dashed line denotes transitional boundary between ripples/dunes and upper-stage plane beds. This figure is based on Figure 3.3 and the genetic stability field plot of Leeder (1983a)

changes in the fluid dynamics and sediment transport characteristics as a function of bed shear stress which dictate the transition between one bedform type and another; and (3) to focus on areas in which research must proceed in order to realize more complete fluid dynamic explanations for the generation and stability of small-scale alluvial bedforms. The plot identifies six bedform fields and their controlling fluid dynamic/sediment transport processes:

(1) *Ripples.* Multiple sweep-induced defects, sweep/streak concentration over flow-parallel sediment ridges; flow separation over bed defects, lee-side scour and downstream bedform propagation.

(2) *Lower-stage plane beds and particle clusters.* Modification of boundary layer structure by grain roughness; particle-induced flow field; mutual particle interference.

(3) *Two-dimensional dunes and bedload sheets.* Particle interactions; bedform amalgamations (kinematic waves?); ineffective flow separation (roughness-dominated flow?). Possible sweep/streak concentration over flow-parallel ridges.

(4) *Dunes.* Macroturbulence generated by enhanced, large-scale flow separation; Kelvin–Helmholtz instabilities shed off free-shear layer dominate flow field, cause sediment suspension and enhanced erosion at reattachment region.

(5) *Upper-stage plane beds.* Modification of near-bed flow (and flow separation) by increased suspended-sediment concentrations. Formation and migration of low-amplitude bedwaves.

(6) *Antidunes and transverse ribs.* Supercritical flow and free-water surface interaction. Standing waves and particle interactions.

Despite the wealth of morphological data upon bedforms, little detailed work has sought to investigate the structure of turbulence over many bed configurations and, until this research is conducted, many ideas concerning the stability of bedforms will remain largely speculative. The critical *transitional* regions between bedform fields would seem the most fruitful areas on which to concentrate. Additionally, the influence of suspended sediment upon the structure of turbulence clearly has many important implications within both higher shear stress flows and in rivers where the sediments are poorly sorted or predominantly fine grained.

ACKNOWLEDGEMENTS

I would like to thank my colleagues for the many stimulating and valuable discussions over the last few years on a range of topics covered in this review . In particular the fruitful input of Mike Leeder, John Bridge, Henry Pantin, Sean Bennett, André Roy, Phil Ashworth, Alistair Kirkbride, John Livesey, Stuart McLelland and Rowena Ditchfield is greatly appreciated. I am also grateful for the detailed reviews of Paul Carling, Sean Bennett and Tim Salter on an earlier draft of this paper. Randy Dinehart kindly supplied the data used in Figure 3.16a. Research in the laboratory, which lies behind the development of many of these ideas, has been funded and supported by grants from the UFC, NERC and University of Leeds for which I am most thankful.

NOMENCLATURE

b_0, b_1, b_2, b_3, b_4	wake heights
B	quadrant 2 event threshold $(= uv/u''v'')$
C_D	coefficient of drag (see Figure 3.10 for definition)
C_L	coefficient of lift (see Figure 3.10 for definition)
C_P	bed pressure coefficient (see Figure 3.10 for definition)
d	height of bed defect
D	particle diameter
D_{50}, D_{84}	particle size at the 50th and 84th percentile of the grain size distribution
f	frequency of eddy shedding
Fr	Froude number $(= U/\sqrt{gY})$
g	acceleration due to gravity
G	distance between clasts
H	bedform height
H_{max}	maximum bedform height
H_t	ripple height
H_e	equilibrium ripple height
L_o, L_i, L	wavelength of eddies (bursts)
m	slope of line
p	pressure
q_s	sediment transport rate
r	correlation coefficient
Re	flow Reynolds number (UY/ν)
Re_d	bed defect Reynolds number $(= du_*/\nu)$
Re_p	particle Reynolds number $(= Du_*/\nu)$
St	Strouhal number $(= fD/U$ for particles; $= fY/U$ for dunes$)$
t	time
t_e	time required for bedform to achieve equilibrium height or wavelength
T	time between bursts
T_b	dimensionless burst period $(= TU/Y)$
u_b	near-bed velocity
u_*	shear velocity
u''	root-mean-square value of downstream, U, velocity
U	mean downstream flow velocity
U_{max}	maximum downstream velocity
v''	root-mean-square value of vertical, V, velocity
y	height of a point in the flow
Y	total flow depth
λ	spanwise streak spacing
$\lambda_a, \lambda_d, \lambda_r$	wavelength of antidunes, dunes and ripples
λ_b	bedform wavelength
λ_{req}	equilibrium ripple wavelength
λ_R	lateral (spanwise) wavelength of ridges

ν kinematic viscosity

σ standard deviation of variable

σ_g geometric standard deviation of grain size distribution ($= D_{84}/D_{50}$)

u, v and w refer to flow velocities in the downstream (x), vertical (y) and spanwise (z) axes, respectively. The superscript $^+$ denotes that a variable has been made dimensionless by the inner wall variables of u_* and ν (i.e. $\lambda^+ = \lambda u_*/\nu$) whilst an overbar indicates the mean value of a variable.

REFERENCES

Acarlar, M.S. & Smith, C.R. 1987. A study of hairpin vortices in a laminar boundary layer. Part 1. Hairpin vortices generated by a hemispherical protuberance. *J. Fluid Mech.*, **175**, 1–41.

Adams, C.E. & Weatherley, G.L. 1981. Suspended sediment transport and benthic boundary-layer dynamics. In: *Sedimentary Dynamics of Continental Shelves* (Ed. C.A. Nittrower), Developments in Sedimentology, **32**, Elsevier, Amsterdam.

Allen, J.R.L. 1968. *Current Ripples: their relation to patterns of water and sediment motion*, North Holland Publishing Company, Amsterdam, 433 pp.

Allen, J.R.L. 1982. *Sedimentary Structures: their character and physical basis*, Elsevier, Amsterdam, 539 pp.

Allen, J.R.L. 1983a. River bedforms: progress and problems In: *Modern and Ancient Fluvial Systems* (Eds J.D. Collinson & J. Lewin), International Association of Sedimentologists Special Publication, **6**, 17–33.

Allen, J.R.L. 1983b. Gravel overpassing on humpback bars supplied with mixed sediment: examples from the lower Old Red Sandstone, Southern Britain. *Sedimentology*, **30**, 285–294.

Allen, J.R.L. 1984. Parallel lamination developed from upper-stage plane beds: a model based on the larger coherent structures of the turbulent boundary layer. *Sedimentary Geol.*, **39**, 227–242.

Allen, J.R.L. 1985. *Principles of Physical Sedimentology.*, Allen & Unwin, London, 272 pp.

Allen, J.R.L. & Leeder, M.R. 1980. Criteria for the instability of upper-stage plane beds. *Sedimentology*, **27**, 209–217.

Antonia, R.A. & Luxton, R.E. 1971. The response of a turbulent boundary layer to a step change in surface roughness. Part 1. Smooth-to-rough. *J. Fluid Mech.*, **48** (4) 721–761.

Antonia, R.A. & Luxton, R.E. 1972. The response of a turbulent boundary layer to a step change in surface roughness. Part 2. Rough-to-smooth. *J. Fluid Mech.*, **53** (4), 737–757.

Antonia, R.A. & Luxton, R.E. 1974. Characteristics of turbulence within an internal boundary layer. In: *Turbulent Diffusion in Environmental Pollution*, Vol. 18A, Advances in Geophysics, 263–285.

Ashley, G.M. 1990. Classification of large-scale subaqueous bedforms: a new look at an old problem. *J. Sed. Petrol.*, **60**, (1), 160–172.

Baas, J.H. 1993. *Dimensional analysis of current ripples in recent and ancient depositional environments*, PhD thesis Department of Geology, University of Utrecht, Netherlands, 199 pp.

Baas, J.H. 1994. A flume study on the development and equilibrium morphology of small-scale bedforms in very fine sand. *Sedimentology*, **41**, 185–209.

Baas, J.H., Oost, A.P., Sztano, O.K., De Boer, P.L. & Postma, G. 1993. Time as an independent variable for current ripples developing towards linguoid equilibrium morphology. *Terra Nova*, **5**, 29–35.

Bagnold, R. A. 1966. *An approach to the sediment transport problem from general physics*. US Geological Survey Professional Paper, **422–1**.

Barwis, J.H. & Hayes, M.O. 1985. Antidunes on modern and ancient washover fans. *J. Sed. Petrol.*, **55**, 907–916.
Bennett, S.J. 1992. *A theoretical and experimental study of bedload transport of heterogeneous sediment*, unpublished PhD thesis, State University of New York at Binghamton, USA.
Bennett, S.J. & Best, J.L. 1994. Structure of turbulence over two-dimensional dunes. In: *Sediment Transport Mechanisms in Coastal Environments and Rivers EUROMECH 310* (Eds M. Bélorgey, R.D. Rajaona & J.F.A. Sleath), World Scientific Publishing Corporation, Singapore, 3–13.
Bennett, S.J. & Best, J.L. 1995. Mean flow and turbulence structure over fixed two-dimensional dunes: implications for sediment transport and dune stability. *Sedimentology*, **42**, 491–513.
Bennett, S.J. & Bridge, J.S. 1995. The geometry and dynamics of low-relief bedforms in heterogeneous sediment in a laboratory channel, and their relationship to water flow and sediment transport. *J. Sed. Res.*, **A65**, 29–39.
Best, J.L. 1992. On the entrainment of sediment and initiation of bed defects: insights from recent developments within turbulent boundary layer research. *Sedimentology*, **39**, 797–811.
Best, J.L. 1993. On the interactions between turbulent flow structure, sediment transport and bedform development: some considerations from recent experimental research. In: *Turbulence: Perspectives on Flow and Sediment Transport* (Eds N.J. Clifford, J.R. French & J. Hardisty), Wiley, Chichester, 61–92.
Best, J.L. & Brayshaw, A.C. 1985. Flow separation: a physical process for the concentration of heavy minerals within alluvial channels. *J. Geol. Soc. Lond.*, **142**, 747–755.
Best, J.L. & Bridge, J.S. 1992. The morphology and dynamics of low amplitude bedwaves upon upper-stage plane beds and the preservation of planar laminae. *Sedimentology*, **39**, 737–752.
Best, J.L. & Leeder, M.R. 1993. Drag reduction in turbulent muddy seawater flows. *Sedimentology*, **40**, 1129–1137.
Blinco, P.H. & Partheniades, E. 1971. Turbulence characteristics in free surface flows over smooth and rough boundaries. *J. Hydraulic Res.*, **9**, 43–69.
Bluck, B.J. 1987. Bed forms and clast size changes in gravel-bed rivers. In: *River Channels: environment and process* (Ed. K.S. Richards), Blackwell, Oxford, 159–178.
Bogard, D.G. & Tiederman, W.G. 1986. Burst detection with single-point velocity measurements. *J. Fluid Mech.*, **162**, 389–413.
Boyer, C. and Roy, A.G. 1991. Morphologie du lit autour d'un obstacle soumis à un écoulement en couche mince. *Géog. phys. Quat.*, **45**(1), 91–99.
Bradshaw, P. & Wong, F.Y.F. 1972. The reattachment and relaxation of a turbulent shear layer. *J. Fluid Mech.*, **52**(1), 113–135.
Brayshaw, A. C. 1983. *Bed microtopography and bedload transport in coarse-grained alluvial channels*, unpublished PhD thesis, Birkbeck College, University of London, 414 pp.
Brayshaw, A.C. 1984. Characteristics and origin of cluster bedforms in coarse-grained alluvial channels. In: *Sedimentology of Gravels and Conglomerates* (Eds E.H. Koster & R.J. Steel) Canadian Society of Petroleum Geologists Memoir, **10**, 77–85.
Brayshaw, A.C., Reid, I. & Frostick, L.E. 1983. The hydrodynamics of particle clusters and sediment entrainment in coarse alluvial channels. *Sedimentology*, **30**, 137–143.
Bridge, J.S. 1981. Bed shear stress over subaqueous dunes, and the transition to upper-stage plane beds. *Sedimentology*, **28**(1) 33–36.
Bridge, J.S. 1982. Bed shear stress over subaqueous dunes, and the transition to upper-stage plane beds: reply. *Sedimentology*, **29**(5), 744–747.
Bridge, J.S. 1985. Paleochannel patterns inferred from alluvial deposits: a critical evaluation. *J. Sed. Petrol.*, **55**(4) 579–589.
Bridge, J.S. & Best, J.L. 1988. Flow, sediment transport and bedform dynamics over the transition from dunes to upper-stage plane beds: implications for the formation of planar laminae: *Sedimentology*, **35**, 753–764.

Bridge, J.S. & Best, J.L. 1990. Flow, sediment transport and bedform dynamics over the transition from dunes to upper-stage plane beds: implications for the formation of planar laminae: reply. *Sedimentology*, **37**, 551–553.

Brown, G.L. & Thomas, A.S.W. 1977. Large structure in a turbulent boundary layer. *Phys. Fluids*, **20**(10), S243–251 (Suppl.).

Carling, P.A. 1990. Particle over-passing on depth-limited gravel bars. *Sedimentology*, **37**, 345–355.

Carling, P.A. & Glaister, M.S. 1987. Rapid deposition of sand and gravel mixtures downstream of a negative step: the role of matrix-infilling and particle-overpassing in the process of bar-front accretion. *J. Geol. Soc. Lond.*, **144**, 543–551.

Chakraborty, C. & Bose, P.K. 1992. Ripple/dune to upper stage plane bed transition: some observations from the ancient record. *Geol. J.*, **27**, 349–359.

Cheel, R.J. 1990a. Horizontal lamination and the sequence of bed phases and stratification under upper-flow regime conditions. *Sedimentology*, **37**, 517–530.

Cheel, R.J. 1990b. Flow, sediment transport and bedform dynamics over the transition from dunes to upper-stage plane beds: implications for the formation of planar laminae: discussion. *Sedimentology*, **37**, 549–551.

Chiew, Y.M. 1991. Bed features in nonuniform sediments. *J. Hydraul. Engrng*, **117**, 116–120.

Coleman, J.M. 1969. Bramahputra river: channel processes and sedimentation. *Sed. Geol.*, **3**, 129–239.

Costello, W.R. 1974. *Development of bed configurations in coarse sands*, unpublished PhD thesis, Massachusetts Institute of Technology, Report 74–1, 120 pp.

Costello, W.R. & Southard, J.B. 1980. Flume experiments on lower-regime bed forms in coarse sand. *J. Sed. Petrol.*, **51**(3), 849–864.

Dal Cin, R. 1968 "Pebble clusters": their origin and utilisation in the study of palaeocurrents. *Sed. Geol.*, **2**, 233–241.

de Jong, C. 1991. A reappraisal of the significance of obstacle clasts in cluster bedform dispersal. *Earth Surf. Process. Landform.*, **16**, 737–744.

Dinehart, R.L. 1989. Dune migration in a steep, coarse-bedded stream. *Water Resources Res.*, **25**(5), 911–923.

Dinehart, R.L. 1992. Evolution of coarse gravel bed forms: field measurements at flood stage. *Water Resources Res.*, **28**(10), 2667–2689.

Ditchfield, R. & Best, J. 1992. Development of bed features: discussion. *J. Hydraul. Engrng*, ASCE, **118**, 647–650.

Drake, T.G. Shreve, R.L., Dietrich, W.E., Whiting, P.J. & Leopold, L.B. 1988. Bedload transport of fine gravel observed by motion-picture photography. *J. Fluid Mech.*, **192**, 193–217.

Engel, P. 1981. Length of flow separation over dunes. *J. Hydraul. Div.*, ASCE., **107**, 1133–1143.

Engelund, F. & Fredsøe, J. 1974. *Transition from dunes to plane bed in alluvial channels*, Institute of Hydrodynamics and Hydraulic Engineering, Technical University of Denmark, Series Paper No. **4**, 56 pp.

Falco, R.E. 1977. Coherent motions in the outer region of turbulent boundary layers. *Phys. Fluids*, **20**(10), S124–132(Suppl.).

Falco, R.E. 1991. A coherent structure model of the turbulent boundary layer and its ability to predict Reynolds number dependence. *Phil. Trans. R. Soc. Lond. Series A*, **336**(1641), 103–129.

Ferguson, R.I., Prestegaard, K.L. & Ashworth, P.J. 1989. Influence of sand on hydraulics and gravel transport in a braided gravel bed river. *Water Resources Res.*, **25**(4), 635–643.

Foley, M.G. & Vanoni, V.A. 1977. Pulsing flow in steep alluvial streams. *J. Hydraul. Div.*, ASCE, **103**(HY8), 843–850.

Gabel, S.L. 1993. Geometry and kinematics of dunes during steady and unsteady flows in the Calamus River, Nebraska, USA. *Sedimentology*, **40**, 237–269.

Grass, A.J. 1970. Initial instability of fine bed sand. *J. Hydraul. Div.*, ASCE, **96**, 619–631.

Grass, A.J. 1971. Structural features of turbulent flow over smooth and rough boundaries. *J. Fluid Mech.*, **50**, 233–255.

Grass, A.J. 1983. The influence of boundary layer turbulence on the mechanics of sediment transport. In: *Mechanics of Sediment Transport, Proceedings of Euromech 156* (Eds B. Mutlu Sumer & A. Müller) Balkema, Rotterdam, 3–18.

Grass, A.J. Stuart, R.J. & Mansour-Tehrani, M. 1991. Vortical structures and coherent motion in turbulent flow over smooth and rough boundaries *Phil. Trans. R. Soc. Lond. Series A*, **336**(1640), 35–65.

Gyr, A. & Schmid, A. 1989. The different ripple formation mechanism. *J. Hydraul. Res.*, **27**(1), 61–74.

Gyr, A., Müller, A. & Schmid, A. 1989. Observation of self stabilisation processes in sediment transport linked to scales of coherent structures. In: *Proceedings of 23rd International Association of Hydraulics Research Congress*, Ottawa, August 1989, Vol. A: Turbulence in Hydraulics, A-31–A-38.

Haque, M.I. & Mahmood, K. 1985. Geometry of ripples and dunes. *J. Hydraul. Engrng, ASCE*, **111**(1), 48–63.

Haque, M.I. & Mahmood, K. 1986. Analytical study on steepness of ripples and dunes. *J. Hydraul. Engrng, ASCE*, **112**(3), 220–236.

Head, M.R. & Bandyopadhyay, P. 1981. New aspects of turbulent boundary layer structure. *J. Fluid Mech.*, **107**, 297–338.

Hoey, T. 1992. Temporal variations in bedload transport rates and sediment storage in gravel-bed rivers. *Prog. Phys. Geog.*, **16**(3), 319–338.

Hubbell, D.W., Stevens, H.H. Skinner, J.V. & Beverage, J.P. 1987. *Laboratory data on coarse-sediment transport for bedload-sampler calibrations*, US Geological Survey Water Supply Paper **2299**, 31 pp.

Ikeda, H. 1983. *Experiments on bedload transport, bed forms, and sedimentary structures using fine gravel in the 4-meter-wide flume*, Environmental Research Center Papers, No. **2**, University of Tsukuba, 78 pp.

Ikeda, S. & Asaeda, T. 1983. Sediment suspension with rippled bed. *J. Hydraul. Engrng*, **109**(3), 409–423.

Iseya, F. 1984. *An experimental study of dune development and its effect on sediment suspension*, Environmental Research Center Papers, No. **5**, University of Tsukuba, 56 pp.

Iseya, F. & Ikeda, H. 1986. Effect of dune development on sediment suspension under unsteady flow conditions. *Proceedings of 30th Japanese Conference on Hydraulics*, Japanese Society of Civil Engineers, 505–510 (in Japanese).

Iseya, F. & Ikeda, H. 1987. Pulsations in bedload transport rates induced by longitudinal sediment sorting: A flume study using sand and gravel mixtures. *Geogr. Ann.*, **69**(A), 15–27.

Itakura, T. & Kishi, T. 1980. Open channel flow with suspended sediment on sand waves. In: *Proceedings of 3rd International Symposium on Stochastic Hydraulics* (Eds H. Kikkawa & Y. Isawa), Tokyo, 599–609.

Jackson, R.G. 1976. Sedimentological and fluid-dynamic implications of the turbulent bursting phenomenon in geophysical flows. *J. Fluid Mech.* **77**, 531–560.

Jackson, R.G. 1978. Mechanisms and hydrodynamic factors of sediment transport in alluvial streams. In: *Research in Fluvial Systems, Proceedings of 5th Guelph Symposium on Geomorphology* (Eds R. Davidson-Arnott, & W. Nickling), 9–44.

Karahan, M.E. & Peterson, A.W. 1980. Visualisation of separation over sand waves. *J. Hydraul. Div., ASCE*, **106**(HY8), 1345–1352.

Kawanisi, K., Maghrebi, M.F. & Yokosi, S. 1993. An instantaneous 3-D analysis of turbulent flow in the wake of a hemisphere. *Boundary-Layer Meteorol.* **64**, 1–14.

Kennedy, J.F. 1961. *Stationary waves and antidunes in alluvial channels*. Report No. *KH-R-2*, W.M. Keck Laboratory of Hydraulics and Water Resources, California Institute of Technology, 146 pp.

Kennedy, J.F. 1963. The mechanics of dunes and antidunes in erodible-bed channels. *J. Fluid Mech.*, **16**, 521–544.

Kirkbride, A. 1993. Observations of the influence of bed roughness on turbulence structure in depth limited flows over gravel-beds. In: *Turbulence: Perspectives on Flow and Sediment Transport* (Eds N.J. Clifford, J.R. French & J. Hardisty), Wiley, Chichester, 185–196.

Kirkgoz, M.S. 1989. Turbulent velocity profiles for smooth and rough open channel flow. *J. Hydraul. Engrng*, **115**(11), 1543–1561.

Klebanoff, P.S., Cleveland, W.G. & Tidstrom, K.D. 1992. On the evolution of a turbulent boundary layer induced by a three-dimensional roughness element. *J. Fluid Mech.*, **237**, 101–187.

Kline, S.J. & Robinson, S.K. 1989. Turbulent boundary layer structure: progress, status and challenges. In: *Structure of Turbulence and Drag Reduction* (Ed. A. Gyr), Hemisphere, New York, 3–22.

Kline, S.J. & Robinson, S.K. 1990. Quasi-coherent structures in the turbulent boundary layer. Part 1: status report on a community-wide summary of the data. In: *Near Wall Turbulence*, Proceedings of 1988 Zoran Zaric Memorial Conference (Eds S.J. Kline & N.H. Agfan, Hemisphere, New York, 200–217.

Kostaschuk, R.A. & Church, M.A. 1993. Macroturbulence generated by dunes: Fraser River, Canada. *Sed. Geol.*, **85**, 25–37.

Koster, E.H. 1978. Transverse ribs: their characteristics, origin and palaeohydraulic significance. In: *Fluvial Sedimentology* (Ed. A.D. Miall), Canadian Society of Petroleum Geologists Memoir **5**, 161–186.

Langord, R. & Bracken, B. 1987. Medano Creek, Colorado, A model for upper-flow-regime fluvial deposition. *J. Sed. Petrol.*, **57**(5) 863–870.

Lapointe, M. 1992. Burst-like sediment suspension events in a sand bed river. *Earth Surf. Process. Landform.*, **17**, 253–270.

Laronne, J.B. & Carson, M.A. 1976. Interrelationships between bed morphology and bed material transport for a small gravel bed channel. *Sedimentology*, **23**, 67–86.

Lasheras, J.C. & Choi, H. 1988. Three-dimensional instability of a plane free shear layer: an experimental study of the formation and evolution of streamwise vortices. *J. Fluid Mech.*, **189**, 53–86.

Leeder, M.R. 1980 On the stability of lower stage plane beds and the absence of ripples in coarse sand. *J. Geol. Soc. Lond.* **137**, 423–429.

Leeder, M.R. 1983a. On the interactions between turbulent flow, sediment transport and bedform mechanics in channelized flows. In: *Modern and Ancient Fluvial Systems* International Association of Sedimentologists Special Publication, (Eds J.D. Collinson & J. Lewin), **6**, 5–18.

Leeder, M.R. 1983b. On the dynamics of sediment suspension by residual Reynolds stresses – confirmation of Bagnold's theory. *Sedimentology*, **30**, 485–491.

Levi, E. 1984. A universal Strouhal law. *J. Engrng. Mech.*, **109**(3), 718–727.

Levi, E. 1991. Vortices in hydraulics. *J. Hydraul. Engrng.*, **117**(4), 399–413.

Ligrani, P.M. & Moffat, R.J. 1986. Structure of transitionally rough and fully rough turbulent boundary layers. *J. Fluid Mech.*, **162**, 69–98.

Lowe, D.R. 1988. Suspended-load fallout rate as an independent variable in the analysis of current structures. *Sedimentology*, **35**, 765–776.

Luchik, T.S. and Tiederman, W.G. 1987. Timescale and structure of ejections and bursts in turbulent channel flows. *J. Fluid Mech.*, **174**, 529–552.

Luchik, T.S. and Tiederman, W.G. 1988. Turbulent structure in low-concentration drag-reducing channel flows. *J. Fluid Mech.*, **190**, 241–263.

Lyn, D.A. 1993. Turbulence measurements in open-channel flows over artificial bedforms. *J. Hydraul. Engrng.*, ASCE, **119**(1) 306–326.

Mantz, P.A. 1978. Bedforms produced by fine, cohesionless, granular and flakey sediments under subcritical water flows. *Sedimentology*, **25**, 83–103.

McDonald, B.C. & Banerjee, I. 1971. Sediments and bedforms on a braided outwash plain. *Can. J. Earth Sci.*, **8**, 1282–1301.

McDonald, B.C. and Day, T.J. 1978. *An Experimental Flume Study on the Formation of Transverse Ribs*, Geological Survey of Canada, Paper **78–1A**, 441–451.

McLean, S.R. & Smith, J.D. 1979. Turbulence measurements in the boundary layer over a sand wave field. *J. Geophys. Res.*, **84**(C12), 7791–7808.

McLean, S.R., Nelson, J.M., & Wolfe, S.R. 1994. Turbulence structure over two-dimensional bedforms: implications for sediment transport. *J. Geophys. Res.*, **99**, 12729–12747.

Mendoza, C. & Shen, H.W. 1990. Investigation of turbulent flow over dunes. *J. Hydraul. Engrng*, **116**(4) 459–477.

Metcalfe, R.W., Orszag, S.A., Brachet, M.E., Menon, S. & Riley, J.J. 1987. Secondary instability of a temporally growing mixing layer. *J. Fluid Mech.*, **184**, 207–243.

Miller, M.C., McCave, I.N. & Komar, P.D. 1977. Threshold of sediment motion in unidirectional currents. *Sedimentology*, **24**, 507–528.

Moss, A.J. 1972. Bed-load sediments. *Sedimentology*, **18**, 159–219.

Müller, A. & Gyr, A. 1982. Visualisation of the mixing layer behind dunes. In: *Mechanics of Sediment Transport, Proceedings of Euromech 156* (Eds B. Mutlu Sumer & A. Müller) Balkema, Rotterdam, 41–45.

Müller, A. & Gyr, A. 1986. On the vortex formation in the mixing layer behind dunes. *J. Hydraul. Res.*, **24**(5), 359–375.

Nelson, J.M. & Smith, J.D. 1989. Mechanics of flow over ripples and dunes. *J. Geophys. Res.*, **94**(C6), 8146–8162.

Nelson, J.M., McLean, S.R. & Wolfe, S.R. 1993. Mean flow and turbulence fields over two-dimensional bed forms. *Water Resources. Res.*, **29**, 3935–3953.

Nezu, I. & Nakagawa, H. 1989. Self forming mechanism of longitudinal sand ridges and troughs in fluvial open-channel flows. In: *Proceedings of 23rd International Association of Hydraulics Research Congress*, Ottawa, August 1989, Volume B: Fluvial Hydraulics, B-65–B-72.

Nezu, I. & Nakagawa, H. 1993. *Turbulence in Open-Channel Flows*, Balkema, Rotterdam, 281 pp.

Nowell, A.R.M. & Church, M. 1979. Turbulent flow in a depth-limited boundary layer. *J. Geophys. Res.*, **84**(C8) 4816–4824.

Onslow, R.J., Thomas, N.H. & Whitehouse, R.J.S. 1993. Vorticity and sandwaves: the dynamics of ripples and dunes. In: *Turbulence: Perspectives on Flow and Sediment Transport* (Eds N.J. Clifford, J.R. French & J. Hardisty), Wiley, Chichester, 279–293.

Paola, C., Gust, G. & Southard, J.B. 1986. Skin friction behind isolated hemispheres and the formation of obstacle marks. *Sedimentology*, **33**, 279–293.

Paola, C., Wiele, S.M. & Reinhart, M.A. 1989. Upper-regime parallel lamination as the result of turbulent sediment transport and low-amplitude bedforms. *Sedimentology*, **36**, 47–60.

Perry, A.E., Lim, T.T. & Teh, E.W. 1981. A visual study of turbulent spots. *J. Fluid Mech.*, **104**, 387–405.

Rao, K.N., Narasimha, R. & Badri Narayanan, M.A. 1971. The "bursting" phenomenon in a turbulent boundary layer. *J. Fluid. Mech.*, **48**, 339–352.

Raudkivi, A.J. 1963. Study of sediment ripple formation. *J. Hydraul. Div.*, ASCE, **89**, 15–33.

Raudkivi, A.J. 1966. Bed forms in alluvial channels. *J. Fluid Mech.*, **26**(3), 597–514.

Reid, I. & Frostick, L.E. 1986. Dynamics of bedload transport in Turkey Brook, a coarse-grained alluvial channel. *Earth Surf. Process. Landform.*, **11**, 143–155.

Reid, I., Brayshaw, A.C. & Frostick, L.E. 1984. An electromagnetic device for automatic detection of bedload motion and its field applications. *Sedimentology*, **31**, 269–276.

Richards, K. & Clifford, N. 1991. Fluvial geomorphology: structured beds in gravelly rivers. *Prog. Phys. Geog.*, **15**(4), 407–422.

Rifai, M.F. & Smith, K.V.H. 1971. Flow over triangular elements simulating dunes. *J. Hydraul. Div.*, ASCE, **97**(HY7), 963–976.

Robert, A. 1990. Boundary roughness in coarse-grained channels. *Prog. Phys. Geog.*, **14**, 42–70.

Robert, A., Roy, A.G. & De Serres, B. 1992. Changes in velocity profiles at roughness transitions in coarse grained channels. *Sedimentology*, **39**, 725–735.

Robinson, S.K. 1990a. Coherent motions in the turbulent boundary layer. *Ann. Rev. Fluid. Mech.*, **23**, 601–639.

Robinson, S.K. 1990b. A review of vortex structures and associated coherent motions in turbulent boundary layers. In: *Structure of Turbulence and Drag Reduction* (Ed. A. Gyr), Hemisphere, New York, 23–50.

Robinson, S.K., Kline, S.J. & Spalart, P.R. 1990. Quasi-coherent structures in the turbulent boundary layer: Part II. Verification and new information from a numerically simulated flat-plate layer. In: *Near Wall Turbulence, Proceedings of 1988 Zoran Zaric Memorial Conference* (Eds S.J. Kline, & N.H. Agfan), Hemisphere, New York, 218–247.

Roe, S.L. 1987. Cross-strata and bedforms of probable transitional dune to upper stage plane bed origin from a Late Precambrian fluvial sandstone, northern Norway. *Sedimentology*, **34**, 89–101.

Rood, K.M. & Hickin, E.J. 1989. Suspended sediment concentration in relation to surface-flow structure in Squamish River estuary, southwestern British Columbia. *Can. J. Earth Sci.*, **26**, 2172–2176.

Saunderson, H.C. & Lockett, F.P. 1983. Flume experiments on bedforms and structures at the dune-plane bed transition. In: *Modern and Ancient Fluvial Systems* (Eds J.D. Collinson, & J. Lewin) International Association of Sedimentologists Special Publication, **6**, 49–58.

Silveira Neto, A., Grand, D., Métais, O. & Lesieur, M. 1993. A numerical investigation of the coherent vortices in turbulence behind a backward-facing step. *J. Fluid Mech.*, **256**, 1–25.

Simons, D.B., Richardson, E.V. and Haushild, W.L. 1963. *Some effects of fine sediment on flow phenomena* US Geological Survey Water Supply Paper **1498-G**, 47 pp.

Simpson, R.L. 1989. Turbulent boundary-layer separation. *Ann. Rev. Fluid Mech.*, **21**, 205–234.

Smith, C.R. 1984. A synthesized model of the near-wall behaviour in turbulent boundary layers. In: *Proceedings of Eighth Symposium on Turbulence* (Eds G.K. & Patterson & J.L. Zakin) University of Missouri-Rolla, 199–327.

Smith, C.R. & Metzler, S.P. 1983. The characteristics of low-speed streaks in the near-wall region of a turbulent boundary layer. *J. Fluid Mech.*, **129**, 27–54.

Smith, C.R. & Walker, J.D.A. 1990. A conceptual model of wall turbulence. In: *Proceedings of NASA Langley Boundary Layer Workshop* (Ed. S. Robinson), 7 pp.

Smith, C.R., Walker, J.D.A., Haidara, A.H. & Sobrun, U. 1991. On the dynamics of near-wall turbulence. *Phil. Trans. R. Soc. Lond. Series A*, **336**(1641), 131–175.

Smith, J.D. 1970. Stability of a sand bed subjected to a shear flow of low Froude number. *J. Geophys. Res.*, **75**(30) 5928–5940.

Smith, J.D. & McLean, S.R. 1977. Spatially averaged flow over a wavy surface. *J. Geophys. Res.*, **86**(12), 1735–1746.

Snishchenko, B.F., Muhamedov, A.M. & Mazhidov, T.Sh. 1989 Bedload composition effect on dune shape parameters and on flow characteristics. In: *Proceedings of 23rd International Association of Hydraulics Research Congress*, Ottawa, August 1989, Volume B: Fluvial Hydraulics, B-105–B-112.

Soulsby, R.L., Atkins, R., Waters, C.B. & Oliver, N. 1991. Field measurements of suspended sediment over sandwaves. In: *Sand Transport in Rivers, Estuaries and the Sea*, Proceedings of Euromech 262 (Eds R.L. Soulsby & R. Bettess), Balkema, Rotterdam, 155–162.

Southard, J.B. & Boguchwal, L.A. 1990 Bed configurations in steady unidirectional water flows. Part 2. Synthesis of Flume data. *J. Sed. Petrol.*, **60**(5) 458–479.

Southard, J.B. & Dingler, J.R. 1971. Flume study of ripple propagation behind mounds on flat sand beds. *Sedimentology*, **16**, 251–263.

Talmon, A.M. Kunen, J.M.G. & Ooms, G. 1986. Simultaneous flow visualisation and Reynolds-stress measurement in a turbulent boundary layer. *J. Fluid Mech.*, **163**, 459–478.

Teissyre, A.K. 1977. Pebble clusters as a directional structure in fluvial gravels: modern and ancient examples. *Geologica Sudetica*, **12**, 79–84.

Tiederman, W.G. 1990. Eulerian detection of turbulent bursts. In: *Near Wall Turbulence, Proceedings of 1988 Zoran Zaric Memorial Conference* (Eds S.J. Kline, & N.H. Agfan), Hemisphere, New York, 874–878.

Tiederman, W.G., Luchik, T.S. & Bogard, D.G. 1985. Wall structure and drag reduction. *J. Fluid Mech.*, **156**, 419–437.

Tsujimoto, T. 1989a. Longitudinal stripes of alternate lateral sorting due to cellular secondary currents. In: *Proceedings of 23rd International Association of Hydraulics Research Congress*, Ottawa, August 1989, Volume B: Fluvial Hydraulics, B-17–B-24.

Tsujimoto, T. 1989b. Longitudinal stripes of sorting due to cellular secondary currents. *J. Hydroscience Hydraul. Engnrng*, **7**(1), 23–34.

Van den Berg, J.H. & Van Gelder, A. 1993. A new bedform stability diagram, with emphasis on the transition of ripples to plane bed in flows over fine sand and silt In: *Alluvial Sedimentation* (Eds M. Marzo & C. Puidefabregas), Special Publication of the International Association of Sedimentologists, **17**, 11–21.

Van der Knaap, F.C.M., Van Mierlo, M.C.L.M. & Officier, M.J. 1991. Measurements and computations of the turbulent flow field above fixed bed-forms. In: *Sand Transport in Rivers, Estuaries and the Sea, Proceedings of Euromech 262* (Eds R.L. Soulsby & R. Bettess) Balkema, Rotterdam, 179–185.

Wan, Z. 1982. *Bed material movement in hyperconcentrated flow*. Technical University of Denmark, Institute of Hydrodynamics and Hydraulic Engineering, Series Paper **31**, 79 pp.

Wan, Z. 1983. Some phenomena associated with hyperconcentrated flow. In: *Mechanics of Sediment Transport, Proceedings of Euromech 156* (Eds B. Mutlu Sumer & A. Müller) Balkema, Rotterdam, 189–194.

Wan, Z. & Niu, Z. 1989. Hyperconcentrated density current in rivers. In: *Proceedings of 23rd International Association of Hydraulics Research Congress*, Ottawa, August 1989, Volume B: Fluvial Hydraulics, B-73–B-80.

Watanabe, K. & Hirano, M. 1988. Effect of suspended sediment on formation of sand waves. *J. Hydroscience Hydraulic Engng.*, **6**(2), 29–43.

Weedman, S.D. & Slingerland, R. 1985. Experimental study of sand streaks formed in turbulent boundary layers. *Sedimentology*, **32**, 133–145.

Wei, T. & Willmarth, W.W. 1991. Examination of v-velocity fluctuations in a turbulent channel flow in the context of sediment transport. *J. Fluid. Mech.*, **223**, 241–252.

Whiting, P.J., Dietrich, W.E., Leopold, L.B., Drake, T.G. & Shreve, R.L. 1988. Bedload sheets in heterogeneous sediment. *Geology*, **16**, 105–108.

Wilcock, P.R. 1992. Experimental investigation of the effect of mixture properties on transport dynamics. In: *Dynamics of Gravel-bed Rivers* (Eds P. Billi, R.D. Hey, C.R. Thorne & P. Taconni), Wiley Chichester 109–139.

Wilcock, P.R. & Southard, J.B. 1989. Bed load transport of mixed size sediment: fractional transport rates, bed forms, and the development of a coarse bed surface layer. *Water Resources Res.*, **25**(7), 1629–1641.

Williams, G.P. 1967. *Flume experiments on the transport of a coarse sand*, US Geological Survey Professional Paper **562-B**, 31 pp.

Williams, G.P. 1970. *Flume width and water depth effects in sediment-transport experiments*, US Geological Survey Professional Paper **562-H**, 37 pp.

Williams, J.J., Thorne, P.D., & Heathershaw, A.D. 1989. Measurements of turbulence in the benthic boundary layer over a gravel bed. *Sedimentology*, **36**, 959–979.

Williams, P.B. & Kemp, P.H. 1971. Initiation of ripples on flat sediment beds. *J. Hydraul. Div.*, ASCE, **97**, 505–522.

Williams, P.B. & Kemp, P.H. 1972. Initiation of ripples by artificial disturbances. *J. Hydraul. Div.*, ASCE, **98**, 1057–1070.

Willmarth, W.W. & Lu, S.S. 1972. Structure of the Reynolds stress near the wall. *J. Fluid Mech.*, **55**, 65–92.

Willmarth, W.W. & Lu, S.S. 1974. Structure of the Reynolds stress and the occurrence of bursts in the turbulent boundary layer. In: *Turbulent Diffusion in Environmental Pollution*, Vol. 18A, Advances in Geophysics, 287–314.

Yalin, M.S. 1976 Origin of submarine dunes. In: *Proceedings of 15th International Coastal Engineering Conference*, Honolulu, July 1976, 2127–2135.

Yalin, M.S. 1977. *Mechanics of Sediment Transport*, Pergamon Press, Oxford, 298 pp.

Yalin, M.S. 1992. *River Mechanics*. Pergamon Press, Oxford, 219 pp.

Yalin, M.S. & Karahan, E. 1979. Steepness of sedimentary dunes. *J. Hydraul. Div.*, ASCE, **105**, 381–392.

4 Entrainment of Sediments from Deposits of Mixed Grain Sizes and Densities

PAUL D. KOMAR

College of Oceanic & Atmospheric Sciences, Oregon State University, USA

INTRODUCTION

The threshold of sediment motion or entrainment of grains by flowing water is of considerable interest to sedimentologists and engineers. This critical condition is meant to signify the first movement of grains from a previously static deposit, a flow stage that produces a very low level of sediment transport. The traditional approach to establish equations for evaluating sediment transport rates generally correlates the quantity of sediment movement to the difference between the measured flow strength (velocity, discharge, mean stress, or flow power) and that required to initiate sediment motion, thereby yielding relationships which predict zero transport at the threshold condition. That approach is still widely used, for example by Hassan *et al.* (1992) in their analyses of the mean rates and distances of travel of individual gravel particles in streams. Einstein (1950) hypothesized that sediment transport is a stochastic process and did not explicitly define a threshold condition, yet sediment movement still is considered to involve repeated occurrences of grain entrainment and stepwise transport. This philosophy is contained in the most recent and sophisticated models of sediment transport, most notably in those developed by Bridge & Bennett (1992) which incorporate evaluations of grain entrainment to assess the transport of different grain sizes, shapes and densities. Geologists have also employed evaluations of grain threshold in interpretations of sediment deposits, for example, to assess discharges of extreme floods from the sizes of cobbles or boulders that were entrained and transported, whether recorded in modern or ancient sediments (e.g. Lord & Kehew, 1987; O'Connor, 1993; Wohl *et al.*, 1994; Kehew & Teller, 1994).

A variety of empirical curves is available for the direct evaluation of sediment threshold. Such curves simply relate the entrainment of a specified grain diameter and density to some measure of the flow such as the mean velocity or bed stress. Those empirical relationships are based almost entirely on laboratory flume data where the sediment consists of a small range of grain sizes, generally one sieve fraction, and the bed is carefully smoothed so as to eliminate any preferential exposure of individual grains to the flow and to remove form resistance produced by

Advances in Fluvial Dynamics and Stratigraphy. Edited by P.A. Carling and M.R. Dawson.
© 1996 John Wiley & Sons Ltd.

bedforms. Although such idealized experiments may properly represent a necessary first step toward an understanding of the grain-threshold condition and its evaluation, one can only wonder about their actual application to natural sediment deposits which consist of mixtures of grains having different densities and wide ranges of sizes. The selective entrainment of grains from deposits of mixed sizes and densities is a process that is extremely important to natural occurrences of sediment sorting. In the case of sorting by grain density, selective entrainment can play a role in heavy-mineral enrichment that ranges from small segregations in the crests of ripples or within scour marks, to the development of large-scale placers containing valuable minerals at high concentrations. The sorting of grains by size is significant in virtually every sediment deposit, but is likely to be particularly important for gravels in streams and in the marine environment due to the extreme ranges of grain sizes typical of such systems.

The overall objective of this chapter is to review our understanding of grain entrainment from natural deposits of mixed sizes and densities. This review will begin with a brief consideration of the empirical threshold curves based on experiments with uniform grains. That consideration is undertaken not so much for the potential application of those curves, but as a point of departure for an examination of the more complex processes and patterns of selective entrainment. Size sorting is most significant in gravels, while density sorting is more important in sands. Therefore, separate sections are devoted to those grain-size ranges, and it will be seen that there is evidence for distinctly different patterns of selective entrainment. Most of the results reviewed in those sections involve empirical analyses based on data from laboratory flume research and the collection of field data. In order to provide a more fundamental understanding of the processes of selective entrainment, this chapter also examines the role of flow turbulence in the initiation of grain movement, and will review the development of mechanical models for grain entrainment by their pivoting or sliding over underlying grains. It will become apparent from this review that although more investigations are required, research in recent years has revealed the patterns of grain sorting due to selective entrainment and has provided empirical and semitheoretical equations that can be used to quantitatively evaluate grain entrainment from deposits of mixed sizes and densities.

THE ENTRAINMENT OF UNIFORM SEDIMENTS

There has been a long history in the development of empirical threshold curves, beginning with the publications of Hjulström (1935, 1939) and Shields (1936). A number of researchers have undertaken flume experiments to collect data relating grain sizes and densities to flow velocities, discharges, and mean stresses needed to initiate particle movement. These data have been reviewed by Miller *et al.* (1977) and compiled into various threshold curves. Their curve for the entrainment of quartz-density grains in water is shown in Figure 4.1, in this case relating the flow's threshold mean stress τ_t to the grain diameter D. As expected, the coarser the grain size the greater the flow stress required for initiation of movement. There is a slight

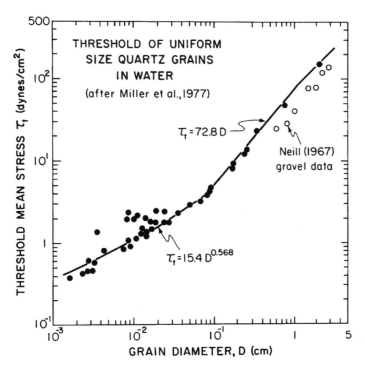

Figure 4.1   The entrainment flow stress τ_t for quartz-density grains of diameter D in water, based on the data compilation of Miller *et al.* (1977)

inflection in the curve at about $D = 0.1$ cm. For grain sizes coarser than that inflection, there is a simple proportionality

$$\tau_t = 72.8D \qquad (4.1)$$

where the units of τ_t and D are respectively dynes/cm^2 and centimetres. For finer grain sizes the empirical relationship is

$$\tau_t = 15.4D^{0.568} \qquad (4.2)$$

There is no minimum in the threshold curve of Figure 4.1 comparable to that found by Hjulström (1935, 1939), with the finest grain sizes (silts and clays) requiring higher flows for entrainment due to their cohesiveness. This results from the artificial conditions of the experiments undertaken to collect the data in Figure 4.1 – flume experiments which utilized clean sieve fractions and so eliminated cohesiveness due to clays and organic matter. The changing slope of the curve and its inflection likely reflect the progressive modification of the boundary layer, with a viscous sublayer existing at low τ_t versus D data combinations, to the development of full turbulence down to the grain level in the experiments with coarser sediments (Inman, 1949; Yalin & Karahan, 1979).

Although the laboratory experiments involved sieve fractions of nearly uniform grains and carefully smoothed sediment beds, the data in Figure 4.1 are still scattered and there are systematic differences between various studies, most notably in the measurements of Neill (1967). This results from the subjectivity of identifying the threshold condition and differences in techniques employed to define that critical condition. It is the experimenter's judgement as to just how much grain movement constitutes "threshold". Therefore, measurements by the individual studies have systematic differences representing varying degrees of sediment movement. Shields (1936) and various subsequent investigators measured sand transport rates under a range of flow conditions, and then extrapolated the results back to a zero-transport level to yield an evaluation of the threshold state. However, most studies of threshold are based on visual assessments of grain movement and therefore represent a small quantity of sediment transport. The threshold criterion has been defined variously as the condition of "weak movement", "general bed movement", the stage for a single stone to be "first displaced", "scattered particle movement", and so on (Miller *et al.*, 1977). There is also a time element in the threshold determination due to the statistical nature of fluid turbulence and bottom stresses, a variability in the fluid flow that is discussed below in the context of grain entrainment. Such problems account for the data of Neill (1967) consistently plotting lower than the other measurements in Figure 4.1. Neill himself criticized his 1967 data on the ground that he applied a less severe criterion for initiation of motion to the large grains in his study than was applied by others to small grains (Neill, 1968). The most objective first movement criterion that has been proposed, dealing directly with the magnitude of transport and the time element, is that of Neill & Yalin (1969) and further refined by Yalin (1972). Their analyses indicate that in experiments with larger grains, one must watch a greater bottom area for first grain movement and must watch that area for a longer period of time than for small grains.

Such problems aside, the resulting curve in Figure 4.1 can still be used to evaluate the mean flow stress required to entrain grains of diameter D. The potential application of this curve is limited, however, as it is suitable only for the entrainment of uniform quartz-density grains in water. The threshold of grains having some other density would require another empirical curve, based on additional series of flume experiments. This requirement has been eliminated largely by the development of "universal" threshold curves such as that of Shields (1936) – the version shown in Figure 4.2 is that derived by Yalin & Karahan (1979) in their data compilation. Shields presented physical arguments to combine the parameters of interest into the dimensionless relationship

$$\theta_t = \frac{\tau_t}{(\rho_s - \rho)gD} = \frac{\rho u_{*t}^2}{(\rho_s - \rho)gD} = f\left(\frac{u_{*t}D}{\nu}\right) \tag{4.3}$$

where ρ_s and ρ are respectively the densities of the sediment grains and fluid, $u_{*t} = \sqrt{\tau_t/\rho}$ is the shear velocity, and ν is the fluid viscosity. The analysis by Shields indicates that θ_t will be a function of the grain Reynolds number $Re_{*t} = u_{*t}D/\nu$, which is proportional to the ratio of the grain diameter as a measure of the bed roughness to the thickness of the potential viscous sublayer, the thin zone of fluid flow closest to the sediment bed. Accordingly, one can interpret the threshold

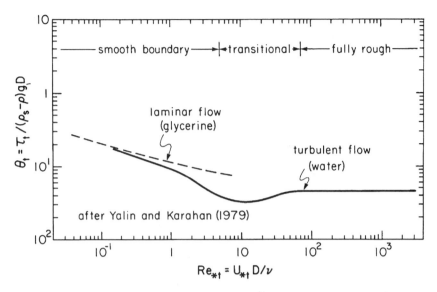

Figure 4.2 The Shields curve established by the data compilation of Yalin & Karahan (1979) for the threshold of grains of different densities within liquids. Also shown is how the left limb of the Shields curve at low Re_{*_t} converges with the line based on grain-threshold measurements within laminar flows of glycerine (after Yalin & Karahan, 1979)

condition in terms of the changing relationship of the viscous sublayer to the grain size which determines the bed roughness. In the present context, however, more important is that the Shields analysis provides a threshold curve that is applicable to many combinations of granular materials and fluids. This "universality" is apparent in that the data compiled to establish the Shields curve in Figure 4.2 included measurements in flume experiments utilizing water and oil as fluids, and grain materials as diverse as amber, lignite, polystyrene and ilmenite, as well as quartz-density sands and glass spheres (grain-density range $\rho_s = 1.05$ to 4.70 g/cm^3). The curve is seen to level off at high Re_{*_t} values; the original Shields (1936) curve gave $\theta_t = 0.06$ at high Re_{*_t}, but Figure 4.2 yields $\theta_t = 0.045$, a value found in the data reviews of both Miller *et al.* (1977) and Yalin & Karahan (1979). From the definition of θ_t given by eqn 4.3, this constant-θ_t portion of the Shields curve corresponds to the simple threshold relationship

$$\tau_t = 0.045(\rho_s - \rho)gD \qquad (4.4)$$

which is dimensionally homogeneous and includes the dependence on the particle's density as well as its diameter. Equation 4.1 derived from the fit to the quartz-density data in Figure 4.1 is a reduced version of eqn 4.4 obtained by substituting the densities of quartz particles ($\rho_s = 2.65$ g/cm^3) and fresh water ($\rho = 1.00$ g/cm^3), as well as $g = 981$ cm/s^2 for the acceleration of gravity. Unfortunately, a comparable dimensionally homogeneous relationship is not derivable for the left limb of the Shields curve where θ_t varies with Re_{*_t}, an equation that might have application to the entrainment of sand and finer sediments. The modification of the empirical eqn 4.2 to include the expected dependence on particle and fluid densities yields

$$\tau_t = 0.0095(\rho_s - \rho)gD^{0.568} \qquad (4.5)$$

where the units of the densities are g/cm^3, and τ_t and D are again, respectively, dynes/cm^2 and centimetres.

Grain shape as well as size and density can affect entrainment, but this has been established only for the extreme case of the threshold of flat mica plates. Mantz (1973, 1977) has undertaken a series of measurements of mica threshold and transport in a laboratory flume, and his results are plotted on a Shields-type graph in Figure 4.3 where the line based on the data of White (1970) represents the entrainment of equant grains. The mica data are plotted three times, the grain "size" being respectively taken as the plate thickness, its face diameter, and its nominal diameter (the diameter of a sphere having the same weight). As might be expected, plotting the results in terms of the nominal diameter provides the closest agreement with the standard Shields curve, and is the preferred comparison. The results indicate that mica achieves threshold at lower Shields values and flow stresses than more equant grains. However, additional experiments by Mantz (1980) demonstrated that imbrication of discoidal grains such as mica greatly increases flow stresses required for entrainment, indicating that the packing arrangement of the flat grains on the bed is of paramount importance to their initiation of movement.

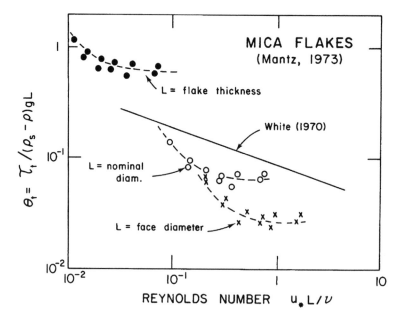

Figure 4.3 Flow-threshold curves for flat mica plates, based on the flume measurements of Mantz (1973). Three curves are given, based on representing the mica size as the flake thickness, face diameter & nominal diameter. The curve from White (1970) is for the entrainment of equant grains (after Mantz, 1973)

The Shields curve of Figure 4.2 can be used to determine the threshold flow stresses of non-cohesive granular materials of uniform density and size in any liquid (but not in air). Combinations of θ_t and Re_{*_t} digitized from the curve can be employed for those grain–fluid combinations. This is illustrated in Figure 4.4 for a series of common sedimentary minerals in water, each curve having been based on the original Shields curve of Figure 4.2. As expected, for a given D, the denser the mineral the greater the flow stress required for its entrainment. However, this sequence of curves is not applicable to considerations of selective entrainment of grains of different sizes and densities from a mixed deposit. Each curve in Figure 4.4 should be used only for deposits consisting of 100% of the mineral. Considerations of selective entrainment by size and density from a mixed deposit require different curves based on different types of data – those data and the resulting selective entrainment relationships will be reviewed later.

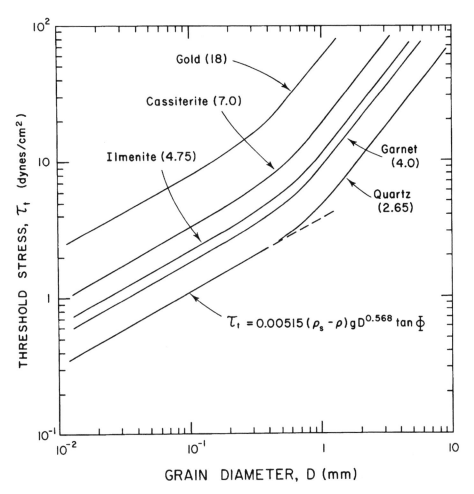

Figure 4.4 Grain-threshold curves for a series of common minerals, the curves having been computed from the Shields curve of Figure 4.2 (after Komar, (1989b)

FLOW TURBULENCE AND THE INITIATION OF GRAIN MOVEMENT

The left limb of the Shields curve at very low Re_{*_t} (Figure 4.2) corresponds to a condition where a viscous sublayer exists between the bed of sediments and the otherwise turbulent flow present at greater distances above the bottom. Of significance is that the grains are well down within this laminar-like flow of the viscous sublayer. Indeed, some of the experiments used to establish this low Re_{*_t} limb of the Shields curve actually employed viscous fluids such as oils and glycerine rather than water, so the flow was entirely laminar (White, 1970). Yalin & Karahan (1979) have undertaken flume experiments with both laminar and turbulent flows, and found that the turbulent-flow curve asymptotically approaches that for laminar flows as Re_{*_t} decreases, shown by the respective curves in Figure 4.2. With turbulent flows, first grain movement occurs under an instantaneous flow stress of a strong eddy, a stress that is greater than the mean stress. This causes the Shields curve, evaluated in terms of mean stresses, to be lower than the curve for purely laminar flow; this in part accounts for the dip in the Shields curve at intermediate Re_{*_t} values. As Re_{*_t} decreases, grains at the threshold condition are increasingly sheltered by the viscous sublayer, a fluid flow that is more laminar-like in character but with periodic bursting events (Chapter 1). This accounts for the asymptotic approach of the Shields curve to the "laminar flow" curve in Figure 4.2 based on viscous liquids such as oils and glycerine.

In view of the importance of turbulent eddies in the entrainment of sediments by flowing water, it has been hypothesized that the initiation of sediment movement and the resulting transport are stochastic processes, a view that brings into question the basic validity of a threshold criterion that separates a condition of no grain movement versus a slightly greater flow where some grain movement and transport occur. Some investigators have argued that a degree of sediment transport exists no matter how low the flow velocity or bed stress, and there is never a condition of "no grain movement" as is sometimes assumed for a threshold criterion. The basis for such a view is illustrated by the analyses of Grass (1970, 1983) who obtained detailed measurements of instantaneous stresses (rather than a time-averaged mean stress) when grain movement occurs. The measurements involved fine-grained particles whose threshold was achieved under sufficiently low-flow conditions that a viscous sublayer existed at the interface between the water and granular bed. Therefore, the initial grain movement involved bursting events, measured by Grass with the hydrogen-bubble technique. As noted by Grass, the initial grain instability results from the interaction between two statistically random variables, that associated with the fluctuating flow turbulence close to the bed and a second produced by the susceptibility of the grains to movement. Even with fairly uniform sediments, the many grains on the bed produce a distribution of degrees of susceptibility to entrainment, a distribution of individual threshold stresses caused by their varying shapes, packing within surrounding grains, and exposure to the fluid-flow. As shown schematically in Figure 4.5, when the mean flow velocity and stress are increased, the distribution of instantaneous flow stresses produced by the turbulence approaches and finally overlaps the distribution of potential grain

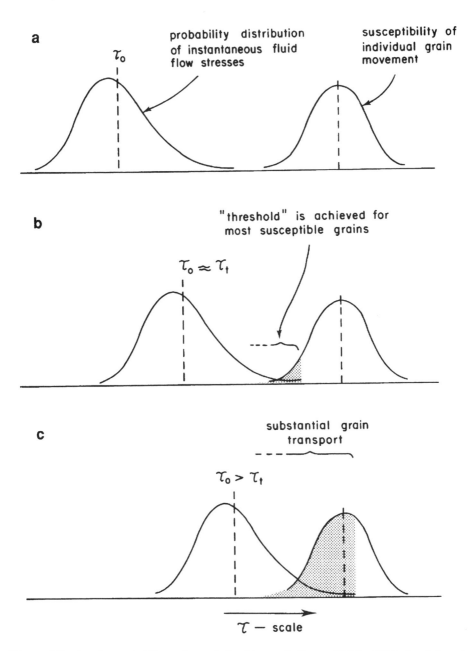

Figure 4.5 A schematic illustration of the ideas of Grass (1970, 1983) involving the existence of a probability distribution of instantaneous fluid-flow stresses and a distribution of the susceptibility of individual grain movement. Threshold is achieved when the two distributions overlap, the degree of overlap governing the quantity of the resulting sediment movement (after Grass, 1970)

threshold stresses. At the actual onset of grain motion, the most susceptible particles of that distribution are moved by the highest instantaneous shear stresses occurring within the shear-stress distribution applied to the bed by the flow. It is seldom possible in flume measurements, where threshold is observed visually, to note this condition of first absolute grain movement; practically speaking, this may not necessarily be the condition in which we are most interested for defining "grain threshold" since so little grain movement is involved. As the mean flow is progressively increased (Figure 4.5) there is more and more overlap of the two distributions and progressively more particles achieve threshold and are transported. As discussed earlier, most observations of grain threshold actually involve a certain degree of grain transport, and it is now seen that the subjectivity involves the amount of overlap of the respective distributions illustrated in Figure 4.5. Even when this is decided upon, the results would generally be reported as the mean value of the flow-stress distribution rather than the instantaneous stress, and as a mean grain size rather than the most susceptible grains that were actually involved in the initial movement.

Critical then to the problem of defining initial grain movement is determining (1) the distribution of flow shear stresses, and (2) the critical shear stress distribution of the bed material as governed by the range of individual grain susceptibilities for movement. An example of the measurements by Grass (1970) of a flow-stress distribution determined during grain-entrainment experiments is shown in Figure 4.6 together with the distribution of the instantaneous critical stresses that produced grain movement, determined with the aid of high-speed movies. The stress distribution for grain entrainment is rather broad, and many more measurements would have been required to fully define it. The results still illustrate that the instantaneous stress for grain entrainment differs from the mean stress exerted by the flow, the actual grain movement having been produced by the overlap of the flow turbulence and grain-susceptibility distributions.

In that at least theoretically the respective distributions shown in Figures 4.5 and 4.6 continue indefinitely to both high and low flow-stress levels, it can be argued that some grain movement (though extremely small) would take place even under the lowest flow

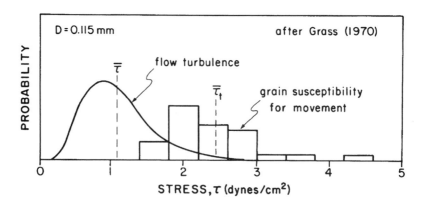

Figure 4.6 Measured distributions by Grass (1970) of instantaneous bed shear stresses associated with the flow turbulence and critical stresses required for individual grain movement (after Grass, 1970)

stages. This is a basic premise in some sediment-transport models, particularly that developed by Einstein (1950) which is based on probabilities of grain movement, the probability not reaching zero until there is no flow. This approach is supported by the direct measurements of Paintal (1971) and Helland-Hansen *et al.* (1974) of gravel transport rates under very low flow conditions where visual observations of the bed gave the impression that it was completely stable with no movement. In the case of the experiments by Helland-Hansen *et al.*, the transport rates involved the movement of only one particle per hour down to as few as one per day. Even so, as seen in Figure 4.7, these low transport rates correlate well with the flow strength as measured by the discharge, and there is a systematic pattern of decreasing grain sizes transported with decreasing flow strength (the typical flow-competence relationship found for gravels, to be discussed later). This indicates that selective size entrainment is in part the cause of there being some transport no matter how low the flow, making it difficult to define a specific threshold condition of no grain movement when a range of sizes is present within the bed materials. With an artificially narrow range of grain sizes as utilized in the flume experiments to establish threshold curves like those in Figures 4.1 and 4.2, the distributions for the susceptibility for grain movement as depicted in Figure 4.5 would narrow considerably, but might still persist due to different packing arrangements of individual grains on the bed, with some projecting higher into the flow.

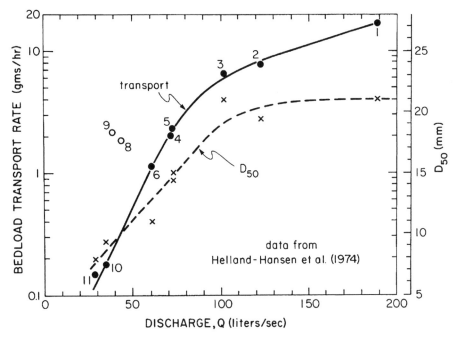

Figure 4.7 The data of Helland-Hansen *et al.* (1974) documenting the existence of some bedload transport even at very low discharges. There is also a correlation between the median grain size D_{50} of the transported bedload and the discharge, a flow-competence relationship. The sample at the time of the highest transport rate represents the movement of only one particle per hour and at the lowest discharge the transport can be represented by the movement of only a single 1 cm diameter particle per day

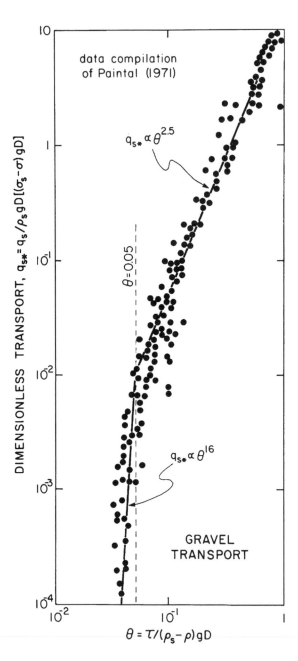

Figure 4.8 Gravel transport measurements compiled by Paintal (1971), showing an inflection in the trend at about $\theta = 0.05$ which corresponds to the $\theta_t = 0.045$ level portion of the Shields curve in Figure 4.2 that defines the threshold condition for gravel. The gravel-transport data shown here indicate that some transport does occur below the standard threshold condition, but is dropping at a rapid rate and represents the movement of only a few grains per day (after Paintal, 1971)

In a series of flume experiments and a compilation of earlier laboratory data, Paintal (1971) demonstrated that with progressively increasing flow discharge and mean bottom stress, there is a stage when an abrupt change occurs in the relationship between the bedload transport and the flow parameters. With low flows (Figure 4.8) the dimensionless bedload transport rate is proportional to the 16th power of the Shields $\theta = \tau/(\rho_s - \rho)gD$, while at higher flow the bedload transport is proportional to $\theta^{2.5}$. Of special interest, the transition between these two dependencies is reasonably abrupt (Figure 4.8), placed at $\theta = 0.05$ by Paintal. This is effectively the same as the $\theta_t = 0.045$ value given by the Shields curve of Figure 4.2 for gravel entrainment (high Re_{*_t}), established by Miller *et al.* (1977) and Yalin & Karahan (1979). From this it would appear that our standard threshold curves (Figures 4.1 and 4.2) tend to place the threshold criterion at the condition where there is an abrupt transition in the rate of bedload transport. With the transport being proportional to θ^{16} below that threshold condition, there is a precipitous reduction in the quantity of transported sediment as the flow stress is decreased further; the transport then involves only a few grains per day. Therefore, there is a basic reality to our standard threshold curves, recognizing however that they do not establish a condition of absolutely no grain movement.

SELECTIVE ENTRAINMENT OF SAND-SIZE GRAINS

Deposits of sand are generally characterized by relatively narrow ranges of grain sizes, but may contain a variety of minerals having different densities. A sand typically contains "light minerals" – particles of quartz and feldspars – plus a small content of "heavy minerals" – hornblende, augite, magnetite, garnet, and so on. The potential exists, therefore, for the occurrence of selective entrainment and sorting controlled by differences in grain densities as well as sizes. The product of the sorting can range from concentrations of heavy minerals on the crests of ripple marks in riverine or marine sediments, to massive black-sand concentrations of heavy minerals on beaches. In some cases the sorting processes can lead to the formation of placers, deposits in which valuable minerals such as gold, platinum, ilmenite and rutile have been concentrated by the natural processes (Slingerland & Smith, 1986; Komar, 1989b; Force, 1991). The role of selective entrainment of grains of different densities and sizes has been studied in particular as one of the factors important to the formation of placers.

The analyses of selective entrainment of sand grains indicate that both density and size are important, and particle shape can also be a factor. Important is that in sands there tends to be an inverse relationship between density and size, the heavy minerals generally being finer grained than the light minerals (Slingerland & Smith, 1986).

This inverse relationship is in part inherited from the source rocks, but it can also be produced when the initial deposition of the sand is from suspension and therefore is governed by particle settling velocities. This is illustrated by the series of minerals found in the beach placer studied by Komar & Wang (1984); Figure 4.9 shows that there is a distinct inverse relationship between densities and median grain diameters

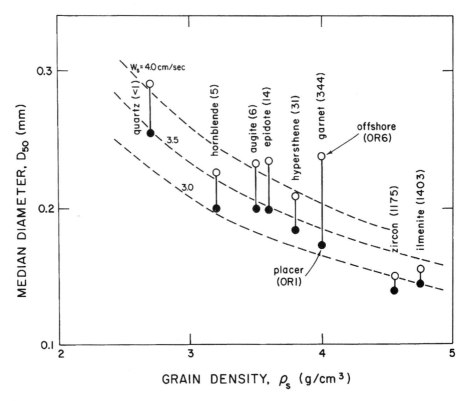

Figure 4.9 The inverse relationship between median diameters and densities for the suite of minerals on the beach studied by Komar & Wang (1984). Cross-shore sorting had taken place during beach face erosion, so medians of both the offshore samples and the placer lag at the landward edge of the beach are given. The dashed curves represent constant settling velocities, suggesting that the inverse relationship between size and density is governed by a narrow range of settling velocities of the grains residing on the beach (after Komar & Wang, 1984)

of the minerals in the deposit. The three dashed curves passing through the data are for the constant settling velocities, $w_s = 3.0$, 3.5 and 4.0 cm/s, establishing that the different minerals representing a large range of densities have only a narrow range of settling velocities. This narrow range was interpreted by Komar & Wang as resulting from grain settling being an important factor in governing which particles remain on the beach, any grains with low settling rates compared to the turbulent velocities of the breaking waves having moved offshore, leaving a sand deposit with grains of nearly constant settling velocity. A similar inverse relationship between density and size with a nearly constant settling velocity has been found by Li & Komar (1992a) in a study of beach placers involving a different suite of minerals from those in the study of Komar & Wang (1984).

The significance of the inverse relationship between mineral density and size to selective entrainment is illustrated schematically in Figure 4.10. Here the small heavy mineral has roughly the same settling rate, and approximately the same

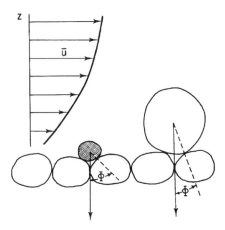

Figure 4.10 Effects of relative grain exposure and variable pivoting angles on the selective entrainment of particles from a deposit of mixed grain sizes and densities. The conditions experienced by a dark heavy mineral are contrasted with those of a larger particle of quartz that is both more exposed to the flow and has a lower pivoting angle Φ (from Komar, 1989b)

immersed weight, as the large light mineral. The observed grain-sorting patterns as well as quantitative evaluations of entrainment stresses demonstrate that in sands, it is the large low-density grains of quartz and feldspars that are more easily entrained than the small, dense heavy minerals. The product of this selective entrainment is a lag of concentrated heavy minerals, with the light minerals having been transported away.

An example of the resulting grain-sorting patterns is shown in Figure 4.11, again from the work of Komar & Wang (1984). The concentration factor (CF) of a mineral found within the beach placer is defined as its weight percentage in the placer divided by its percentage in a sample from the outer surf zone of the beach, the sorting having involved the erosion of the beach face and the cross-shore transport of the eroded sand and its deposition in the outer surf zone. A series of samples along a beach profile established that the quartz and feldspar were selectively entrained and transported offshore, leaving behind a lag of heavy minerals. Accordingly, the concentration factors plotted in Figure 4.11 become a direct evaluation of the degree to which a mineral has been concentrated in the placer as opposed to having moved offshore with the quartz and feldspar; CF is therefore a measure of the sorting efficiency. The highest concentration factor is that of ilmenite (CF = 1403), with zircon a close second (CF = 1175), signifying that both remained in the placer and virtually none moved offshore during beach face erosion. Of the heavy minerals, hornblende has the lowest concentration factor (CF = 5), indicating that it tended to remain in the placer but the concentrating processes were inefficient so that much of it moved offshore. Quartz/feldspar has a concentration factor <1 since it preferentially moved offshore rather than remaining in the placer. This series demonstrates that for the several minerals found in the beach sand, the concentration factor of the mineral in the placer increases with increasing grain

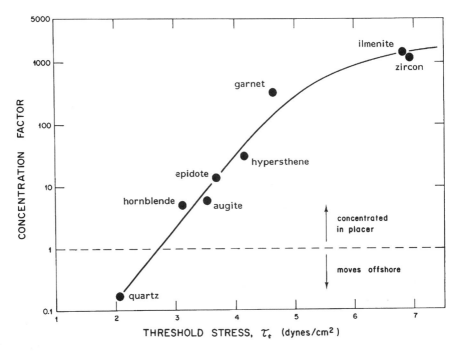

Figure 4.11 Concentration factors of minerals in the beach placer studied by Komar & Wang (1984), versus the flow stresses required for their entrainment as calculated with eqns 4.6a and 4.6b (from Komar & Wang, 1984)

density and decreasing median diameter (Figure 4.11). Apparently, ilmenite is most efficiently concentrated in the placer because it is the densest of the minerals and is also the finest grained. Within the series of heavy minerals, the processes were least efficient in concentrating the hornblende because it has the lowest density and the highest median diameter of the heavy minerals. The pattern of sorting leading to the formation of the placer is therefore one where the processes have selectively removed the grains of lower density and coarser sizes, leaving grains of high density and small diameters in the placer, much as shown schematically in Figure 4.10.

The concentration factors are compared in Figure 4.11 with calculated entrainment stresses for the individual minerals within the initial deposit of mixed sizes and densities. The strong trend establishes that within the mixed sediment, particles of ilmenite were most difficult to entrain while the quartz/feldspar grains were most easily moved by the flowing water. This led to the conclusion that selective entrainment was of primary importance to the observed sorting and formation of the placer, with differential rates of subsequent transport also having played a role (Komar & Wang, 1984). The studies by Frihy & Komar (1991) and Li & Komar (1992a) have further established the importance of selective entrainment and transport in beach placer formation, but where there is sorting in the longshore direction as well as cross-shore sorting.

In calculating the selective-entrainment stresses presented in Figure 4.11, Komar & Wang (1984) utilized the relationships

$$\tau_t = 0.00515(\rho_s - \rho)gD^{0.568} \tan \Phi \tag{4.6a}$$

$$\Phi = 61.5\left(\frac{D}{K}\right)^{-0.3} \tag{4.6b}$$

based in part on grain-pivoting models that will be reviewed later in this chapter. Equation 4.6a is equivalent to the empirical eqns 4.2 and 4.5 established in Figure 4.1, modified to include the expected dependence on the grain-pivoting angle Φ as illustrated in Figure 4.10. Equation 4.6b for the pivoting angle is from Miller & Byrne (1966) for natural beach sands. Important is that the pivoting angle decreases with the relative grain size, the ratio of the diameter D of the pivoting grain which is to be entrained to the diameter K of the underlying grains, a geometric dependence shown schematically in Figure 4.10 and established by data to be reviewed later. In practice, K is taken as the median diameter of the deposit as a whole.

These entrainment relationships, eqns 4.6a and 4.6b, have been tested by Li & Komar (1992b) in laboratory experiments which involved the movement of uniform grains of diameter D over a fixed bed consisting of grains of diameter K glued to the bottom of the flume. The measured versus predicted entrainment stresses are compared in Figure 4.12, showing nearly perfect agreement. The relative grain size D/K ranged from 0.59 to 1.41 in the series of experiments, as labelled in Figure 4.12, where it is seen that the data confirm that the entrainment stress decreases with an increase in D/K. These experiments, therefore, confirm that the relative grain size is important to the selective entrainment of sand-size grains, with the coarser

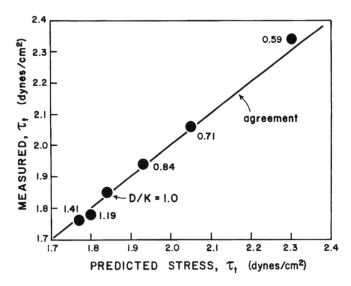

Figure 4.12 Measurements of threshold stresses versus those predicted by eqns 4.6a and 4.6b, covering a range of D/K relative grain sizes where D is the diameter of the moving grain and K is the grain size of the fixed bed roughness (from Li & Komar, 1992b)

sizes in the mixture being preferentially entrained. With respect to the field observations of Komar & Wang (1984) discussed above, the results reaffirm that the larger sizes of the quartz and feldspar minerals are important to their having been more easily entrained than the heavy minerals, acting in concert with differences in mineral densities.

The series of threshold curves in Figure 4.13 is based on the calculations of Komar & Wang (1984) using the combined eqns 4.6a and 4.6b for the series of minerals found in the placer involved in that study. The selective entrainment curve for quartz obliquely crosses the standard threshold curve for beds of uniform quartz grains. The crossing point is at the median grain size of the beach sand as a whole (indicated by the cross on the curve), which is initially dominated by the quartz fraction. The curves for the other minerals were calculated on the basis of this same median grain size. Each selective entrainment curve slopes downward to the right,

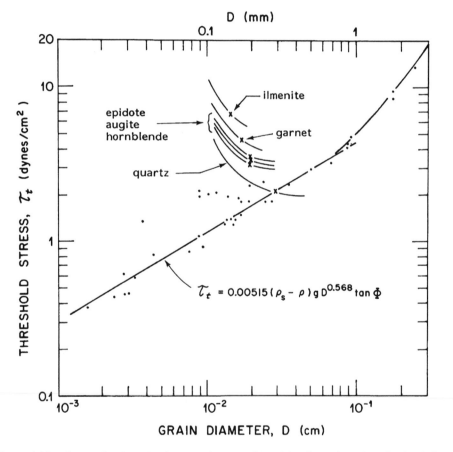

Figure 4.13 Curves for the selective entrainment of particles from deposits of mixed sizes, calculated with eqns 4.6a and 4.6b for the series of minerals found on the beach studied by Komar & Wang (1984). The standard threshold curve for deposits of uniform quartz grains is shown for comparison (from Komar & Wang, 1984)

signifying that the stress required for grain entrainment from a bed of mixed sizes decreases with increasing grain diameter within that mixture, just as established in Figure 4.12 from the experimental tests of Li & Komar (1992b). The series of curves in Figure 4.13 predicts that of the several minerals found within the beach sand, grains of ilmenite would be most difficult to entrain due in part to their having the highest density, but also due to ilmenite being the finest grained of the mineral series. Of the heavy minerals, hornblende is predicted to be most easily entrained, being the least dense and coarsest. As established in Figure 4.11, this predicted order of resistance to entrainment is matched by the degrees of concentration of these minerals in the beach placer.

A number of studies have demonstrated the importance of grain density in the sorting of sands (for example, Trask & Hand (1985) and Frihy & Komar (1991), as well as those discussed in detail above). The review presented here indicates that differences in grain sizes can also be important, and generally act in support of density sorting due to the inverse relationship between density and size found in most mineral suites. Of interest, the results demonstrate that for a fixed density, within the range of sand sizes, the larger the particle the more easily it is entrained from a deposit of mixed sizes. This is opposite to the size-sorting pattern found in gravel deposits.

SELECTIVE ENTRAINMENT OF GRAVEL

Gravel in streams and in marine sediments is generally characterized by wide ranges of particle sizes, while densities are roughly the same. The problem then is one of entrainment from deposits of mixed sizes, with grain shape and packing arrangements also being important to the process.

In deposits of mixed sizes, the focus is on the initial movement of a particular size fraction within the total distribution of sizes, and on the very largest particles the flow can entrain and transport. The relative entrainment stresses of different size fractions and their rates of transport are reflected in samples of bedload derived from the deposits. In the case of gravel-bed streams, bedload samples are generally collected with small, portable samplers such as those developed and tested by Helley & Smith (1971) and Hubbell (1964, 1987), with enlarged versions of otherwise portable samplers (Bunte, 1992), or with permanent samplers that span the complete stream width and capture the entire bedload (Milhous, 1973; Leopold & Emmett, 1976, 1977; Carling, 1983). It is generally found that the greater the stream discharge, velocity or mean stress, the coarser grained the entire bedload grain-size distribution (Ashworth & Ferguson, 1989; Shih & Komar, 1990a; Kuhnle, 1992). With such data, the critical condition for entrainment is associated with the maximum particle size D_m in the bedload sample. In most data sets it is found that there is a shift in D_m toward larger sizes with an increase in flow discharge, velocity or mean bed stress; representative relationships of this type are listed in Table 4.1, derived from a number of field and laboratory investigations. An example correlation is shown in Figure 4.14 for the data of Milhous (1973) collected with the total-bedload trap operated in Oak Creek, Oregon. The basic response is one of

Table 4.1 Flow-competence relationships for the entrainment of large particle sizes

Source	Location	$\bar{D}_B$ (cm)	D_m range (cm)	$\tau_c = rD_m'$	$\theta_c = a(D_m/\bar{D}_B)^b$	$\bar{u}_c = cD_m^d$
Hooker (1896)	various rivers		5–49			$c = 64$; $d = 0.47$
Grimm & Leupold (1939)	laboratory		5–52			$c = 102$; $d = 0.34$
Lane & Carlson (1953)	San Luis Valley, Col.		2.5–12	$r = 44$; $s = 0.73$		
Fahnestock (1963)	White River, Wash.		6–48	$r = 193$; $s = 0.48$		$c = 128$; $d = 0.18$
Egiazaroff (1965)	laboratory				$a = 0.050; b = .44$	
Scott & Gravlee (1968)	Rubicon River, Calif.		46–329	$r = 8.1$; $s = 1.56$		
Helley (1969)	Blue Creek, Calif.		15–52			$c = 96$; $d = 0.26$
Milhous (1973)	Oak Creek, Oregon	2	0.8–11.7	$r = 108; s = 0.57$		$c = 38.7; d = 0.55$
Day (1980)	laboratory	0.042	0.015–0.286		$a = 0.045; b = .43$	
		0.044	0.015–0.203		$a = 0.047; b = .53$	
		0.175	0.015–1.11		$a = 0.045; b = .66$	
		0.155	0.015–0.406		$a = 0.026; b = .29$	
					$a = 0.029; b = .66$	
Andrews (1983)	East Fork, Snake and Clearwater Rivers				$a = 0.083; b = .87$	
Carling (1983)	Great Eggleshope Beck, England	2	1–20	$r = 288; s = 0.185$	$a = 0.039; b = -0.82$	
Hammond et al. (1984)	English Channel	0.75	0.5–4	$r = 55$; $s = 0.42$	$a = 0.045; b = -0.58$	
Andrews & Erman (1986)	Sagehen Creek, Calif.				$a = 0.11$; $b = -1$	
Ashworth & Ferguson (1989)	Lyngdalselva, Dubhaig and Feshie Rivers	2.3–9.8			$a = 0.089; b = -0.74$	
Ashworth et al. (1992)	Sunwapta River, Alberta, Canada	2–3			$a = 0.049; b = .69$	
Petit (1994)	laboratory	1.28			$a = 0.058; b = -0.66$	
		1.96			$a = 0.049; b = -0.68$	
		2.42			$a = 0.047; b = -0.73$	
		3.92			$a = 0.045; b = -0.81$	

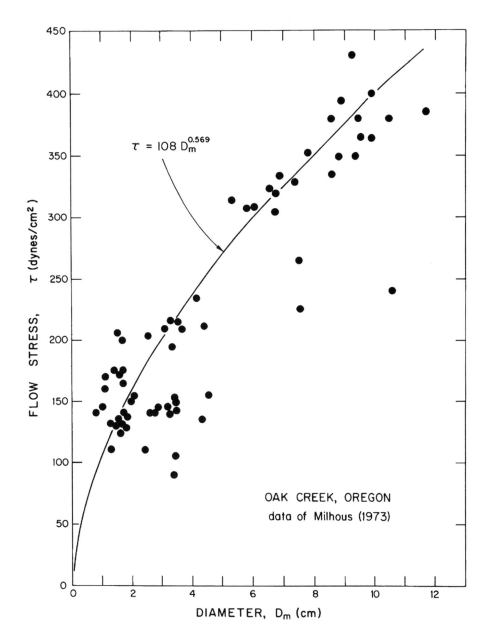

Figure 4.14 The data of Milhous (1973) from Oak Creek, Oregon, for the D_m of the largest gravel particles found in bedload samples at various mean-flow bed stresses (from Komar, 1987b)

increasing D_m with increasing flow stress, but in the correlation the stress is made the dependent variable so as to obtain

$$\tau_c = 108 D_m^{0.57} \qquad (4.7)$$

by regression, where the units of D_m are centimetres and the flow stress is dynes/cm^2. This is a type of flow-competence relationship, of potential use in evaluating the mean flow stress of a flood from the maximum diameter particles transported (Baker & Ritter, 1975; Costa, 1983; Komar, 1987b).

There is some concern in using maximum particle diameters as a measure of the flood's magnitude. As discussed by Costa (1983), these extreme sizes could in some instances have been contributed by debris flows entering the channel, or introduced by bank erosion with minimal actual transport. Costa recommends using an average of the five or ten largest particle sizes present in the flood deposit to provide a more representative sample reflecting the flood's magnitude. In a study of the Bonneville flood deposits of Idaho, O'Connor (1993) used boulders from large bars to ensure that they had actually been transported. Wilcock (1992) has more fundamentally argued against the use of maximum particle sizes in competence evaluations. One argument he presented was specific to the trend seen in Figure 4.14 for the Oak Creek data of Milhous (1973); Wilcock noted that larger samples of bedload were collected at higher flow stresses, and suggested that this may have biased the sampling in favour of capturing larger particles at the higher flow stages. Expressed another way, he argues that if equal volumes of samples had been collected at all flow stages, the bedload grain-size distributions would have been essentially the same and there would not have been a τ_c versus D_m trend as seen in Figure 4.14 for the Milhous data. However, a detailed analysis of the Oak Creek data by Komar & Carling (1991) refutes this interpretation. They demonstrate that the pattern of flow competence is reflected in the entire bedload grain-size distribution, with systematic increases in the median grain size (D_{50}), the 60th percentile (D_{60}), and so on up to D_{95} and finally D_m. The results demonstrate that D_m is behaving as part of the entire distribution of grain sizes in shifting to coarser sizes as the flow discharge and stress increase. Furthermore, as the bedload grain-size distributions in Oak Creek coarsen with increasing flow stage, they systematically approach the grain-size distribution of the bed material (Shih & Komar, 1990b; Komar & Shih, 1992). A comparable coarsening of bedload grain-size distributions has been found by Kuhnle (1992, fig. 7.5) in Goodwin Creek, Mississippi, with a near congruence between the bedload samples and the grain-size distribution of the bed material at high flow stages. Finally, as seen in Table 4.1, there is a considerable number of data sets that demonstrate flow-competence relationships of increasing D_m with increasing bed stress and mean velocity, a trend found in gravel-bed streams and in marine gravels, and also derived from experiments in the controlled conditions of the laboratory (Day, 1980; Bathurst, 1987; Petit, 1994). Although Wilcock's arguments are insufficient to reject the concept of flow-competence evaluations based on D_m, his concerns regarding sample sizes and their possible influence on grain-size distributions are valid, especially when using small portable samplers, and must be borne in mind when designing field sampling programmes. A probable example is provided by the study of Ashworth et al. (1992) on the Sunwapta River, Alberta, employing Helley–Smith portable samplers during marginal transport conditions. Although they statistically found a trend of increasing D_m with increasing flow stress, the data are scattered and there was no accompanying shift in the median diameters of the bedload samples. Analyses of the fractional transport rates did show size-selective entrainment in the coarser fractions, so Ashworth *et al.* concluded that some of the increase in D_m may still reflect a flow-competence response, even though the correlation was affected by sample sizes.

Representative τ_c versus D_m flow-competence relationships from the list in Table 4.1 are plotted in Figure 4.15. The data of Carling (1983) also came from a gravel-bed stream using a total bedload trap, while those of Day (1980) are derived from flume experiments. The data of Hammond *et al.* (1984) came from marine continental shelf gravels, the movement of D_m having been determined from camera observations while boundary-layer flow measurements were simultaneously obtained to evaluate the mean flow stress. Each flow-competence curve in Figure 4.15 obliquely crosses the dashed line in the graph for the standard threshold of uniform grains, equivalent to that given in Figure 4.1. The series of D_m curves can be viewed as selective entrainment relationships, the maximum particle sizes that are entrained from the deposits of mixed sizes at different flow-stress levels. For each curve the larger sizes are entrained at flow stresses that are lower than required for their initial movement from beds of uniform sizes, while the smaller

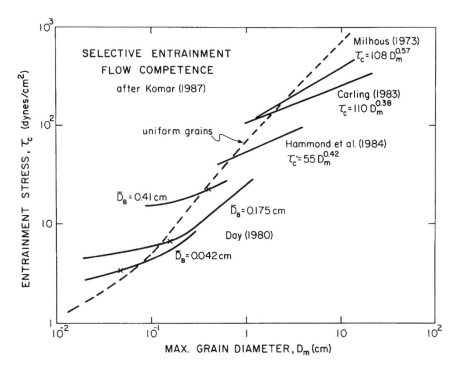

Figure 4.15 Flow stress versus D_m flow-competence curves from various environments and studies (Table 4.1). The dashed curve is that for uniform sediments, equivalent to that given in Figure 4.1 (from Komar, 1987b)

size fractions require higher flow stresses than in the uniform-bed case. The reason for this is that the largest sizes within the mixed sediments are more exposed to the flow, while the smallest size fractions tend to be sheltered from the flow by the larger particles. As will be seen in a later section of this chapter, where mechanical models of grain entrainment are reviewed, differences in pivoting angles of small versus large grains may also be important to the size-sorting patterns observed in Figure 4.15. A significant conclusion is that the larger size fractions within the bed of mixed sizes are substantially more mobile than if they formed a bed of uniform sizes. Within the deposit of mixed sizes, however, larger flow stresses are still required to entrain the larger particles, the typical flow-competence dependence for gravels.

The flow-competence curves in Figure 4.15 cross the standard threshold curve (dashed) for uniform grains at approximately their respective median grain diameters (Komar, 1987a). Therefore, the echelon arrangement of curves reflects the overall coarseness of the sediment mixtures, with the Oak Creek gravels in the study of Milhous (1973) being coarsest, and the flume studies of Day (1980) having used the finest gravel-size mixtures. This suggests that the series of curves might converge if normalized as $D_m/\bar{D}_B$ where $\bar{D}_B$ is the median diameter of the bed material (Andrews, 1983). The entrainment stress can be normalized as $\theta_c = \tau_c/(\rho_s - \rho)gD_m$, the Shield θ of eqn 4.3 but here expressed in terms of flow-competence evaluations

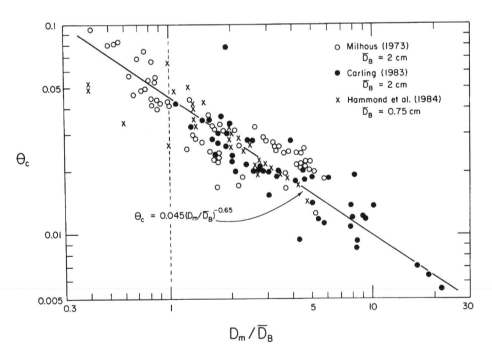

Figure 4.16 The dimensionless Shields θ_c versus $D_m/\bar{D}_B$ for field measurements of gravel movement, a normalization by the median diameter $\bar{D}_B$ of the bed material that results in a reasonable convergence of the data (from Komar, 1989a)

for the movement of the maximum particle size. This yields the non-dimensional plot given in Figure 4.16, and the correlation

$$\theta_c = 0.045\left(\frac{D_m}{\overline{D}_B}\right)^{-0.6} \tag{4.8}$$

as an average representation of the data (Komar, 1989a). The results show that θ_c decreases as D_m increases relative to the median diameter $\overline{D}_B$ of the deposit. When $D_m/\overline{D}_B = 1$, that is D_m corresponds to the median diameter, then $\theta_c = 0.045$ which is the value of the Shields curve of Figure 4.2 for uniform coarse grains. Using $\theta_c = \tau_c/(\rho_s - \rho)gD_m$, the non-dimensional eqn 4.8 can be converted into the equivalent dimensional form

$$\tau_c = 0.045(\rho_s - \rho)g\overline{D}_B^{0.6}D_m^{0.4} \tag{4.9}$$

for a direct computation of the flow-competence stress. This form of the relationship shows the strong dependence of τ_c on the median diameter of the bed material as well as on the maximum particle size entrained at some particular flood stage. The form of eqn 4.9 is compared with field data in Figure 4.17, including the cobble entrainment measurements of Fahnestock (1963) obtained in the White River,

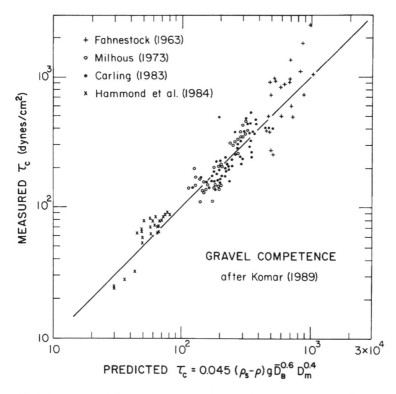

Figure 4.17   The measured flow-competence stress τ_c for the movement of D_m versus that predicted by eqn 4.9 (from Komar, 1989a)

Washington. The studies of Wohl *et al.* (1994) and Kehew & Teller (1994) provide examples of the use of eqn 4.9 in flow-competence evaluations, respectively, to cobble deposits within bedrock channels in Israel and to glacial spillways which transmitted extreme discharges in the northern United States.

Strong support for eqns 4.8 and 4.9, which were based on field data (Komar, 1989a), is provided by the comprehensive series of flume experiments undertaken by Petit (1994). The experiments included series with a fixed bed and a few movable grains, and series where the bed materials were completely mobile. The median grain sizes of the bed materials ranged from 1.2 to 3.9 cm, while the range of marked particles was 1.0 to 6.5 cm. Nearly one hundred marked pebbles were used in each experiment, and Petit documented the local bed stresses for those pebbles that did not move as well as for the pebbles that were entrained. Graphs of shear stresses versus particle sizes and also of θ versus $D_m/\overline{D}_B$ were developed for each experiment, and lines for the initial threshold and for generalized movement were established. The empirical equations so derived for the threshold condition are listed in Table 4.1, where it is seen that the exponents for the θ versus $D_m/\overline{D}_B$ relationship ranged from -0.66 to -0.81, while the proportionality coefficients ranged from 0.045 to 0.058; Petit recommended use of -0.70 and 0.050 based on the complete series of experiments, values which are very close to those given by eqn 4.8 based on the field data.

The -0.6 exponent in eqn 4.8 and the $D_m^{0.4}$ dependence of the competence stress in eqn 4.9 reflect the degree of selective entrainment within deposits of mixed sizes. Those coefficients are products of the combined data, but Table 4.1 reveals that individual data sets yield somewhat different values for the empirical coefficients. Of interest is how much of this range results from different patterns of grain sorting in different streams, which might in turn depend on the grain-size distributions and fabrics of the bed materials. For example, the segregation of bed material into a coarse pavement and finer-grained subpavement (sometimes referred to as the armour and subarmour layers) should tend to equalize the entrainment of the different grain-size fractions, in the extreme leading to equal mobility where all grain sizes are entrained at the same flow stress (Parker *et al.*, 1982). Perfect equal mobility is sometimes interpreted as corresponding to a -1 exponent in eqn 4.8 (Andrews & Erman, 1986). By that criterion, Oak Creek should depart significantly from equal mobility since its exponent is -0.43 (Table 4.1), yet its bed material comes close to the classical segregation of sizes into pavement and subpavement layers, and the data of Milhous (1973) from that site were employed by Parker *et al.* (1982) to formulate the equal-mobility model. The range of empirical coefficients in Table 4.1 derived from the many investigations instead appears to result largely from systematic differences in sampling techniques employed in the studies, particularly the use of portable samplers versus a total bedload trap. Another problem is the possibility of collecting random data for the stress versus D_m. In the extreme, a series of samples at the same flow stage (constant τ) will yield a range of D_m values. Plotted as θ_c versus $D_m/\overline{D}_B$ as in Figure 4.16, the data would yield a -1 exponent in eqn 4.8 since one is simply plotting $\theta_c \propto 1/D_m$ versus D_m. As discussed by Komar & Shih (1992), this is almost certainly the cause of the -1 exponent for the measurements of Andrews & Erman (1986) in Squaw Creek, California. Data sets could also contain subsets of random measurements that would tend to pull the empirical exponent in eqn 4.8 closer to -1. This has been illustrated for the Oak Creek data of Milhous (1973) as analysed by Komar & Shih (1992). At a discharge of about $1 \, \text{m}^3/\text{s}$ the pavement of the bed material begins to break up, and data collected above that flow stage yield a strong trend of increasing D_{m} with increasing flow stress (Figure 4.14); those data alone yield the -0.43 exponent in a relationship of the form of eqn 4.8. Measurements of D_{m} collected during flow stages lower than $1 \, \text{m}^3/\text{s}$ are random, and if plotted in a graph such as Figure 4.16 they form a separate trend having a -1 slope (Komar & Shih, 1992, fig. 7). The entire data set yields a -0.64 exponent, intermediate between the significant -0.43 value and -1 from the random data. It is apparent that a -1 exponent in a graph of θ_{c} versus $D_{\text{m}}/\bar{D}_{\text{B}}$ cannot alone be interpreted as a demonstration of equal mobility of different grain-size fractions found within the bed material.

The concept of equal mobility was introduced by Parker *et al.* (1982) as a *first-order approximation* in developing relationships for evaluating bedload transport rates of individual grain-size fractions within a gravel-bed stream. The equal-mobility approximation yielded reasonable results in determinations of gravel transport rates in Oak Creek, but did not account for observed variations in bedload grain-size distributions in that it predicts every bedload sample has the same grain-size distribution, approximately the same as the distribution of the bed material. It was seen in Figure 4.14 that there actually are significant shifts in D_{m} with increasing flow stage in Oak Creek, and the studies of Shih & Komar (1990a) and Komar & Carling (1991) have established the existence of a systematic coarsening of the entire grain-size distributions for the transported bedload as the flow stress increases. Parker *et al.* recognized this failure of the equal-mobility assumption, and made initial attempts to develop higher-order solutions for the transport computations that would account for the observed shifts in grain-size distributions; Diplas (1987) and Shih & Komar (1990b) present analyses that provide improved calculations of bedload transport rates as well as accounting for the coarsening grain-size distributions.

The original equal-mobility analysis of Parker *et al.* (1982) made use of a reference transport rate for the grain-size fractions which can be viewed as a threshold condition, although, as will be seen below, the level of transport represents more grain movement than is inherent in traditional threshold evaluations. Their use of a reference transport rate to converge the data for the different size fractions yielded the relationship

$$\tau_{\text{ir}}^* = 0.0876\left(\frac{D_{\text{i}}}{\bar{D}_{\text{B}}}\right)^{-0.982} \tag{4.10}$$

where D_{i} represents the size fraction (in practice a sieve-size fraction) and the Shields $\tau_{\text{ir}}^* = \tau_{\text{ir}}/(\rho_{\text{s}} - \rho)g D_{\text{i}}$ is in terms of D_{i} as well, with τ_{ir} being the reference stress for the size fraction. In using τ_{ir}^* rather than θ, I am employing the symbols used by Parker *et al.* (1982) and others who have analysed gravel transport rates, in this way emphasizing that their reference transport is inherently different from direct assessments of grain entrainment used to establish flow-competence relationships in terms of D_m. Although eqns 4.8 and 4.10 are similar in form, their derivations and supporting data are entirely different, and it is important not to confuse their applications. Relationships in the form of eqn 4.10 are listed in Table 4.2, derived from measured gravel transport rates in streams and also in flume experiments. The exponents vary widely, but many are close to -1, in which case the investigators generally interpreted the results as a demonstration that all grain-size fractions are entrained at the same flow stress, one aspect of equal mobility. The -0.982 exponent derived by Parker *et al.* (1982) is equivalent to the simple proportionality $\tau_{ir} \propto D_i^{0.018}$, which does imply that all size fractions have nearly the same "threshold" τ_{ir}, or at least the same reference stress used in calculations of the bedload transport. However, Komar & Shih (1992) have shown that eqn 4.10 in this instance resulted from the analysis approach of Parker *et al.*, specifically the use of dimensionless ratios of gravel transport rates and fluid-flow parameters. The -0.982 exponent of eqn 4.10 is close to -1 owing to the dimensionless form of the basic gravel transport relationship and cannot necessarily be used to infer that all size fractions are entrained at the same flow stress. In their laboratory experiments and reanalysis of other available data sets, Wilcock & Southard (1988) employed the same dimensional analysis approach as Parker *et al.*, and similarly derived exponents close to -1 (Table 4.2).

Komar & Shih (1992) have shown that if it is assumed that the exponent in eqn 4.10 is -1 so as to correspond to perfect equal mobility, then the reference transport rate defined by Parker *et al.* (1982) is equivalent to

$$q_{sir} = 0.0000519 f_i \left(\frac{(\rho_s - \rho)g}{\rho} \right)^{1/2} \bar{D}_B^{3/2} \qquad (4.11)$$

Table 4.2 The $\tau_{ir}^* = a' (D_i / \bar{D}_B)^{b'}$ relationship for a reference "threshold" condition for the grain-size fraction D_i based on a fixed level for the dimensional transport rate as defined by Parker *et al.* (1982)

Source	Location	$\bar{D}_B$ (mm)	a'	b'
Parker *et al.* (1982)	Oak Creek, Oregon	20	0.0876	-0.982
Day (1980)[*]	laboratory	1.82	0.0368	-0.809
		5.28	0.0368	-0.953
Dhamotharan *et al.* (1980)[*]	laboratory	2.16	0.0708	-1.091
Misri *et al.* (1984)[*]	laboratory	2.36	0.0475	-0.997
		3.81	0.0415	-0.953
		4.00	0.0371	-0.920
Wilcock & Southard (1988)	laboratory	1.83	0.0301	-1.006
		1.83	0.0356	-0.970
		0.670	0.0226	-0.984
		5.28	0.0371	-1.064
Ashworth & Ferguson (1989)	Dubhaig R., Scotland	2.3–9.8	0.072	-0.65
	Feshie R., Scotland	5.2–6.3	0.054	-0.67
	Lyngsdalselva, Norway	6.9	0.087	-0.92
Kuhnle (1992)	Goodwin Creek, Miss.	8.31	0.0856	-0.805
Ashworth *et al.* (1992)	Sunwapta River, Alberta	20–30	0.061	-0.79

[*] Data sets reanalysed by Wilcock & Southard (1988)

for the actual transport rate, having eliminated its dimensionless form. It is seen that q_{sir} lacks any direct dependence on individual D_i grain-size fractions, so there can be no selective entrainment, but is proportional to f_i of the size fractions as found in the bed material and also depends on the $\bar{D}_B$ median of the bed material. These latter dependencies are reasonable for a reference threshold condition, especially for those connected with evaluations of gravel transport rates. As noted earlier, the use of a small level of transport as a threshold criterion has been used in previous studies. In formulating such a threshold criterion for a size fraction within the bedload, it is reasonable to have it proportional to the availability of that size in the stream bed (i.e. to f_i) and to the overall grain size as represented by the median of the bed material. This is in keeping with the quantitative definitions of grain threshold as developed by Neill & Yalin (1969).

This can be explored further using the Oak Creek data of Milhous (1973) by directly plotting the measurements of q_{si} and q_{si}/f_i versus τ for the individual grain-size fractions of the bedload samples (Komar, 1992a); an example is shown in Figure 4.18a for the -4.25ϕ sieve fraction. A good straight-line fit can be established between the transport rate of the size fraction and the flow stress, a condition that is found for all of the size fractions analysed. A compilation of the series of lines for the several sieve fractions is given in Figure 4.18b. The reference transport level in the equal-mobility analysis of Parker et al. (1982) corresponds to $q_{sir}/f_i = 6.38$ kg/hm as calculated with eqn 4.11 (Komar & Shih, 1992). This reference level is marked on the graphs of Figure 4.18 where it is seen that it must represent a significant amount of sediment movement. Furthermore, from Figure 4.18b it is apparent that the selection of a lower value than 6.38 kg/hm for the reference transport would result in a wider range of reference "threshold" stresses for the series of sieve fractions. At low stresses the curves for the size fractions are widely spaced and a great deal of selective entrainment and transport clearly occurs. At higher flow stresses the curves converge, and one can then think in terms of an approximate condition of equal mobility in that the different size fractions have roughly comparable transport rates when normalized to their availability in the bed material.

A threshold condition representing a small amount of bedload transport in Oak Creek might be better taken as $q_{si}/f_i = 1$ kg/hm, shown as the short-dashed horizontal line in Figure 4.18B. The corresponding threshold flow stresses for the series of sieve fractions are plotted in Figure 4.18c, showing the expected increase in threshold flow stresses for progressively coarser size fractions and establishing a condition of non-equal mobility in terms of grain entrainment.

Wilcock & McArdell (1993) have followed this approach in analysing the threshold stresses of grain-size fractions in terms of a reference transport rate, having abandoned the dimensionless analysis used earlier in Wilcock & Southard (1988). Equally important, Wilcock & McArdell employed wider ranges of grain sizes than used in earlier flume experiments, grain-size distributions that are more like those found in natural streams. With such changes in analysis procedures and bed materials, the results now yielded patterns of increasing reference threshold stresses with increasing coarseness of the grain-size fraction. The results of Wilcock

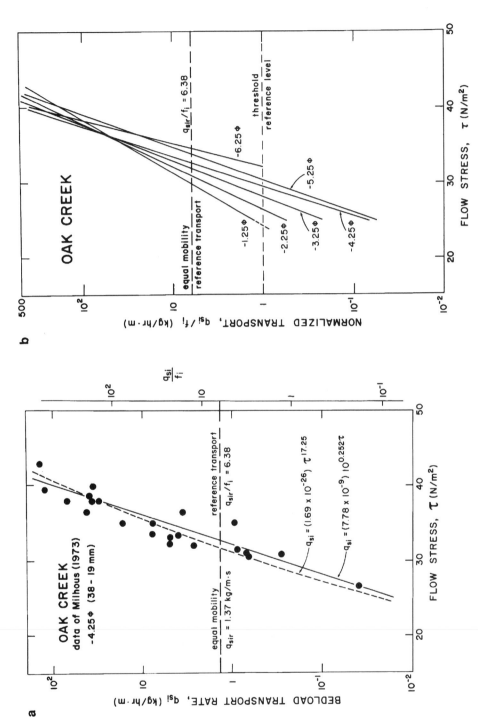

Figure 4.18 *For caption see facing page*

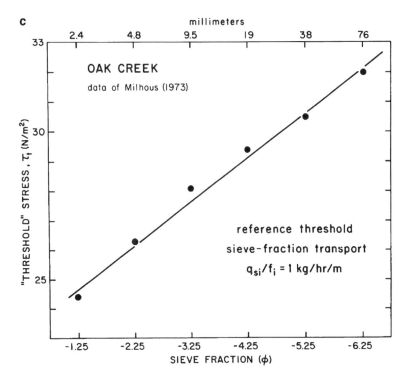

Figure 4.18   (a) The bedload transport rate of the $-4.25\,\phi$ sieve fraction in Oak Creek for the measurements of Milhous (1973). Two regressions are shown, log–log and log–linear. The reference transport rate for the equal mobility analysis of Parker *et al.* (1982) for this size fraction is given as q_{sir} and q_{sir}/f_i where f_i is the frequency of that fraction in the subpavement bed material. (b) A compilation of curves such as that in (a) for the bedload transport of various grain-size fractions in Oak Creek. The equal mobility reference transport of Parker *et al.* is given by $q_{sir}/f_i = 6.38$ kg/hm, a relatively high quantity of transport. Also indicated is the $q_{sir}/f_i = $ kg/hm transport level, arbitrarily taken to represent a threshold condition. (c) The resulting "threshold" curve of the flow stress required to produce the 1 kg/hm transport level as given in (b) for the series of grain-size fractions (after Komar, 1992a)

& McArdell are shown in Figure 4.19 where the "initial threshold" curve is based on analyses similar to those used in Figure 4.18 for the Oak Creek data. Wilcock & McArdell also derived data in support of their "initial threshold" curve from observations of the movement of individual particles in the coarsest fractions, made possible in their experiments by having painted each size fraction a different colour. Also given in Figure 4.19 is a curve for the "complete mobilization" of the grain-size fraction, which on average occurs at a stress that is approximately twice that of the "initial threshold". Wilcock & McArdell define a condition of "partial transport" where the coarser grains remain largely immobile on the bed, with their fractional transport rates being produced by the rare movement of only a few grains. This partial transport condition contrasts with the complete mobilization of the finer grain-size fractions where the fractional transport rates are a function of only their

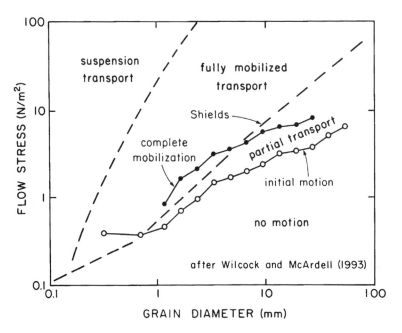

Figure 4.19 Curves for the "initial motion" and "complete mobilization" of grain-size fractions where the bed material has a wide distribution of sizes, based on the flume experiments of Wilcock & McArdell (1993) and the definition of initial motion in terms of a reference transport rate as illustrated in Figure 4.18 for Oak Creek (after Wilcock & McArdell, 1993)

proportion on the bed surface and on the total transport rate. Such patterns of partial transport of the coarser grain-size fractions have been observed in gravel-bed streams (Ashworth & Ferguson, 1989; Kuhnle, 1992) as well as in the flume experiments of Wilcock & McArdell.

Comparable analyses to define the threshold condition of a grain-size fraction can be undertaken in terms of numbers of particles being transported rather than expressed as mass transport rates (q_{sir}). Analyses in terms of numbers of particles in motion have been undertaken by Bunte (1992) based on bedload samples from Squaw Creek, Montana. The results are given in Figure 4.20 for the seven coarsest sieve-size fractions (0.50ϕ intervals), each showing the expected rapid increase in numbers of particles transported per minute per metre of channel width with increasing flow discharge. The final graph of Figure 4.20 presents the regression lines for the series of size fractions. The threshold condition could be placed at 0.01 particle/m min, equivalent to one particle transported in 100 min. Particles smaller than 45 mm (lines 1–4) have a common threshold of motion at about 3.5 m³/s, and their transport rates increase exponentially with the discharge above that level. Particles larger than 45 mm do not start to move unless the discharge exceeds 4.5 m³/s, and there is an orderly progression of coarser fractions requiring greater flow discharges for their threshold. At discharges over 5.5 m³/s, the transport rates of the different size fractions appear to be converging, perhaps again suggesting an

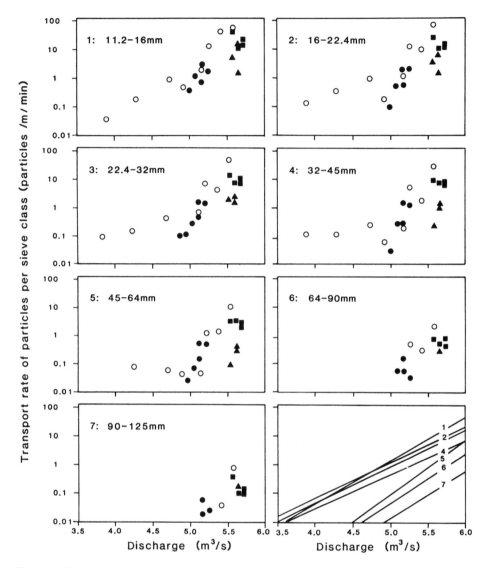

Figure 4.20 Particle transport rates in Squaw Creek, Montana, expressed in terms of the numbers of particles transported in the series of sieve fractions, derived from bedload samples collected in a large trap. The "threshold" condition for any size fraction can be represented by a 0.01 particle/m/min transport rate, which is seen in the last graph to depend on the flow discharge (from Bunte, 1992)

approach toward equal mobility. It is apparent that the overall patterns seen here in terms of numbers of particles transported are similar to those in Figure 4.18 for mass-transport rates of different size fractions in Oak Creek, Oregon. Komar (1992b) has analysed the Oak Creek data of Milhous (1973) in terms of numbers of particles transported and similarly to Bunte (1992), found a regular progression of increasing threshold discharges for the series of sieve fractions, having defined the threshold condition as the transport of 1 particle/m h. The results are consistent with the flow-competence relationships found for Oak Creek (Figure 4.14) representing a shift in the largest particles transported in the bedload (D_m) with increasing mean bed stress.

Thus far in dealing with the entrainment of gravel, the analyses have been based on the mean bed stress of the flowing water. In the case of flow-competence evaluations for the transport of the maximum particle sizes (D_m), the corresponding mean-flow velocity can be calculated from the evaluated stress if a suitable friction coefficient for the flow as a whole is selected, and this in turn could yield an estimate of the flood discharge. Alternatively, some investigators have attempted to develop flow-competence equations for gravel that directly yield the flow velocity or discharge from the sizes of the largest particles transported. It has been argued that in the case of the entrainment of extreme-sized particles such as an individual boulder, the velocity or discharge is a more relevant measure of flow magnitudes acting on the large particle.

Bathurst *et al.* (1982) and Bathurst (1987) have attempted to establish a flow-competence equation that directly yields the discharge. Based on data from flume experiments with bed materials having relatively uniform sizes, the empirical relationship

$$q_t = 0.15 g^{0.5} D^{1.5} S^{-1.2} \qquad (4.12)$$

was derived for the critical discharge per unit channel width required to entrain grains of diameter D in a channel having a slope S. The analysis was then expanded to an examination of selective entrainment from deposits of mixed sizes, the competence discharge q_c required to entrain the maximum diameter D_m. Data from gravel-bed streams were used to examine correlations between q_c and D_m, and it was found that the individual data sets have trends that obliquely cross the curve of eqn 4.12 for uniform grains, much like the series of curves in Figure 4.15 for the stress-based analysis. The cross-over points were found to be approximately at the median diameters of the distributions, so that the data could be normalized as

$$q_c = q_t \left(\frac{D_m}{\bar{D}_B}\right)^k \qquad (4.13)$$

where q_t is the value from eqn 4.12 evaluated for the median diameter $\bar{D}_B$ of the deposit. Based on data from two rivers, Bathurst determined that k is on the order of 0.2 to 0.4, and further suggested that its value might depend on the sorting of the gravel within the bed material as evaluated by the D_{84}/D_{16} ratio. More data from a greater variety of field sites are required to better establish the discharge-based flow competence equation, and analyses are needed to explore its relationship to the stress-based competence equations.

Attempts have also been made to develop flow-competence equations for the mean velocity and for the flow power, the velocity times the stress (Baker & Ritter 1977; Costa, 1983; Williams, 1983; O'Connor, 1993). These investigations have established scatter diagrams for the velocity or power versus the sizes of the transported clasts, with regressions of the data to yield flow-competence relation-ships. Examples are shown in Figure 4.21 from the study of O'Connor where the

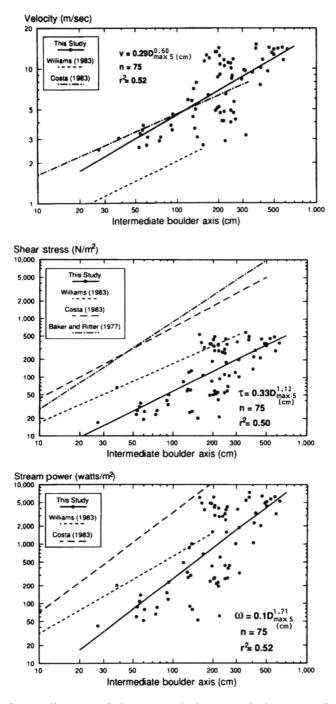

Figure 4.21 Scatter diagrams of the mean velocity, mean bed stress, and stream power versus the intermediate diameters of the five largest boulders found in the Bonneville flood deposits of Idaho. The data are compared with the empirical relationships of Baker & Ritter (1977), Costa (1983) and Williams (1983) (from O'Connor 1993)

data are derived from the Bonneville flood deposits of Idaho. In this instance the flood velocities, stresses and powers are estimated from hydraulic calculations, rather than representing direct measurements of flow parameters as was the case for data employed to establish the gravel-entrainment competence relationships discussed above. It is seen in Figure 4.21 that the empirical curves proposed by the several investigations diverge considerably, yielding markedly different assessments of the flow conditions for the movement of a specific clast diameter. Kehew & Teller (1994) have applied these relationships to boulder transport in glacial spillways in the northern United States, and concluded that the empirical equations proposed by Costa and Williams yield estimated stresses and flow powers that are too large by magnitudes which in turn yield excessive estimates for flow depths and discharges. Better results were obtained with eqns 4.8 and 4.9 derived by Komar's (1989a) analyses where the evaluations are based on both the maximum clast sizes transported and median diameters of the deposits. Kehew & Teller did not employ the curves and empirical competence relationships established by O'Connor based on boulder transport in the Bonneville flood deposits. The stress-based relationship of O'Connor, in yielding much lower estimates than the competence equations of Costa and Williams, does yield more reasonable estimates of flow parameters, values that are consistent with those obtained with eqns 4.8 and 4.9.

The divergence of the flow-competence curves in Figure 4.21 results from the contrasting types of data employed to establish them. For example, Costa (1983) relied on measurements of gravel transport in active rivers, the same data employed by Komar (1987a,b, 1989a) to establish the selective-entrainment eqns 4.8 and 4.9 that can be applied in flow-competence evaluations. The boulder-transport data of O'Connor (1993) represent the movement of larger clasts, generally with higher $D_m/\bar{D}_B$ ratios. This difference is apparent in Figure 4.22. Equation 4.8 can be rearranged to

$$\tau_c = 0.045(D_m/\bar{D}_B)^{-0.6}(\rho_s - \rho)gD_m \tag{4.14}$$

which reveals that it can be plotted as a series of lines of τ_c versus D_m as in Figure 4.22, with one line for each value of $D_m/\bar{D}_B$. The line for $D_m/\bar{D}_B = 1$ corresponds to the constant θ_t portion of the original Shields curve in Figure 4.2. The empirical flow-competence equation of Costa approximately corresponds with eqn 4.14 when $D_m/\bar{D}_B < 5$, approaching a ratio of 1 (the Shields curve) for D_m on the order of 1 metre. The curve of Williams plotted in Figure 4.21 was established as the lower limit of the data scatter in a τ_c versus D_m diagram (Williams, 1983, figs 3 and 4), and corresponds to $D_m/\bar{D}_B = 13$ for equivalence with eqn 4.14. This explains why Kehew & Teller (1994) found that the flow-competence relationships of those studies yielded estimated flow stresses that were unrealistically high, the application being to flood deposits where the $D_m/\bar{D}_B$ ratios were significantly greater. It is seen in Figure 4.22 that the boulder-transport data of O'Connor correspond to eqn 4.14 when $D_m/\bar{D}_B$ is in the range 5 to greater than 100. The regression equation of O'Connor given in Figure 4.21 is approximately equivalent to eqn 4.14 and thus with eqns 4.8 and 4.9 when $D_m/\bar{D}_B = 80$. It is uncertain whether the boulder-transport data of O'Connor actually reflect continued agreement with eqns 4.8 and 4.9, which were empirically based on gravel-transport data where $D_m/\bar{D}_B < 30$ (Figure 4.16). The

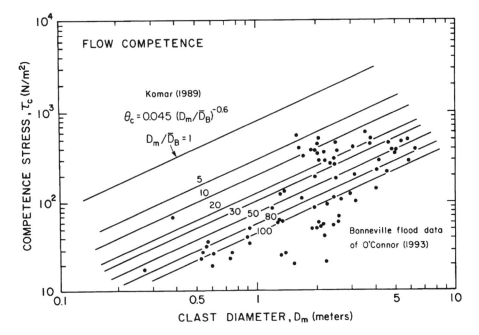

Figure 4.22   A series of lines for a range of $D_m/\bar{D}_B$ ratios calculated with eqn 4.14 which is equivalent to eqn 4.8 from Komar (1989a). Agreement with the boulder transport data of O'Connor (1993) would require high $D_m/\bar{D}_B$ ratios, implying that many of the boulders are"erratics" in being much coarser than the dominant bed material

transport of the larger boulders, when $D_m/\bar{D}_B$ is in excess of 50 to 100, must certainly represent the movement of "erratics", that is, individual clasts that are significantly coarser than the main distribution of grain sizes within the bed material. This distinction has been made by Krumbein & Lieblein (1956) with a precise definition of what constitutes an erratic within natural, coarse-grained deposits. Here the distinction should eventually be in terms of $D_m/\bar{D}_B$, the ratio of the largest particles transported to the median grain size of the bed material as a whole. When $D_m/\bar{D}_B$ is comparatively small, less than 30 to 50, the transported clasts (D_m) are likely to be part of the main distribution of grain sizes of the bed materials, and their entrainment is empirically predicted with eqns 4.8 and 4.9 while the actual process of entrainment involves the pivoting of particles over one another as analysed in the physical models developed below. At some stage when $D_m/\bar{D}_B$ is large the clast becomes an erratic, and its initiation of movement becomes more complex due to the high mobility of the supporting sediment which is much finer. Entrainment then depends on whether scour around the erratic results in its burial before the flow can roll it along the bed. Fahnestock & Haushild (1962) experimented with clasts having diameters ranging from 3 to 15 cm moving over sand beds. They found that the clasts would move downstream only if the flow was in the upper regime, that is, where the bedforms were plane bed or antidunes. With the lower-flow regime, the clasts were consistently buried within scour holes that developed around them. More research is needed that focuses on the $D_m/\bar{D}_B$ limits to which eqns 4.8 and 4.9 can be applied, and on the conditions under which erratics can be moved by floods.

PIVOTING MODELS FOR GRAIN ENTRAINMENT

The processes involved in the entrainment of sediments by a current are best considered through mechanical analyses which focus on the forces of fluid flow and particle resistance. The earliest models were developed by Rubey (1938) and White (1940), respectively analysing the fluid forces required to slide a grain up and over neighbouring particles and the condition of a grain pivoting over underlying particles. Recent analyses of this type have been conducted by Slingerland (1977), Phillips (1980), Naden (1987), Wiberg & Smith (1987), Komar & Li (1988), James (1990), Carling *et al.* (1992) and Bridge & Bennett (1992). The focus here will be on analytical relationships that have been developed to predict the entrainment of individual clasts from deposits of mixed sizes, with potential applications to flow-competence evaluations.

The mechanical analysis of grain entrainment involves balancing the moments of the forces acting on the grains as illustrated in Figure 4.23. First movement occurs when the fluid forces of drag (F_d) and lift (F_L) overcome the grain's immersed weight (W_t), pivoting the grain over the underlying particles through the pivot angle Φ so that the grain rotates out of its resting position. Most analyses have considered the mean stress exerted by the flow on the bed. However, in examinations of selective entrainment due to differences in particle sizes and densities, it is necessary to focus on the fluid forces acting directly on the individual grains. This requires a

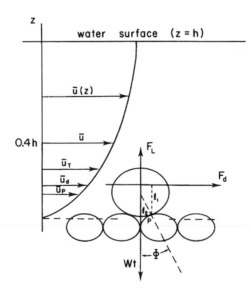

Figure 4.23 Analysis of a grain being entrained by a current of depth h and mean velocity $\bar{u}$, the flow producing lift (F_L) and drag forces (F_d) that overcome the grain"s immersed weight (Wt), pivoting it over underlying grains through the angle Φ

derivation of the threshold condition in terms of flow velocities and stresses acting at the grain level. These grain-level velocities can then be converted into equations for the mean-flow velocity or stress.

Through such analyses, Komar & Li (1988) obtained the following relationship for the velocity $\bar{u}_p$ at the level of the grain's centre (Figure 4.23):

$$\bar{u}_p = \Omega \left[\frac{4}{3} \frac{\rho_s - \rho}{\rho} gD_b \frac{\tan \Phi}{A + B(D_b/D_c)\tan \Phi} \right]^{1/2} \tag{4.15a}$$

where

$$A = C_d \Psi k_d^2 \tag{4.15b}$$

and

$$B = C_L k_L^2 \tag{4.15c}$$

Here D_b and D_c are respectively the intermediate and smallest axial diameters of the overall ellipsoidal-shaped particle, the Ω parameter was introduced to account for turbulence in the flow, C_d and C_L are drag and lift coefficients, k_d and k_L are factors of proportionality between the velocities $\bar{u}_d$ and $\bar{u}_T$ with $\bar{u}_p$ shown in Figure 4.23, and Ψ is a factor introduced to account for sheltering of the particle being entrained by neighbouring particles such that the drag force is reduced. Using the logarithmic relationship for the velocity profile for a fully rough boundary, the equation for the mean flow stress required to entrain the particle becomes

$$\tau_p = \frac{4\Omega^2/3}{[5.75 \log(30z_p/\zeta\bar{D}_B)]^2} \left[\frac{4}{3} \frac{\rho_s - \rho}{\rho} gD_b \frac{\tan \Phi}{A + B(D_b/D_c)\tan \Phi} \right]^{1/2} \tag{4.16}$$

where z_p is the elevation of the particle's centre above the bed (i.e. above the $u = 0$ level of the velocity profile), and $\bar{D}_B$ is the median diameter of the bed material, representing the overall bed roughness (i.e. the roughness coefficient $k_s = \zeta\bar{D}_B$). In applications, $z_p \approx 0.6(0.8D_b)$ based on a representative particle shape (Komar & Li, 1988).

Equations 4.15 and 4.16 are typical of relationships derived in grain-pivoting models for particle entrainment, and show the expected dependence of the flow velocity and stress on the density difference between the particle and fluid, $\rho_s - \rho$, on the overall size of the particle as represented by its intermediate axial diameter, D_b, on the grain's shape or flatness, D_b/D_c, on the particle's projection distance above the bed (z_p), and on its pivoting angle Φ over the underlying grains (in the case of entrainment by sliding, $\tan\Phi$ becomes the friction coefficient). These factors include those important to the density and size-sorting of grains during entrainment. The larger the particle perched atop other particles, the further it extends up into the flow (z_p) and in general the greater its exposure to the flow since the velocity within the boundary layer increases upward (Figure 4.23). A parallel effect is the decrease in the pivoting angle Φ as the ratio of the diameter of the perched grain to the underlying grains increases. The greater exposures and reduced pivoting angles of the larger particles act in concert to enhance the entrainability of the larger grains.

However, the value of the diameter D_b for the larger particles would increase their resistance to movement.

Miller & Byrne (1966) and Li & Komar (1986) have undertaken laboratory measurements to examine how the pivoting angle depends on grain size, shape and factors such as imbrication. Their measurements were used to establish the relationship

$$\Phi = \alpha \left(\frac{D}{K}\right)^{-\beta} \tag{4.17}$$

where D is the diameter of the pivoting grain, K is the diameter of the underlying grains over which the pivoting occurs, and α and β are empirical coefficients. This equation shows that Φ is inversely proportional to D/K, the ratio of the pivoting grain to the underlying grains. This dependence means that for a fixed K, the magnitude of the pivoting angle decreases as D increases, apparent from the geometry of the grains in Figures 4.10 and 4.23. Miller & Byrne (1966) and Li & Komar (1986) provide values for the α and β coefficients for spheres, ellipsoidal grains with and without imbrication, and for angular particles. Figure 4.24 shows the results for ellipsoidal particles, where the grain diameters are expressed as intermediate axial diameters (D_b and K_b). The analyses further indicated that grain movement as pivoting versus sliding depends on the ratio of the particle's smallest axial diameter to its intermediate diameter (D_c/D_b), a measure of the grain's "flatness".

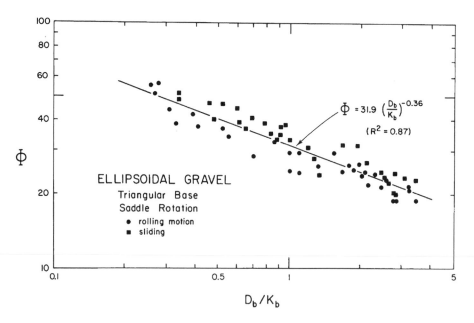

Figure 4.24 Measured values of the pivoting angles Φ for ellipsoidal gravel particles, where the grain diameters are expressed as intermediate axial diameters (D_b and K_b) (from Li & Komar, 1986)

Komar & Li (1988) compared the model of eqn 4.16, utilizing eqn 4.17 to evaluate the pivoting angle, with the data of Milhous (1973), Carling (1983) and Hammond *et al.* (1984) for the movement of the largest particle size D_m. The series of curves in Figure 4.25 represent two pivoting analyses with $\beta = 0.2$ and 0.3 in eqn 4.17, an analysis involving pure sliding, and one for pure pivoting. The comparison is most favourable for the pivoting model with $\beta = 0.2$, but given the scatter of the data and uncertainties in choosing values for drag coefficients (C_d and C_L) and the other parameters in the relationships, the seeming agreement in Figure 4.25 can hardly be taken as confirmation of the models, only perhaps suggestive of their use in flow-competence evaluations.

Accurate tests of the grain-pivoting, process-based models require laboratory flume data where one can determine exactly the pivoting conditions of the moving particle and the instant when first movement occurs. Fenton & Abbott (1977) undertook such experiments that focused on documenting the importance of grain protrusion. The experiments were with a fixed bed of regularly spaced grains, with one movable grain that could be pushed upward, progressively increasing its protrusion distance until movement was initiated by the flowing water. Their data are plotted in Figure 4.26 as the Shields entrainment function, expressed in terms of the diameter D of the moving particle versus P/K, where P represents the height of the top of the entrained particle over the tops of the supporting particles of uniform

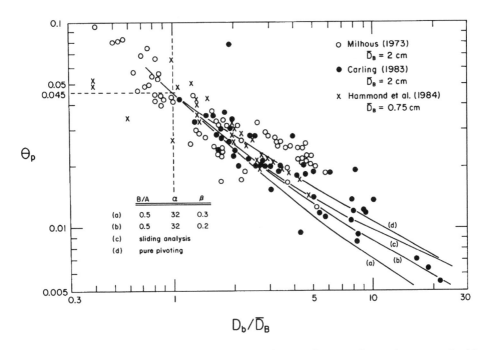

Figure 4.25 The model of eqn 4.16 for the selective entrainment of gravel, compared with field data. In the case of grain pivoting, the pivoting angles Φ were derived from eqn 4.17 and therefore depend on $D_b/K \approx D_m/\overline{D}_B$. The sliding analysis assumes a fixed value for $\tan\Phi$ which becomes a friction coefficient (after Komar & Li, 1988)

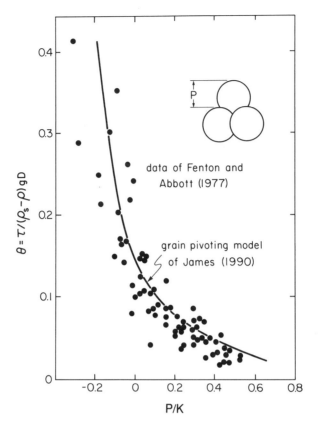

Figure 4.26 The data of Fenton & Abbott (1977) for the Shields entrainment parameter θ as a function of the ratio of the protrusion distance P of the moving grain above the level of the base grains of uniform diameter K. The curve is based on the theoretical grain-pivoting model derived by James (1990) (after James, 1990)

diameter K. As expected, the greater the ratio P/K the lower the θ required for entrainment due to the greater projection and exposure of the particle to the flow.

The curve in Figure 4.26 compared with the data of Fenton & Abbott (1977) was computed by James (1990) based on grain-pivoting models and relationships similar to eqns 4.15 and 4.16. James has developed the most detailed analysis to date that yields analytical rather than numerical solutions, and provides a thorough review of what is known regarding the relative importance of drag versus lift forces, and other factors important to the analysis. He also undertook a detailed theoretical analysis of the pivoting of moving spheres over base-layer spheres, a case of special interest to him as this condition corresponded to idealized experiments he undertook in a flume. Figure 4.27 compares model predictions with laboratory results for a single sphere on a bed of spheres, showing excellent agreement. Also included is the empirical Shields curve for natural sediments, derived from the data compilations by Miller *et al.* (1977) and Yalin & Karahan (1979) and shown in Figure 4.2. Again, good agreement is found between the empirical curve and the pivoting model of James

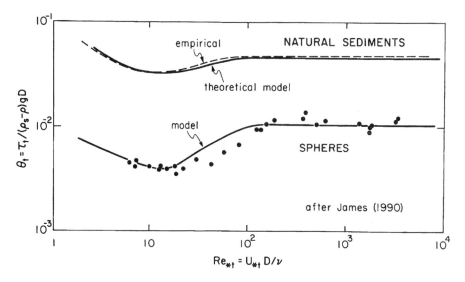

Figure 4.27 Comparisons between the grain-pivoting model of James (1990) and the laboratory results for the entrainment of single spheres on a bed of fixed spheres, and the empirical Shields curve for natural sediments (Figure 4.1) (after James, 1990)

(1990), the model having used a semiempirical coefficient that represents a combined measure of the effective grain protrusion and mean pivoting angle, reflecting different degrees of bed packing. This packing factor accounts for the higher θ values required for the entrainment of natural sediments versus the isolated spheres on a bed of spheres. Use of that same semiempirical coefficient in the model yielded the curve in Figure 4.26 compared with the data of Fenton & Abbott (1977).

Figure 4.28 shows the experimental results of James (1990) for the movement of spheres of variable sizes D over a bed of fixed spheres of diameter K. This graph is analogous to Figure 4.16 which yielded the flow-competence relationships, eqns 4.8 and 4.9. However, here the slope of the data trend for the entrainment of spheres is close to 45°. This would yield a -1 exponent in a regression of θ versus D/K, and in this case does represent a condition of equal mobility in terms of all sphere sizes being entrained at effectively the same flow stress. The curve shown in Figure 4.28 for spheres was computed by James from his grain-pivoting model, and shows good agreement with the data and in effect predicts that equal mobility should approximately exist in this idealized condition for the movement of spheres over base spheres. This was also the conclusion of Wiberg & Smith (1987) based on their models of grain entrainment. However, the theoretical curve of Figure 4.28 calculated by James for natural sediments represents a condition of non-equal mobility, and when $D/K > 1$ his theoretical curve is essentially congruent with the empirical curve shown in Figure 4.16 established by Komar (1989) for selective entrainment within gravels.

Similar experiments to those of James (1990) have been conducted by Carling *et al.* (1992), but employing regularly shaped particles (rods, ellipsoids, discs and

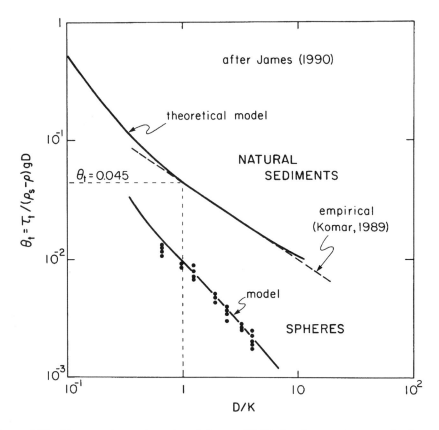

Figure 4.28   The experimental results of James (1990) for the entrainment of spheres of diameter D over a fixed bed of spheres of diameter K, yielding a condition of equal mobility $(-1$ slope) and agreement with his grain-pivoting model. Also shown are the model results for natural sediments, predicting a condition of non-equal mobility and close agreement with the empirical results in Figure 4.16 (after James, 1990)

cuboids) as well as spheres. The bed consisted of fixed roughness elements, gravel to cobbles cemented in place. The results were not compared with grain-pivoting models, but do illustrate that grain shape is important to entrainment and that the grains often vibrate and readjust their positions within bed pockets prior to actually being entrained. Such processes will have to be considered in further refinements of pivoting models. Carling *et al.* stressed the importance of particle-shape variability and grain orientations such as imbrication in yielding a range of entrainment stresses for particles that otherwise have the same weight.

Bridge & Bennett (1992) have developed a mathematical model for the entrainment and bedload transport of sediments, a model which accounts for different grain sizes, shapes and densities. The entrainment portion of the model is based on a grain-pivoting analysis, and includes considerations of turbulent fluctuations in fluid drag and lift forces acting on the particles. The entrainment analysis is first compared with the standard Shields curve (Figure 4.2) for the threshold of uniform grains, and then with available data sets for the entrainment of different grain sizes from deposits of mixed sizes. The model shows good agreement with the data, and Bridge & Bennett provide analyses which test the sensitivity of the numerical results to the various parameters included in the entrainment calculations (e.g. roughness coefficients and pivoting-angle coefficients). Their paper is particularly interesting in demonstrating how grain-entrainment analyses can be incorporated into a broader model of sediment transport, one that calculates transport rates of the grains depending on their sizes, shapes and densities.

Kirchner *et al.* (1990) and Buffington *et al.* (1992) have stressed the importance of the total distribution of pivoting-angle values, rather than just the mean value given by eqn 4.17, noting that it will be the particles having the lowest Φ values within the distribution that will tend to move first. The measurements of Kirchner *et al.* were made as part of a flume study of gravel transport, working with a poorly sorted mixture (range 1 to 12 mm). Buffington *et al.* made their measurements in Wildcat Creek, California, the first attempt to determine pivoting angles for a naturally formed gravel in a stream. In both studies a portion of the bed was fixed with a glue that did not alter the surface texture, and this bed surface was then transferred intact to the laboratory. Direct measurements of pivoting angles involved the placement of test particles of various sizes and shapes randomly on the surface, and then tilting the bed until the grain pivoted out of position. Due to the irregularity of the surface, any test particle of a specific size would have a range of pivoting angles depending on where it fell onto the surface. The medians of the pivoting angles largely agreed with eqn 4.17 and the results of Miller & Byrne (1966) and Li & Komar (1986); Kirchner *et al.* (1990) found $\beta = 0.31$ for the water-worked deposit in the flume, while the field values of Buffington *et al.* (1992) ranged β from 0.21 to 0.28. However, more important are the total ranges of pivoting angles for a specific experiment. Replicate measurements for a given test grain/bed surface combination typically span a range of 40 to 60°. Figure 4.29 shows the results of Kirchner *et al.* for the series of percentiles n within the distributions of measured pivoting angles. It is important that for a specific D/K_{50} ratio, where K_{50} is the median diameter of the fixed bed, there is a large range of pivoting angles that forms a distribution around the median Φ_{50}. Kirchner *et al.* derived the relationship

$$\Phi_n = (30 + 0.5n)\left(\frac{D}{K_{50}}\right)^{-0.3} \tag{4.18}$$

for the series of curves shown in Figure 4.29 for the individual percentiles. Based on their field data, Buffington *et al.* modified this to

$$\Phi_n = (25 + 0.57n)(\sigma)^{-(0.21 + 0.0027n)}\left(\frac{D}{K_{50}}\right)^{-(0.16 + 0.0016n)} \tag{4.19}$$

which indicates that the proportionality depends on the sorting coefficient σ of the bed material, and the exponent of D/K_{50} changes for the series of n percentiles.

Kirchner *et al.* (1990) also analysed the statistics of the protrusions and exposures of grains on the fixed bed derived from their flume experiments. This involved the "placement" of ideal circular discs atop the measured streamwise bed surface

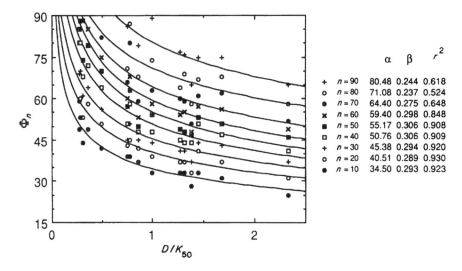

Figure 4.29   The measurements of Kirchner *et al.* (1990) for the series of percentiles *n* within the distributions of measured pivoting angles. For a specific D/K_{50} ratio, where K_{50} is the median diameter of the fixed bed, there is a large range of pivoting angles that forms a distribution around the median Φ_{50}. The series of curves for the several percentiles correspond to the empirical eqn 4.18 (from Kirchner *et al.*, 1990)

topography as illustrated in Figure 4.30. As expected, the grain projection and exposure vary markedly with both grain size and packing geometry. Large grains project further above the surface and have a greater exposure above upstream obstacles, both directly because they are larger and indirectly because their size makes them less likely to fall into gaps between other grains. They found that nearly half of the small grains do not project above the mean bed level, and nearly all are sheltered by large grains upstream. One surprising result is that the measurements show no relationship between the pivoting angle and the protrusion of individual grains for a given bed type and grain size. This is counter-intuitive in view of the schematic Figure 4.10 where the larger the grain the greater its protrusion above the bed and the smaller its pivoting angle.

Due to the variability in pivoting angles and grain protrusion within a given bed surface, there will be a corresponding distribution of critical entrainment stresses for a certain grain size. Kirchner *et al.* (1990) and Bufffington *et al.* (1992) developed models comparable to eqns 4.15 and 4.16, but where the pivoting angle Φ and protrusion distance z_p are represented by probability distributions rather than mean values. Figure 4.31 from Buffington *et al.* illustrates the computed entrainment stresses for a series of D sizes on a fixed K_{50}; similar graphs are presented by Buffington *et al.* for other K_{50} values, and Kirchner *et al.* present comparable results based on the flume bed material. The computed distributions of Figure 4.31 indicate that the smallest grain sizes ($D = 4.5$ mm) require the highest entrainment stresses; this result being due to their higher pivoting angles than the larger grains and because they are more sheltered. Theré is also a greater range of entrainment stresses for the smaller grain sizes, whereas the coarsest fraction ($D = 33$ mm) has a comparatively

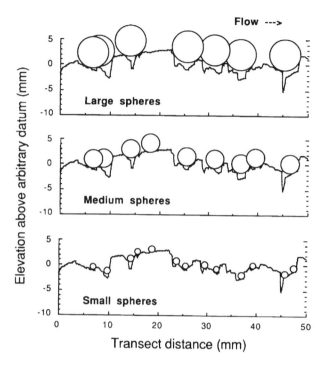

Figure 4.30 The "placement" of ideal circular discs atop the measured streamwise bed surface topography from a flume experiment, the approach used to analyse the statistics of the protrusions and exposures of grains on the fixed bed (from Kirchner *et al.*, 1990)

narrow range of critical entrainment stresses. For each fraction, first movement is represented by the percentage close to 0, that having the lowest required entrainment stress (presumably being the grains with the smallest pivoting angles, greatest projections, and minimal sheltering by nearby grains). As discussed by Kirchner *et al.* and Bufffington *et al.*, the fact that the distribution curves for the series of *D* sizes tend to converge at their low stress ends suggests an equal mobility with respect to all sizes being entrained at approximately the same flow stress. However, as noted by Buffington *et al.*, this apparent equal mobility conflicts with their on-going measurements of grain entrainment in Wildcat Creek, experiments with tagged particles which show that the larger sizes require higher flow stresses and discharges for initiation of movement, the typical flow-competence relationship. They suggest that this discrepancy between their analysis based on theoretical pivoting models and direct measurements with tagged particles results from the effects of packing, partial burial of the larger clasts, and local grain protrusion-induced flow accelerations. Because of such effects, they further recommend that field-based studies which attempt to define particle pivoting angles and protrusion should be done, if possible, on grains already resting on the bed, rather than on grains placed on the bed.

The use of tagged particles in natural gravel-bed streams and in the marine environment has the potential for further testing and advancing the use of pivoting models for grain entrainment. There have already been several investigations to

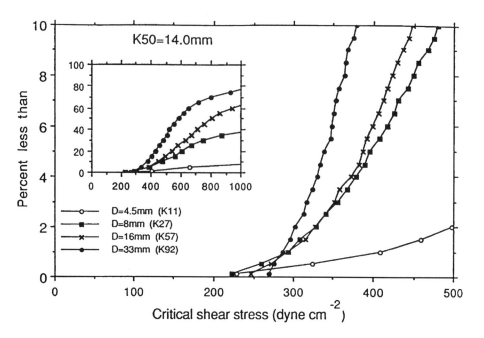

Figure 4.31 Calculated distributions of the critical entrainment stresses for test grains (size D) on a surface having the median diameter $K_{50} = 14.0$ mm (from Buffington *et al.*, 1992)

document the entrainment of individually tagged gravel and cobbles placed on the bed of a river. Ritter (1967) used painted cobbles in the Middle Fork of the Eel River, California, placed across a river section at a gauging station. An approximate relationship was found between the cobble diameter and average velocity of flow required for entrainment, and a reasonable correlation was established with other data sets on particle threshold. The investigation that has made the greatest effort thus far to use tagged cobbles to test entrainment models such as eqns 4.15 and 4.16 was that of Helley (1969). The measurements were made in Blue Creek, a small tributary to the Klamath River in northern California. Rocks of various sizes, shapes and lithologies were used as test particles, and local water velocities were measured with portable current meters. Unfortunately, the ranges of grain sizes used and their resulting entrainment velocities were relatively narrow and therefore did not establish good trends. However, Helley demonstrated that the data show good agreement with a grain-pivoting model that expressed the threshold condition in terms of the "bed velocity", that is, the velocity of flow at the level of the cobble (at $0.6D_a$ above the mean level of the bed, where D_a is the grain's shortest axial diameter oriented perpendicular to the bed). The data of Helley also show an order-of-magnitude agreement with a simplified version of eqn 4.16 that assumes a fixed average value for the pivoting angle Φ.

As part of his research on gravel transport in Great Eggleshope Beck, Carling (1983) employed tagged cobbles as well as catching the transported gravel in a total bedload trap. The combined data have been employed in Figures 4.16 and 4.17 to establish the empirical eqns 4.8 and 4.9 for flow-competence evaluations, and in

Figure 4.25 as a test of eqn 4.16 for grain entrainment by pivoting or sliding. Of significance is that the correlations established by the D_m largest particles captured in the bedload trap are consistent with the tagged cobbles which are still larger and moved only during the most extreme flow stages that occurred during the field study of Carling. The advent of new technology for monitoring the movement of naturally magnetic or artificially magnetized gravel particles increases the potential for obtaining measurements required to better establish the equations for evaluating grain entrainment and flow competence (Ergenzinger & Custer, 1983; Reid et al., 1984; Custer et al., 1987; Bunte, 1992; Schmidt & Ergenziner, 1992).

SUMMARY

Considerations of the threshold of sediment motion have focused mainly on the entrainment of particles from uniform deposits consisting of grains having the same density and a narrow range of sizes. Considerable effort has gone into the collection of data from laboratory flumes where the sediments and flows can be carefully controlled. The products of these efforts have been our standard grain-threshold curves such as Figure 4.1 for quartz-density sand and the more broadly applicable Shields curve of Figure 4.2. However, the relevance of these empirical curves to natural sediments is limited since deposits found in rivers and marine environments typically contain particles with wide ranges of sizes and densities. Important in these natural sediments are the processes of selective entrainment, where grains having contrasting sizes and densities are entrained at different flow velocities or bed stresses. Selective entrainment, particularly by size, is an important factor in producing observed variations in the grain-size distributions of bedload in rivers and in sorting patterns of the bed material. In sands, selective entrainment due to contrasting particle densities (and sizes) commonly leads to concentrations of heavy minerals as lag deposits since the low density quartz and feldspar grains are most easily picked up by the flowing water; in the extreme, this sorting process can lead to the development of black-sand placers.

In sand, the contrasting densities of the particles largely control the sorting of the various minerals within the deposit, there being a selective entrainment of the light minerals (quartz and feldspars) versus the series of heavy minerals. However, sorting by grain size also appears to be important, particularly due to the inverse relationship between the mineral densities and grain sizes commonly found in sand. Field data for the observed sorting patterns and laboratory experiments on grain entrainment for sand-size particles indicate that with a range of grain sizes, the larger particles are more easily entrained.

The sorting pattern by grain size in gravel is opposite to that found in sand, illustrated by the schematic diagram in Figure 4.32. Field and laboratory data demonstrate that in the entrainment of gravel from deposits of mixed sizes, the greater the flow strength (velocity, discharge or bed stress) the coarser the material entrained from the bed. This is generally seen as a progressive coarsening of the bedload grain-size distributions, with a shift in the diameter D_m of the maximum particle sizes found in the bedload being part of that general coarsening. The

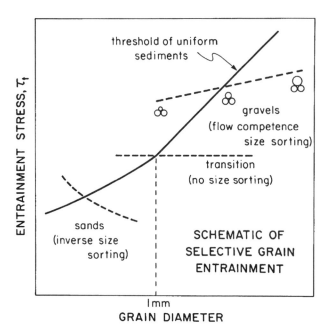

Figure 4.32 A schematic diagram illustrating the contrasting patterns of selective entrainment from deposits of mixed sizes for sand versus gravel

threshold consideration then becomes a flow-competence evaluation, relating the flow's discharge or stress to D_m. Much attention has been devoted to establishing flow competence relationships due to their important application in the evaluation of the hydraulics of extreme floods from the largest sizes of cobbles and boulders transported by the flood.

The contrasting patterns of selective size entrainment in sand versus gravel (Figure 4.32) have been attributed to different dynamics of water flow near the sediment bed, with fully turbulent flow extending to the interface in the case of the entrainment of gravel, while a viscous sublayer becomes important to the entrainment of fine-grained sediments. This needs to be further investigated, particularly in detailed flume experiments. Such experiments could also provide an improved understanding of pivoting models for grain entrainment, models that focus on the actual physics of the initiation of first-grain movement. In addition to flume experiments, there is a need for field programmes utilizing tagged gravel and cobbles to better establish the grain-pivoting models under natural conditions, and to better define the movement of "erratics" that are much larger than the dominant grain sizes of bed materials.

ACKNOWLEDGEMENTS

This work has been supported by the Planetary Geology and Geophysics Program of NASA, my research having the objective of better establishing flow-competence equations which can be used to quantitatively evaluate the hydraulics of floods on Mars that eroded the large outflow channels. I would like to thank the anonymous reviewer of this manuscript who provided many helpful suggestions for its improvement.

REFERENCES

Andrews, E.D. (1983). "Entrainment of gravel from naturally sorted riverbed material", *Geological Society of America Bulletin*, **94**, 1225–1231.

Andrews, E.D. & Erman, D.C. (1986). "Persistence in the size distribution of surficial bed material during an extreme snowmelt flood", *Water Resources Research*, **22**, 191–197.

Ashworth, P.J. & Ferguson, R.I. (1989). "Size-selective entrainment of bed load in gravel bed streams", *Water Resources Research*, **25**, 627–634.

Ashworth, P.J., Ferguson, R.l., Ashmore, P.E., Paola, C., Powell, D.M., & Prestegaard, K.L. (1992). "Measurements in a braided river chute and lobe: 2. Sorting of bed load during entrainment, transport and deposition", *Water Resources Research*, **28**, 1887–1896.

Baker, V.R., & Ritter, D.F. (1975). "Competence of rivers to transport coarse bedload materials", *Geological Society of America Bulletin*, **86**, 975–978.

Bathurst, J.C. (1987). "Critical conditions for bed material movement in steep, boulder-bed streams", In *Erosion and Sedimentation in the Pacific Rim* (eds, R.L. Beschta, T. Blinn, G.E. Grant, G.G. Ice & F.J. Swanson) IAHS Publ. No. **165**, pp. 309–318.

Bathurst, J.C., Graf, W.H. & Cao, H.H. (1982). "Initiation of sediment transport in steep channels with coarse bed material", *Euromech 156: Mechanics of Sediment Transport* (eds B.M. Sumer & A. Müller), Balkema, Rotterdam 207–213.

Bridge, J.S. & Bennett, S.J. (1992). "A model for the entrainment and transport of sediment grains of mixed sizes, shapes, and densities", *Water Resources Research*, **28**, 337–363.

Buffington, J.M., Dietrich, W.E. & Kirchner, J.W. (1992). "Friction angle measurements on a naturally-formed gravel streambed: Implications for critical boundary shear stress", *Water Resources Research*, **28**, 411–425.

Bunte, K. (1992). "Particle number grain-size composition of bedload in a mountain stream", In: *Dynamics of Gravel-Bed Rivers* (eds P. Billi, R.D. Hey, C.R. Thorne & P. Tacconi) John Wiley, Chichester, 55–72.

Carling, P.A. (1983). "Threshold of coarse sediment transport in broad and narrow natural streams", *Earth Surface Processes and Landforms*, **8**, 1–18.

Carling, P.A., Kelsey, A. & Glaister, M.S. (1992). "Effect of bed roughness, particle shape and orientation on initial motion criteria", In: *Dynamics of Gravel-Bed Rivers* (eds P. Billi, R.D. Hey, C.R. Thorne & P. Tacconi) John Wiley, Chichester, 23–37.

Costa, J.E. (1983). "Paleohydraulic reconstruction of flash-flood peaks from boulder deposits in the Colorado Front range", *Geological Society of America Bulletin*, **94**, 986–1004.

Custer, S.G., Bugosh, N., Ergenzinger, P.E. & Anderson, B.C. (1987). 'Electromagnetic detection of pebble transport in streams: A method for measurement of sediment transport waves", *Recent Developments in Fluvial Sedimentology*, Special Publication **39**, Soc. Econ. Paleontol. and Mineral., Tulsa, OK, 21–26.

Day, T.J. (1980). "A study of initial motion characteristics of particles in graded bed material", *Geological Survey of Canada Current Research*, Part A, Paper **80–1A**, 281–286.

Dhamotharan, S., Wood, A., Parker, G. & Stefan, H. (1980). "Bedload transport in a model gravel stream," *Project Report 190*, St. Anthony Falls Hydraulic Lab., Univ. of Minnesota.

Diplas, P. (1987). "Bedload transport in gravel-bed streams", *Journal of Hydraulic Engineering*, ASCE **113**, 277–292.

Egiazaroff, l.V. (1965). "Calculation of nonuniform sediment concentrations", *Journal of Hydraulic Engineering*, **91** (HY4), 225–247.

Einstein, H.A. (1950). *The bedload function for sediment transport in open channel flows*, Technical Bulletin **1026**, Soil Conservation Service, US Dept. of Agriculture, Washington, DC.

Ergenzinger, P.J., & Custer, S.G. (1983). "Determination of bedload transport using naturally magnetic tracers: First experiences at Squaw Creek, Gallatin County, Montana", *Water Resources Research*, **19**. 187–193.

Fahnestock, R.K. (1963). *Morphology and hydrology of a glacial stream – White River, Mount Rainier, Washington*, US Geological Survey Professional Paper **422-A**, 70 pp.

Fahnestock, R.K. & Haushild, W.L. (1962). "Flume studies of the transport of pebbles and cobbles on a sand bed," *Geological Society of America Bulletin*, **73**, 1431–1436.

Fenton, J.D. & and Abbott, J.E. (1977). "Initial movement of grains on a stream bed: the effect of relative protrusion," *Proceedings of the Royal Society of London, Series A*, **352**, 523–537.

Force, E.R. (1991). "Placer deposits", In: *Sedimentary and Diagenetic Mineral Deposits: A Basin Analysis Approach to Exploration* (eds E.R. Force, J.J. Eidel & J.B. Maynards), Reviews in Economic Geology, Society of Economic Geologists, **5**, 131–140.

Frihy, O.E. & Komar, P.D. (1991). "Patterns of beach-sand sorting and shoreline erosion on the Nile delta," *Journal of Sedimentary Petrology*, **61**, 544–550.

Grass, A.J. (1970). "Initial instability of fine bed sand", *Journal of the Hydraulics Division*, ASCE **96** (HY3), 619–632.

Grass, A.J. (1983). "The influence of boundary layer turbulence on the mechanics of sediment transport", *Euromech 156: Mechanics of Sediment Transport*, Balkema, Rotterdam, 3–17.

Grimm, C.l. & Leupold, N. (1939). *Hydraulic data pertaining to the design of rock revetments*, U.S. Engineering Office, North Pacific Division, Portland, Oregon, 32 pp (unpublished mimeograph).

Hammond, F.D.C., Heathershaw, A.D. & Langhorne, D.N. (1984). "A comparison between Shields' threshold criterion and the movement of loosely packed gravel in a tidal channel", *Sedimentology*, **31**, 51–62.

Hassan, M.A., Church, M. & Ashworth, P.J. (1992). "Virtual rate and mean distance of travel of individual clasts in gravel-bed channels," *Earth Surface Processes and Landforms*, **17**, 617–627.

Helland-Hansen, E., Milhous, R.T. & Klingeman, P.C. (1974). "Sediment transport at low Shields parameter values," *Journal of the Hydraulics Division, ASCE*, **100** (HY1), 261–265.

Helley, E.J. (1969). Field measurement of the initiation of large particle motion in Blue Creek near Klamath, California, *US Geological Survey Professional Paper 562-G*, 19 pp.

Helley, E.J. & Smith, W. (1971). *Development and calibration of a pressure-difference bedload sampler*, Open-File Report, US Geological Survey, Menlo Park, California.

Hjulström, F. (1935). "Studies of the morphological activity of rivers as illustrated by the River Fyris," *Geological Institute of the University of Upsala Bulletin*, **25**, 221–527.

Hjulström, F. (1939). "Transportation of detritus by moving water," In: *Recent Marine Sediments* (ed. P.D. Trask), American Association of Petroleum Geologists, Tulsa, 5–31.

Hooker, E.H. (1896). "Suspension of solids in flowing water," *Transactions of the American Society of Civil Engineers*, **36**, 239–324.

Hubbell, D.W. (1964). *Apparatus and techniques for measuring bedload*, US Geological Survey Water Supply Paper, **1748**, 74 pp.

Hubbell, D.W. (1987). "Bed load sampling and analysis," In: *Sediment Transport in Gravel-Bed Rivers* (eds C.R. Thorne, J.C. Bathurst & R.D. Hey), John Wiley, Chichester, 89–118.

Inman, D.L. (1949). "Sorting of sediments in the light of fluid mechanics", *Journal of Sedimentary Petrology*, **19**, 51–70.

James, C.S. (1990). "Prediction of entrainment conditions for nonuniform noncohesive sediments", *Journal of Hydraulic Research*, **28**, 25–42.

Kehew, A.E. & Teller, J.T. (1994)."Glacial-lake spillway incision and deposition of a coarse-grained fan near Watrous, Saskatchewan", *Canadian Journal of Earth Sciences* **31**, 544–553.

Kirchner, J.W., Dietrich, W.E., Iseya, F. & Ikeda, H. (1990)."The variability of critical shear stress, friction angle, and grain protrusion in water-worked sediments," *Sedimentology*, **37**, 647–672.

Komar, P.D. (1987a)."Selective grain entrainment by a current from a bed of mixed sizes: A reanalysis", *Journal of Sedimentary Petrology*, **57**, 203–211.

Komar, P.D. (1987b)."Selective gravel entrainment and the empirical evaluation of flow competence", *Sedimentology*, **34**, 1165–1176.

Komar, P.D. (1989a)."Flow-competence evaluations of the hydraulic parameters of floods: an assessment of the technique", In: *Floods: Hydrological, Sedimentological and Geomorphological Implications* (eds K. Beven & P. Carling, John Wiley, Chichester. 107–134.

Komar, P.D. (1989b)."Physical processes of waves and currents and the formation of marine placers", *CRC Critical Reviews in Aquatic Sciences*, **1**, 393–423.

Komar, P.D. (1992a)."Reply" to discussions of Komar & Shih (1992), in *Dynamics of Gravel-Bed Rivers* (ed P. Billi, R.D. Hey, C.R. Thorne & P. Tacconi), John Wiley, Chichester, 100–106.

Komar, P.D. (1992b)."Discussion" of Bunte (1992) in *Dynamics of Gravel-Bed Rivers* (eds P. Billi, R.D. Hey, C.R. Thorne & P. Tacconi) John Wiley, Chichester, 69–71.

Komar, P.D. & Carling, P.A. (1991)."Grain sorting in gravel-bed streams and the choice of particle sizes for flow-competence evaluations", *Sedimentology*, **38**, 489–502.

Komar, P.D. & Li, Z. (1988)."Applications of grain-pivoting and sliding analyses to selective entrainment of gravel and to flow-competence evaluations", *Sedimentology*, **35**, 681–695.

Komar, P.D. & Shih, S.-M. (1992)."Equal mobility versus changing grain sizes in gravel-bed streams," In: *Dynamics of Gravel-Bed Rivers* (eds P. Billi, R.D. Hey, C.R. Thorne & P. Tacconi) John Wiley, Chichester, 73–93.

Komar, P.D. & Wang, C. (1984)."Processes of selective grain transport and the formation of placers on beaches", *Journal of Geology*, **92**, 637–655.

Krumbein, W.C. & Lieblein, J. (1956)."Geological application of extreme-value methods to interpretation of cobbles and boulders in gravel deposits", *Transactions of the American Geophysical Union*, **37**, 313–319.

Kuhnle, R.A. (1992)."Fractional transport rates of bedload on Goodwin Creek", In: *Dynamics of Gravel-Bed Rivers* (eds P. Billi, R.D. Hey, C.R. Thorne & P. Tacconi) John Wiley, Chichester, 141–155.

Lane, E.W. & Carlson, E.J. (1953)."Some factors affecting the stability of canals constructed in coarse granular materials", *Proceedings International Association of Hydraulic Research*, 37–48.

Leopold, L.B. & Emmett, W.W. (1976)."Bedload measurements, East Fork River, Wyoming", *Proceedings of the National Academy of Sciences*, **73**, 1000–1004.

Leopold, L.B. & Emmett, W.W. (1977)."1976 bedload measurements, East Fork River, Wyoming", *Proceedings of the National Academy of Sciences*, **74**, 2644–2648.

Li, Z. & Komar, P.D. (1986)."Laboratory measurements of pivoting angles for applications to selective entrainment of gravel in a current", *Sedimentology*, **33**, 413–423.

Li, Z. & Komar, P.D. (1992a)."Longshore grain sorting and beach placer formation adjacent to the Columbia River", *Journal of Sedimentary Petrology*, **62**, 429–441.

Li, Z. & Komar, P.D. (1992b)."Flume experiments on the selective entrainment of mixed size and density sands", *Journal of Sedimentary Petrology*, **62**, 584–590.

Lord, M.L. & Kehew, A.E. (1987)."Sedimentology and paleohydrology of glacial-lake outburst deposits in southeastern Saskatchewan and northwestern North Dakota," *Geological Society of America Bulletin*, **99**, 663–673.

Mantz, P.A. (1973)."Cohesionless, fine graded, flaked sediment transported by water", *Nature (Physical Sciences)*, **246**, 14–16.

Mantz, P.A. (1977)."Incipient transport of fine grains and flakes by fluids – Extended Shields diagram", *Journal of the Hydraulics Division, ASCE*, **103**(HY6), 601–615.

Mantz, P.A. (1980). "Low sediment transport rates over flat bed", *Journal of the Hydraulics Division, ASCE*, **106**(HY7), 1173–1190.

Milhous, R.T. (1973). *Sediment transport in a gravel-bottomed stream*, PhD thesis, Oregon State Univ., Corvallis, 232 pp.

Miller, M.C., McCave, I.N. & Komar, P.D. (1977). "Threshold of sediment motion in unidirectional currents", *Sedimentology*, **24**, 507–528.

Miller, R. L. & Byrne, R.J. (1966). "The angle of repose for a single grain on a fixed rough bed," *Sedimentology*, **6**, 303–314.

Misri, R.L., Garde, R.J. & Ranga Raju, K.G. (1984). "Bed load transport of coarse nonuniform sediment", *Journal of Hydraulic Engineering*, **110**(HY3), 312–328.

Naden, P. (1987). "An erosion criterion for gravel-bed rivers", *Earth Surface Processes and Landforms*, **12**, 83–93.

Neill, C.R. (1967). "Mean-velocity criterion for scour of coarse uniform bed material," *International Association of Hydraulic Research, 12th Congress*, **3**, 46–55.

Neill, C.R. (1968) "Note on initial movement of coarse uniform bed material," *Journal of Hydraulic Research*, **6**, 173–176.

Neill, C.R. & Yalin, M.S. (1969). "Quantitative definition of beginning of bed movement", *Journal of the Hydraulics Division, ASCE*, **95**(HY1), 585–587.

O'Connor, J.E. (1993). *Hydrology, hydraulics, and geomorphology of the Bonneville Flood*, Geological Society of America Special Paper **274**, 83 pp.

Paintal, A.S. (1971). "Concept of critical shear stress in loose boundary open channels", *Journal of Hydraulic Research*, **9**, 91–113.

Parker, G., Klingeman, P.C. & McLean, D.G. (1982). "Bedload and size distribution in paved gravel-bed streams", *Journal of the Hydraulics Division*, ASCE, **108**(HY4), 544–571.

Petit, F. (1994). "Dimensionless critical shear stress evaluation from flume experiments using different gravel beds", *Earth Surface Processes and Landforms*, **19**, 565–576.

Phillips, M. (1980). "A force balance model for particle entrainment into a fluid stream", *Journal of Physics, D. Applied Physics*, **13**, 221–233.

Reid, I., Brayshaw, A.C. & Frostick, L.E. (1984). "An electromagnetic device for automatic detection of bedload motion and its field applications", *Sedimentology*, **31**, 269–276.

Ritter, J.R. (1967). *Bed-material movement, Middle Fork Eel River, California*, Geological Survey Professional Paper **575-C**, C219–C221.

Rubey, W.W. (1938). *The force required to move particles on a stream bed*, Geological Survey Professional Paper **189-E**, 121–141.

Schmidt, K.-H. & Ergenzinger, P. (1992). "Bedload entrainment, travel lengths, step lengths, rest periods – studied with passive (iron, magnetic) and active (radio) tracer techniques", *Earth Surface Processes and Landforms*, **17**, 147–165.

Scott, K.M. & Gravlee, G.C. (1968). *Flood surge on the Rubicon River, California: hydrology, hydraulics and boulder transport*, U.S. Geological Survey Professional Paper **422-M**, 38 pp.

Shields, A. (1936). "Anwendung der Ahnlichkeitsmechanik und der Turbulenzforschung auf die Geschiebebewegung Mitteilungen der Preuss", *Versuchanst fur Wasserbau und Schiffbau*, **26**, 26 pp (translated by W.P. Ott & J.C. van Uchlen, US Dept. Agriculture, Soil Conservation Service Coop. Lab., California Institute of Technology).

Shih, S.-M. & Komar, P.D. (1990a). "Hydraulic controls of grain-size distributions of bedload gravels in Oak Creek, Oregon, USA", *Sedimentology*, **37**, 367–376.

Shih, S.-M. & Komar, P.D. (1990b). "Differential bedload transport rates in a gravel-bed stream: A grain-size distribution approach", *Earth Surface Processes and Landforms*, **15**, 539–552.

Slingerland, R.L. (1977). "The effects of entrainment on the hydraulic equivalence relationships of light and heavy minerals in sands", *Journal of Sedimentary Petrology*, **47**, 753–770.

Slingerland, R. & Smith, N.D. (1986). "Occurrence and formation of water-laid placers", *Annual Reviews of Earth and Planetary Sciences*, **14**, 113–147.

Trask, C.B. & Hand, B.M. (1985). "Differential transport of fall-equivalent sand grains, Lake Ontario", *Journal of Sedimentary Petrology*, **55**, 226–234.

White, C.M. (1940). "The equilibrium of grains on the bed of a stream", *Proceedings of the Royal Society of London, Series A*, **174**, 332–338.

White, S.J. (1970). "Plane bed thresholds of fine grained sediments", *Nature*, **228**, 152–153.

Wiberg, P.L. & Smith, J.D. (1987). "Calculations of the critical shear stress for motion of uniform and heterogeneous sediments", *Water Resources Research*, **23**, 1471–1480.

Wilcock, P.R. (1992). "Flow competence: A criticism of a classic concept", *Earth Surface Processes and Landforms*, **17**, 289–298.

Wilcock, P.R. & McArdell, B.W. (1993). "Surface-based fractional transport rates: Mobilization thresholds and partial transport of a sand-gravel sediment", *Water Resources Research*, **29**, 1297–1312.

Wilcock, P.R. & Southard, J.B. (1988). "Experimental study of incipient motion in mixed-size sediment", *Water Resources Research*, **24**, 1137–1151.

Williams, G.P. (1983). "Paleohydrological methods and some examples from Swedish fluvial environments, 1 – Cobble and boulder deposits", *Geografiska Annaler*, **65A**, 227–243.

Wohl, E.E., Greenbaum, N., Schick, A.P. & Baker, V.R. (1994). "Controls on bedrock channel incision along Nahal Paran, Israel", *Earth Surface Processes and Landforms*. **19**, 1–13.

Yalin, M.S. (1972). *Mechanics of Sediment Transport*, Pergamon Press, Oxford, 290 pp.

Yalin, M.S. & Karahan, E. (1979). "Inception of sediment transport", *Journal of the Hydraulics Division, ASCE 105*(HY11), 1433–1443.

5 Unsteady Transport of Sand and Gravel Mixtures

ROGER A. KUHNLE
National Sedimentation Laboratory, Oxford, Mississippi, USA

INTRODUCTION

Unsteadiness in the transport of sand and gravel sediment mixtures in alluvial channels is known to occur over a wide range of time scales. Periods of fluctuation from minutes to days have been documented in studies of field and laboratory channels. Longer term fluctuations caused by factors external to river systems undoubtedly also exist, yet are more difficult to quantify. Several periods of fluctuation may be present in the same system at the same time with each controlled by a different process. Knowledge of the processes causing unsteadiness in alluvial channels is necessary to design effective sampling strategies which will yield reliable estimates of the mean transport rate for a given set of sediment and flow conditions.

Unsteadiness in the transport of sand and gravel mixtures has been shown to be present in a variety of alluvial channels. Large fluctuations in transport rate with time have been measured by a number of researchers in field streams (Meade, 1985; Bunte, 1992; Reid *et al.*, 1985; Kang, 1982; Whiting *et al.*, 1988; Dinehart, 1989; Tacconi & Billi, 1987; Kuhnle *et al.*, 1989), in model studies of braided channels (Hoey and Sutherland, 1991; Ashmore, 1988), and in flume studies with single channels (Hubbell, 1987; Iseya & Ikeda, 1987; Kuhnle & Southard, 1988; Wilcock & Southard, 1989). When transport rates are measured at closely spaced time intervals in streams with gravel and sand beds, the presence of large fluctuations in bedload transport rate appears to be the rule rather than the exception.

The unsteady transport of sand and gravel sediment mixtures will be addressed by this chapter in three parts: a survey of measured unsteadiness of transport, the processes that cause unsteadiness in transport, and probability distributions associated with bedload transport.

STATEMENT OF THE PROBLEM

The transport of cohesionless sediment by flowing water in an alluvial channel is a complicated process. Quantifying the resisting and entraining forces on the sediment

Advances in Fluvial Dynamics and Stratigraphy. Edited by P.A. Carling and M.R. Dawson.
© 1996 John Wiley & Sons Ltd.

grains is complicated by the uncertainty of the flow field at the level of the sediment grains, the differing exposures of the grains to the flow, and the heterogeneity of the size and packing of grains on the bed surface. When transport of the sediment begins the size distribution of the bed surface sediment may change. With increasing flows bedforms of different types may appear on the bed which affects the flow field and may also affect the size distribution of the bed surface.

Complicating the process of fluvial transport of sand and gravel mixtures is the fact that the sediment mixture will often have a bimodal size distribution (Pettijohn, 1975, p. 158). The presence of a strongly bimodal size distribution in the bed material has been shown to significantly affect the transport of the bed sediment (Wilcock, 1992, 1993; Kuhnle, 1993a,b). For strongly bimodal sediments the two modes tend to be transported independently of one another at low flows and tend to interact with each other during transport at high flows.

Although it is clear that the motion of the bed material in an alluvial channel is driven by the force of the flow on the grains, the best way to measure and characterize the flow strength is less clear. For this study the rate of bedload transport (q_b) will be assumed to be a function of the bed shear stress (τ_0):

$$q_b = f(\tau_0) \tag{5.1}$$

With other things being equal, eqn 5.1 predicts that greater transport rates will result from greater shear stresses. A fluctuation in the rate of sediment transport in a channel in which the average flow and sediment quantities are constant can happen in at least two ways. There could be some cyclical increase and decrease in flow velocity and bed shear stress that occurs on the same time scale as the fluctuations in sediment transport rate, or there could be a repetitive change in the bed surface (in either elevation or roughness) which would cause the shear stress to increase and decrease in response to the changes in the depth of the flow or the roughness of the bed surface. Actually these processes are related and can serve to increase each other. Some theories on the formation of dune bedforms rely on an initial bed disturbance and the large scale structures of flow turbulence to explain the scale of the resulting bedforms (Yalin, 1972). A wealth of data from sandbed streams relates the fluctuation of sediment transport rates to the migration of bedforms (Gomez et al., 1989). Recent data from studies of channels with sand and gravel sediments also relate fluctuations in sediment transport to bedforms (Iseya & Ikeda, 1987; Kuhnle & Southard, 1988; Whiting et al., 1988; Dinehart, 1992). As a working hypothesis some type of bed or channel form will be assumed to be responsible for the fluctuations that have been shown to occur in channels with bed sediments composed of sand and gravel. In the studies considered here, with the exception of the studies by Emmett et al. (1983) and Meade (1985), unsteady flow was likely not the cause of the fluctuations in bedload transport.

Adequately characterizing the sizes of sediment available for transport is a diffi-cult task. This problem is further compounded for sediments with size distributions other than unimodal. A measure of mean sediment diameter is commonly assumed to be sufficient to characterize a unimodal bed material size distribution for sediment transport predictions. For sediments with bimodal or multimodal size distributions, the mean sediment size may constitute only a very small percentage of the total sediment mix and will probably not adequately characterize the sediment. Recent work (Wilcock, 1992; Kuhnle, 1993b) suggests that information on the absolute size of the total sediment mix, the size of each fraction relative to the whole mix, the degree of separation in size of the two modes, the amount of material in the size fractions between the modes, and the proportion in each mode is necessary to characterize the transport of strongly bimodal sediments. Wilcock (1993) proposed a bimodality parameter (B):

$$B = \left(\frac{D_c}{D_f}\right)^{1/2} \Sigma P_m \qquad (5.2)$$

where D_c is grain size of the coarse mode, D_f is grain size of the fine mode, and ΣP_m is the proportion of the sediment mixture contained in the two modes. Several data sets have shown that sediments with a bimodality parameter (B) greater than 1.7 are entrained as bimodal sediments, while those with $B < 1.7$ are entrained as unimodal sediments (Wilcock, 1993).

Collecting bedload samples that yield accurate mean transport rates with a high probability is also a problem when the transport rate is fluctuating (Kuhnle & Southard, 1988). Bedload transport rates vary spatially as well as temporally and this must be taken into account when sampling programmes are designed (Hubbell, 1987).

SURVEY OF TRANSPORT DATA

A list of studies that found evidence for fluctuations in sediment transport rates is contained in Table 5.1. The studies are divided according to whether they were field or laboratory studies, and also whether the channel was single- or multiple-thread on the basis that different processes may be acting according to the scale and type of channel pattern in each study. A further division of the studies may be made as to whether the flow was steady or unsteady and whether the sediment transport samples were collected at a point or over the whole cross-section. The flow was steady in all of the laboratory studies listed in Table 5.1, and with only one exception the fluctuations in transport rate in the field studies were apparently not related to the flow unsteadiness. In the study of Meade (1985) the rising and falling of the flow stage was closely related to the sediment transport fluctuations associated with riffles and pools (see below). Whether the bedload transport rate was sampled at one location or integrated over the entire cross-section of the channel is probably important for the type of distribution the samples of transport rate possess (see below).

Experimental studies of sediment transport allow independent variables of the channel to be controlled and thus offer an opportunity to isolate the shorter scales of transport fluctuations in an alluvial channel. Of the five laboratory studies listed in Table 5.1, three had single-thread and two had multiple-thread channels. Comparisons among the studies would have been facilitated if sample intervals and the duration of sampling had been comparable in the five different investigations.

Table 5.1   Survey of transport data

Reference	Study type[a]	Flow (S steady, U unsteady)	Sampling (WC whole cross-section, P point)	Sample interval	Total sample time	Fluctuation period
Hoey & Sutherland (1991)	LMC	S	WC	15 min	50 h	2–5 h
Kang (1982)	FSC, FMC	U	WC	1, 2, 5 min	11 h	13–30 min
Ashmore (1988)	LMC	S	WC	15 min	60 h	2–3 h 6–8 h
Tacconi & Billi (1987)	FSC	U	WC	1, 5 min	8.5 h	30 min
Reid et al. (1985)	FSC	U	WC, P	continuous	14 h	1.4–2 h
Whiting et al. (1988)	FSC	S	P	1.3 min	80 min	8–12 min
Hubbell (1987)	LSC	S	P	2 min	4 h	12 min 1.3 h
Iseya & Ikeda (1987)	LSC	S	WC	10 s	20 min	4 min
Bunte (1992)	FSC	U	WC	5 min	72 h	1.5 h[b]
Kuhnle & Southard (1988)	LSC	S	WC	30 s	2.5 h	6, 14, 26 min
Kuhnle et al. (1989)	FSC	S	P	3 min	4.5 h	25 min
Meade (1985)	FSC	U	WC	24 h	80 days	5–20 days[c]
Schick et al. (1987)	FMC	U	P	8.3 min	3.3 h	30 min
Dinehart (1989)	FSC	U	P	10 s	4.4 h	2–5 min 10–30 min[d]

[a] LSC laboratory single channel, LMC laboratory multiple-thread channel, FSC field single channel, FMC field multiple-thread channel
[b] From magnetic tracer technique
[c] From channel survey data
[d] From bed form migration data

However, the focus of each type of study was different. In the multiple-thread channel studies, the effect of channel formation and destruction on sediment transport was investigated using long sample intervals (15 min) and long total sampling times (50 and 60 h). The single-channel studies concentrated on shorter-term in-channel processes and therefore sampled transport rates more frequently and for shorter durations. The sampling design of each study controlled the time scale of the fluctuations that could be resolved.

Do transport fluctuations with periods of minutes, similar to those identified in the single-channel studies, occur in multiple-thread channels? Also, do fluctuations with periods of hours or longer occur in single-thread channels? The study by Kang

(1982) suggests that short-term fluctuations similar to those identified in single-channel studies do in fact occur in multiple-thread channels. Although Kang (1982) was not able to determine unambiguously the cause of the observed 10–30 min transport fluctuations, correlations with water surface slope data and a lack of correlation with changes in the flow strongly suggest that bedforms similar to the bedload sheets described by Iseya & Ikeda (1987), Kuhnle & Southard (1988) and Whiting *et al.* (1988) were the cause of the transport fluctuations. Longer-term fluctuations in single-thread channels were identified by Reid *et al.* (1985), and very long fluctuations (on the order of days) associated with unsteady flows were identified by Meade (1985). What relation these have with the long-term fluctuations identified in the multiple-thread channel studies is unclear. Longer-term fluctuations were also suggested by the findings of Kuhnle & Southard (1988). Clast jams were identified which reduced transport for periods of 20–40 min. This suggests that peaks in transport rates from the destruction of the clast jams could occur up to 80 min apart, placing their period of fluctuation in the same order of magnitude as some of the long-term fluctuations measured in multiple-thread channels.

In summary, quasi-periodic fluctuations in the transport of sand and gravel mixtures have been identified at periods ranging from a few minutes, to several hours, to days. Evidence suggests that the transport fluctuations with periods on the order of minutes are present and are caused by similar processes (bedload sheets or dunes, see below) in single- and multiple-thread channels. Fluctuations that occur over hours or longer are present in both types of channels but the processes causing them are not well known and may be different for different channel types.

PROCESSES CAUSING UNSTEADY TRANSPORT

While the presence of quasi-periodic variability in transport rates of alluvial channels with gravel and sand sediment mixtures has been shown by a number of studies (Table 5.1), the cause of this variability is often not clear (e.g. Kang, 1982; Reid *et al.*, 1985). Bedforms have been identified in some gravel-bed streams which affect the initiation of motion of the bed particles but apparently do not cause fluctuations in transport rate (cluster bedforms, e.g. Reid *et al.*, 1984, 1992). In other studies an apparently different class of bedforms have been identified as the cause of the fluctuations in sediment transport rates (Hubbell, 1987; Iseya & Ikeda, 1987; Kuhnle & Southard, 1988; Whiting *et al.*, 1988). These bedforms have been shown to sometimes be very subtle and difficult to recognize under less than ideal conditions (Iseya & Ikeda, 1987; Kuhnle & Southard, 1988; Whiting *et al.*, 1988). In studies where fluctuations in transport rate were identified but bedforms were not, it was assumed that the lack of identification resulted from less than ideal observing conditions. Another type of feature, termed riffles and pools, has been identified as being associated with unsteady sediment transport. These features scale with the width of the channel (here called channel forms) and apparently require unsteady flows to maintain them. In this section the processes which are known to cause fluctuations in the transport of sand and gravel sediment mixtures will be reviewed.

Dunes

Dunes may be defined as bedforms, with a gentle upstream slope and an abrupt downstream slope near the angle of repose of the sediment, which migrate in the direction of the flow, and whose sizes are generally believed to be strongly dependent on the depth of the flow (Yalin, 1972). Most studies of dunes have been in sand sediment, probably because of the difficulties associated with the high velocities necessary for the formation of dunes in gravel. It is likely that dunes do not routinely occur in many gravel-bed streams because the flow strengths tend to be near the critical stress for entrainment of the sediment (Parker, 1978) and are too low to form dunes. Dunes have recently been observed in gravel–sand-bed rivers (Dinehart, 1989, 1992; Kuhnle et al., 1989) and laboratory flumes (Hubbell, 1987; Kuhnle & Southard, 1988; Wilcock & Southard, 1989), and may be more common in natural sand- and gravel-bed streams than has previously been recognized. Fluctuations in transport rate that have been measured in gravel-bed channels with dunes have been shown to be similar to those that occur in sand-bed channels with dunes (Hubbell, 1987; Dinehart, 1992).

Bedload sheets

The term bedload sheets was used by Whiting et al. (1985, 1988) to describe the long and low bedforms that have been recognized by several workers (Iseya & Ikeda, 1987; Kuhnle & Southard, 1988) in channels with a mixture of sand and gravel bed sediment. Bedload sheets have been measured to be approximately one to two coarse grain diameters (D_{90}) high and to have spacings in the range 0.5–2.0 m at Duck Creek (Whiting et al., 1988), 0.5–3.0 m in the flume study of Kuhnle & Southard (1988) and 1.3–2.5 m in the flume study of Iseya & Ikeda (1987).

Bedload sheets appear to result from interactions between the fine and coarse fractions of the sediment. In a bed composed of a mixture of sizes, the local roughness of the bed has a large effect on the shear stress needed for entrainment of each grain size. Increasing the percentage of sand causes the roughness of the bed to decrease and the gravel sediment to be entrained at lower shear stresses. Increasing the percentage of gravel causes the roughness of the bed to increase and the sand to be entrained at higher shear stresses. Iseya & Ikeda (1987) demonstrated this convincingly by experimenting with bed sediments ranging from all-gravel to mixtures of sand and gravel. Data collected in the hydraulic laboratory at the National Sedimentation Laboratory (Figure 5.1) also show the effect of different percentages of sand on the shear stress necessary for initiation of motion of sand–gravel bed sediments.

The following scenario is a likely explanation for the processes that are acting as a bed load sheet migrates down a channel. It is necessary that there is enough sand in the system such that the sand is able to move without any gravel in motion. In other words not all of the sand of the mixture is trapped in the voids between the gravel grains. Because of the difficulty of the sand moving through the coarser gravel, the sand tends to pile up and move as a long and low wave. As the wave of sand moves down the channel the local roughness of the bed is lessened and the near-bed flow

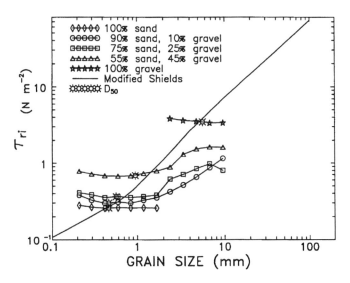

Figure 5.1 Critical bed shear stress for initiation of motion of bimodal sediments. Sediment beds were 100% sand; 90% sand, 10% gravel (SG10); 75% sand, 25% gravel (SG25); 55% sand, 45% gravel (SG45); and 100% gravel. Modified Shields curve (Miller *et al.*, 1977) is shown for comparison

velocity is locally increased. This decrease in roughness and increase in near-bed flow velocity causes the local bed shear stress to increase enough that gravel grains that were not in motion at the front of the sheet are entrained as the wave of sand migrates down the channel. Most of the gravel grains entrained by the migration of the bedload sheet move only a short distance past the downstream end of the sheet

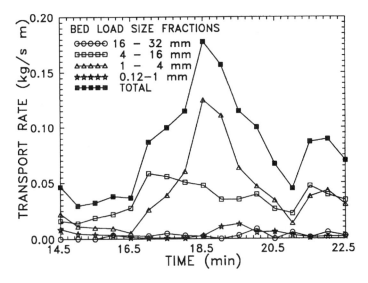

Figure 5.2 Fractional transport rate versus time for run H3 of Kuhnle & Southard (1988). The transport rate of four size fractions and the total rate is shown

because of the increase in bed roughness and decrease in near-bed flow velocity in the region downstream of sheet. This explains the peak in the transport of coarse grains followed by a peak in the transport of the finer grains (see Figure 5.2) reported by Iseya & Ikeda (1987), Kuhnle & Southard (1988), and Whiting *et al.* (1988).

To test the hypothesis that the gravel sizes were at or below their critical shear stress before the passage of a bedload sheet, calculations of the critical shear stress using Miller *et al.*'s (1977) initiation relation were made for the fine and coarse fractions of the sediment beds used by Whiting *et al.* (1988), Kuhnle & Southard (1988), and Iseya & Ikeda (1987). The idea that the gravel is at or below its critical shear stress before the passage of the bedload sheet is approximately supported by these calculations (Table 5.2).

To summarize, bedload sheets are formed in mixtures of sand and gravel where the percentages of the sand and gravel are such that each affects the shear stress at initiation of motion of the other. The critical shear stress of the sand is increased by the presence of the gravel and the critical shear stress of the gravel is decreased by the sand. If the proportion of sand in the bed material is too high dunes rather than bedload sheets will form. If the proportion of gravel in the bed is too high bedload sheets will not form because the sand will not be available for transport until the coarser sizes of the bed are in motion. In channels where bedload sheets do form they are replaced by dunes at higher bed shear stresses (Kuhnle & Southard, 1988; Whiting *et al.*, 1988; Dinehart, 1989, 1992).

Bedform stability fields

What are the relationships among flow strength, sediment size, and bedform type and ultimately how do they relate to fluctuations in the transport of bedload? For sand-bed channels, data from many laboratory experiments and field studies have allowed the types of bedforms and their stability fields to be identified (Simons & Richardson, 1961; Guy *et al.*, 1966; Southard, 1971; Boguchwal & Southard, 1990; Southard & Boguchwal, 1990a,b). On the other hand, data on bedforms in gravel, and mixed gravel and sand beds are rare. To the author's knowledge, no systematic studies of bedforms in sand- and gravel-bed channels have been completed in which flow strength and sediment size distribution were varied.

Table 5.2 Parameters for bedload sheets

Reference	Bed shear stress ($N\ m^{-2}$)	Bed material sizes (mm)	Critical shear stress ($N\ m^{-2}$)
Whiting *et al.* (1988)	5.5, 8.8	$D_{50} = 5$ $D_{max} = 16$	3.4 for 5 mm 10.4 for 15 mm
Kuhnle & Southard (1988)	5.4–7.3	$D_{50} = 3$ $D_{90} = 10$	1.8 for 3 mm 7.8 for 10 mm
Iseya & Ikeda (1987)[a]	1.5–2.9	$D_{50s} = 0.4$ $D_{max} = 4$	0.2 for 0.4 mm 2.7 for 4 mm

[a] Only runs 6–9 considered, D_{50s} is for sand fraction only

It is clear, at least for some mixtures of sand and gravel, that bedload sheets are present at low flows and dunes occur at higher flows (Whiting *et al.*, 1988; Kuhnle & Southard, 1988). The spread of sediment sizes in the bed material may control the flow strength at which the transition from bedload sheets to dunes occurs. At the present state of knowledge, the bedforms that will occur for a given sand and gravel sediment bed and flow strength cannot be reliably predicted.

Riffles and pools

Riffles and pools are channel forms which have been recognized in streams with gravel and sand beds and occur at scales comparable to the size of the channel. Riffles and pools are characterized by a high elevation coarse-grained riffle and a lower elevation finer-grained pool. The spacing of riffles and pools tends to be of the order of five to seven channel widths (Leopold *et al.*, 1964). At low flows, the water surface slope over the riffle tends to be steep while the water surface slope tends to be low over the pool. The slopes tend to reverse during large runoff events, possibly around bankfull, with greater slopes over the pool than over the riffle (Keller, 1971; Richards, 1976; Emmett *et al.*, 1983). In a study spanning several years Leopold *et al.* (1964) found that the position of riffles and pools changed very slowly or not at all.

The fluctuation in bedload transport caused by the migration of dunes and bedload sheets does not require unsteady flows; however, changes in flow are an integral part of the process by which unsteadiness in sediment transport occurs over riffles and pools. During low flows the high water surface slopes in riffles and the low water surface slopes in pools results in the pool acting as a sediment storage zone. When flows increase, the relatively high water surface slopes in the pools causes a net degradation of the bed in the pools and aggradation of the bed in the riffles. As the flow decreases, sand and fine gravel sediment is eroded from the riffles and stored in the pools. For a given flow discharge this means the sediment transport just downstream of a pool will be higher during the rising stage of a flood event than during the falling stage of an event. The opposite is true for positions just downstream of a riffle: for a given flow discharge sediment transport will be lower during the rising stage than during the falling stage of a flood event. The result is that relationships between the bedload transport and shear stress just downstream of the riffles and pools is described by loops with different rates of transport for the same flow during rising and falling stages. Loops in bedload transport were found from measurements on the East Fork River by Emmett *et al.* (1983) and Meade (1985). These loops in bedload transport were found to collapse into a single relation if stream power is used instead of flow discharge. However, the relation between bedload transport rate and stream power was not general for different parts of the East Fork River (Meade, 1985).

Evidence that a reversal in flow velocity is not necessary for maintaining equilibrium riffles and pools has been given by Carling (1991). Flow measurements from three riffles and pools on the River Severn show that neither the sectionally averaged velocity nor the near-bed shear velocity was greater in the pools than over the riffles during bankfull or near-bankfull flow. Rather than a velocity reversal

Carling (1991) found that the flow over riffles and pools on the River Severn became nearly equal as the flow rate increased. From continuity Carling (1991) concluded that the riffle needs to be considerably wider than the pools for velocity reversals to occur under stable conditions.

It appears that while flow reversal may occur on some riffles and pools, on others it does not occur. A velocity reversal at high flows is apparently not a necessary requirement for the maintenance of riffles and pools.

Similar loops in bedload transport have also been found by Kuhnle (1992) on two small gravel-bed streams. No evidence for riffles and pools on these streams has been found, however. It is possible that these loops were caused by a lag in the formation and destruction of bedforms on the channel, although no data on bedforms have been collected.

PROBABILITY DISTRIBUTIONS OF TRANSPORT RATES

With the realization that the transport of sand and gravel sediment mixtures is highly variable a generalized probability distribution function that would describe the variability of the transport process would be valuable. To accomplish this goal many data sets of sequential transport samples taken under known flow conditions and with a wide range of flows and sediment mixes are required. Data sets with enough values to calculate representative distribution functions are rare. The few that do exist tend to be from laboratory channels in which the flow was steady and uniform.

A few studies have been completed which consider the statistical properties of sediment transport records. Willis & Bolton (1979) found that measurements of sediment concentration closely followed a gamma probability distribution; however, this study considered both suspended load and bedload in laboratory sand-bed channels. Carling et al. (1993) found that 10 min averages of bedload count data from Squaw Creek were well described by skewed normal distributions. These data lacked corresponding data on the size distribution of the bedload, however, and are difficult to compare to data from other studies where the bedload was measured in mass per time. Other studies (Carey, 1985; Carey & Hubbell, 1986; Gomez et al., 1989) have found that the distributions of relative bedload transport rates from flume studies and a field study closely followed the distribution derived by Hamamori (1962).

Probability distributions were calculated for laboratory flume data from Kuhnle & Southard (1988) and recently collected but unpublished flume data from the National Sedimentation Laboratory (NSL). In these two studies total bedload transport rates were measured at 0.5 and 1 min intervals for several different flow strengths for periods of 2.5 or 4 h (Table 5.3). When plotted, it was apparent that the distributions of transport data from Kuhnle & Southard (1988) and from NSL resembled a normal distribution. The data were standardized as follows:

$$t = \frac{q_b - \bar{q}_b}{\sigma} \tag{5.3}$$

where q_b is the individual bedload transport rate, $\bar{q}_b$ is the mean transport rate, and σ

Table 5.3 Range of experimental conditions of Kuhnle & Southard (1988) and NSL

Study	Sediment	Number of exp. runs	Range of flow discharge $(m^3 \, s^{-1} \, m^{-1})$	Range of flow depths (m)	Range of sediment transport rate $(kg \, s^{-1} \, m^{-1})$
Kuhnle & Southard (1988)	$D_{50} = 3.0$ mm range: 0.12–32 mm	7	0.028–0.089	0.036–0.074	0.08–1.0
NSL	Three mixtures of 0.44 mm sand and 5.6 mm gravel	18	0.029–0.083	0.095–0.108	0.000004–0.17

is the standard deviation. Distributions of the standardized transport data from three experimental runs (H1, L1, L2) from Kuhnle & Southard (1988) (Figure 5.3) and the composited data from seven runs conducted at the NSL were found to closely resemble the normal distribution (Figure 5.4):

$$f(t) = \frac{1}{\sqrt{2\pi}} \, e^{-t^2/2} \tag{5.4}$$

Figures 5.3 and 5.4 also show that these data do not follow the standardized form of Hamamori's distribution:

$$f(t) = -\frac{\sqrt{7}}{12} \ln\left(\frac{\sqrt{7}}{12} t + \frac{1}{4}\right), \quad -\frac{3}{\sqrt{7}} \leqslant t \leqslant \frac{9}{\sqrt{7}} \tag{5.5}$$

Hamamori's distribution was derived from the assumption that primary and secondary two-dimensional triangular-shaped bedforms generate a distribution which ranges from near zero to four times the mean rate, with a large fraction of the transport rates (60% less than the mean) at zero or low transport rates (Carey, 1985; Carey & Hubbell, 1986). This has been shown to closely resemble distributions of relative transport rates by Carey (1985), Carey & Hubbell (1986), and one of the data sets considered by Gomez *et al.* (1989). One common characteristic of the transport data sets used in these three studies is that they were collected at essentially a single point in a laboratory or natural channel. On the other hand, the data used for Figures 5.3 and 5.4 was total bedload transport from sampling devices that collected the sediment from the whole channel cross-section. One of the data sets from Gomez *et al.* (1989), in which the whole cross-section was sampled, had a distribution that did not resemble Hamamori's distribution.

It is possible that the differences in the distributions of bedload transport rate are the result of different sampling periods; however, it is the opinion of the author that averaging over the cross-section was more critical than the differences in sampling periods used in the different studies. Bedforms are usually three-dimensional. When sampling the whole cross-section it is unlikely that the trough (or low transport zone)

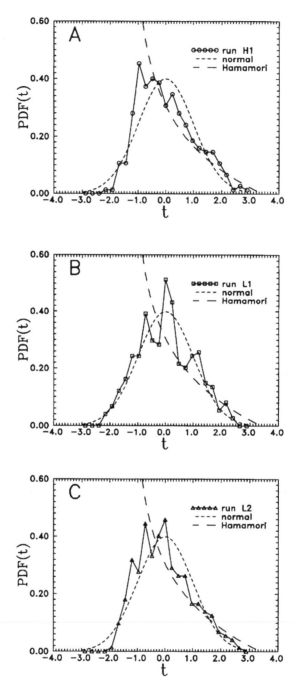

Figure 5.3   Probability distribution functions for data of (A) run H1, (B) run L1, and (C) run L2 from Kuhnle & Southard (1988)

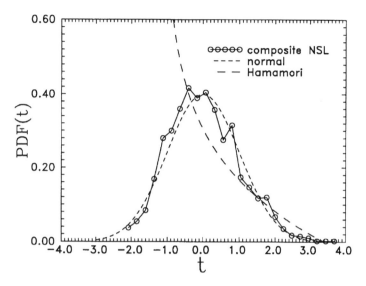

Figure 5.4 Probability distribution function for the standardized transport rate distributions of the NSL data

of a bedform will cover the whole width of the channel. As a result the number of zero or near-zero samples will be much fewer than for samples taken over just a fraction of the cross-section. It is also possible that the processes of sediment transport were distinctly different in the different experiments. This is thought not to be the case, however, as similar processes associated with the migration of bedforms were responsible for the variability of transport rates in the data of Gomez *et al.* (1989), Kuhnle & Southard (1988), and NSL. It may be the case that point samples of bedload taken for short periods of time would yield a probability distribution similar to that derived by Hamamori (1962), while bedload samples taken for longer periods of time or over a greater width of a channel cross-section would yield a probability distribution which closely resembles the normal distribution. A data set which had both point and total cross-section data could shed more light on this subject.

Implications for bedload sampling

The accurate sampling of bedload transport for a given flow condition is a difficult task. In addition to the temporal variability of the transport rate, there is spatial variability across the channel. Hubbell & Stevens (1986) dealt with the effect of lateral variability on the accuracy of bedload sampling and concluded that a few strategically chosen sample locations with multiple samples at each was the best approach.

But for how long (or how frequently) should samples be taken at a location to assure that the average transport rate of the samples is within a reasonable limit of the actual rate? McLean & Tassone (1987) used a Monte Carlo simulation to calculate the number of samples necessary to obtain a given coefficient of variation. Their calculations showed that approximately 20 samples were needed to obtain a coefficient of variation (CV) of approximately 20% using Hamamori's distribution and that 20 samples were required to obtain a CV of 30% using an empirical distribution from Fraser River data. If bedload transport data are assumed to be normally distributed, the number of samples needed to estimate the mean transport rate with a given probability can be calculated if the mean and standard deviation of the transport are known:

$$|\mu - \bar{q}_b| < k \frac{\sigma}{\sqrt{n}}$$ (5.6)

where μ and σ are the population mean and standard deviation, respectively, $\bar{q}_b$ is the mean of the samples, n is the number of samples, and k is a multiplier whose value depends on the chosen distribution and probability level. If the term on the right-hand side of eqn 5.6 is set equal to a fraction (c) times the mean transport rate (μ) and n is solved for:

$$n = \left(\frac{k\sigma}{c\mu}\right)^2$$ (5.7)

then the number of samples needed depends on the value of the ratio of the standard deviation to the mean. In practice the population mean and standard deviation are never known. However, if the population mean and standard deviation are estimated from a sample mean (m) and standard deviation (s), for a probability of 0.95 $(k = 1.96)$, with $s/m = 0.5$, and $c = 0.1$, then the number of samples (n) needed would be 96. If the mean of the samples only needed to be within ±20% $(c = 0.2)$ of the population mean for the same probability then $n = 24$. However, to calculate the number of samples necessary for a given probability level, the mean and standard deviations need to be known or estimated before the samples are collected.

The sampling period necessary to predict accurately the mean transport rate of the cross section was also addressed by Kuhnle & Southard (1988) and the same technique was used for the NSL data (Figures 5.5 and 5.6). The data contained in Figures 5.5 and 5.6 were calculated using a computer program to combine the transport samples from the six experimental runs into samples with successively longer sampling intervals and then to compute the mean and standard deviation of each "new" data set. The coefficient of variation (CV) $((s/m) \times 100)$ was then computed to indicate the probability of obtaining an accurate estimate of the actual mean transport rate for a sample taken for a given period of time. If the data are assumed to be normally distributed, the probability is 0.68 that a sample taken for a given length of time will be within one CV of the mean, or 0.95 that the sample will be within two CVs.

As shown by Figure 5.5 the sampling period necessary to obtain a coefficient of variation of 10% is 7.5 min for run L1 and 35.5 and 32.5 min for runs H1 and L2, respectively. The much shorter sample time needed for L1 was due to the absence of longer-term fluctuations which were present in runs H1 and L2 (Table 5.4). The

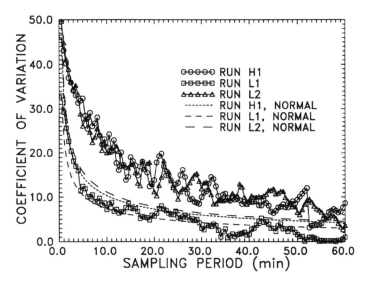

Figure 5.5 Coefficient of variation versus sampling period for data from Kuhnle & Southard (1988)

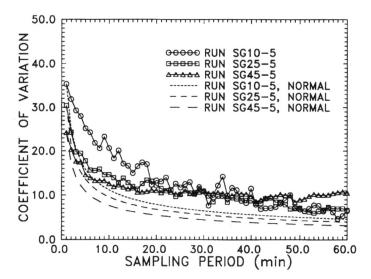

Figure 5.6 Coefficient of variation versus sampling period for data from NSL

three other lines in Figure 5.5 were generated assuming that the standard deviation of the "new" data sets decreased by

$$s_i = \frac{s}{\sqrt{n}}$$
(5.8)

as the sampling period of the combined samples increased, where s_i is the standard deviation of the "new" data set, s is the standard deviation of the original data set,

Table 5.4 Significant periods of transport fluctuations from experiments of Kuhnle & Southard (1988)

Run	Significant periods (min)	Sampling period to obtain CV of 10%
H1	9.8, 14.2, 26.1	35.5
L1	6.1, 6.4	7.5
L2	6.0, 11.5, 24.9	32.5

and n is the number of samples combined to form the "new" data set. Equation 5.8 is from the Central Limit Theorem and the deviation of the three lines from the calculated CV data may be taken as an indicator of the deviation of the data set from the normal distribution. The probability distribution of run L1 appears to be closer to the normal distribution than the distributions of the other two runs plotted in Figure 5.5.

The NSL data (Figure 5.6) show a similar trend except that the initial coefficient of variation values are lower. The data from the three runs in Figure 5.6 indicate that a sample length of about 30 min is needed to obtain a CV of 10%. The three lines generated using eqn 5.8 also show that the deviations from a normal distribution of the distributions of the three runs cause the CV values to drop more slowly with increasing sampling period than they would if they were precisely normal.

Sampling periods of 7 to 35 min are not possible with most conventional bedload samplers. Many samplers (e.g. Helley-Smith) do not have the capacity to hold the volume of sediment required for samples collected over periods of tens of minutes. Also, flows in many streams do not remain steady for periods of 30 min. The collection of a series of samples at the same flow rate for more than one runoff event may be necessary on some streams.

SUMMARY AND CONCLUSIONS

Unsteadiness in the transport of sand and gravel mixtures in alluvial channels has been shown to be the rule rather than the exception. Periods of fluctuation from minutes to days have been measured in laboratory and natural channels. Bedforms including bedload sheets and dunes have been shown to cause variations in bedload transport rates during steady flows at periods from minutes to tens of minutes. Riffle and pool channel forms have been shown to cause bedload transport fluctuations that have periods of hours to days and require unsteady flows.

From the limited data available, probability distributions of bedload transport rates appear to be well described by Hamamori's distribution for samples taken at a point, and appear to resemble the standardized normal distribution if the whole cross-section is sampled.

Sample strategies to determine accurately mean cross-section bedload transport rates show that sampling periods (or number of samples) must cover at least one period of the longest fluctuation periods of the transport rate to have a high probability of accurately predicting the mean transport rate. The fluctuation periods of the bedload transport and the type of bedforms for a given flow and a given sediment bed on a sand- and gravel-bed stream cannot be reliably predicted at the present.

REFERENCES

Ashmore, P.E. (1988) Bed load transport in braided gravel-bed stream models. *Earth Surface Processes and Landforms* **13**, 677–695.

Boguchwal, L.A. & Southard, J.B. (1990) Bed configurations in unidirectional water flows. Part 1. Scale model study using fine sands. *Journal of Sedimentary Petrology* **60**, 649–657.

Bunte, K. (1992) Particle number grain-size composition of bedload in a mountain stream. *Dynamics of Gravel-Bed Rivers* (eds P. Billi, R.D. Hey, C.R. Thorne & P. Tacconi), John Wiley & Sons, Chichester, 55–68.

Carey, W.P. (1985) Variability in measured bedload-transport rates. *Water Resources Bulletin* **21**, 39–48.

Carey, W.P. & Hubbell, D.W. (1986) Probability distributions for bedload transport. *Proceedings of 4th Federal Interagency Sedimentation Conference*, 4–131–4–140.

Carling, P.A. (1991) An appraisal of the velocity-reversal hypothesis for stable pool–riffle sequences in the River Severn, England. *Earth Surface Processes and Landforms* **16**, 19–31.

Carling, P.A., Williams, J.J., Glaister, M.G. & Orr, H.G. (1993) *Particle dynamics and gravel-bed adjustments*. Final Technical Report, European Research Office of the US Army, Contract Number DAJA45–90-C–0006, London, 46 pp.

Dinehart, R.L. (1989) Dune migration in a steep coarse-bedded stream. *Water Resources Research* **25**, 911–923.

Dinehart, R.L. (1992) Evolution of coarse gravel bed forms: field measurements at flood stage. *Water Resources Research* **28**, 2667–2689.

Emmett, W.W., Leopold, L.B. & Myrick, R.M. (1983) Some characteristics of fluvial processes in rivers. *Proceedings, 2nd International Symposium on River Sedimentation*, 730–754.

Gomez, B., Naff, R.L. & Hubbell, D.W. (1989) Temporal variations in bedload transport rates associated with the migration of bedforms. *Earth Surface Processes and Landforms* **14**, 135–156.

Guy, H.P., Simons, D.B. & Richardson, E.V. (1966) *Summary of alluvial channel data from flume experiments, 1956–1961*. US Geological Survey Professional Paper **462-I**.

Hamamori, A. (1962) *A theoretical investigation on the fluctuations of bedload transport*. Delft Hydraulics Laboratory Report **R4**.

Hoey, T.B. & Sutherland, A.J. (1991) Channel morphology and bedload pulses in braided rivers: a laboratory study. *Earth Surface Processes and Landforms* **16**, 447–462.

Hubbell, D.W. (1987) Bed load sampling and analysis. *Sediment Transport in Gravel-Bed Rivers* (eds C.R. Thorne, J.C. Bathurst & R.D. Hey), John Wiley & Sons, Chichester, 89–106.

Hubbell, D.W. & Stevens, H.H. Jr (1986) Factors affecting accuracy of bedload sampling. *Proceedings of the 4th Federal Interagency Sedimentation Conference*, 4–20–4–29.

Iseya, F. & Ikeda, H. (1987) Pulsations in bedload transport rates induced by a longitudinal sediment sorting: a flume study using sand and gravel mixtures. *Geografiska Annaler* **69** A, 15–27.

Kang, S. (1982) *Sediment transport in a small glacial stream: Hilda Creek, Alberta*. Unpublished MS thesis, University of Illinois Chicago, 265 pp.

Keller, E.A. (1971) Areal sorting of bed-load material: the hypothesis of velocity reversal. *Geological Society of America Bulletin* **82**, 753–756.

Kuhnle, R.A. (1992) Bed load transport during rising and falling stages on two small streams. *Earth Surface Processes and Landforms* **17**, 191–197.

Kuhnle, R.A. (1993a) Fluvial transport of sand and gravel mixtures with bimodal size distributions. *Sedimentary Geology* **85**, 17–24.

Kuhnle, R.A. (1993b) Experiments on the incipient motion of sand-gravel sediments with bimodal size distributions. *Journal of Hydraulic Engineering* **119**, 1400–1415.

Kuhnle, R.A. & Southard, J.B. (1988) Bed load transport fluctuations in a gravel bed laboratory channel. *Water Resources Research* **24**, 247–260.

Kuhnle, R.A., Willis, J. C. & Bowie, A.J. (1989) Variations in the transport of bed load sediment in a gravel-bed stream, Goodwin Creek, Mississippi, U.S.A. *Proceedings, 4th International Symposium on River Sedimentation*, 539–546.

Leopold, L.B., Wolman, M.G. & Miller, J.P. (1964) *Fluvial Processes in Geomorphology.* W.H. Freeman, San Francisco, 522 pp.

McLean, D.G. & Tassone, B. (1987) Discussion of bed load sampling and analysis. *Sediment Transport in Gravel-Bed Rivers* (eds C.R. Thorne, J.C. Bathurst & R.D. Hey), John Wiley & Sons, Chichester, 109–113.

Meade, R.H. (1985) Wavelike movement of bedload sediment, East Fork River, Wyoming. *Environmental Geology and Water Science* **7**, 215–225.

Miller, M.C., McCave, I.N. & Komar, P.D. (1977) Threshold of sediment motion under unidirectional currents. *Sedimentology* **24**, 507–527.

Parker, G. (1978) Self-formed straight rivers with equilibrium banks and mobile bed. Part 2. The gravel river. *Journal of Fluid Mechanics* **89**, 127–146.

Pettijohn, F.J. (1975) *Sedimentary Rocks*, 3rd edn. Harper & Row, New York, 628 pp.

Reid, I., Brayshaw, A.C. & Frostick, L.E. (1984) An electromagnetic device for automatic detection of bedload motion and its field applications. *Sedimentology* **31**, 269–276.

Reid, I., Frostick, L.E. & Layman, J.T. (1985) The incidence and nature of bedload transport during flood flows in coarse-grained alluvial channels. *Earth Surface Processes and Landforms* **10**, 33–44.

Reid, I., Frostick, L.E. & Brayshaw, A.C. (1992) Microform roughness elements and the selective entrainment and entrapment of particles in gravel-bed rivers. *Dynamics of Gravel-Bed Rivers* (eds P. Billi, R.D. Hey, C.R. Thorne & P. Tacconi), John Wiley & Sons, Chichester, 253–266.

Richards, K.S. (1976) The morphology of riffle–pool sequences. *Earth Surface Processes* **1**, 71–88.

Schick, A.P., Lekach, J. & Hassan, M.A. (1987) Bed load transport in desert floods: observations in the Negev. *Sediment Transport in Gravel-Bed Rivers* (eds C.R. Thorne, J.C. Bathurst & R.D. Hey), John Wiley & Sons, Chichester, 617–636.

Simons, D.B. & Richardson, E.V. (1961) Forms of bed roughness in alluvial channels. *Journal Hydraulics Division* **87**, 87–105.

Southard, J.B. (1971) Representation of bed configurations in depth–velocity–size diagrams. *Journal of Sedimentary Petrology* **41**, 903–915.

Southard, J.B. & Boguchwal, L.A. (1990a) Bed configurations in steady unidirectional water flows. Part 2. Synthesis of flume data. *Journal of Sedimentary Petrology* **60**, 658–679.

Southard, J.B. & Boguchwal, L.A. (1990b) Bed configurations in steady unidirectional water flows. Part 3. Effects of temperature and gravity. *Journal of Sedimentary Petrology* **60**, 680–686.

Tacconi, P. & Billi, P. (1987) Bed load transport measurements by the vortex-tube trap on Virginio Creek, Italy. *Sediment Transport in Gravel-Bed Rivers* (eds C.R. Thorne, J.C. Bathurst & R.D. Hey), John Wiley & Sons, Chichester, 583–606.

Whiting, P.J., Dietrich, W.E., Leopold, L.B. & Collins, L. (1985) The variability of sediment transport in a fine-gravel stream (abstract). *Proceedings, 3rd International Fluvial Sedimentology Conference*, 38.

Whiting, P.J., Dietrich, Leopold, L.B., Drake, T.G. & Shreve, R.L. (1988) Bedload sheets in heterogeneous sediment. *Geology* **16**, 105–108.

Wilcock, P.R. (1992) Experimental investigation of the effect of mixture properties on transport dynamics. *Dynamics of Gravel-Bed Rivers* (eds P. Billi, R.D. Hey, C.R. Thorne & P. Tacconi), John Wiley & Sons, Chichester, 109–130.

Wilcock, P.R. (1993) Critical shear stress of natural sediments. *Journal of Hydraulic Engineering* **119**, 491–505.

Wilcock, P.R. & Southard, J.B. (1989) Bed load transport of mixed size sediment: fractional transport rates, bed forms, and the development of a coarse bed surface layer. *Water Resources Research* **25**, 1629–1641.

Willis, J.C. & Bolton, G.C. (1979) Statistical analysis of concentration records. *Journal of the Hydraulics Division* **105**, 1–15.

Yalin, M.S. (1972) *Mechanics of Sediment Transport*. Pergamon Press, Oxford, 290 pp.

6 Sediment Sorting over Bed Topography

PETER J. WHITING

Department of Geological Sciences, Case Western Reserve University, USA

INTRODUCTION

Flow over a mobile streambed not only transports sediment and creates topography, but in the processes of erosion, transport and deposition, sediment is sorted by size, shape and density. We can see this sorting in the development of clusters of large grains on flat beds, or in the accumulation of small heavy mineral grains on ripple crests. At intermediate scales we see coarse gravels in the troughs of dunes or along the outer bank in a river bend. At the largest scale, we see rivers progressively fine toward the sea.

Sorting in the context of streamflow can be defined as the process by which particles having particular characteristics (such as size, shape or specific gravity) are naturally separated from associated but dissimilar particles. This definition is modified only slightly from that proposed by the American Geological Institute (1984). A detailed mathematical analysis is not my goal here, but a mathematical definition of the sorting process must be based on a grain specific conservation of mass equation

$$\nabla Q_i = 0$$

Such a mathematical formulation is schematically illustrated in Figure 6.1. Any changes in the discharge of each sediment size fraction Q_i in the downstream (s) or cross-stream (n) directions must be accommodated by changes in the bed elevation (z) or by changes in the bed composition. The composition of the bed adjusts depending upon the influx of each constituent sediment fraction. Sorting processes act, and the differences in material properties are expressed in three directions: vertical, longitudinal and lateral. Vertical sorting is segregation between layers that are adjacent. Longitudinal sorting is segregation in the downcurrent direction in the plane of the transport surface. Lateral sorting is segregation in the cross-stream direction perpendicular to the longitudinal direction but in the plane of the transport surface. Sorting often involves the combination of directions – outward rolling of coarser clasts on a point bar slipface while being rolled downstream, or the vertical stacking of longitudinally sorted bedforms. For a more complete mathematical

Advances in Fluvial Dynamics and Stratigraphy. Edited by P.A. Carling and M.R. Dawson.
© 1996 John Wiley & Sons Ltd.

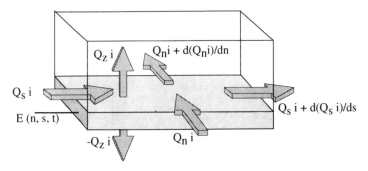

Figure 6.1 Sketch of the components of a mass balance describing sediment sorting

derivation of equations describing the sorting process, see Parker (1991a) among others.

The separation of particles with shared properties from other particles occurs for a variety of reasons. Segregation may occur as material is preferentially added or as material is preferentially left behind because of a particle's mobility or immobility relative to another class of particles. Segregation may occur because of different transport mechanisms or transport directions compared to other classes. Local differences in size may arise as the supply of material to the channel varies. Sorting also occurs in avalanching grainflows. Table 6.1 summarizes the major sorting mechanisms.

Table 6.1 Major sorting mechanisms

Differential entrainment
 winnowing
 mobile armours
 longitudinal corridors
Differential distrainment
 clusters
 overpassing
 transverse ribs
 topsets
 channel sorting
 downstream fining
Self-mediated entrainment/distrainment
 bedload sheets
 longitudinal corridors
Differential transport modes/trajectories
 sideslope tipping
 secondary flow
 suspended-load fallout
Grain flows
 dispersive pressures
Changes in supply
 tributary loading
 attrition

Sorting depends upon the transport of the bed's constituent particles; in different settings this can include the motion of a certain class of particles, or the lack thereof, the style of motion, and the transport direction. The transport process can be thought of as having three parts: entrainment, displacement and distrainment, following Drake *et al.* (1988). Entrainment is the initiation of movement of individual grains on the surface. Since grains may jiggle in their resting sites in pockets between other grains, the definition of entrainment should require that particles shift more than a grain diameter. Grains may move initially by rollover, sliding, liftoff, or ejection when hit by other grains. Displacement describes the translation of grains and modes of motion include suspension, saltation, rolling and sliding. Distrainment is the, at least temporary, cessation of motion as grains return to the bedsurface. This occurs as the moving grains lose the momentum necessary to continue moving through the fluid and across a rough surface.

Segregation processes are greatly enhanced by topography and particularly by various scales of superimposed topography. Riffle and pool sequences in rivers and streams create sloping surfaces that alter a grain's mobility and trajectory. Flow acceleration and associated shear stress variation forced by such topography allow for selective movement and deposition. Dunes and ripples, both two- and three-dimensional, similarly create topographic and flow variation that can act during entrainment, displacement, or distrainment. When flow separates over steep topography, sorting mechanisms can be unusually effective. A steep slipface is reworked by small failures in the granular material. Furthermore, the stratigraphic preservation of sorting is favoured when the segregation is developed over undulating topography.

Sorting is important not only because it is so ubiquitous in fresh deposits and in sedimentary rocks, not only because fluvial segregation has created economic placers as well as toxic concentrates, but also because an understanding of the segregation process is fundamental to the understanding of sediment transport in general. A better understanding of sorting can be applied to solve vexing problems at scales ranging from the entrainment of a single grain on a heterogeneous surface to the resolution of the seemingly conflicting observations of both downstream fining and near-equal mobility. Sorting is a critical hydraulic clue in environmental reconstructions, in delineating old surfaces or even in recognizing which direction is up in outcrops or hand samples. Sorting by size can introduce spatial variation in the suitability of the substrate for organismal feeding, hiding or spawning, and in so doing can determine biological patterns. The hydraulic resistance to flow is set by the size of constituent elements making up the bedsurface so that sorting can determine local friction.

This chapter will be organized to discuss each of the major mechanisms of segregation of sediments within the channel banks (Table 6.1). The topic of floodplain sorting must await another book or at least Chapter 8 of this volume. The discussion here will highlight the current understanding of the physics of each segregation process across the range of scales and environments where this mechanism has been observed and will focus on surface segregation rather than stratigraphic preservation.

SORTING FROM DIFFERENTIAL ENTRAINMENT

The progressive stripping of the most mobile of grains in a mixture is seemingly the easiest to comprehend of the mechanisms for generating segregation on a streambed. As the more mobile grains are entrained and carried away they leave a surface that becomes enriched in less mobile grains – a lag. This lag may be composed of the largest grains from the former mixture – less mobile because of their greater volume and weight – or composed of the densest grains (Chapter 4). In some particular cases the lag may be composed of oddly shaped grains that make them less prone to transport (Carling *et al.*, 1992).

A surface of many grains has a complex geometry with a great variety of pockets created between adjacent grains. The depth and breadth of these pockets, taken with the size of the particles sitting in these pockets, determines the fluid drag that must be applied to roll grains up and out of the pockets (Li & Komar, 1986; Komar & Li, 1986; Buffington *et al.*, 1992), or to otherwise dislodge grains. The shear stress required to entrain a gravel particle sitting on a bed of similar size particles increases linearly with the particle diameter (Figure 6.2). However, in mixtures of various grain sizes and densities, the shear stress required to move a grain smaller than the median can be expected to be substantially greater than if it were on a surface of other small grains (Figure 6.3). The smaller grain is hidden from the full force of the flow as it sits deep in steep pockets of the bed. On the other hand, the shear stress required to move a grain larger than the median can be expected to be substantially less than if it were on a surface of large grains. The larger grain is exposed to more of the flow as it projects out of shallow-angle pockets of the bed (Figure 6.3).

The much discussed implication of the gross equivalence of critical shear stress is that a bed of various grain sizes should become mobile at approximately the same

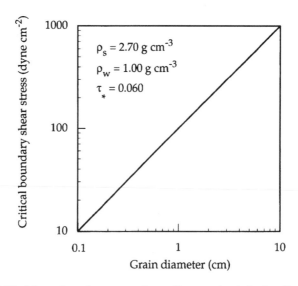

Figure 6.2   Critical boundary shear stress for uniform grains in hydraulically rough flow

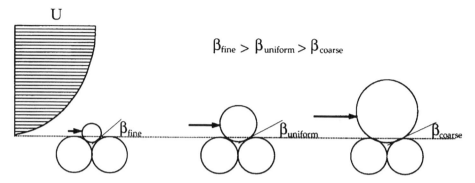

Figure 6.3 Sketch of the effect of pocket geometry on mobility. Smaller than median grains sit deep in pockets of the bed (large β angles) and are relatively hidden from the flow compared to a median size grain or a large grain sitting on a finer bed (smaller β angles). The magnitude and height of the average fluid velocity impinging upon the grain is shown by arrows

shear stress: the "equal mobility" hypothesis (Parker *et al.*, 1982). This phenomenon has been used to explain the common observation in mobile gravel-bed channels of a surface layer that is distinctly coarser than the substrate. As proposed and further developed by Parker and his colleagues (Parker *et al.*, 1982; Parker & Klingeman, 1982; Andrews & Erman, 1986), the mobile surface coarsening can be viewed as regulating the size distribution of the bedload such that there is an approximate equivalence between sediment supply and bedload size distributions. The subsurface is generally considered to be most representative of the sediment supply. Far more detailed and authoritative discussions are available in Parker's papers and the papers of others (e.g. Wilcock, 1992) and in Chapter 4 of this volume.

If in fact equal mobility was exactly realized, we would not observe lags associated with selective entrainment. Calculations of the force needed to entrain particles on heterogeneous beds (i.e. Wiberg & Smith, 1987) predict that critical boundary shear stress approaches "equal mobility" (Figure 6.4), but does not exactly match such a condition. Grains smaller than the median more closely approach the condition of invariant critical shear stress than do the coarser grains, but there is in general a progressive increase in critical boundary shear stress with size on a non-uniform bed. Unimodal and weakly bimodal sediments approach equal mobility more closely than strongly bimodal sediments (Wilcock, 1992). It appears that while equal mobility serves as a useful paradigm that reminds us that the surface texture has a significant effect on entrainment stresses, precise equal mobility is not achieved (Ashworth & Ferguson, 1989; Komar, 1987), at least across the full range of boundary shear stress typically encountered in streams. Consequently, as boundary shear stress is raised, progressively larger and denser grains are moved. And conversely, as shear stress drops, progressively finer grains drop out of the load.

Recent work by Dietrich *et al.* (1989), Kuhnle (1989), and Wilcock & Southard (1989) has shown that the degree of surface armouring depends on the supply of sediment compared to the channel's ability to transport sediment and that the armour layer disappears as the load matches the capacity. Thus ·it would seem that static

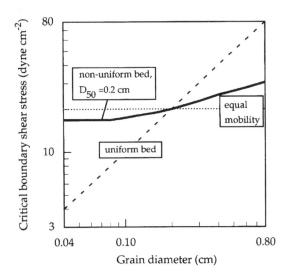

Figure 6.4   Critical boundary shear stress for mixed-size beds as calculated from Wiberg & Smith (1987) compared to equal mobility and linearly increasing critical boundary shear stress. The calculation is performed for a median size of 0.2 cm

armours (lags) and mobile armours both lie along a continuum between two distinct end members – one of active transport of all fractions over a completely mixed surface equivalent to the substrate and one of vanishing transport over a surface distinctly coarser than the substrate. The mobility of the various size fractions adjusts as the supply and surface texture adjust.

On flat beds, deflation lags have been reported by a number of workers (Gessler, 1965; Kellerhals, 1967; Little & Mayer, 1976). Common to all these studies is selective entrainment of the finer fraction which led to surface coarsening compared to the substrate. Progressive removal of the more mobile fraction left a residue of generally larger grains. But it is not always the coarse grains that form lags. Kuhnle & Southard (1990) developed fine-grained concentrates of heavy minerals at the surface during degradation without sediment feed. Brady & Jobson (1973) similarly found heavy mineral layers on a bed being eroded. They linked the concentration to the greater density of the fine grains and to hiding of the small heavy grains in pockets of the bed. To illustrate the effect both greater density and smaller size have on the mobility of these "heavies" in a mixture with quartz-density grains, see Figure 6.5.

Innumerable examples of surface coarsening have been reported from channels downstream of dams. The dams act to restrict, if not cut off, the supply of sediment, particularly coarse sediment, and the nearly clearwater flows winnow the bed below the dam of finer sediment. The streambed sediment size may be coarsened for tens of kilometres, but the surface gradually approaches the initial composition downstream (Figure 6.6). This effect is most dramatic just below dams and in reaches without major tributaries. Williams & Wolman (1984) give a good compilation of effects below dams.

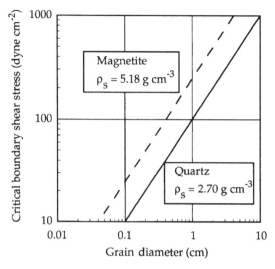

Figure 6.5 Critical boundary shear stress for uniform grains of differing densities

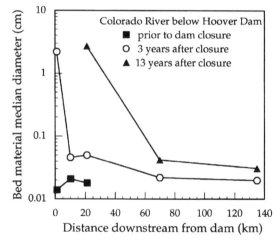

Figure 6.6 Grain size downstream of the Hoover Dam on the Colorado River before and after closure (from Williams & Wolman, 1984)

Size differentiation between the surface and subsurface, which occurs inevitably with differential entrainment, does not require a variation in shear stress – only a critical shear stress difference between grains. On the other hand, longitudinal and lateral segregation due to entrainment differences does require a spatial variation in boundary shear stress. This variation is tied to accelerations forced by curvature, or to topographic accelerations forced by ripples, dunes and bars. While stress divergence on a mobile bed is commonly thought to be balanced by erosion or deposition or lateral transport of sediment, a response to spatial variation in the boundary shear stress can be almost entirely through bed grain size adjustment (Dietrich, 1987) accomplished largely through selective entrainment.

Topographic variation over dunes provides a mechanism to introduce spatial variation in the stress applied to the bed. At Muddy Creek, Wyoming, USA, bedsurface size increases toward the crest (Whiting & Dietrich, 1990) presumably due to increasing local boundary shear stress. The same general increase in sediment size has been noted over aeolian dunes (Barndorff-Nielsen et al., 1982). Bridge & Best (1988) and Dinehart (1992) as well as others have noted that bedload is typically coarsest near the crest. While Bridge and Best (1988) linked this observation to increasing shear stress over the back of dunes, a coarser bedload need not mean that the bedsurface is itself coarser – higher velocities of more exposed large grains and/or a lack of distrainment (as for example in overpassing, to be discussed later) may raise the size of the load without altering the stoss surface size.

Another class of sorted surfaces produced by selective entrainment of more mobile grains is "bottom set placers". The less entrainable grains are left in the troughs of migrating ripples and dunes. In a mixture of fine heavy minerals and larger quartz grains, it is the fine heavies that form the lag (Brady & Jobson, 1973). The heavier grains are left because, in addition to their density, their steep climb out of pockets in the bed and lesser exposure to the flow make them less erodible than their larger quartz companions (Figure 6.5). On a nearly flat bed with a sediment feed containing 3% by weight heavy minerals (Kuhnle & Southard, 1990), "heavies" prograded downstream forming a continuous sublayer below a discontinuous sheet of lighter grains. Exposure of the heavier sublayer occurred in the lee of cluster of the lighter grains, and in the troughs of very low bedforms described as being similar to bedload sheets. Kuhnle & Southard argue that the reduced mobility of the heavies explains the formation of the sublayer. They also suggest that the fine heavy minerals may infiltrate the surface of coarse grains.

Schiller & Rowney (1984) described armouring that occurred in a flume filled with sand and gravel. Gravel accumulated in the troughs of migrating sand ripples and eventually formed a continuous layer under the ripples. Such a mechanism is limited to situations where bedform height is sufficiently small so as to preclude break-up of the armour in the bedform trough.

At the scale of channel-wide architecture, it has been noted that the median size of bedsurface sediment in a bend tracks the corridor of maximum boundary shear stress (Dietrich et al., 1979; Dietrich & Smith, 1984) or flow velocity (Jackson, 1975). This appears to be due to the preferential entrainment of the finer sediment sizes or their entrainment and suspension. Van Alphen et al. (1984) also noted a correspondence between stress and bed material size in the upstream part of bends. In the downstream part of bends, where effects of sediment supply became important, the linkage was less clear. Lisle & Madej (1992) and others have proposed that some spatial variation in bed material size results from lower flow winnowing of fine sediment from riffles and its deposition in pools. Winnowing of the head of braid bars was envisioned by Leopold & Wolman (1957) to lead to the downstream fining along the bar.

In other situations the selective entrainment of finer grains does not lead to a coarse residue but rather the selective entrainment leads to the formation of a locally fine mobile bed. Jackson & Beschta (1982), Mosley (1988), and Whiting & Dietrich (1991) have described sand that moved in narrow longitudinal corridors over a coarse bed. Ferguson *et al.* (1989), calling these sand ribbons, suggested that caving banks and reworking of gravels as stage rose were the sources of the sand. Jackson & Beschta (1982) proposed that finer material is stored in pools, the near-bank regions, and behind obstructions. Because the sand and finer material is segregated, the finer sediment is mobile at lower shear stresses than the mixed material elsewhere in the channel. If all the sediment sizes were mixed, the pockets would act to hide the fine sediment which would become mobile only with the motion of the coarser, pocket-forming grains. At Solfatara Creek, Wyoming, USA (Whiting & Dietrich, 1991), sand began to move off the downstream part of an upstream point bar and over a gravel bar as stage climbed. Sand filled pockets between gravel grains and in so doing extended a narrow (~15 cm wide) corridor of sand on an otherwise gravel bed (Figure 6.7). The longitudinal orientation of the sand corridor and its persistence may be tied to flow patterns set up past the edge of the bar or to differences in the friction between the gravel and sand beds.

When the mobile sand is formed into dunes or ripples there may be spectacular variation in the bedsurface size. Coarse sediment of the immobile bed may be exposed in the troughs of sand dunes. At a larger scale, Meade (1985) described sand waves up to 30 channel widths long that migrated over the gravel bed of the East Fork River, Wyoming, USA.

The spatial segregation of sizes described above has implications that extend beyond the reach. The collection of the finer grains among other fine sediment allows the intrinsically greater mobility of the finer smaller particles to be realized –

Figure 6.7 View looking downstream showing fine sediment in longitudinal corridors infilling a coarse gravel bed, Solfatara Creek, Yellowstone National Park, Wyoming, USA

at least compared to a coarse grain or to a fine grain on a coarse bed. Thus local segregation, especially in longitudinally continuous strips, may be sufficient to allow for downstream fining even while near-equal mobility conditions prevail in areas where coarse and fine sediment are mixed – a point made by Lisle & Madej (1992).

Further discussion of the effects of entrainment on surface and load characteristics can be found in Chapter 4.

SORTING FROM DIFFERENTIAL DISTRAINMENT

In other situations, it is the preferential distrainment of grains that leads to local concentrations of grains and hence to surface segregation. As with preferential entrainment, the primary controls on whether a particle class drops out of the transported load are grain size, density and shape, particularly as these compare to the bedsurface. Distrainment occurs because either the fluid forces acting to keep grains in motion decrease, or because the geometry of the bedsurface is such as to impede the downstream motion of a certain class of grains. Kuenen (1966) linked selective distrainment to the tendency of particles moving along the bottom to join stationary ones of equal weight, density and shape.

Since transport is the repeated entrainment and distrainment of grains, a distinction between selective entrainment and selective distrainment might appear to depend upon one's perspective. Selective distrainment literally can mean one grain stops on the bed and another keeps going, or in a probabilistic sense, in the great number of liftoffs and returns, one set of grains is less likely to be re-entrained. To distinguish between these views, with preferential distrainment a segregated area is built by local deposition of grains. In the case of preferential entrainment, a segregated area is built by the removal of a class of grains – the segregation is a residue. This suggests that we should find preferentially deposited segregations better sorted, in a sedimentological sense, than residual segregations. Examples of selective distrainment are copious over the range of bed states.

At the smallest scale, local accumulations of a few clasts may form "particle clusters" (Brayshaw, 1985). Their formation is linked to the availability of stable sites for deposition. As the "birds-of-a-feather" expression (McBride et al., 1975) suggests, clusters form because the constituent grains find very stable pockets among their neighbours (Figure 6.8). In some cases, the grains can remain on the bed only because these coarse grains sit adjacent to one another. Clusters in turn can act to modify the flow field and themselves lead to the distrainment of other size fractions (Reid et al., 1992). Current shadows of finer sediment form in the lee of these obstacles. Fine grains accumulate in the complex wake until the bed in the lee of the clast has been so smoothed that additional grains are either unable to find a stable pocket, or flow is altered by this shadow and finer grains are moved past the obstacle. In other cases, sediment may accrete to the front of the clusters. These static accumulations are possibly seeds for the development of other sorted features such as migrating bedload sheets.

Whether triggered by clusters or by a single obstructing grain, Moss (1963, 1972) coined a term "traction clogging" to describe a hypothetical situation where an

Figure 6.8 Sketch of clusters of coarse grains projecting above the average bed. In the lee of the cluster finer sediment has collected to form a tail

assemblage of rolling or sliding clasts is essentially frozen in place to form a coarse patch. The moving mass becomes jammed by a local obstruction and stops moving. The obvious kinematic analogy is to a car stopped on a highway, or to too many people all at once trying to pass through a doorway. While other workers have suggested traction clogging as a possible mechanism to create collections of coarse clasts, bedload in clearwater flows is not sufficiently concentrated for this to occur: perhaps it is a viable mechanism when debris flows run down channels.

On the basis of the geometry of surface layers and trenches cut into gravel bars, Dunkerley (1990) recognized accumulations of coarse grains several grains thick that appear to have been deposited preferentially, as opposed to having been left as a lag. In one picture included with the text, a coarse layer is draped over topset and foreset beds of a bar built out of the sandy material, and the bedding does not appear truncated by erosion. Milhous & Klingeman (1973) have described coarse layers that they believed resulted from deposition of less mobile clasts.

Coarse sediment moving over a finer bed covered with dunes or bars has been observed to pass over the stoss side of the feature without stopping, and to deposit in the trough or along the foreset (Hooke, 1975; Carling, 1990). Calling this "overpassing" after Everts (1973), Allen (1983) suggested it could explain features in the Old Red Sandstone of southwestern Ireland in which gravelly foresets passed upward into sandy topsets. He linked the preferential collection of the coarsest grains in the foresets to their being passed to the trough after being rejected for deposition on the finer stoss bed. He cited as corroboration the results of Brady & Jobson (1973) and McQuivey & Keefer (1969). Carling (1990) has shown that overpassing can ultimately lead to the formation of alternating layers of openwork and matrix-filled gravels. Overpassing coarse particles deposit matrix-free gravel on the leeside. Fine suspended sediment is trapped in the separated flow zone in the lee and also in the framework of the dune crest. Consequently an openwork gravel-layer develops sandwiched by two fine-sediment layers. In an accretionary setting this leads to alternating openwork and matrix-filled layers (see also Chapter 9).

Transverse ribs of regularly spaced clusters of coarse clasts have been observed in a number of settings, particularly braided channels and very steep mountain channels (Boothroyd 1970; McDonald & Banerjee, 1971; Gustavson, 1978). Transverse ribs appear to develop in association with hydraulic jumps (Allen, 1984) and the largest grains deposit under the decelerating flow of the rising part of the standing waves. While the ribs are composed of the coarsest grains in the mixture, size variation across a set of ribs is often made obvious by waning flow deposition of finer sediment between the coarse-grained ribs (Allen, 1984).

Clusters of gravel one grain thick were created in supercritical flow from coarse material dropped on the upstream side of breaking antidunes by the flow disruptions accompanying the collapse of the standing wave (Foley, 1977). Foley called these lenses of gravel "dropout armour". One difference between features described as transverse ribs and dropout armour is that ribs are developed from the coarsest few grains in a generally coarse bed, whereas dropout armour represents the coarsest grains on an otherwise fine bed.

In other situations it is not the large, less mobile grains that are distrained, but instead it is the dense fine fraction that drops out. On the crest of symmetrical waves and nearly flat beds, Cheel (1984) described accumulations of heavy minerals he called "heavy mineral shadows". These accumulations had a sharp upstream edge and a diffuse downstream edge. The formation of these shadows was linked to the development of in-phase water waves as the Froude number climbed to near 1.0. In a later paper, Cheel (1990) described the possibility of forming subhorizontal laminae of heavy minerals with the downstream migration of heavy mineral-dominated wave crests. Another way to form a subhorizontal layer of heavy minerals comes from the results of Kuhnle & Southard (1990). A number of recent articles have discussed the formation of parting laminations. I point to several papers including Paola *et al.* (1989), Cheel (1990) and Best & Bridge (1992) for descriptions of sorting at the finest scale, but sorting that is primarily obvious in the vertical direction.

Both McQuivey & Keefer (1969) and Brady & Jobson (1973) experimentally developed accumulations of heavy minerals upstream of the brinkpoint of dunes and ripples. These topset deposits were believed to form as the heavy grains were preferentially deposited past the dune crest as turbulent intensity and maximum instantaneous shear stress decreased. Selective accumulation of the coarse fraction on the dune crest has not been reported presumably because, even if stress locally decreased, the substantial projection of the large grains into the flow precludes their deposition.

At the scale of the channel itself, longitudinal as well as lateral variation in topography and shear stress that occur in association with channel bends, create many opportunities for selective distrainment. A number of workers have confirmed Bluck's (1971) observation that the downstream portion of a point bar is finer than the bar head (e.g. Jackson, 1975; Dietrich *et al.*, 1979; Van Alphen *et al.*, 1984). This pattern, which will be discussed in more detail later, is at least partially due to decreasing boundary shear stress along the bar. As boundary shear stress drops, systematically smaller grains come to rest on the bed (Figure 6.9a). Lewin (1976) described coarse bar heads on alternate bars (Figure 6.9b) and Bluck (1974), Leopold & Wolman (1957) and recently Ashworth *et al.* (1992a,b) have described coarse heads and fine tails on braid bars (Figure 6.9c).

Riffles in many coarse gravel streams are distinctly coarser than pools (Leopold *et al.*, 1964; Keller, 1971; Lisle, 1979; Milne, 1982; Grant *et al.*, 1990). The accumulation of sand and finer material as flow is ponded behind the riffle at low discharges accentuates pre-existing size segregation. Below the superficial fines the pools are floored by gravels finer than the gravels or cobbles of the riffle. The

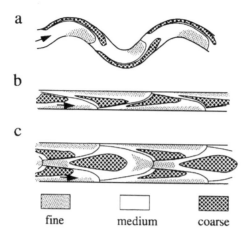

fine   medium   coarse

Figure 6.9 Size variation in meandering channels (a), as for instance seen at Muddy Creek (Dietrich & Smith, 1984), straight channels with alternate bars (single-row bars) (b), as for instance seen by Lisle & Madej (1992), and braid bars (double-row bars) (c), as for instance seen by Bluck (1974)

coarser riffle has been linked to decreasing shear stress in the downstream portion of the pool at formative stages. Coarser particles are believed to deposit over the riffle while finer particles are carried downstream. Earlier in this discussion, some of the coarsening of the riffle was tied to low flow winnowing (Lisle & Madej, 1992). I do not care to expound upon the hypothesis of "velocity reversals" from pool to riffle as stage varies (Keller, 1971; Andrews, 1979; Lisle, 1979), but suggest the recent reviews by Clifford & Richards (1992) and Carling & Wood (1994) for additional discussion.

The observations by Hill (1987) are relevant to the significance of these features. In a sinuous flume, Hill observed coarse "lumps" that formed several channel widths downstream of an outer bank pool in a bend. He linked their formation to decreasing boundary shear stress as flow spread out once past the point bar margin. Hill noted that the coarse lumps in turn caused an additional scour hole to develop along the opposite bank thus producing alternate-bar topography. Lisle *et al.* (1991), in their flume experiments on alternate-bar behaviour with mixed-size sediment, also described "coarsening shoals" that formed as flow diverged after emerging from pools along the outer bank. These coarse accumulations then inhibited erosion of the bars and led to bar stabilization. In braided channels, Leopold & Wolman (1957) and Ashworth (1992a,b) described mid-channel bars formed by the deposition of less mobile coarse bedload. The accumulation was not always grain by grain; Ashmore (1991) observed mid-channel bars initiated by the stacking of coarse-fronted bedload sheets. The bar heads continued to catch coarse material until they filled the channel chute and caused flow to bifurcate around the new bar. In very steep channels, stepped-bed topography (Grant *et al.*, 1990) is built from the coarsest clasts available. Grant and others believe these clasts organize because the largest grains deposit only amongst other similar grains; the large grains traverse smooth reaches without stopping.

So it seems that the selective deposition of coarse grains has effects other than establishing sorting – the selective distrainment can lead to the creation of the basic channel architecture. In this sense, the "coarse lumps", "coarsening shoals", and "coarse patches" described by Hill (1987), Lisle *et al.* (1991), Leopold & Wolman (1957) and Ashworth *et al.* (1992a,b), respectively, are a manifestation of the tendency for a deformable bed to develop bars, whether alternate or row (braid) bars. Apparently channel-wide flow patterns initiate the local deposition of the most difficult to move coarse clasts and the local texture of the surface then acts to accumulate more coarse grains, thus accentuating the channel-scale effects.

Selective distrainment may lead to downstream fining along rivers. Paola *et al.* (1992) developed longitudinal sorting in a long flume with a bimodal feed of gravel and sand. The input of load caused a progressively prograding wedge of sediment to form. The aggradation in the upstream portion of the flume was balanced by a net decrease in sediment transport. The grains preferentially distrained were the coarse ones.

Finally one might view falling stage as a mechanism for preferential deposition of less mobile material. While the decreasing stress accompanying the falling stage means that transport rate decreases and deposition occurs more often than erosion, the time scale of water-level fluctuation is such that segregation only occurs at the scale of the channel network.

SORTING FROM SELF-MEDIATED ENTRAINMENT/ DISTRAINMENT

We have seen that selective entrainment and distrainment can act independently to create spatial segregation. It also appears that selective entrainment and selective distrainment can act together to create distinctive sorted surfaces. Self-mediated distrainment and entrainment occurs as the coarse and fine fractions in a mixture alternately modify the mobility of the other end member of the size distribution. Coarse sediment moves easily over a finer bed and stops only amongst other coarse grains. Fine sediment moves easily over a bed of similar material, but becomes trapped in deep pockets of a coarse bed. Downstream progression of fine sediment requires infilling of the pockets formed by coarse particles. Coarse sediment is selectively entrained and transported over finer beds while selectively distrained over coarse beds. Fine sediment is selectively distrained over coarse beds compared to fine beds, and is transported in appreciable quantities over fine beds. The surface texture modification, in addition to changing the geometry of bed surface pockets, may modify the flow itself through the friction of the surface (Whiting & Dietrich, 1990).

Corridors of fine sediment snaking downstream through a bed of gravel were discussed earlier in the context of selective entrainment of the more mobile sediment. While in those cases the sand was the only mobile fraction, in other settings coarser sediment is weakly mobile. When coarser grains are entrained and move onto the locally smooth sand bed, they travel rapidly in a manner analogous to overpassing. They may be carried to the downstream end of the longitudinal sand corridor or rolled off to the side onto the gravel bed. Once on the rougher gravel, they are quickly distrained. Lisle *et al.* (1991) observed that coarse sediment accumulated outside the band of active transport in areas of rougher microtopography. Dietrich *et al.* (1989) showed that a finer corridor, less armoured than the surrounding areas, narrowed as the sediment feed was reduced. Another effect that may act to maintain segregation is the consequent streaming of flow over the geometrically and hence hydraulically smoother bed (Ferguson *et al.*, 1989). Secondary flow is established by the variation in near-bed friction and finer material is swept toward the corridor from both sides.

Transverse bands of sorted sediment reflecting repeated longitudinal segregation have also been recognized. Among some of the first descriptions are those of Gessler (1965) and Gustavson (1978). Whiting *et al.* (1985, 1988) described thin, longitudinally sorted, transverse bedforms that they called bedload sheets. Bedload sheets are one to two grains thick at their leading edge and sediment is coarsest at the downstream edge of the feature and finest at the upstream edge (Figure 6.10). In planform, the bedload sheets may be nearly straight-crested to lobate. Since the initial report, sheets have been recognized by many workers in both the field (Germanoski, 1989; Ashmore, 1991; Ashworth *et al.*, 1992a,b; Dinehart, 1992) and laboratory studies (Iseya & Ikeda, 1987; Kuhnle & Southard, 1988, 1990; Dietrich *et al.*, 1989; Wilcock, 1992). There is further evidence for such features in the sorted pulses of bedload summarized by Church & Jones (1982). Wavelengths of 0.2 to 0.6 m have been reported in sand while wavelengths of 0.5 to 2.0 m have been

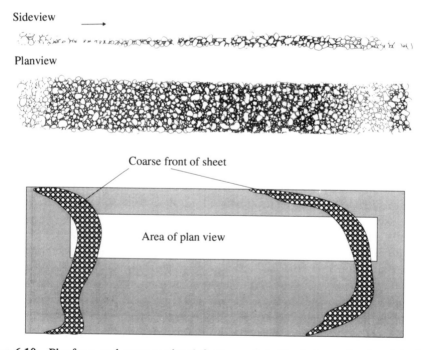

Figure 6.10 Planform and cross-sectional features of bedload sheets developed in Duck Creek, Wyoming. The arrow is about 5 cm long

reported in gravel. The bed surface texture due to the variation in size and degree of sorting was described by Iseya & Ikeda (1987) as congested, transitional and smooth bed states. The leading edge of a bedload sheet corresponds to a congested texture, while the finer part of a bedload sheet is either transitional or smooth.

Iseya & Ikeda's (1987) description of the distinct bed states points to the cause of the segregation: namely the differential mobility of size fractions in a poorly sorted mixture, particularly when moving over a surface composed mainly of another size fraction. Bedload sheets are unknown from uniform-size sediments. In mixed-size sediment, the mobility of different sizes is controlled in part by the geometry of bed surface pockets (Wiberg & Smith, 1987) and the size of the grains on the bed over which these mobile grains travel (Iseya & Ikeda, 1987; Drake et al., 1988). The coarsest fractions in the mixture are distrained and deposit only near other large grains because only here are there stable pockets for resting on the bed and only here do the collective wakes of the coarse grains reduce the drag on these grains which project relatively far into the flow. The finer fraction cannot move over the accumulated coarse grains and collects in the large and relatively deep pockets between coarse grains. This smoothes the bed and prevents incoming coarse grains from depositing in this area because pockets in which to deposit the largest grains are very shallow (Figure 6.3). Furthermore, higher flow velocities are found close to the bed (Whiting et al., 1988; Ferguson et al., 1989). Large grains roll past the smooth reach and deposit near other coarse grains where pockets are deeper and conditions more favourable for distrainment. Re-entrainment of the coarse grains may occur when advancing fines alter the bed or when the coarse grains re-emerge after being buried. Another process that may increase the entrainment of coarse particles might be called "infiltration jacking". As described by Whiting et al. (1988) fine particles may drop under jiggling coarse grains and lift the coarse grains higher into the flow. Once a collection of a few coarse grains is developed, possibly as clusters, the inherent mobility differences seem to easily form longitudinal patches of relatively coarser and finer sediment. This phenomenological explanation is generally held by Whiting et al. (1988) and Iseya & Ikeda (1988) and is the basis of a theoretical analysis by G. Seminara (pers. comm.) to predict the conditions where such features should arise and their expected wavelength. Tsujimoto (1989) linked this same instability to secondary currents initiated by variation in surface roughness. The variation in hydraulic roughness due to variation in surface size (e.g. Whiting & Dietrich, 1990) may act as a feedback mechanism (Ferguson et al., 1989) that could stabilize features, but a hydraulic difference seems unnecessary to explain the initial formation of waves of sorted material. Additional discussion of the dynamics of bedload sheets can be found in Chapter 3 and Chapter 5.

Dietrich et al. (1989) observed that bedload sheets disappeared as the bed armoured. Bedload sheets were most common when sediment feed was at the maximum transport rate permissible without aggradation and were less common in runs where the feed was reduced and an armour developed. As the supply was cut, the corridor of active transport narrowed. This corridor was distinctly finer than the rest of the bed and was close to the load in its composition. Apparently, sediment transport could only occur in these narrow reaches where the finer part of the load smoothed the bed – both locally increasing the near-bed fluid drag on particles and reducing the friction angles of bed pockets. Gessler (1965) described similar features that were most common in the first part of the experiment and eventually disappeared as a coarse surface layer completely covered the bed. Arnott & Hand (1989) studied the formation of horizontal laminations by migration of thin bedforms, coarsest at their leading edge, that they identified as bedload sheets. To mimic the rapid sedimentation of turbidity currents, sand was fed into the top of the flow. They found that sheets were suppressed at high feed rates, presumably because the infall served to obliterate surface texture variation.

Sorted features with a scale much larger than bedload sheets have been described by a number of workers. These low-amplitude features approach channel scale, but are probably formed in a manner analogous to sheets. Depositional sites are limited by the availability of stable pockets hence gravels deposit only near other coarse grains. As mentioned earlier, Moss (1963, 1972) described a process he called traction clogging where coarse bedload grains accumulated in expanding clumps which could eventually cover much of a channel section. While he believed the traction clogging represented the collapse of some dilated sediment layer, the features probably were something like bedload sheets. Leopold & Wolman (1957), based on field and flume observation, suggested that braid bars develop from deposits of the coarse fraction of the bedload moving down the centre of the channel. Finer sediment collected in the lee of the coarse accumulation. Gustavson (1978) described gravel sheets which appear to be equivalent to bedload sheets except that they fine downstream. It may very well be that the direction of fining is only with respect to what he called the feature's leading edge; if the coarse clasts are designated the front, the feature fines upstream, and if the coarse clasts are designated the tail, the feature fines downstream. Hein & Walker (1977) observed diffuse gravel sheets that were nearly as wide as braided channels and only several coarse clasts in thickness. They argued that these features form lags that may act as the nucleus for bar growth. While they themselves did not discuss in much detail the segregation of sizes, it appears that these features are coarser than the other bed material.

SORTING FROM DIFFERENTIAL TRANSPORT MODES/ TRAJECTORIES

The longitudinal and transverse variation in topography and boundary shear stress in channels provides additional mechanisms for sorting sediment. Here we consider the factors that determine paths and modes of transport once grains are in motion. In this sense we are only secondarily concerned with entrainment and distrainment differences; we are concerned primarily with shifts in styles of sediment transport and with how transverse slopes and flow directions alter grain trajectories.

The steep slope of bar fronts and the more gently sloped bar top afford opportunities to sort bedload and alter bed material size. The path a particle takes on the bed is determined by the various forces applied on the particle. A particle's trajectory reflects the sum of the drag exerted by the flow on the exposed area of the grain and the gravitational force acting on the mass of the grain. On a two-dimensional bed without transverse topography, two quartz grains of differing diameter may be entrained at different stresses but the transport of both grains is directly downstream parallel to the flow. When however there is a transversely sloped surface, grains of different size and weight will be deflected from the direction of the applied shear stress. The downslope gravitation force will turn grains toward the base of the slope. The gravitational force applied to a grain depends upon the grain weight, hence larger diameter grains will be turned more directly downslope than smaller diameter grains (since the gravitational force is proportional to D^3 while the drag is proportional to D^2) (Figure 6.11). Dunes superimposed on the bar may further steepen the outward slope. The point bar top is also tipped outward but much more gently than the point bar slope and as a result the gravitational effect is substantially less and coarse particle paths are close to the direction of flow and similar to the paths of the finer fraction.

In meandering alluvial channels, the pool at the base of the point bar slope is typically coarse compared to the rest of the channel bed (Figure 6.9a). This is particularly true in the upstream part of the pool along the outer concave bank (Bluck, 1971, 1976; Hooke, 1975; Jackson, 1975; Bridge & Jarvis, 1976; Dietrich et al., 1979; Dietrich & Smith, 1984; Van Alphen et al., 1984; Dietrich & Whiting, 1989). One cause of this spatial variation in size is the transverse rolling of coarse grains into the pool and their collection when boundary shear stress is insufficient to roll them downstream, and up and out of the pool. The transverse slopes of alternate bars and braid bars (double-row bars) similarly sort sediment over their tops and sideslope margins. Lisle et al. (1991) described the rolling of the coarser fraction off alternate-bar tops and into the pools while fines were carried downstream over the bar tops (Figure 6.9b). The chute and lobe sequence making up braid bars selectively routes bedload, altering the surface size over the features (Ashworth et al., 1992a,b) (Figure 6.9c).

Flow in curved channels establishes a cross-stream pressure gradient that acts to turn near-bed flow inward. While not spanning the entire channel width, and typically restricted to areas beyond the bar front (Dietrich & Smith, 1984), such inward flow preferentially directs finer sediment toward the inner convex bank compared to the coarser sediment. Smaller diameter grains have more surface area per unit of mass than

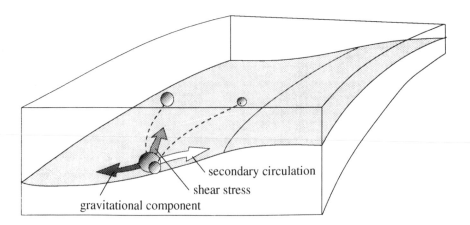

Figure 6.11 Forces on differing size grains on a tipped surface

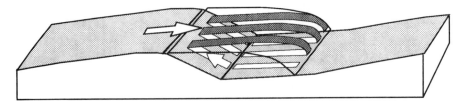

Figure 6.12 Troughwise flow in the lee of oblique topographic steps. Near-bed flow is upstream toward the base of the step. The upper part of the recirculating zone is approximately parallel to the streaming downstream flow

Sediment may also be segregated over a bar or dune front because of the different settling velocities of each grain-size fraction. Larger particles settle more rapidly than smaller particles so that in a mixture of various sizes swept off the top of a dune or bar, the largest grains will have the shortest trajectories. A systematic fallout pattern of proximal coarse and distal fines results (Figure 6.13). The accumulation of the coarsest grains near the crest is followed often by avalanching that buries the finer toe of the slope, producing coarse–fine couplets. The displacement of the finer fraction may be enhanced by their starting slightly higher in the bedload layer at the crest of the dune (D. Mohrig, pers. comm.) Fine sediment may even bypass the trough and land on the stoss side of the downstream dune while coarse material rolls into and collects in the trough (McBride *et al.*, 1975). Slingerland (1984) and Jopling (1967) have also discussed selective transport and deposition related to differential settling velocities.

Another mechanism that can lead to sorting is a change in the style of transport of a certain size fraction. Dietrich & Whiting (1989) described a situation where sand moving as suspended load was swept by secondary currents into a region of decreasing boundary shear stress. This formerly suspended sand settled to the bed and moved as bedload along the downstream part of a point bar of the Rio Grande del Frijoles. Mud drapes may form in this manner in some settings.

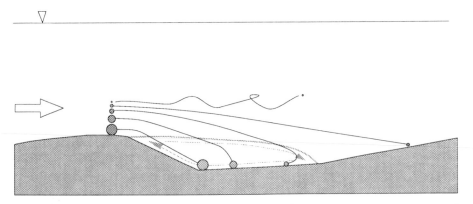

Figure 6.13 Downstream displacement of grains of differing settling velocity swept off a topographic step. The largest grains are deposited closest to the crest because of their large settling velocity whereas finer sediment is carried further downstream

SORTING FROM GRAIN FLOWS

Shear of a concentrated granular dispersion has been observed to generate sorting. In the motion of many individual grains, as for instance in an avalanche, collisions between the grains serve to partially dilate the collective mass of grains (Bagnold, 1954). This dispersive pressure, as it is called, is proportional to the square of the size and density of the constituent particles. In a mass of grains of various size and density, larger and denser particles are preferentially driven toward the top of the flow.

Geologic examples, where dispersive pressures have been proposed to explain segregation, are rather limited due to the high concentrations required to generate the effect. It has been suggested as a mechanism generating beach-swash laminae (Clifton, 1969), but in channels it appears to be important only when avalanching occurs on the foresets of dunes or bars. As the dune brink built from bedload accumulating just past the dune lip collapses, the coarsest grains rise to the top of the shearing layer and are preferentially carried to the base of the slope. Segregation by size occurs along the foreset and between the base and surface of each foreset package (Sallenger, 1979). Allen (1963) pointed out that the avalanche deposits reverse the size pattern that would be expected from grains settling on the slope. On flat beds there do not appear to be sufficient concentrations of bedload to generate dispersive pressures that would lead to dilated masses, and then segregation. While Todd (1989) proposed that a high-density traction carpet supported by dispersive pressures could explain portions of the Lower Old Red Sandstone, modern analogues have not been clearly proven.

SORTING FROM CHANGES IN SUPPLY

A few other situations can be mentioned that lead to size variation in a channel. Mud drapes, as for instance shown by Jackson (1975), are deposited in areas of locally reduced boundary shear stress and as the flow wanes. These materials are often the first to be entrained with the next flood. Size differences can also result from tributary streams that supply a different size sediment. Muddy Creek supplies sand to the gravel-bedded East Fork River in Wyoming, USA. This sand moves in waves over the less mobile gravel. Bank erosion can introduce new sediment sizes to a stream, thus leading to sorting.

Longitudinal sorting as represented by progressive downstream fining has been well known since Sternberg (1875), who proposed a clast diminution model that gave an exponentially decreasing grain size with distance travelled. If abrasion is important, the material carried downstream becomes progressively smaller because the supply to downstream areas is finer. The relative importance of abrasion and selective transport of the finer grains has been debated extensively; however, the recent results of Parker (1991b) suggest that except for grains of material like limestone, selective transport of the finest fraction is most important in leaving a concentration of large grains upstream and providing a large supply of fine sediment downstream.

CONCLUSIONS

A variety of mechanisms give rise to spatial segregation in channels: among these mechanisms there is sorting due to preferential entrainment, preferential distrainment, self-mediated transport, differential transport modes or different trajectories, changes in supply, and grain flows (Table 6.1). The scale of sorting ranges from clusters of a few grains, to coarse bar heads, to downstream fining along the channel length. The sorting mechanisms are enhanced by topography including ripples, dunes and bars. Topography generates spatial variation in shear stresses, provides opportunities to express different transport modes and to reorient transport vectors, establishes zones of separated flow, and creates slopes steep enough to avalanche. Furthermore, topography, and particularly the superposition of bedforms and bars, creates the opportunity to preserve, somewhat selectively, spatial patterns of sorting. Among the most important consequences of the sorting process are economic and pollutant concentrates, substrate alteration that influences the biotic systems, and flow and transport modification that can in turn initiate bedform and even bar development. In closing, a deeper insight into sorting mechanisms taken with the discussions in the other chapters shows that we are making progress toward a more complete understanding of channel processes.

REFERENCES

Allen, J.R.L. (1963). Sedimentation to the lee of small underwater sand waves: an experimental study. *Journal of Sedimentary Petrology*, **23**, 95–116.

Allen, J.R.L. (1970). *Physical Processes in Sedimentation*, London, Allen and Unwin, 248 pp.

Allen, J.R.L. (1983). Gravel overpassing on humpback bars supplied with mixed sediment, examples from the Lower Old Red Sandstone, southwestern Britain. *Sedimentology*, **30**, 285–294.

Allen, J.R.L. (1984). *Sedimentary Structures – their Character and Physical Basis*. Elsevier Science Publishers, Amsterdam, 593 pp.

American Geological Institute (1984). *Dictionary of Geologic Terms*, 3rd edn (eds R.L. Bates & J.A. Jackson), Anchor Press, Garden City, NY, 571 pp.

Andrews, E.D. (1979). *Scour and fill in a stream channel, East Fork River, Western Wyoming*. US Geological Survey Professional Paper **1117**, 49 pp.

Andrews, E.D. & Erman, D.C. (1986). Persistence in the size distribution of surficial bed material during an extreme snowmelt flood. *Water Resources Research*, **22**, 191–197.

Arnott, R.W. & Hand, M. (1989). Bedforms, primary structures, and grain fabric in the presence of suspended sediment rain. *Journal of Sedimentary Petrology*, **59** (6), 1062–1069.

Ashmore, P.E. (1991). How do gravel-bed rivers braid? *Canadian Journal of Earth Science*, **28**, 326–341.

Ashworth, P.J. & Ferguson, R.I. (1989). Size selective entrainment of bedload in a gravel bed stream. *Water Resources Research*, **25**(4), 627–634.

Ashworth, P.J., Ferguson, R.I. & Powell, M.D. (1992a). Bedload transport and sorting in braided channels. In *Dynamics of Gravel-bed Rivers* (eds P. Billi, R.D. Hey, C.R. Thorne & P. Tacconi), John Wiley and Sons, Chichester, 497–513.

Ashworth, P.J., Ferguson, R.I., Ashmore, P.E., Paola, C., Powell, M.D. & Prestegaard, K.L. (1992b). Measurements in a braided river chute and lobe, 2, Sorting of bedload during entrainment, transport, and deposition. *Water Resources Research*, **28** (7), 1887–1896.

Bagnold, R.A. (1954). Experiments on a gravity-free dispersion of large solid spheres in a Newtonian fluid under shear. *Proceedings Royal Society London*, **225**, 49–63.

Bandorf-Nielsen, O., Dalsgaard, K., Halgreen, C., Kuhlman, H., Møller, J.T. & Schou, G. (1982). Variation in particle size distribution over a small dune. *Sedimentology*, **29**, 53–65.

Best, J. & Bridge, J. (1992). The morphology and dynamics of low amplitude bedwaves upon upper stage plane beds and the preservation of planar lamellae. *Sedimentology*, **39**, 737–752.

Bluck, B.J. (1971). Sedimentation in the meandering River Endrick. *Scottish Journal of Geology*, **7**, 93–138.

Bluck, B.J. (1974). Structure and depositional properties of some valley sandur deposits in southern Iceland. *Sedimentology*, **21**, 533–554.

Bluck, B.J. (1976). Sedimentation in some Scottish rivers of low sinuosity. *Trans. Roy. Soc. Edin.*, **69**, 425–455.

Boothroyd, J.C. (1970). Recent braided stream sedimentation (abs.). *Bulletin American Association of Petroleum Geologists*, **54**, 836.

Brady, L.L. & Jobson, H.E. (1973). *An experimental study of heavy mineral segregation under alluvial flow conditions.* US Geological Survey Professional Paper, **562-K**, 38 pp.

Brayshaw, A.C. (1985). Bed microtopography and entrainment thresholds in gravel-bed rivers. *Geological Society America Bulletin*, **96**, 218–233.

Bridge, J.S. (1977). Flow, bed topography, grainsize and sedimentary structure in bends, a three dimensional model. *Earth Surface Processes*, **2**, 401–416.

Bridge, J.S. & Best, J. (1988). Flow, sediment transport and bedform dynamics over the transition from dunes to upper-stage plane beds: implications for the formation of planar laminae. *Sedimentology*, **39**, 737–752.

Bridge, J.S. & Jarvis, J. (1976). Flow and sedimentary processes in the meandering River South Esk, Glen Clova, Scotland. *Earth Surface Processes*, **1**, 303–336.

Buffington, J.M., Dietrich, W.E. & Kirchner, J.W. (1992). Friction angle measurements on a naturally formed gravel stream bed, implications for critical boundary shear stress. *Water Resources Research*, **28**(2), 411–425.

Carling, P.A. (1990). Particle over-passing on depth limited gravel bars. *Sedimentology*, **37**, 345–355.

Carling, P.A., Kelsey, A. & Glaister, M.S. (1992). Effect of bed roughness, particle shape, and orientation on initial motion criteria. In *Dynamics of Gravel-bed Rivers* (eds Billi, R.D. Hey, C.R. Thorne & P. Tacconi), John Wiley and Sons, Chichester, 23–40.

Carling, P.A., & Wood, N. (1994). Simulation of flow over pool-riffle topography: a consideration of the velocity reversal hypothesis. *Earth Surface Process and Landforms*, **19**, 319–332.

Cheel, R.J. (1984). Heavy mineral shadows, a new sedimentary structure formed under upper-flow-regime conditions, its directional and hydraulic significance. *Journal of Sedimentary Petrology*, **54**, 1175–1182.

Cheel, R.J. (1990). Horizontal lamination and the sequence of bed phases and stratification under upper-flow-regime conditions. *Sedimentology*, **37**, 517–527.

Church, M. & Jones, D. (1982). Channel bars in gravel-bed rivers. In *Gravel-bed Rivers* (eds R.D. Hey, J.C. Bathurst & C.R. Thorne), John Wiley, pp. 291–338.

Clifford, N.J. & Richards, K.S. (1992). The reversal hypothesis and the maintenance of riffle pool sequences; a review and field appraisal. In *Lowland Floodplain Rivers: Geomorphological Perspectives* (eds P.A. Carling & G.E. Petts), John Wiley and Sons, Chichester, 43–70.

Clifton, H.E. (1969). Beach lamination: nature and origin. *Marine Geology*, **7**, 553–559.

Dietrich, W.E. (1987). Mechanics of flow and sediment transport in river bends. In *River Channels; Environment and Process* (ed. K.S. Richards), Basil Blackwell, Oxford, 179–227.

Dietrich, W.E. & Smith, J.D. (1984). Bedload transport in a river meander. *Water Resources Research*, **20**(10), 1355–1380.

Dietrich, W.E. & Whiting, P.J. (1989). Boundary shear stress and sediment transport in river meanders of sand and gravel. In *River Meandering* (eds S. Ikeda & G. Parker), Water Resources Monograph **12**, American Geophysical Union, 1–50.

Dietrich, W.E., Smith, J.D. & Dunne, T. (1979). Flow and sediment transport in a sand bedded meander. *Journal of Geology*, **87**, 305–315.

Dietrich, W.E., Kirchner, J.W., Ikeda, H. & Iseya, F. (1989). Sediment supply and the development of the coarse surface layer in gravel-bedded rivers. *Nature*, **340**, 215–217.

Dinehart, R.L. (1992). Evolution of coarse-gravel bedforms, field measurements at flood stage. *Water Resources Research*, **28**(10), 2667–2689.

Drake, T.G., Shreve, R.L., Dietrich, W.E., Whiting, P.J. & Leopold, L.B. (1988). Bedload transport of fine gravel observed by motion-picture photography. *Journal of Fluid Mechanics*, **192**, 193–217.

Dunkerley, D.L. (1990). The development of armour in the Tambo River, Victoria, Australia. *Earth Surface Processes and Landforms*, **15**, 405–412.

Everts, C.H. (1973). Particle overpassing on flat granular boundaries. *Journal of the Waterways and Harbors Division, Proceedings of the American Society of Civil Engineers*, **99**, 425–438.

Ferguson, R.I., Prestegaard, K.L. & Ashworth, P.J. (1989). Influence of sand on hydraulics and gravel transport in a braided gravel bed river. *Water Resources Research*, **25**(4), 635–643.

Foley, M.G. (1977). Gravel-lens formation in antidune-regime flow – a quantitative hydrodynamic indicator. *Journal of Sedimentary Petrology*, **47**, 738–746.

Germanoski, D. (1989). Development of planar horizontal laminae by the migration of low-amplitude linguoid bars under lower-regime flow (abs.). *Geological Society of America Annual Meeting, Abstracts with Programs*, **21**(7), 97.

Gessler, J. (1965). *The beginning of bedload movement of mixtures investigated as natural armouring in channels*. Report **69**, Laboratory of Hydraulic Research and Soil Mechanics of the Swiss Federal Institute of Technology (translated by E.A. Prych).

Grant, G.E., Swanson, F.J. & Wolman, M.G. (1990). Pattern and origin of stepped-bed morphology in high-gradient streams, Western Cascades, Oregon. *Geological Society America Bulletin*, **102**, 340–352.

Gustavson, T.C. (1978). Bedforms and stratification types of modern gravel meander lobes, Nueces River, Texas. *Sedimentology*, **25**, 401–426.

Hein, F.J. & Walker, R.G. (1977). Bar evolution and development of stratification in the gravelly, braided, Kicking Horse River, British Columbia. *Canadian Journal of Earth Science*, **14**, 562–570.

Hill, R. (1987). *Sediment sorting in meandering river*. Unpublished MSc thesis, University of Minn., 75 pp.

Hooke, R.L. (1975). Laboratory study of the influence of granules on flow over a sand bed. *Geological Society America Bulletin*, **79**, 495–500.

Ikeda, H. (1983). *Experiments on bedload transport, bed forms, and sedimentary structures using fine gravel in the 4-meter-wide flume*. Environmental Research Center Papers, number **2**, Tsukuba, Japan, 78 pp.

Iseya, F. & Ikeda, H. (1987). Pulsations in bedload transport rates induced by a longitudinal sediment sorting; a flume study using sand gravel mixtures. *Geografiska Annaler*, **69**, 15–27.

Jackson, R.G. (1975). Velocity–bed-form–texture patterns of meander bends in the lower Wabash River of Illinois and Indiana. *Geological Society America Bulletin*, **86**, 1511–1522.

Jackson, W.L. & Beschta, R.L. (1982). A model of the two-phase bedload transport in an Oregon Coast Range stream. *Earth Surface Processes and Landforms*, **7**, 517–527.

Jopling, A.V. (1967). Origin of laminae deposited by the movement of ripples along a stream bed, a laboratory study. *Journal of Geology*, **75**, 287–305.

Keller, E.A. (1971). Areal sorting of bed-load material; the hypothesis of velocity reversal. *Geological Society America Bulletin*, **82**, 753–756.

Kellerhals, R. (1967). Stable channel with gravel-paved beds. *Journal of the Hydraulics Division, Proceedings of the American Society of Civil Engineers*, **93**, 63–64.

Komar, P.D. (1987). Selective grain entrainment by current from a bed of mixed sizes, a reanalysis. *Journal of Sedimentary Petrology*, **57**(2), 203–211.

Komar, P.D. & Li, Z. (1986). Pivoting analyses of the selective entrainment of sediments by shape and size with application to gravel thresholds. *Sedimentology*, **33**, 425–436.

Kuenen, P.H. (1966). Experimental turbidite lamination in a circular flume. *Journal Geology*, **74**, 523–545.

Kuhnle, R.A. (1989). Bed surface size changes in gravel bed channel. *Journal of Hydraulic Engineering*, **115**(6), 731–743.

Kuhnle, R.A. & Southard, J.B. (1988). Bedload transport fluctuations in a gravel bed laboratory channel. *Water Resources Research*, **24**, 247–260.

Kuhnle, R.A. & Southard, J.B. (1990). Flume experiments on the transport of heavy minerals in gravel-bed streams. *Journal of Sedimentary Petrology*, **60**, 687–696.

Leopold, L.B. & Wolman, M.G. (1957). *River channel patterns – braided, meandering, and straight*. US Geological Survey Professional Paper, **282-B**, 39–85.

Leopold, L.B., Wolman, M.G. & Miller, J.P. (1964). *Fluvial Processes in Geomorphology*. Freeman, San Francisco, 522 pp.

Lewin, J. (1976). Initiation of bed forms and meanders in coarse-grained sediment. *Geological Society America Bulletin*, **87**, 281–285.

Li, Z. & Komar, P.D. (1986). Laboratory measurements of pivoting angels for applications to selective entrainment of gravel in a current. *Sedimentology*, **33**, 413–423.

Lisle, T.E. (1979). A sorting mechanism for a riffle pool sequence. *Geological Society America Bulletin*, **90**, 1142–1157.

Lisle, T.E. & Madej, M.A. (1992). Spatial variation in armouring in a stream channel with high sediment supply. In *Dynamics of Gravel-bed Rivers* (eds P. Billi, R.D. Hey, C.R. Thorne & P. Tacconi), John Wiley and Sons, Chichester, 277–296.

Lisle, T.E., Ikeda, H. & Iseya, F. (1991). Formation of stationary alternate bars in a steep channel with mixed-size sediment; a flume experiment. *Earth Surface Processes and Landforms*, **16**, 463–469.

Little, W.C. & Mayer, P.G. (1976). Stability of channel beds by armouring. *Journal of the Hydraulics Division, Proceedings of the American Society of Civil Engineers*, **102**(11), 1647–1661.

McBride, E.F., Shepherd, R.G. & Crawley, R.A. (1975). Origin of parallel near-horizontal laminae by migration of bed forms in a small flume. *Journal of Sedimentary Petrology*, **45**, 132–139.

McDonald, B.C. & Banerjee, I. (1971). Sediments and bedforms on a braided outwash plain. *Canadian Journal of Earth Science*, **8**, 1282–1301.

McQuivey, R.S. & Keefer, T.N. (1969). *The relation of turbulence to deposition of magnetite over ripples*. US Geological Survey Professional Paper, **650-D**, D244–D247.

Meade, R.H. (1985). Wavelike movement of bedload sediment, East Fork River, Wyoming. *Environmental Geology and Water Science*, **7**, 215–225.

Milhous, R.T. & Klingeman, P.C. (1973). Sediment transport in a gravel-bottom-stream. *Proceedings of the 21st Annual Specialty Conference of the American Society of Civil Engineers, Hydraulics Division*, 293–303.

Milne, J.A. (1982). Bed-material size and the riffle–pool sequence. *Sedimentology*, **29**, 267–278.

Mosley, M.P. (1988). Bedload transport and sediment yield in the Onyx River, Antarctica. *Earth Surface Processes and Landforms*, **13**, 51–67.

Moss, A.J. (1963). The physical nature of the common sandy and pebbly deposits, Part II. *American Journal of Science*, **261**, 297–343.

Moss, A.J. (1972). Bedload sediments. *Sedimentology*, **18**, 159–219.

Paola, C., Wiele, S.M. & Reinhart, M.A. (1989). Upper regime parallel laminations as the result of turbulent sediment transport and low-amplitude bedforms. *Sedimentology*, **36**, 47–60.

Paola, C., Parker, G., Seal, R., Sinha, S.K., Southard, J.B. & Wilcock, P.R. (1992). Downstream fining by selective deposition in a laboratory flume. *Science*, **258**, 1757–1760.

Parker, G. (1991a). Selective sorting and abrasion of river gravel, I, Theory. *Journal of Hydraulic Engineering*, **117**, 131–149.

Parker, G. (1991b). Selective sorting and abrasion of river gravel, II, Applications. *Journal of Hydraulic Engineering*, **117**, 150–171.

Parker, G. & Andrews, E.D. (1985). Sorting of bed load sediment by flow in meander bends. *Water Resources Research*, **21**, 1361–1373.

Parker, G. & Klingeman, P.C. (1982). On why gravel-bed streams are paved. *Water Resources Research*, **18**(5), 1409–1423.

Parker, G., Dhamotharan, S. & Stefan, H. (1982). Model experiments on mobile, paved gravel bed streams. *Water Resources Research*, **18**(5), 1395–1408.

Reid, I., Frostick, L.E. & Brayshaw, A.C. (1992). Microform roughness elements and the selective entrainment and entrapment of particles in gravel bed rivers. In *Dynamics of Gravel-bed Rivers* (eds P. Billi, R.D. Hey, C.R. Thorne & P. Tacconi), John Wiley and Sons, Chichester, 253–266.

Sallenger, A.H. (1979). Inverse grading and hydraulic equivalence in grain-flow deposits. *Journal of Sedimentary Petrology*, **49**, 553–562.

Schiller, E.J. & Rowney, A.C. (1984). Stream-bed armouring under known conditions of upstream sediment input. *Canadian Journal of Earth Science*, **21**, 1061–1066.

Slingerland, R. (1984). Role of hydraulic sorting in the origin of fluvial placers. *Journal of Sedimentary Petrology*, **54**, 137–150.

Sternberg, H. (1875). Untersuchungen uber Langen- und Quer-profil geschiebefuhrende Flusse. *Zeitschrift Bauwesen*, **25**, 483–506.

Todd, S.P. (1989). Stream-driven, high-density gravelly traction carpets; possible deposits in the Trabeg Conglomerate Formation, SW Ireland, and some theoretical considerations of their origin. *Sedimentology*, **36**, 513–530.

Tsujimoto, T. (1989). Instability of bed-surface composition due to sorting process in a stream composed of sand and gravel. In *Proceedings of the International Symposium Sediment Transport Modeling*, ASCE, New Orleans, 302–307.

Van Alphen, J.S., Bloks, P.M. & Hoekstra, P. (1984). Flow and grainsize pattern in a sharply curved river bend. *Earth Surface Processes and Landforms*, **9**, 513–522.

Whiting, P.J. & Dietrich, W.E. (1990). Boundary shear stress and roughness over mobile alluvial beds. *Journal of Hydraulic Engineering*, **116**(12), 1495–1511.

Whiting, P.J. & Dietrich, W.E. (1991). Convective accelerations and boundary shear stress over a channel bar. *Water Resources Research*, **27**(5), 783–796.

Whiting, P.J., Dietrich, W.E., Leopold, L.B. & Collins, L. (1985). The variability of sediment transport in a fine-gravel stream (abs.). In *Proceedings, Third International Fluvial Sedimentology Conference*, Ft. Collins, CO, Colorado State University Press, 38.

Whiting, P.J., Dietrich, W.E., Leopold, L.B., Drake, T.G. & Shreve, R.L. (1988). Bedload sheets in heterogeneous sediment. *Geology*, **16**, 105–108.

Wiberg, P.L. & Smith, J.D. (1987). Calculations of the critical shear stress for motion of uniform and heterogeneous sediments. *Water Resources Research*, **23**, 1472–1480.

Wilcock, P.R. (1992) Experimental investigation of the effect of mixture properties on transport dynamics. In *Dynamics of Gravel-bed Rivers* (eds P. Billi, R.D. Hey, C.R. Thorne & P. Tacconi), John Wiley and Sons, Chichester, 109–131.

Wilcock, P.R. & Southard, J.B. (1989). Bedload transport of mixed size sediment, fractional transport rates, bedforms, and the development of a coarse bed-surface layer. *Water Resources Research*, **25**, 1629–1641.

Williams, G.P. & Wolman, M.G. (1984). *Downstream effects of dams on alluvial rivers*, US Geological Survey Professional Paper, **1286**, 83 pp.

7 Modelling the Sediment Transport Process

ADRIAN KELSEY

Department of Geography, Lancaster University, UK

INTRODUCTION

The range of scales across which the sediment transport process occurs, the number of constituent processes acting and the presence of stochastic elements in the processes occurring at these scales have been described in preceding chapters. When modelling the sediment transport process a balance must be made between the detail with which the process is modelled and the resolution and scale required of the final result. This balance occurs because the necessary information to describe a system at a given resolution will increase as the scale, spatial or temporal, under consideration increases. This increase occurs before any attempt is made to model the behaviour of the system and will only be reduced, not eliminated, if a system can in some sense be described as self-similar. Thus if a model is required to give results at a certain scale this will affect the resolution at which a description of the system can be made or a calculation performed. However, though a process may be occurring at a smaller scale than the resolution of a calculation and can therefore not be represented directly, its effects may still influence the transport process at a larger scale.

The range of scales of the processes described in the previous chapters indicates the consideration of scales that must be included in any description of sediment transport. The mass movement of sediment is the sum of the movement of individual particles; this defines the smallest scales that must be considered when describing sediment transport. The sediment particles move under the influence of turbulent flow; this can be considered to consist of a mean flow component and fluctuating turbulent components of flow. The fluctuating components of flow allow particle movement to occur when mean flow conditions are less than those required for particle movement. The structures present in the turbulence also influence sediment transport, the quantity transported and fluctuations in the transport. These influences on sediment transport are described in Chapter 1. In a model of sediment transport the effects of the mean flow can be described using a deterministic model, while those due to turbulent fluctuations can be described using distributions of velocity fluctuations (Naden, 1987a), shear stress (Bridge & Bennett, 1992) or of transport rate (Williams & Tawn, 1991).

Advances in Fluvial Dynamics and Stratigraphy. Edited by P.A. Carling and M.R. Dawson.
© 1996 John Wiley & Sons Ltd.

The flow around a particle determines the fluid forces acting on it. The forces required to entrain a particle are influenced by the particle and its neighbours. The size and density of a particle influence the forces required to start movement but so does its size relative to other particles forming the bed. In addition the position of particles within a bed, whether embedded or exposed, affects the conditions of entrainment. These influences on the initial movement of particles are described in Chapter 4. A deterministic calculation of the entrainment of a particle due to a flow can be performed for a defined flow, though the flow itself may include stochastic elements. The possible range of particle positions within a bed can be seen in Kirchner *et al.* (1990) and Buffington *et al.* (1992) leading to the use of distributions to define particle positions within the bed (Wiberg & Smith, 1985; Sekine & Kikkawa, 1992).

During the sediment transport process sorting of sediment particles by size and density can occur. The range of conditions for the initial motion of particles is one sorting mechanism but sorting processes are also at work during movement and deposition of particles, described in Chapter 6. The influence of turbulent flow and the local bed geometry on particle movement is such that a range of particle movements is possible for the same conditions, introducing the need for stochastic elements in any description of particle movement. The number of possible processes involved in the movement of particles also means that the same movement may occur due to a number of different processes. When modelling the movement of sediment particles as part of the transport process, consideration must be given to which processes must be described for a particular application and to appropriate descriptions of these processes.

Increasing the scale under consideration from individual particles to groups of particles, the movement of particles modifies the bed geometry leading to bedforms, which modify and interact with the flow (see Chapter 3). The presence of these bedforms and their movement act to make the sediment transport process unsteady at the scale of movement of groups of particles (see Chapter 5).

Since the sediment transport process occurs across a range of scales, models of the process have also been developed across a range of scales. The models of the sediment transport process described in this chapter are based on mathematical descriptions of processes. The first part of the chapter describes models of the movement of individual sediment particles, deterministic and stochastic. Approaches to increasing the scale under consideration from the movement of individual particles to sediment transport and the effects of this transport on the flow and the bed are then considered. Finally models of sediment transport describing duneform and change in bed height and composition are described.

MOVEMENT OF INDIVIDUAL SEDIMENT PARTICLES

Deterministic calculations of the movement of sediment particles due to a mean flow can be performed. The interaction of particles with turbulent fluctuations in the flow requires the introduction of stochastic elements and a realistic description of the bed will also include stochastic elements.

Deterministic and stochastic descriptions of the movement of sediment particles have been developed. The deterministic models solve equations of particle motion and include stochastic elements to describe flow and bed. The stochastic models attempt to describe observed patterns of particle behaviour.

Deterministic models

Deterministic models of sediment particle movement as bedload have been developed to study the behaviour of particles and to obtain descriptions of the movement of particles. The results from these calculations have then been used to develop models of sediment transport rate (Wiberg & Smith, 1989, Sekine & Kikkawa, 1992).

In deterministic models of particle movement the equations of particle motion are solved numerically

$$V_p(\rho_s + c_M\rho)\,\frac{du_{pi}}{dt} = F_i + \rho_s V_p g_i$$

where the particle position and velocity at time t are

$$x_{pi} = (x_{p1}, x_{p2}, x_{p3}) = (x_p, y_p, z_p)$$
$$u_{pi} = (u_{p1}, u_{p2}, u_{p3}) = (u_p, v_p, w_p)$$

the flow velocity is

$$u_i = (u_1, u_2, u_3) = (u, v, w)$$

and ρ and ρ_s are the density of the fluid and the sediment, V_p is the volume of the sediment particle, C_M is the coefficient of added mass and g_i is the acceleration due to gravity. The added mass force is a pressure force opposing the acceleration of a particle through a fluid. The term included on the left-hand side of the equation is due to the particle acceleration. An added mass term is contained in the force term, F_i, on the right-hand side; this term is due to fluid shear. The force F_i represents the forces due to the flow acting on the sediment particle. These can be taken to be:

(i)   buoyancy force
(ii)   drag force due to rectilinear motion
(iii)   lift force
(iv)   Basset history term

The relative contributions of each of these to the total force acting on a particle, the accuracy with which they have been determined and the analysis on which the equation of particle movement is based affect whether they are included in the calculation of particle movement. In Reizes (1978) only drag force, added mass and terms due to gravity were included in the description of particle movement. Murphy & Hooshiari (1982), van Rijn (1984), Wiberg & Smith (1985) and Sekine & Kikkawa (1992) included a lift force in their initial equations of particle movement. The treatment of lift force varied and these terms were not necessarily included when the equations were solved.

The drag force term is evaluated from the expression

$$F_{\mathrm{D}i} = \frac{C_{\mathrm{D}}}{2} \rho A_{\mathrm{p}} \left| (u_{\mathrm{p}j} - u_j) \right| (u_{\mathrm{p}i} - u_i)$$

where A_{p} is the projected area of the particle and C_{D} is the coefficient of drag. The values used in the models for the coefficient of drag were for isolated spheres. The values were calculated for the appropriate Reynolds number from fits made to observations. The value of added mass coefficient, C_{M}, used was 0.5, that for a sphere moving in steady or slowly varying flow. The expressions used for the lift force differ, as do the coefficients of lift used within these expressions. In Wiberg & Smith (1985) an analysis of the lift force due to the differing flow velocities at top and bottom of a particle was presented. The value of the coefficient of lift used in this expression was set empirically, using the data of Chepil (1958). Sekine & Kikkawa (1992) included the expression of Wiberg & Smith (1985) in their initial description of particle movement but found the effects of this term to be small and did not include it in their solutions of particle movement. Murphy & Hooshiari (1982) and Van Rijn (1984) used analytical expressions for the lift force. Murphy & Hooshiari (1982) included lift terms due to particle spin (Barkla & Auchterlonie, 1971) and wall shear (Einstein & El-Samni, 1949) in their description of particle motion; from examination of the tracks of particles in saltation they concluded that the effects of any lift forces were not significant. Van Rijn (1984) considered lift due to shear (Saffman, 1965) and particle spin (Rubinow & Keller, 1961) though both expressions were for viscous flow. Saffman (1965) showed lift due to shear was an order of magnitude greater than lift due to spin, therefore van Rijn (1984) only includes lift due to shear. In the solutions of particle motion it was assumed that the expression for lift due to shear in viscous flow could be applied to turbulent flow. The coefficient of lift in this expression was used to match calculated and observed behaviour resulting in a high coefficient of lift. The Basset history term is usually not included in solutions of particle motion; this term includes the history of the particle acceleration. The theoretical analysis of Basset (1888) omitted convective acceleration and was therefore limited to the condition of low velocity and rapid acceleration, the effects of convective acceleration have been studied and were described in Odar & Hamilton (1964).

The inclusion or otherwise of these terms demonstrates that the equations of particle movement that are solved are approximation, other approximations in these models are the treatment of particles as spheres of equivalent diameter. The numerical methods which are used to solve the equations introduce another possible source of error.

The equations of particle movement describe the motion of particles under the influence of a flow away from the bed. This calculation requires initial conditions to be specified for the particle movement, the flow causing the movement to be specified and the conditions under which the particle movement will cease.

Conditions for initial motion and cessation of motion

The models of particle motion used different descriptions for the start of particle motion; cessation of motion was determined from the result of the impact at the end of each saltation. The outcome of impact with the bed determined whether motion ceased or continued; where motion continued the conditions immediately after impact were used as the initial conditions for the next saltation.

The model of Wiberg & Smith (1985) used a deterministic condition to describe the initial particle velocities at the start of particle motion, based on a pivoting analysis. Van Rijn (1984) used an empirical initial condition; the particles were given initial velocities at the start of the saltation which were a fraction of the shear velocity. In Reizes (1978) and Sekine & Kikkawa (1992) calculations were started with particles possessing zero initial velocity; the particle motion was allowed to develop due to the influence of the flow and the effects of impact.

The results of the impact of particles with the bed were calculated from preservation of momentum; fractions of the components of momentum normal and tangential to the line between particle centres at impact were conserved (Figure 7.1). In Wiberg & Smith (1985) a single value was used for the fraction of momentum conserved for both fractions; this was set by matching calculated values to the observations of Abbott & Francis (1977). Cessation of motion occurred when particles could not clear the next particle. Van Rijn (1984) only calculated single saltations and therefore no conditions for impact and cessation of motion were described. Both Reizes (1978) and Sekine & Kikkawa (1992) included three-dimensional impact models. Reizes (1978) described a model conserving translational and rotational particle momentum in three dimensions, allowing for particle movement with and without slip. In Sekine & Kikkawa (1992) only translational momentum was considered; a single coefficient was used for the fraction of all three components of momentum conserved, and the value of this coefficient was set empirically. The condition of cessation of motion in Reizes (1978) was for particle motion to cease after an arbitrary number of impacts with the same bed particle. In Sekine & Kikkawa (1992) particle motion ceased if too much energy was dissipated for motion to continue. There is a tendency in these models for motion to continue, even if very slowly, hence the need for the imposition of conditions to determine the position at which particle motion ceases.

Stochastic elements

The models of particle movement described above were deterministic, the only stochastic element included in any of them was the description of the bed. The interactions of the particle with the bed were sufficient to give a range of particle movements. The flow over the bed would be turbulent; the mean flow component was usually described by the logarithmic law for turbulent flow over a rough boundary. Observations show that turbulence affects the transport of sediment as bedload (see Chapter 1 and Drake *et al.*, 1988). The effects of turbulence are usually ignored by defining saltation of particles as being unaffected by turbulence. However, flume observations show that particle movement in saltation can be affected by turbulent fluctuations. Particle movement can be affected by turbulence at low transport stages, close to one, the transport stage being defined as the ratio of the mean bed shear velocity to the critical bed shear velocity for initiation of movement of particles (Abbott & Francis, 1977).

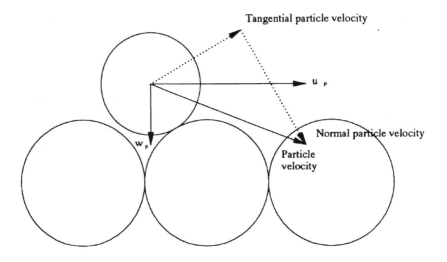

a   Particle velocity and components immediately before impact

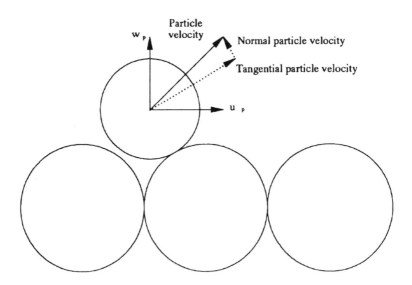

b   Particle velocity and components immediately after impact

Figure 7.1 Conservation of particle momentum on impact with the bed: (a) particle velocity immediately before impact, expressed as horizontal, u_p, and vertical, w_p, velocity components and as normal and tangential velocity components: (b) particle velocity immediately after impact, showing conservation of fractions of the normal and tangential components of velocity

For sediment particles moving as bedload the particles are supported by the bed, either directly, by sliding or rolling, or indirectly, by impacts between saltations. The inclusion of the effects of turbulence on particle movement has been modelled by Naden (1987b), who modelled particle movements based on the observations described in Abbott & Francis (1977). This model included the effects of turbulence on initial motion of particles, by using a probabilistic model of entrainment, and on particle trajectories, by allowing the possibility of a longer particle trajectory than the mean condition, due to the action of turbulence suspending the particle. Deterministic treatments of the effects of turbulent velocity fluctuations on particle movements are described in Yvergniaux & Chollet (1989) for particles moving fully in suspension and in Kelsey *et al.* (1994) for particles moving as bedload, with initial motion and trajectories influenced by turbulence.

The deterministic models reviewed here can be used to calculate particle movements, initial motion, impacts and cessation of movement and the influence of turbulence on these processes. Similar though not identical models exist for the aeolian environment (Rumpel, 1985; Ungar & Haff, 1987; Anderson, 1987; Anderson & Haff, 1988; 1991; Werner, 1990; McEwan & Willetts, 1991). The models are not identical due to the difference in relative densities of sediment particles and the fluid through which they are moving. In water the relative density of sediment is of order one, in air of order one thousand. When calculating particle movement in air, terms which include the fluid density are small compared to those which include the sediment density, and the terms including the fluid density are therefore usually ignored. The behaviour of particles is also modified leading to a different emphasis in the aeolian and fluvial models. In air two distinct thresholds for initiation of particle movement have been observed, one due to impact and another, higher threshold for initial particle movement due to fluid forces. In water entrainment due to impact can occur, but is infrequent (Drake *et al.*, 1988) and is therefore not usually considered. Initial movement is not included in many of the models of fluvial sediment particle movement because of the difficulty in modelling the transition from initial motion to particle movement.

Results

In the model of van Rijn (1984) the calculated saltations were fitted to observations of saltations, calculated quantities were then used to supply parameters for use in a model of bedload transport rate of sediment. Murphy & Hooshiari (1982) only show a comparison with a single observed saltation, with a reasonable fit (Figure 7.2a). Wiberg & Smith (1985) (Figure 7.2b) and Sekine & Kikkawa (1992) (Figure 7.2c) make more comparisons with observed data; plots of calculated against observed values show a reasonable fit.

Stochastic analysis

The deterministic models of particle movement of Wiberg & Smith (1985) and Sekine & Kikkawa (1992) included a stochastic element to enable the model output to match observed behaviour. Analysis of the vertical particle distributions and the

a

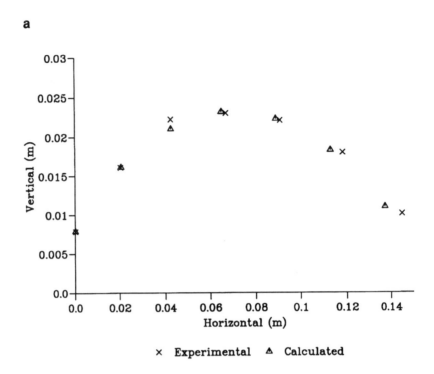

× Experimental   △ Calculated

b

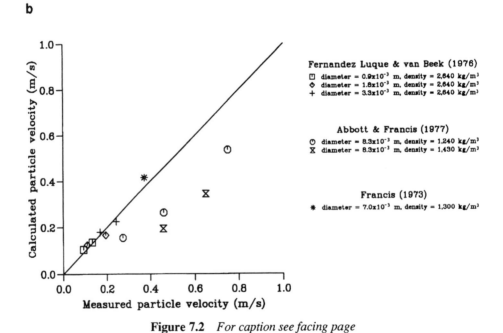

Fernandez Luque & van Beek (1976)
☐ diameter = 0.9x10⁻³ m, density = 2,640 kg/m³
◇ diameter = 1.8x10⁻³ m, density = 2,640 kg/m³
+ diameter = 3.3x10⁻³ m, density = 2,640 kg/m³

Abbott & Francis (1977)
○ diameter = 8.3x10⁻³ m, density = 1,240 kg/m³
X diameter = 8.3x10⁻³ m, density = 1,430 kg/m³

Francis (1973)
✳ diameter = 7.0x10⁻³ m, density = 1,300 kg/m³

Figure 7.2   *For caption see facing page*

c

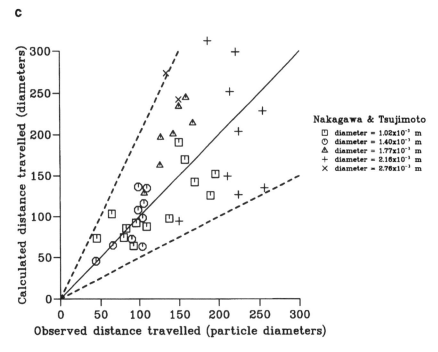

Figure 7.2 Comparison of results of deterministic calculations of particle movement with observations. (a) Comparison of observed and calculated particle trajectory (Murphy & Hooshiari, 1982). (b) Comparison of calculated and measured particle velocities (Wiberg & Smith, 1985). (c) Comparison of observed and calculated distances travelled in a single movement by particles (Sekine & Kikkawa, 1992)

resulting momentum extraction from the flow (Raupach, 1991) shows that particles cannot follow a single "mean" saltation, rather a range of saltations must occur.

An alternative approach to the calculation of particle motion is to use stochastic distributions to characterize the range of particle behaviour. In this method observed distributions of particle movement are reproduced either by the direct fitting of distribution functions or by parameterizing existing distribution functions.

The approach of using stochastic distributions to characterize particle movements is described in Einstein (1937). In this work an attempt was made to find distributions that could explain the range of observed particle movements during flume experiments. Based on observations of the movement of sediment particles in these experiments Einstein (1937) proposed describing the movement of particles as consisting of two components, instantaneous particle movements, interspersed by periods of rest. Einstein (1937) suggested the use of negative exponential distributions to describe the distributions of particle movement and the rest periods. The distributions were characterized by the mean distance travelled and the mean time at rest. These distributions were found to be able to give a good fit to data.

This approach was also used by Hubbell & Sayre (1964) to describe the dispersion of radioactive sand, and a similar description was used by Crickmore & Lean (1962) for the same purpose. These experiments involved the dispersion of radioactive sand and were mainly performed in flumes with steady flow conditions being used. A generalization of this description of particle movement can be found in Shen & Todorovic (1971). However, in all these cases the mean quantities used to define the distributions of movement and rest periods were based on fitting curves to the observed distributions as direct measurements of particle movements could not be made. Measurements of the behaviour of radio tagged particles, enabling these quantities to be determined directly, have been made and are described in Schmidt & Ergenzinger (1992). The information is difficult to obtain, requiring considerable equipment to be installed at a measuring site.

An alternative approach to describing the movement of particles is to attempt to fit distributions directly to the data from observations of the behaviour of tagged particles. This approach is useful when examining movements of particles over events, where the duration and magnitudes of flows are not known. The derivation and calculation of such distributions were described in Hassan & Church (1992) and Kirkby (1991). The distributions fitted were presented in terms of numbers of events rather than movements within events. The observed distributions were fitted using gamma distributions, and different order distributions were calculated to find the best fit to observed behaviour. The order of the distribution was assumed to be related to the number of individual movements making up the total distance travelled during an event.

Summary

The models described demonstrate that the movement of individual particles can be described using both deterministic and stochastic models. The deterministic models attempt to reproduce observed particle behaviour by modelling the processes acting on particles. The stochastic models are derived from the observed behaviour of particles and attempt to model this behaviour without describing the exact physical processes causing the movement.

The movements of individual particles are affected by the composition and structure of the bed over which they move and the turbulent flow with its fluctuations. The possible ranges of these interactions, with no deterministic models to describe either the variation of the bed or fluctuations in the flow, lead to the introduction of stochastic elements into the deterministic models, for example, the determination of the position and height of impact between moving and bed particles from a random distribution (Wiberg & Smith, 1985, Sekine & Kikkawa, 1992). When a deterministic calculation of particle movement is made, the conditions have to be specified and are therefore known precisely. The descriptions of particle movement based on observations of particle movement can be made for known conditions (Einstein, 1937; Hubbell & Sayre, 1964) or for events (Hassan *et al.* 1991; Kirkby, 1991) which will include a wider range of conditions which are probably not known precisely. The use of the latter as predictive tools for particle movement without knowledge of the conditions under which these movements occurred would be difficult.

The surface over which sediment transport occurs and the turbulent flow causing the transport mean that stochastic influences on particle movement must be included in deterministic models of particle movement. The range of scales and pervasiveness of these influences mean that deterministic models are limited in their ability to include all such effects. There will therefore be situations where stochastic models based on observations are more suitable to describe the movement of particles than the deterministic models.

Calculation of the movement of single particles allows the interaction of particles and the bed to be examined but the bed cannot be modified due to the movement of single particles. To model sediment transport and associated modification of the bed and the flow, the scale under consideration must: be increased from the calculation of the movement of single particles to mass movement of particles.

SINGLE PARTICLE MOVEMENT TO SEDIMENT TRANSPORT

Increasing the quantity of particle movement under consideration from single to multiple particles and so sediment transport increases the processes that must be considered when developing the description of sediment transport. The mass movement of sediment particles allows the bed to be modified by this transport. The possibility of interactions between moving particles and so modification of the movement of individual particles is also introduced. The flow can be modified by change in the surface height of the bed and the composition of the bed. The flow can also be modified directly by momentum transfer from moving particles to the bed.

The models of particle movement described in the previous section have been used to examine sediment transport and its effects. When a model of sediment transport is derived from a description of the behaviour of single particles the physical processes described above must be considered. The scale at which sediment transport is modelled and the detail of that description must also be considered.

Rate of sediment transport

Deterministic models

The calculations of particle movement can be used to supply data for models of the rate of transport of sediment. The models of van Rijn (1984), Wiberg & Smith (1985) and Sekine & Kikkawa (1992) have been used for this purpose. For the calculation of the rate of sediment transport this means that rather than using empirical expressions, based on dimensional analysis, or semi-empirical expressions, based on physical reasoning, the expression can be based on deterministic, though approximate, expressions. Rather than being related to a particular set of data, as with the empirical results, the results of such expression should be of general application. Such a use of the results of calculations of particle movement allows the calculation of particle movement to be separated from calculations of sediment transport; the data from a calculation can be used many times without repeating the calculation. The calculation of rate of sediment transport only makes use of mean quantities, or distributions, from the calculations of particle movement, a reduction in the resolution of the result. The effects to be considered when calculating the rate of transport of sediment are interaction of sediment particles with the flow and with other moving particles. Modification of the bed by the transport of sediment is not considered in developing expressions for the rate of transport of sediment.

The interaction of sediment particles with the flow can be considered as a control on the quantity of sediment in motion. In van Rijn (1984), the quantity of sediment in motion was calculated using an empirical expression for the concentration of sediment in the bedload layer. The expression for the bedload concentration was derived from analysis of values of transport rate, q_B. The concentration of bedload sediment was calculated

$$C_B = \frac{q_B}{U_B \delta_B}$$

where U_B and δ_B are values of mean bedload particle velocity and the thickness of the bedload layer. The values for these quantities were calculated using expressions fitted to the results of the calculations of movements of single particles.

Wiberg & Smith (1989) and Sekine & Kikkawa (1992) calculate the quantity of sediment in motion from consideration of the momentum extracted from the flow by the movement of the sediment particles. This approach uses a hypothesis put forward by Owen (1964) that "the concentration of particles within the saltation layer is governed by the condition that the shearing stress borne by the fluid falls, as the surface is approached, to a value just sufficient to ensure that the surface grains are in a mobile state". The saltation layer is the layer in which the majority of particle movement is in saltation and therefore as bedload, and can be regarded as the same as the bedload layer. The total shear stress due to the flow, τ, is regarded as being partitioned between sediment particles and fluid in the bedload layer

$$\tau = \tau_f + \tau_g$$

where τ_f is the shear stress carried by the flow and τ_g is the shear stress carried by sediment particles. At the bed the shear stress carried by the flow falls to the critical value for the entrainment of sediment particles, τ_{cr} (Figure 7.3). If the fluid shear stress at the bed was above the critical value for entrainment, particles would continue to be entrained and momentum extracted from the flow would increase until the critical value was reached, when entrainment would cease. If the fluid shear stress at the bed was below the critical value for entrainment then no particles would be entrained and the momentum extracted from the flow would fall until the critical value was reached, when entrainment would start. The quantity of sediment in motion is determined by a balance between shear stress acting at the bed and momentum extracted from the flow by the sediment in motion in a self-equilibrating balance. When the bedload layer is thin in comparison with the total depth of the flow, the total shear stress within the layer can be regarded as constant and equal to the shear stress exerted by the flow at the top of the saltation layer. The behaviour of sediment particles is assumed to be unaffected by any changes in flow structure within the bedload layer due to the momentum extracted from the flow.

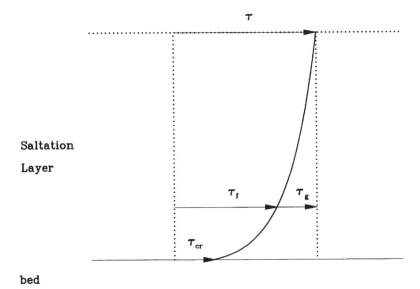

Figure 7.3 Variation in shear stress close to the bed in the saltation layer. The shear stress at the top of the layer, τ, is that due to the flow. Within the saltation layer the shear stress is partitioned between that carried by the fluid, τ_f, and that by the saltating grains, τ_g, due to the momentum extracted from the flow by the grains. The fluid shear stress falls to the critical shear stress for initiation of particle movement at the bed

As the rate of sediment transport increases, the number of particles in motion increases and the likelihood of interactions between particles in motion, modifying particle trajectories, also increases. Interaction of moving particles is not treated in these models, but is offered as a reason for divergence of observed and calculated transport rates (Sekine & Kikkawa, 1992). Limits for the assumption of no interaction of particles have been offered at a transport stage of 2 (Leeder, 1979) and 1.5 (Sekine & Kikkawa, 1992). The models of transport rate give a good fit to observations; however, the data selected at present are for a single size or size range of particles, rather than a range of values as would be found in a natural river.

The use of single-particle models to develop expressions for transport rate allow expressions to be developed and parameterized based on known descriptions of the sediment transport process with a minimum of empirical terms involved. The use of the results of calculations of particle movement in calculations of rate of transport moves away from the spatial element in the particle calculations, though this can be reintroduced in calculations involving sediment continuity. An alternative approach is to calculate the movement of the individual sediment particles forming the sediment transport (Jiang & Haff, 1993).

Stochastic models

The rate of transport of sediment can also be derived from stochastic models of the movement of particles. The expressions for the rate of transport of sediment described here are of two types: those requiring data describing the dispersion of sediment particles, and expressions which have been derived from more general consideration of the transport of sediment.

The expressions for the rate of sediment transport which require information on the dispersion of sediment particles are from the work of Crickmore & Lean (1962) and Hubbell & Sayre (1964). The distribution curves for sediment particles were obtained by the use of radioactive sand as a tracer. The experiments were carried out in flumes using steady flows, or in rivers with steady flow regimes. The steady nature of the flow and uniformity of the channel are assumed in the derivation of the expressions for the rate of transport of sediment. The application of these calculations to other situations is limited by the assumptions involved in their derivation and by the data requirements, distribution curves of streamwise tracer dispersal and information on the vertical distribution of tracer within the bed.

Crickmore & Lean (1962) calculated the rate of transport based on the distribution of tracer material a time t after it had been introduced into the system. The quantity of material that had passed a section x could be written

$$w \sum_{x}^{\infty} p(s)$$

where $p(s)$ was the fraction of material from a unit strip a distance s upstream and w was the original quantity of material entrained. The total quantity of material from all the unit strips upstream of x can be written

$$w \sum_{x=0}^{x=\infty} \sum_{x}^{\infty} p(s) = w \sum_{0}^{\infty} xp(x)$$

so the rate of transport of sediment was

$$\frac{w}{t} \sum_{0}^{\infty} xp(x)$$

For the radioactive tracer used in this work

$$wp(x) = \frac{w_0 R}{R_0}$$

where w_0 was the original quantity of tracer, R_0 was the original level of radioactivity and R was the observed level of radioactivity at x at time t. This method only acknowledges the stochastic nature of the sediment transport process in the use of distribution curves to describe sediment dispersion. The calculated transport rate was 15% greater than the actual transport rate; however, some of the sand being used as tracer was recirculated and could account for this discrepancy.

A simpler approach was used in Hubbell & Sayre (1964)

$$q_B = g\rho_s(1 - p_b)Wd\,\frac{\bar{x}}{t}$$

where g is the acceleration due to gravity, ρ_s is the density of the bed material, p_b is the bed porosity, W is the width of the channel and d is the average depth through which particle movement occurs. This gives a continuity equation assuming that the projected area, Wd, through which the sediment particles are passing corresponds with the mean distance travelled by particles, $\bar{x}$, and the time, t. The values for $\bar{x}$ and t can be supplied from observations or, where the tracer matches the behaviour of the sediment, from the theoretical distributions of sediment movement described in Einstein (1937). These distributions were characterized by a mean step length $1/k_1$ and a mean rest period $1/k_2$ giving a ratio of $\bar{x}$ to t of k_2/k_1. A number of approaches were used to calculate the mean depth of tracer: averaging dune profiles, observations of multiple cores and from concentration curves. The fit of calculated to observed transport rates was explained but the measurements required a long period of time with a steady flow.

The other stochastic expressions for the rate of sediment transport were developed from the idea of intermittent particle movements. The distance travelled and periods of rest were both described by distributions rather than single values, the same description of particle movement found in the stochastic description of the movements of individual particles. The description has also been compared with the calculated behaviour of particles in deterministic models by Sekine & Kikkawa (1992), which calculated particle movements of multiple saltations before a particle returned to rest. This description of the sediment transport process describes observed behaviour and identifies this behaviour with components of the sediment transport process. The process of particle erosion is based on physical models, fitted using empirical terms, and particle movement by a distribution matching observations, without describing process.

In the model of Einstein (1950) the probability of particle movement was calculated based on the probability of the lift force exceeding the submerged weight of the particle. Movement of particles was regarded as consisting of a series of steps of constant length. The expression for the rate of transport was developed from equating deposition and erosion of particles. All particles move a distance LD, where L is the number of particle diameters, D, moved between periods of rest. The weight of bedload per unit time per unit width, q_B, gives the number of particles deposited per unit time per unit bed area

$$N_D = \frac{q_B}{LDA_2D^3\rho_s g}$$

where A_2 is the constant of grain volume and D is the particle diameter. The erosion of particles is related to the number of exposed particles per unit area and to the probability of erosion per unit time, p_E. The number of particles eroded per unit area per unit time is therefore

$$N_E = \frac{p_E}{A_1D^2}$$

where A_1 is a constant of grain area. If the time, t_E, in which one particle is replaced by another is known, the absolute probability of erosion, p, can be written

$$p = p_E t_E$$

The term t_E is a particle property and was assumed to be the time required for a particle to fall its own length in the fluid

$$t_E = A_3 \frac{D}{V_s} = A_3 \sqrt{\frac{D\rho}{g(\rho_s - \rho)}}$$

where A_3 is a constant of the grain timescale and V_s is the settling velocity of the particle, equating the rates of erosion and deposition

$$N_D = N_E$$

$$\frac{q_B}{LDA_2 D^3 \rho_s g} = \frac{p}{A_3 A_1 D^2} \sqrt{\frac{g(\rho_s - \rho)}{D\rho}}$$

For a low probability of erosion, particle movements will occur in single steps

$$L = \lambda$$

where λ is the step length. For higher probabilities particle movements consist of more than one step

$$LD = \sum_{n=0}^{\infty} (1 - p)p^n(n + 1)\lambda D = \frac{\lambda D}{1 - p}$$

Substituting, we obtain

$$\frac{q_B(1 - p)}{A_2 \lambda D g \rho_s D^3} = \frac{p}{A_1 A_3 D^2} \sqrt{\frac{g(\rho_s - \rho)}{D\rho}}$$

which can be rearranged as

$$\frac{p}{1 - p} = \left(\frac{A_1 A_3}{A_2 \lambda}\right)\left[\frac{q_b}{\rho_s g}\left(\frac{\rho}{\rho_s - \rho}\right)^{1/2}\left(\frac{1}{gD^3}\right)^{1/2}\right]$$

$$= A_* \phi$$

The term ϕ is a dimensionless measure of bedload and Einstein regarded A_* as a constant. Yalin (1977) showed that the assumption of a constant value for λ in particular is not true and showed various other errors in the derivation along with an alternative derivation, while stating that the basic approach was sound. The derivation of the probability of entrainment, p, related the lift force due to the fluctuating flow to the probability of entrainment. The calculations of lift force were based on the measurements of Einstein & El-Samni (1949) with the addition of further measurements described in Einstein (1950). The concept of a hiding factor, reducing the probability of erosion of smaller particles, was introduced into the probability of erosion. The final result reduced the calculation of rate of transport to a curve relating non- dimensional flow and transport rate.

A similar model to that of Einstein (1950) is described in Paintal (1971); here, the probability of a particle being eroded was determined by a pivoting analysis of particle entrainment. This included the effects of particle exposure and variation in forces acting due to a fluctuating flow. The distance moved in a single step was represented by a negative exponential distribution. The bedload transport rate was determined by consideration of the number of particles moving past a section in unit time. The number of particles, N_{dx}, eroding from an area of unit width and length dx in unit time can be written

$$N_{dx} = \frac{dx}{A_1 D^2} p_E$$

A particle crosses a line at x with probability $p_x(s > = x)$

$$p_x = \int_x^\infty f(s)\,ds = \exp\left[-\frac{(1-p)x}{\lambda Dp}\right]$$

where $f(s)$ is the mean density function for the distance travelled by a particle

$$f(s) = \frac{1}{\lambda Dp}(1-p)\exp\left[-\frac{(1-p)s}{\lambda Dp}\right]$$

This was developed from the sum of repeated steps, where the distribution function for the steps was assumed to obey a negative exponential distribution for the step length. The total number of particles crossing a section can then be written

$$N = \int_0^\infty \left[\frac{dx}{A_1 D^2} p_E p_x\right] = \frac{p_E \lambda Dp}{A_1 D^2(1-p)}$$

Rearranging gives

$$N A_1 D = \frac{p p_E \lambda}{1-p}$$

The mean probability of entrainment $p = p_E t_p$ where t_p is a characteristic time. Paintal related this to the time the fluid forces act on a particle to move it a step. The probability of erosion also affects this time giving an expression

$$t_p = \frac{A_3 D}{p U_*}$$

where A_3 is a time constant, though not the same as that of Einstein (1950), and U_* is the mean bed shear velocity. So

$$N A_1 D = \frac{p^3 U_* \lambda}{A_3(1-p)D}$$

Converting to a transport rate, $q_B = N A_2 D^3 g \rho_s$ and rearranging gives

$$\frac{q_B}{\rho_s g}\left(\frac{\rho}{\rho_s - \rho}\right)^{1/2}\left(\frac{1}{gD^3}\right)^{1/2} = \frac{A_2 \lambda}{A_1 A_3}\frac{p^3}{(1-p)}\tau_{o_*}^{1/2}$$

$$\phi = A_* f(B\tau_{o_*})$$

where $f(\tau_{0*})$ is an expression developed for the probability of entrainment of a spherical particle and B is a constant representing the behaviour of non-spherical particles. The values of A. and B were determined empirically. The final version of the equation is similar though not identical to that of Einstein (1950); the difference is due to the slightly different assumptions made in the derivation.

Comparison of the expression developed in Einstein (1950) with observations and the results of other expressions for rate of bedload transport (Gomez & Church, 1989; Yang & Wan, 1991) showed that in some circumstances the Einstein formula performed as well as other more recent expressions. Gomez & Church (1989) recommended its use for cases where the local transport rate needs to be calculated for known local hydraulic conditions. Yang & Wan (1991) found it could predict sediment transport in large rivers though not small rivers and flumes. The hiding and lifting factors introduced for calculation of transport of non-uniform sediments were found to overcorrect, predicting coarser material in transport than that forming the bed.

The basic assumption used in the stochastic models – that particle motion is not continuous but consists of particle movement followed by a period of rest – is confirmed by observations. The interpretation of the mechanisms which cause this behaviour varies: in Einstein (1950) the probability of entrainment was related purely to lift force acting, while in Paintal (1971) a pivoting analysis was used.

Individual assumptions can also be incorrect: using the same probability to describe the probability of initial motion due to flow and continued motion after impact relates a flow-based motion criterion and an impact-based motion criterion. The use of empirically derived constants allows the observed fit of models and data, though the use of constants is also incorrect (Yalin, 1977) since at least some of these terms are in fact variables.

Mobile bed models

The expressions for rate of sediment transport based on calculations of particle movement simplify the problem of scaling from single particle to sediment transport by concentrating on rate, while not modelling the bed over which the transport occurs. Calculations of particle movement including the bed and its modification by sediment transport have been made by Naden (1987b), Jiang & Haff (1993) and Haff *et al.* (1993).

The model described in Naden (1987b) was two-dimensional. The bed was described on a two-dimensional grid, 1000 grains long by 50 grains deep, overlying a fixed layer. The simulation used two sizes of grain, the smaller of a diameter equal to the grid size, the larger of a diameter equal to twice the grid size (Figure 7.4a). At the start of a simulation the bed was flat with random perturbations of $+/-$ one grain diameter. This bed of particles provided the supply of material for sediment transport and was modified by this transport.

The model of sediment transport used was a queuing model which included processes of entrainment, transport and deposition. The model of entrainment of particles was stochastic, combining a pivoting analysis of the forces required to initiate movement of particles with a description of forces acting due to turbulent

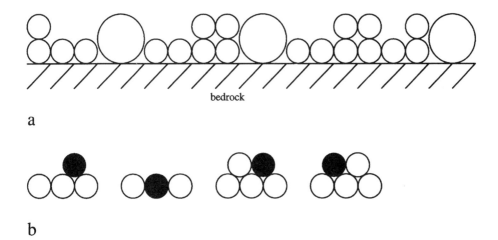

a

b

Figure 7.4 Representation of the bed in the model of sediment transport of Naden (1987b): (a) bed formed from two diameters of particles over a bedrock surface; (b) particle geometries considered for entrainment, shaded particle is the particle being considered

flow. A range of possible particle geometries was considered, based on the possible configurations that could form on the surface of the grid that formed the bed (Figure 7.4b). A stochastic element was introduced by considering fluid forces acting on particles due to turbulent flows. The horizontal and vertical turbulent velocity fluctuations were assumed to follow Gaussian distributions (Naden, 1987a). The mean velocity component was determined from a logarithmic law velocity profile for turbulent flow over a rough boundary. The standard deviations of the turbulent velocity fluctuations were determined from empirical expressions based on the data of McQuivey (1973). This information was used to assign probabilities to the entrainment of particles. The transport of particles in saltation was described by lengths and heights of saltations, based on analysis of the data of Abbott & Francis (1977), and particle velocity from the data of Abbott & Francis (1977) and that of Meland & Norrman (1966). The movements included the possibility of particles being suspended by velocity fluctuations due to turbulence. The effect of suspension was to prolong a particle trajectory, the number of possible occurrences being related to the duration of the trajectory. The distance travelled in a saltation determined the position of particle impact; the angle of impact was calculated from the height and length of the saltation. The occurrence or otherwise of deposition was determined by a balance of the tangential components of momentum on impact, the normal component being transferred to the bed and the tangential component being conserved.

The flow over the bed was calculated assuming steady gradually varied flow, with the flow being recalculated at intervals determined by the modification of the bed. The model was only used to calculate low rates of sediment transport. Direct modification of the flow due to the momentum extracted by particle movement and interaction of moving particles was therefore not included, only modification of the flow due to the change in surface height and composition.

the larger diameter grains. Consequently, the secondary flow-induced drag on the smaller grains is more easily able to reorient the path inward. The general importance of this effect has been recognized by Bridge & Jarvis (1976), Dietrich *et al.* (1979), Van Alphen *et al.* (1984) and Parker & Andrews (1985), among others, and a number of early models to predict sorting depended upon the cross-stream flow and the gravitational redirection to establish the transverse force balance on particles (Allen, 1970; Bridge, 1977). In meandering channels, cross-stream flow develops more slowly than the topography hence is most important in the downstream part of bends (Van Alphen *et al.*, 1984). The net effect of the gravitational redirection of the sediment, which is more important for the large grains, and the secondary flow-induced redirection, which is more important for the smaller grains, is that coarse and fine sediment transport paths cross in the bend (Figure 6.11). Coarse sediment is found typically along the outer bank and along the inside bank well upstream of the bend apex whereas fine sediment is found along the inside bank downstream of the bend apex (Figure 6.9a).

In channels without planform curvature, but with curved streamlines, the same process occurs. The topography of alternate bars and braid bars forces streamline curvature that can selectively route finer sediment. Ashworth *et al.* (1992a,b) observed braid-bar sorting in flume and field experiments where coarse particles rolled up onto the bar head while finer material was steered by the flow around the bar head and then toward the bar tail. They speculated that this steering was associated with roughness effects at the bar head and secondary flow further downstream in each of the flanking channels.

Another process that acts to sort sediment by size in bends or over dunes is flow separation. In the lee of steps or steep slopes a recirculation zone forms that is not associated with the flow streaming above. In situations where flow approaching the step is perpendicular to the lip of the step, sorting occurs because of settling velocity differences. We will treat that later. In many situations flow does not approach the crest of dunes or bars normal to the step, but rather approaches obliquely. When flow is oblique to such a bedform, a current is established in the lee of the crest that is parallel to the crest. Flow can be viewed as like a corkscrew lying parallel to the lip of the step in the lee of the step. Near-bed flow is directly back toward the base of the step, while flow near the top of the recirculating zone is aligned with the flow streaming over the step and oblique to the crest (Figure 6.12).

Dietrich and co-workers have described the important role of troughwise flow in the lee of oblique bedforms in routing sediment of different size through a bend and in creating the spatial pattern of bedsurface size (Dietrich *et al.*, 1979; Dietrich & Smith, 1984). Over an oblique dune tipped over the side of a bar, both coarse and fine sediment may avalanche outward, but some of the finer grains will be carried back inward by troughwise flow. The oblique flow over the margins of bars often routes the fine sediment and organic material along the base of the front (e.g. Whiting & Dietrich, 1991). Flow in these recirculating zones is typically too weak to move the coarser fraction. Ikeda (1983) shows a striking figure of sorting over alternate bars in a flume experiment where the bar top and curving bar edge is composed of an openwork gravel while the pool has a cover of sand which has infiltrated the gravel base. It would appear that the sand collects in the deeper flow and in the lee of the gravel-bar edge.

This model reproduced observed behaviour for the rate of sediment transport, transport rate reducing over time for the same flow, repeated events of similar magnitude causing less transport and pulses of sediment transport, associated with infrequent movement of large particles. The surface topography generated was also found to bear similarities to observations of river surfaces, forming waves of short and long wavelength (Naden & Brayshaw, 1987). The short wavelengths forms were five to ten grains long and were associated with the formation of clusters. These short bedforms were superimposed on longer bedforms with a wavelength of 30–40 grains, which were associated with the step-pools of Whittaker & Jaeggi (1982) and antidunes of Bathurst *et al.* (1982). A reduction in transport rate from that occurring over the initial random surface was also observed, and was caused by the formation of clusters. The clusters reduced the availability of grains; as the flow was increased clusters were destroyed and the transport rate increased again but not up to the initial rate.

The model described in Naden (1987b) was a grain-scale model of the mass movement of sediment. It was assumed that the rate of sediment transport was such that no interactions occurred between moving particles. The system was constrained so that only particles exposed at the surface were able to move, thus any modification of the surface could only occur by entrainment of surface particles exposing underlying material; interactions of particles within the bed were not calculated. This approach reduced the computational requirements; only the behaviour of surface grains and grains in motion needed to be calculated. The calculations themselves were also simplified, since the types of interaction were limited.

Calculations have been made in which the interactions between all the sediment particles in a system have been performed; this approach is described in Jiang & Haff (1993) and Haff *et al.* (1993). The method used was a particle dynamics method (PDM); these methods have their origin in molecular dynamics methods of physical chemistry. Their initial application to problems of sediment transport was for aeolian transport of sediment; a review of the method and applications can be found in Haff & Anderson (1993). This approach to calculation of sediment transport was used because it allowed multiple particle calculations to be performed at the grain scale, for which no system of differential equations exists. The data from these calculations can be used at a larger scale (Haff *et al.*, 1993).

The application of PDN to fluvial sediment transport requires consideration of forces dependent on the relative density of fluid and sediment. These can be ignored in calculations of aeolian sediment transport, but the small relative density of sediment and water requires consideration of buoyancy, forces due to fluid acceleration caused by particle movement and lubrication effects. The buoyancy force was treated as a reduction in weight due to the fluid displaced by a particle, while the effects of local acceleration of the fluid were treated by assuming an increased particle mass. The effects of lubrication of particles nearly in contact were treated by reducing the coefficient of restitution. The description of the contact forces between particles was a simplification of the actual forces occurring; the system itself is also simplified by modelling all the particles as spheres with their centres constrained to move in two dimensions, horizontal and vertical. The mass of particles is therefore based on that of a sphere of the diameter of the particle but the interactions between particles are calculated between circles of the diameter of the particle. These simplifications enabled calculations to be performed at a small scale: the bed was formed by 100 particles with a periodic boundary condition being used, the duration of calculations was 2 s in prototype.

The vector equation for the forces acting on each particle i in a system can be written

$$F_i = F_{gi} + \sum_j F_{ij} + F_{Di}$$

The three terms on the right-hand side of the equation represent the forces due to gravity, due to particle contacts and the flow acting on particle i. Once these forces have been calculated the acceleration of the particle can be calculated

$$F_i = m_i a_i$$

where m_i is the mass of particle i and a_i is its acceleration. The movement of the particle due to this acceleration can then be calculated.

The force term due to gravity is

$$F_{gi} = m_i g_i$$

where g_i is the acceleration due to gravity. The term due to contact between particles is the sum of all the forces due to the j particles in contact with particle i. The contact forces were represented by a normal and a tangential component of force; the force between the particles i and j was described by

$$F_{ij} = F_{n\ ij} + F_{t\ ij}$$

The normal component of force, $F_{n\ ij}$, was modelled as a stiff, damped spring

$$F_{nij} = -k\delta_{ij} - b\ \frac{\mathrm{d}\delta_{ij}}{\mathrm{d}t}$$

where δ_{ij} is the overlap of particle i and j, k is a spring constant and b is a damping coefficient which is related to the coefficient of restitution. The tangential component of force, $F_{t\ ij}$ was determined from the product of a coefficient of friction, μ, and the instantaneous value of the applied normal load, $F_{n\ ij}$. The only fluid force acting on particles considered directly was a drag force. This was calculated assuming that the drag force acting was the same as if the particle was isolated moving in an unperturbed flow, the same assumption as in the calculation of the movement of individual particles. The effects of other fluid forces, added mass and buoyancy, on particles were included by modifying the particle mass and weight.

The flow itself was modelled as a slab, driven by a shear stress on the upper surface and retarded by the drag forces acting due to particles

$$\tau A_{slab} - \left(\sum_i F_{Di}\right)_x = M_{slab} a_{slab}$$

where τ is the shear stress acting on the area A_{slab} of the upper surface of the slab, the term in brackets represents the x component of the drag forces, M_{Slab} is the mass of the slab and a_{slab} is the acceleration of the slab in the x direction. This is a simplified model of flow, with no vertical structure; however, it shows similarities to the hypothesis of Owen (1964) described earlier. The interaction of the flow with the bed is determined by the fraction of the surface particles with their centres initially within the slab; this must be set somewhere between the extremes of all particle centres within the slab and no particle centres within the slab. In the calculations performed the bottom of the fluid slab was positioned so that the centres of 30% of the surface particles lay within the slab.

The bed used in these calculations was formed by allowing particles of a range of sizes to drop under the influence of gravity; their final positions were determined by the rest position after the kinetic energy of each particle had dissipated. The surface produced by this was characterized by the range of friction angles present at the surface. For particles perched on the surface a reduction in friction angle occurred after the surface had been worked by the flow. Observations performed for the surface used in flume experiments showed similar behaviour for particles dropped on a surface (Kirchner *et al.*, 1990). The values of friction angle from the simulation were always less than for similar observations. The results compare a two-dimensional calculation with observations for three dimensions, which eliminates many of the possible geometries in which particles could rest. In addition, the circular cross-sections of the simulation cannot interlock and the calculated values of friction angles are for all possible particle rest positions. The observations were based on random drops of particles onto a surface which might preferentially sample certain rest positions. An alternative calculation of the friction angle based on the particles forming the surface of the bed found an increase in friction angle after the surface had been worked. Thus while over-passing particles see a worked surface as smoother, the particles forming the bed see a rougher surface and the bed surface becomes more stable.

Comparison of the bedload transport rate calculated from the simulation with the bedload transport rate calculated using the Einstein–Brown formula (Vanoi, 1977) shows a similar magnitude and variation with dimensionless shear stress. The use of simulations for this purpose would not be a sensible application due to the amount of computation required. The expressions developed by van Rijn (1984), Wiberg & Smith (1989) and Sekine & Kikkawa (1992) are based on particle dynamics and give good results for transport rate, though they do not include all the effects included in the PDM model, notably the effects of collisions between moving particles.

The movements of individual particles calculated using the model of Jiang & Haff (1993) increased with increasing shear stress, both in quantity of movements and distance travelled in the movements. In any one simulation particles could move more than once, increasing the total distance moved. The calculated behaviour of particles consisted of general particle movement, rather than movement of individual particles, even close to the critical shear stress for motion. The model also predicted rapid mixing of particles within the traction layer; the traction layer itself increased in thickness with increasing shear stress.

The particle dynamics method and results from it show that a grain-scale calculation can be performed and used to calculate the transport rate of sediment directly. However, such a use of the technique would be wasteful due to the computation required for the calculation. It is in the examination of structure or mechanisms that it is worth using the detailed models of PDM with their large computational overheads. The continuity in the treatment of particle behaviour rather than breaking particle movements into entrainment, transport and deposition is also an advantage in these calculations. Other than the small scales, spatial and temporal, of the PDM calculations another problem is that the number of parameters to which values must be assigned is large. Of these, Jiang & Haff (1993) described sensitivity analysis for the spring constant and coefficient of friction, but the coefficient of restitution and the height of the base of the flow with respect to the surface of the bed must also be set.

In aeolian sediment transport, impact rather than fluid forces provides the main mechanism by which particles are entrained. Studies of impact, both experimental (Mitha *et al.*, 1986; Willetts & Rice, 1985) and numerical (Rumpel, 1985, Werner & Haff, 1988), have therefore been a major component in attempts to model aeolian sediment transport. It was to model the impact process that particle dynamics methods were first developed for calculations of aeolian sediment transport (Werner & Haff, 1988; Haff & Anderson, 1993). The results of these calculations have been used to reduce the many degrees of freedom present in the original problem to a simpler (possibly stochastic) description, called for particles impact the "splash function". These functions give a statistical description linking the impact speed and impact angle with the ejection speed and number of particle ejected. Once such calculations of splash function have been made they can be used in models of aeolian sediment transport (Anderson & Haff, 1988, Werner, 1990, McEwan & Willetts, 1991) or in more complicated models of sediment transport over bedforms, such as that described in Anderson & Bunas (1993).

Summary

The deterministic models of sediment transport either reduce the resolution with which the calculation is performed, as with the use of calculations of particle movements to develop expressions for the rate of sediment transport, or are small in scale, spatial and temporal. The results show that particle-based calculations can successfully be used to supply parameters for rate of transport expressions which are almost completely physically based. The PDM work described is limited by computation, but simple scaling of these calculations without examining the assumptions made for both the forces acting and the parameters describing these forces would not necessarily lead to useful results. Appropriate experimental data sets to compare calculated with observed results are needed, as with the work on particle impact in aeolian sediment transport. Appropriate values for the parameters used in these models also need to be determined experimentally to ensure correct usage.

Both deterministic and stochastic models of particle movement were able to be used as the basis for calculations of the rate of transport. The deterministic models were used to supply parameters describing the movement of sediment particles, while the stochastic models were used to derive expressions for the rate of transport which were then fitted to data using empirical parameters. In both these approaches a detailed description of particle movement was replaced by mean quantities describing particle behaviour. The use of a technique like PDM allows detailed studies of components of the sediment transport process, with the requirement of suitable experimental results for comparison. Once such models have been developed they can be used to widen the understanding of processes or to enable the development of descriptions of these processes to use in larger scale models. These make possible increases in scale from calculations of movements of individual grains to calculations of sediment transport, allowing effective use of available data and computing resources.

CONTINUUM MODELS

The models of sediment transport described up to this point have been based on grain-scale calculations. The direct use of grain-scale calculations to calculate mass movement of sediment has been shown to be limited to small spatial and temporal scales by the quantities of computation involved in these calculations. At least for the present, larger scale descriptions of sediment transport and its effects will depend on expressions for the rate of transport and the applications of sediment continuity to determine the effects of sediment transport on bedforms and bed composition. The calculations performed will be continuum calculations.

Only in the model of Naden (1987b) was fluid flow and hence the bed shear stress calculated rather than regarded as a given. In the other models a shear stress, related to a flow, was always assumed. Calculations of sediment transport which include the effects of the sediment transport on the bed allow modification of the flow. The flow must be recalculated to obtain the shear stress to recalculate the rate of sediment transport. The calculations can either be coupled, where a single system of equations, including sediment transport and its effects, is solved at each step, or uncoupled, where flow and sediment transport are calculated independently. As with the calculations of grain-scale quantities, continuum calculations are performed at different scales and resolutions depending on the purpose of the calculation. The choice of whether to use a coupled or an uncoupled solution depends on the time interval over which the flow and sediment transport solutions are being calculated. If the degree of modification of the bed within this interval could affect the flow then a coupled solution must be used, otherwise an uncoupled solution can be used.

Two groups of models of the effects of sediment transport will be reviewed here, bedform development and one-dimensional mobile bed models. The first of these requires a description of the flow in two or three dimensions. The calculation of flow is further complicated by the possibility of flow separation on the lee face of dunes. The expressions used to calculate the rate of transport of sediment all involve raising the calculated shear stress to a power. Any error in the calculated shear stress will have a greater effect on the calculated transport rate and so on the calculated development of the bedform. The result of these considerations is that for models of this type much of the effort in development has gone into describing the flow rather than modelling the sediment transport, since without a good description of the flow the calculated shear stress distribution and magnitude will be incorrect and hence all the quantities relating to sediment transport.

The one-dimensional mobile bed models involve calculations at much larger scales; bedforms, if present, would only be represented by a roughness scale. The flow calculation is reduced to a one-dimensional solution, usually either steady, gradually varied flow or the St. Venant's equations. One result of this is that since the flow description is relatively well known and a number of different solution methods are already known, more emphasis is placed on the modelling of sediment transport and its effects.

In all these models a form of sediment continuity equation is used to describe the effects of the rate of sediment transport on the bed. The changes in bed height due to the rate of sediment transport are calculated using the equation of sediment continuity; in two dimensions this can be written

$$(1 - p_{b}) \frac{\partial h}{\partial t} = -\frac{\partial q_v}{\partial x}$$

where p_b is the porosity, h is the bed height above a datum, t is the time and q_v is the volumetric sediment transport rate per unit width.

Bedform development

Modelling sediment transport over bedforms requires the calculation of shear stress distribution over the bedform. The shear stress distribution can either be determined empirically or calculated from a flow solution. Fredsoe (1982) used the shear stress distribution measured downstream of the reattachment point of a rearward-facing step. This distribution was then scaled for different shear stresses. Where a calculation of flow over duneforms has been made two approaches have been used: building the flow field from components (Nelson & Smith, 1989) and using a single turbulence model to close the equations being solved throughout the flow. The flow solutions themselves are usually made in two dimensions, horizontal and vertical, consideration of the influence of three-dimensional flow effects can be found in Johns & Xing (1993).

Flow calculations

The approach of breaking the flow into components was used in the models of flow over dunes of McLean & Smith (1986), Nelson & Smith (1989) and in the model of flow over ripples of Wiberg & Nelson (1992). This approach was used because the flow over these bedforms involves effects at different scales and it was thought that these scales were so different that a flow solution based on a single turbulence closure model would not successfully be able to predict the flow field. The components of flow described in Nelson & Smith (1989) were an inner boundary layer region and an outer wake region. The inner boundary layer started at the reattachment point on the stoss slope of a dune and grew along this slope to the separation point at the crest of the dune. The wake region started at the flow separation point at the crest of the dune. The calculation of wake regions from each dune was continued downstream, stacking them over the wakes of following dunes.

The flow field was calculated throughout the depth of the flow; velocities and shear stress were matched between the inner boundary layer and the wake above it and between each wake above, allowing these quantities to be continuously defined through the depth. The flow field in the separation region downstream of the crest of the dune was calculated using the same flow description for the boundary layer as over the stoss slope of the dune. The calculation started at the reattachment point and continued upstream to the separation point at the crest of the dune, matching boundary layer thickness and velocities at the separation surface (Figure 7.5).

There are a number of different models of flow solving two-dimensional hydrodynamic flow equations using finite difference methods and a single turbulence closure model. The turbulence models used to close the system of equations vary in complexity from a mixing length model (Johns, 1991), through $k - \varepsilon$ models (Termes, 1988), and algebraic stress models (Mendoza & Shen, 1990). All these models predict velocity over the dune and the presence of a region of separated flow downstream of the crest of the dune. The predictions of bed shear stress and the comparison of this quantity with observations vary.

Sediment transport

All the models of duneform development use uncoupled solutions of flow and sediment transport. The flow calculations converge to a steady-state solution through time and the duration of each iteration is such that the modification of the bedform in any iteration will be small. In Johns *et al.* (1990) two different time intervals were used, a shorter time interval for flow solutions and a longer time interval at which the sediment transport was calculated and the bedform updated. The results of calculation of bedform were only described for the flow model described in Dawson *et al.* (1983), a hydrostatic version of the model described in Johns (1991) for marine flows. A hydrodynamic version used to calculate unidirectional flow is described in Johns *et al.* (1993). The hydrostatic flow solution did not predict flow separation so the flow quantities in the region of the lee slope of the dune differ from measurements. Bedload movement in the region of the lee slope would be dominated by the influence of gravity since this surface will approach the angle of

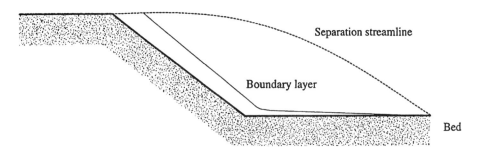

Figure 7.5 Representation, of region of flow separation in model of flow over dunes of Nelson & Smith (1989). Separation streamline marks edge of region of separated flow. The boundary layer in the region starts at the point of reattachment and grows upstream, matching the height of the stoss slope boundary layer at the dune crest

repose for the dune, so the inability to predict the flow in this region may well not affect the ability of the model to calculate duneform development. The combined model of flow and sediment transport described in Johns *et al.* (1990) proved capable of qualitative prediction of changes in bedform due to both suspended and bedload transport of sediment. Quantitative prediction of duneform development from prediction of flow and sediment transport rate is still limited by the accuracy of prediction of bed shear stress and the expressions used to calculate the rate of sediment transport. The expression for sediment transport rate used in Johns *et al.* (1990) supplies a mean rate of transport of a representative grain size; the model is therefore incapable of predicting any change in composition over the surface of the dune or within the dune. An alternative approach using cellular automata to calculate the development of aeolian duneforms, described in Anderson & Bunas (1993), has the advantage of representing these aspects of dune development though at the expense of a simplified description of flow.

One-dimensional mobile bed modelling

The problems examined at this scale are much larger spatially and temporally than those described so far. The flow models used therefore tend to be one-dimensional flow models, modelling only the downstream variation in flow depth. The aims of such models are to evaluate long-term evolution of rivers, aggradation and degradation, particularly in response to changing conditions, which may well be superimposed. For example, damming of a river would introduce the possibility of sedimentation upstream of the dam and degradation due to lack of supply of material downstream of the dam.

The models described are continuum models with the rate of transport of sediment and sediment continuity used to describe sediment transport and its effects. Though calculated in this way variation of surface composition is often included in these models, particularly to allow calculation of the formation of armour layers.

The system of equations used to describe one-dimensional mobile beds has been solved as an uncoupled system of equations (Bennett & Nordin, 1977; Borah *et al.*, 1982; Willetts *et al.*, 1987; van Niekerk *et al.*, 1992). More recently coupled systems of equations have been developed and used (Lyn, 1987, Rahuel *et al.*, 1989; Correia *et al.*, 1992).

In uncoupled systems of equations each iteration consists of independent calculations of flow over the bed, rate of sediment transport due to this flow and finally the modification due to the sediment transport. This approach is justified by assuming negligible change in bed height during an iteration or approximating the discharge to a constant value during an iteration. The solution of the equations is also simpler, making the development of models using this approach easier.

Solution of coupled systems of equations is more complicated and the development of models of this type is more recent. A number of justifications for the use of these more complicated solutions have been made. Examination of the systems of equations used in uncoupled and coupled solutions (Lyn, 1987) shows that assumptions made in using the uncoupled solutions are not generally met, though they can be met for certain conditions. The assumptions made in describing other components of the systems can also invalidate the use of an uncoupled solution. In most of the models, material available for transport can only be supplied from an active layer at the surface of the bed. For uncoupled models the material in this layer is fixed at the start of an iteration. If the supply is exhausted during an interval, the transport is regarded as being availability limited; this is a numerical availability rather than a physical condition. If the thickness of the active layer is fixed the only way to avoid this happening is to reduce the time interval over which the solution is calculated. Problems due to large changes in the height of the bed during an iteration can also only be solved by reducing the time interval over which the solution is calculated; the coupled solutions should therefore be capable of producing solutions using longer time intervals.

The coupled solutions make less assumptions than uncoupled solutions about the effects of changing bed height on the flow, though results from uncoupled solutions show good fit with data (Vogel *et al.*, 1992) can be achieved. The accuracy of the equations used in these models is affected by the values of the terms within these equations, the rate of sediment transport and bed roughness, and even the structure of the bed and availability of material are semi-empirical. Thus even where a better numerical solution, containing less assumptions, is available the choice of semi-empirical terms or even the description of processes necessary for such models could lead to worse results.

In comparison with the descriptions of the bed by particle position and size used in Naden (1987b) and Jiang & Haff (1993) the material present in the bed is described by the fraction of each size present in the bed. This change is a function of the change from the discrete particle models to the continuum models. A further difference is the description of the bed by a number of layers, used by all the models. The layers consist of an active layer at the interface of the bed and the flow, possibly mixing layers below this, and beneath these layers undisturbed bed material. The surface described by particles in Naden (1987b) and Jiang & Haff (1993) is replaced by a layer of known volume and fractional composition. The treatment and definition of the thickness of the different layers representing the bed varies. In most models the number of layers is reduced to two, an active surface layer and underlying undisturbed bed material. This simplification arises from the definition of the layers: the active layer, in contact with the flow during any particular interval; the mixing layer, material that can be worked by the flow; and under these undisturbed bed material, which is never worked by the flow. The result of these definitions is that the thickness of the active layer is related to the time-scale under consideration or the time interval on which the calculation is performed. For a short time interval the active layer corresponds only to those particles on the surface which are liable to immediate entrainment; over longer periods it corresponds to the depth through which dunes and ripples move. This relation between thickness of active layer and time interval relates to the possibility of the sediment becoming exhausted in uncoupled solutions.

The definitions used for the thickness of the active layer vary from simple fractions of sediment size (Bennett & Nordin 1977; Willetts *et al.*, 1987) to descriptions involving the smallest immobile particle size and the quantity of material necessary to form an active layer from such material (Borah *et al.*, 1982; Park & Jain, 1987) to a function of flow depth (Rahuel *et al.*, 1989; see Table 7.1).

Table 7.1 Expressions used to calculate the thickness of the active layer of the bed in one-dimensional mobile bed models

Model	Active layer thickness[a]	Constant or variable thickness
Bennett & Nordin (1977)	ND_{50}	Constant
Willetts *et al.* (1987)	D_{100}	Constant
Van Niekerk *et al.* (1992)	$2D_{50} \dfrac{\bar{\tau}'}{\tau_{c50}}$	Variable
Borah *et al.* (1982)	$\dfrac{1}{\displaystyle\sum_{i=L}^{M} P_i} \dfrac{D_{\mathrm{L}}}{(1-\lambda_{\mathrm{L}})}$	Variable
Park & Jain (1987)	$\dfrac{1}{(1-\lambda)} \dfrac{\displaystyle\sum_{i=L}^{M} P_{\mathrm{p}i}D_i}{\displaystyle\sum_{i=L}^{M} P_{\mathrm{p}i}}$	Constant
	equilibrium $c_{\mathrm{d}}h$	
	erosion	
Rahuel *et al.* (1989)	$\dfrac{D_{\mathrm{L}}}{(1-\lambda_{\mathrm{L}})} \dfrac{1}{\displaystyle\sum_{i=L}^{M} P_i}$	Variable

[a] D_{50} is the particle diameter for which 50% of the bed material by mass is smaller and D_{100} is the particle diameter for which 100% of the bed material is smaller, N is a calibration constant, $\bar{\tau}'$ is the skin friction component of the mean bed shear stress, τ_{c50} is the critical bed shear stress to entrain particles of diameter D_{50}, λ is the porosity of the bed, D_i and P_i are the ith grain size of the bed material and the fractional volume of that grain size in the bed, subscript p refers to the parent bed material, L is the first grain size that the flow cannot entrain, where there are M grain size intervals, c_{d} is a coefficient between 0.10 and 0.20 and h is the flow depth

All of the models use different equations to calculate the rate of transport of sediment. Since mixing and sorting effects of transport are to be studied using the models, the transport rates of different size fractions and the different entrainment rates of the size fractions forming the bed must be calculated. The equilibrium rates of transport for each size fraction were calculated using expressions for rate of bedload transport modified to account for the presence of a range of size fractions (and densities). The availability of different fractions in the surface combined with the material input determined the rate of either pick-up or deposition of material. When material was not available in the surface due to either the calculation method or armouring, another source of material, usually finer grains, was considered until an armoured surface was created.

The exact method used to determine availability and which fraction was used to make up deficits varied from model to model. The representations of these processes were based on observations rather than a physical description of the system.

Results

The coupled models described (Lyn, 1987, Holly & Rahuel, 1990; Correia *et al.*, 1992) have only been tested using theoretical cases and against each other. Their performance against field data has not been reported since the solution methods used are still being tested and developed. Once this stage of development has been achieved, and with advances in computing power, they will be available for calculations. The uncoupled solutions by contrast have been tested against field data with reasonable results and can be used for calculations (Vogel *et al.*, 1992).

CONCLUSIONS

The availability of computing power has enabled development of models of the sediment transport process across a range of scales. Increases in computing power and new computer architectures have allowed the development of approaches to the calculation of sediment transport processes based on grain-scale calculations rather than mean quantities; these calculations can include the stochastic effects present in the sediment transport process. The development of these grain-scale calculations allows the detailed examination of sediment transport processes and the development of descriptions of these processes suitable for use in larger scale models. The scales at which these models can be used will continue to increase allowing alternatives to empirical expressions that are used at present. The present models and further developments require detailed measurements of flow and sediment transport processes across a range of scales to validate their results.

REFERENCES

Abbott, J.E. & Francis, J.R.D. 1977 Saltating and suspension trajectories of solid grains in a water stream. *Philosophical Transactions of the Royal Society of London* **284A**, 225–254.

Anderson, R.S. 1987 Eolian sediment transport as a stochastic process: the effects of a fluctuating wind on particle trajectories. *Journal of Geology* **95**, 497–512.

Anderson, R.S. & Bunas, K.L. 1993 Grain size segregation and stratigraphy in aeolian ripples modelled with a cellular automaton. *Nature* **365**, 740–743.

Anderson, R.S. & Haff, P.K. 1988 Simulation of eolian saltation. *Science* **241**, 820–823.

Anderson, R.S. & Haff, P.K. 1991 Wind modification and bed response during saltation of sand in air. *Acta Mechanica* (suppl.) **1**, 21–51.

Barkla, H.M. & Auchterlonie, L.J. 1971 The Magnus or Robins effect on rotating spheres. *Journal of Fluid Mechanics* **47**, 437–447.

Basset, A.B. 1888 On the motion of a sphere in a viscous fluid. *Philosophical Transactions of the Royal Society London* **179A**, 43–63.

Bathurst, J.C., Graf, W.H. & Cao, H.H. 1982 Bedforms and flow resistance in steep gravel-bed channels. In Sumer, B.M. & Muller, A. (eds.) *Mechanics of Sediment Transport*, A. A. Balkema, Rotterdam, 215–221.

Bennett, J.P. & Nordin, C.F. 1977 Simulation of sediment transport and armouring. *Hydrological Sciences Bulletin* **22**, 555–569.

Borah, D.K., Alonso, C.V. & Prasad, S.N. 1982 Routing graded sediment in streams: formulations. *Journal of the Hydraulics Division ASCE* **108**, 1486–1503.

Bridge, J.S. & Bennett, S.J., 1992 A model for the entrainment and transport of sediment grains of mixed sizes, shapes and densities. *Water Resources Research* **28**, 337–363.

Buffington, J.M, Dietrich, W.E. & Kirchner, J.W., 1992 Friction angle measurements on a naturally formed gravel streambed: implications for critical boundary shear-stress. *Water Resources Research* **28**, 411–425.

Chepil, W.S. 1958 Use of evenly spaced hemispheres to evaluate aerodynamic forces on soil surfaces. *Eos. Transactions of the AGU* **39**, 397–404.

Correia, L.R.P., Bommanna, G., Krishnappan, G. & Graf, W.H. 1992 Fully coupled unsteady mobile boundary flow model. *Journal of Hydraulic Engineering* **118**, 476–494.

Crickmore, M.J. & Lean, G.H. 1962 The measurement of sand transport by means of radioactive tracers. *Proceedings of the Royal Society of London* **266A**, 402–421.

Dawson, G.P., Johns, B. & Soulsby, R.L. 1983 A numerical model of shallow-water flow over topography. In: Johns, B. (ed.) *Physical Oceanography of Coastal and Shelf Seas*, Elsevier, Amsterdam, 267–320.

Drake, T.G., Shreve, R.L., Dietrich, W.E., Whiting, P.J. & Leopold, L.B. 1988 Bedload transport of fine gravel observed by motion-picture photography. *Journal of Fluid Mechanics* **192**, 193–217.

Einstein, H.A. 1937 Bedload transport as a probability problem. English translation by Sayre, W.W. (1972) in Shen, H.W. (ed.) *Sedimentation*, H.W. Shen, Fort Collins, Colorado.

Einstein, H.A. 1950 *The bed-load function for sediment transportation in open channel flows*. Technical Bulletin No. **1026**, United States Department of Agriculture, Soil Conservation Service.

Einstein, H.A. & El-Samni, E. 1949 Hydrodynamic forces on a rough wall. *Review of Modern Physics* **21**, 520–524.

Fernandez Luque, R. & Van Beck, R. 1976 Erosion of transport of bed-load sediment. *Journal of Hydraulic Engineering* **14**, 127–144.

Francis, J.R.D. 1993 Experiments on the motion of solitary grains along the bed of a water stream. *Proceedings of the Royal Society of London* **A332**, 443–471.

Fredsoe, J. 1982 Shape and dimensions of stationary dunes in rivers. *Journal of the Hydraulics Division ASCE* **108**, 932–947.

Gomez, B. & Church, M 1989 An assessment of bed load sediment transport formulae for gravel bed rivers. *Water Resources Research* **25**, 1161–1186.

Haff, P.K. & Anderson, R.S. 1993 Grain scale simulations of loose sedimentary beds: the example of grain-bed impacts in aeolian saltation. *Sedimentology* **40**, 175–198.

Haff, P.K., Jiang, Z. & Forrest, S.B. 1993 Transport of granules by wind and water: micromechanics to macromechanics in geology and engineering. *Mechanics of Materials* **16**, 173–178.

Hassan, M.A. & Church, M. 1992 The movement of individual grains on the streambed. In Billi, P., Hey, R.D., Thorne, C.R. & Taconi, P. (eds) *Dynamics of Gravel-bed Rivers*, Wiley, Chichester, 159–173.

Hassan, M.A., Church, M. & Schick A.P. 1991 Distance of movement of coarse particles in gravel bed streams. *Water Resources Research* **27**, 503–511.

Holly, F.M. & Rahuel, J.-L. 1990. New numerical/physical framework for mobile-bed modelling. Part 1: numerical and physical principles. *Journal of Hydraulic Research* **28**, 401–416.

Hubbell, D.W. & Sayre, W.W. 1964 Sand transport studies with radioactive tracers. *Journal of the Hydraulics Division ASCE* **90**, 39–68.

Jiang, Z. & Haff, P.K. 1993 Multiparticle simulation methods applied to the micromechanics of bed load transport. *Water Resources Research* **29**, 399–412.

Johns, B. 1991 The modelling of the free surface flow of water topography. *Coastal Engineering* **15**, 257–278.

Johns, B. & Xing, J. 1993 Three dimensional modelling of the free surface turbulent flow of water over a bedform. *Continental Shelf Research* **13**, 705–721.

Johns, B., Soulsby, R.L. & Chesher, T.J. 1990 The modelling of sandwave evolution resulting from suspended and bed load transport of sediment. *Journal of Hydraulic Research* **28**, 355–374.

Johns, B., Soulsby, R.L. & Xing, J. 1993 A comparison of numerical model experiments of free surface flow over topography with flume and field observations. *Journal of Hydraulic Research* **31**, 215–228.

Kelsey, A., Allen, C.M., Beven, K.J. & Carling, P.A. 1994 Particle tracking model of sediment transport. In: Beven, K.J., Chatwin, P.C., & Millbank, J.H. (eds) *Mixing and Transport in the Environment*, Wiley, Chichester, 419–442.

Kirchner, J.W. Dietrich, W.E., Iseya, F. & Ikeda, H. 1990. The variability of critical shear stress, friction angle and grain protrusion in water-worked sediments. *Sedimentology* **37**, 647–672.

Kirkby, M.J. 1991 Sediment travel distance as an experimental and model variable in particulate movement. *Catena* suppl. **19**, 111–128.

Leeder, M.R. 1979 Bedload dynamics: grain-grain interactions in water flows. *Earth Surface Processes* **4**, 229–240.

Lyn, D.A. 1987 Unsteady sediment-transport modelling *Journal of Hydraulic Engineering* **113**, 1–15.

McEwan, I.K & Willetts, B.B. 1991 Numerical model of the saltation cloud. *Acta Mechanica* (suppl.) **1**, 53–66.

McLean, S.R. & Smith, J.D. 1986 A model for flow over two-dimensional bed forms *Journal of Hydraulic Engineering* **112**, 300–317.

McQuivey, R.S. 1973 *Summary of turbulence data from rivers, conveyance channels and laboratory flumes – Turbulence in water.* US Geological Survey Professional Paper **802-B**.

Meland, N. & Norrman, J.O. 1966 Transport velocities of single particles in bed load motion. *Geografiska Annaler* **48**, 165–182.

Mendoza, C. & Shen, H.W. 1990. Investigation of turbulent flow over dunes. *Journal of Hydraulic Engineering* **116**, 459–477.

Mitha, S., Tran, M.Q., Werner, B.T. & Haff, P.K 1986. The grain-bed impact process in aeolian saltation. *Acta Mechanica* **63**, 267–278.

Murphy, P.J. & Hooshiari, H. 1982 Saltation in water dynamics. *Journal of the Hydraulics Division ASCE* **108**, 1251–1267.

Naden, P. 1987a An erosion criterion for gravel-bed rivers. *Earth Surface Processes and Landforms* **12**, 83–93.

Naden, P. 1987b Modelling gravel-bed topography from sediment transport. *Earth Surface Processes and Landforms* **12**, 353–367.

Naden, P. & Brayshaw, A.C. 1987 Small- and medium-scale bedforms in gravel-bed rivers. In Richards, K. (ed.) *River Channels, Environment and Process*, Blackwell, Oxford, 249–271.

Nakagawa, H. & Tsujimoto, T. 1982 Closure to discussion of sand bed instability due to bed load motion. *Journal of the Hydraulics Division ASCE* **108**, 1402–1405.

Nelson, J.M. & Smith, J.D. 1989 Mechanics of flow over ripples and dunes. *Journal of Geophysical Research* **94**, 8146–8162.

Odar, F. & Hamilton, W.S. 1964 Forces on a sphere accelerating in a viscous fluid. *Journal of Fluid Mechanics* **18**, 302–314.

Owen, P.R. 1964 Saltation of uniform grains in air. *Journal of Fluid Mechanics* **20**, 225–242.

Paintal, A.S. 1971. A stochastic model of bed load transport. *Journal of Hydraulic Research* **9**, 527–554.

Park I. & Jain, S.C. 1987 Numerical simulation of degradation of alluvial channel beds *Journal of Hydraulic Engineering* **113**, 845–859.

Rahuel, J.L., Holly, F.M., Chollet, J.P., Belleudy, P.J. & Yang, G., 1989 Modeling of riverbed evolution for bedload sediment mixtures. *Journal of Hydraulic Engineering* **115**, 1521–1542.

Raupach, M.R. 1991 Saltation layers, vegetation canopies and roughness lengths. *Acta Mechanica (suppl.)* **1**, 83–96.

Reizes, J.A. 1978 Numerical study of continuous saltation. *Journal of the Hydraulics Division ASCE* **104**, 1305–1321.

Rubinow, S.I. & Keller, J.B. 1961 The transverse force on a spinning sphere moving in a viscous fluid. *Journal of Fluid Mechanics* **11**, 447–459.

Rumpel, D.A. 1985 Successive aeolian saltation: studies of idealized collisions. *Sedimentology* **32**, 267–280.

Saffman, P.G. 1965 The lift on a small sphere in a slow shear flow. *Journal of Fluid Mechanics* **22**, 385–400.

Schmidt, K-H. & Ergenzinger, P. 1992 Bedload entrainment, travel lengths, step lengths, rest periods-studied with passive (iron, magnetic) and active (radio) tracer techniques. *Earth Surface Processes and Landforms* **17**, 147–165.

Sekine, M. & Kikkawa, H. 1992 Mechanics of saltating grains. *Journal of Hydraulic Engineering* **118**, 536–558.

Shen, H.W. & Todorovic, P. 1971 A general stochastic model for the transport of sediment bed material. In Chiu, C.L. (ed.) *Stochastic Hydraulics*, University of Pittsburgh Press, Pittsburgh, 489–503.

Termes, A.P.P. 1988 *Rivers: Application of mathematical models for a turbulent flow field above artificial bed forms*. Delft Hydraulics, Netherlands. Report **TOW A56 Q787**.

Ungar, J.E. & Haff, P.K. 1987 Steady state saltation in air. *Sedimentology* **34**, 289–299.

Van Niekerk A., Vogel, K.R., Slingerland, R.L. & Bridge, J.S. 1992 Routing of heterogeneous sediments over moveable bed: model development. *Journal of Hydraulic Engineering* **118**, 246–262.

Van Rijn, L.C. 1984 Sediment transport, part I: bed load transport. *Journal of Hydraulic Engineering*, **110**, 1431–1456.

Vanoi, V.A. 1977 *Sedimentation Engineering*. ASCE, New York.

Vogel, K.R., van Niekerk A., Slingerland, R.L. & Bridge, J.S. 1992 Routing of heterogeneous sediments over movable bed: model verification. *Journal of Hydraulic Engineering* **118**, 263–279.

Werner, B.T. 1990 A steady-state model of wind-blown sand transport. *Journal of Geology* **98**, 1–17.

Werner, B.T. & Haff, P.K. 1988 The impact process in aeolian saltation: two-dimensional simulations. *Sedimentology* **35**, 189–196.

Whittaker, J.G. & Jaeggi, M.N.R. 1982 Origin of step- pool systems in mountain streams. *Journal of Hydraulics Division ASCE* **108**, 758–773.

Wiberg, P.L. & Nelson, J.M. 1992 Unidirectional flow over asymmetric and symmetric ripples. *Journal of Geophysical Research* **97**, 12 745–12 761.

Wiberg, P.L. & Smith, J.D. 1985 A theoretical model for saltating grains in water. *Journal of Geophysical Research* **90**, 7341–7354.

Wiberg, P.L. & Smith, J.D. 1989 Model for calculating bed load transport of sediment. *Journal of Hydraulic Engineering* **115**, 101–123.

Willetts, B.B. & Rice, M.A. 1985 Inter-saltation collisions. In: Barnsdorff-Nielsen, O.E., Moller, J.T., Rasmussen, K.R. & Willetts, B.B. (eds) *Proceedings of the International Workshop on Physics of Blown Sand* Department of Theoretical Statistics, Aarhus University, Denmark, Memoirs **8**, Vol. **I**, 83–100.

Willetts, B.B., Maizels, J.K. & Florence, J. 1987 The simulation of stream bed armouring and its consequences. *Proceedings of the Institute of Civil Engineers Part I* **82**, 799–814.

Williams, J.J. & Tawn, J.A. 1991 Simulation of bedload transport of marine gravel *WRDA Coastal Sediments '91 Proceedings*, 703–716.

Yalin, M.S. 1977 *Mechanics of Sediment Transport*, 2nd edn, Pergamon Press, Oxford.

Yang, C.T. & Wan, S.G. 1991 Comparisons of selected bed-material load formulas. *Journal of Hydraulic Engineering* **117**, 973–989.

Yvergniaux, P. & Chollet, J.-P. 1989 Particle trajectories modelling based on a Lagrangian memory effect. In *IAHR XXIII Congress, Hydraulics and the Environment, Turbulence in Hydraulics*, 301–313.

8 Channel Morphology and Element Assemblages: A Constructivist Approach to Facies Modelling

GARY J. BRIERLEY

School of Earth Sciences, Macquarie University, Australia

The degree of completeness with which we can deduce the conditions of deposition of ancient strata is in direct proportion to our knowledge of recent sediments and the factors determining their attributes. (J.R.L. Allen, 1965, p. 89)

INTRODUCTION

Facies models provide general summaries of sedimentary environments (Walker & James, 1992), often in schematic forms, in which observed sediment sequences are seen to result from particular sets of processes operating under certain environmental conditions. By definition, individual models must synthesize information from a range of examples; otherwise, each case study could be considered a model in itself. The challenge to fluvial sedimentologists is to generate a suite of facies models which covers the spectrum of river styles, while each model must retain specific characteristics which define a particular river type. To respond successfully to this challenge, principles from fluvial geomorphology must be integrated with fluvial sedimentology, for it is geomorphic factors which control the geometry, preservation and stacking arrangement of depositional units.

Individual river reaches are classified primarily on the basis of their planform type, defined as the configuration of a river in plan view. It is now generally accepted that river styles demonstrate a continuum of morphologic complexity (see reviews in Bridge, 1985; Miall, 1985). This presents a range of problems for facies modelling, as individual planform-scale facies models must have discrete characteristics, while the continuum of variability between these facies models cannot be neglected. Facies patterns viewed to be intermediate between two or more models, reflecting a mixture of components of various models, cannot necessarily be inferred to be indicative of an intermediary environmental setting. Indeed, as noted for deltas by McPherson *et al.* (1988), the continuum of river styles may be obscured by rigid compartmentalization of channel morphologies on the basis of planform style (Reading & Orton, 1991). As geomorphic research on river styles continues, knowledge of the complexity of river depositional styles continues to

Advances in Fluvial Dynamics and Stratigraphy. Edited by P.A. Carling and M.R. Dawson.
© 1996 John Wiley & Sons Ltd.

expand, and more intermediate and even end-member planform styles are identified, thus expanding the range of potential models with which ancient examples can be compared.

In this chapter the application of planform-scale fluvial facies models is critiqued and an alternative conceptual framework to facies modelling is proposed, based on the geomorphic units that make up river depositional environments. These three-dimensional units, termed elements, are viewed as the building blocks of any river reach. Individual field situations are analysed as assemblages of elements rather than as similar or dissimilar to any particular planform-scale facies model. This is termed a "constructivist" approach to facies modelling. This approach essentially constitutes a readoption of the three-dimensional building-block approach to alluvial sediment inventory summarized in Allen (1965), set in the context of the broader range of depositional units and river styles that have been recognized in the past 25 years.

The layout of this chapter is as follows. First, conceptual approaches to river sediment analysis are summarized at a range of differing scales. Channel planform sedimentology is then briefly reviewed. Limitations of a planform-based approach to facies modelling are evaluated in terms of problems with planform definition, modelling considerations, and the lack of geomorphic insight. The constructivist approach is then proposed and the character of various elements is briefly described. Finally, the application and implications of a constructivist approach to facies modelling are outlined.

SCALES OF RIVER SEDIMENT ANALYSIS

Hierarchical scales of depositional units have been identified which seek to explain form–process associations at a range of spatial and temporal scales (Table 8.1, Figure 8.1; see Miall, 1985, 1988a, 1991; DeCelles *et al.*, 1991; Brierley *et al.*, 1993). No specific dimensions can be attached to units in this hierarchy: an individual bar on a major river may extend over several kilometres, whereas an equivalent bar type in a smaller river may extend only metres. In Figure 8.1, one set of conditions at a particular instant determines the style of bedform formation, while subsequent events dictate the deformation and/or preservation of these deposits and the stacking of additional bedforms. Incorporation of broader-scale bar and channel-fill sequences into the floodplain is conditioned by the position of these deposits and later flow history. Various longer-term controls determine whether these deposits ultimately become part of the larger valley or basin fill. The character of these differing scales of depositional unit – bedform, bedform assemblage, bar/ channel unit, floodplain sequence, and valley fill – is conditioned by differing sets of processes and environmental controls, ranging from instantaneous discharge events in response to particular meteorologic events to longer-term system evolution dictated by climate change, base-level change and tectonic history.

In recent years emphasis in fluvial sedimentology has been placed on development of facies models which describe channel morphology in terms of channel planform type. These models have typically been defined in terms of assemblages of bedform-scale facies (e.g. Miall, 1977). However, since bedforms merely record

Table 8.1 Hierarchical arrangement of sedimentary scales

Contemporary field unit	Conceptual unit	Preservational control
Individual grain		Autocyclic
Bedforms (e.g. ripples, dunes)	Sedimentary structure Facies (*sensu* Miall, 1977)	Autocyclic
Bedsets/cosets/composite beds	Facies assemblage/model (*sensu* Miall, 1977)	Autocyclic
Field geomorphic unit (e.g. Levee, ox-bow)	Macroform (Jackson, 1975) Morphostratigraphic unit (Brierley, 1991a) Architectural element (Miall, 1985, 1988a) Element (Brierley, 1991b) Allomember (NACSN, 1983)	Autocyclic
River reach (e.g. planform type)	Element assemblage (Brierley, 1991b) Alloformation (NACSN, 1983) Facies model (*sensu* Walker, 1992) Depositional system (Galloway, 1992; Blum 1993, pers. comm.)	Autocyclic
Change in boundary condition, including potential for long-term preservation		
Abandoned segments of river systems (e.g. buried geomorphic unit, terrace)	Reworked/preserved element assemblage Allostratigraphic unit (NACSN, 1983)	Allocyclic
Valley fill/terrace complex	Alloformation (NACSN, 1983)	Allocyclic
Catchment	Basin Stratigraphic group Allogroup	Allocyclic

flow/sediment interactions determined by the passage of bodies of water of particular velocity, depth and duration over certain materials on a given slope, there is not necessarily any direct association to a specific depositional environment. As such, applicability of facies models based on stacked bedforms may not be restricted to particular river styles (e.g. Collinson, 1978; Jackson, 1978; Bridge, 1985; Brierley, 1989).

The interpretative significance of any sediment inventory depends primarily on the scale of observation (Table 8.2). Indeed, field techniques employed to analyse these differing scales of depositional unit are themselves different (Table 8.3). Reliable interpretation of depositional units is best applied at the scale which most clearly reflects form – process associations in river systems. This scale lies between the scale of bedforms and planform types, and is referred to as the element scale, i.e. the scale of geomorphic units. Appreciation of the critical nature of form/ process associations in analysis of river sediments has been recognized by sedimentologists and geomorphologists alike, but unfortunately these different groups have not always worked in unison, and the sedimentological literature has become riddled with jargon and idiosyncratic terminology. Nowhere is this better reflected than in the readoption of sediment classification procedures based on three-

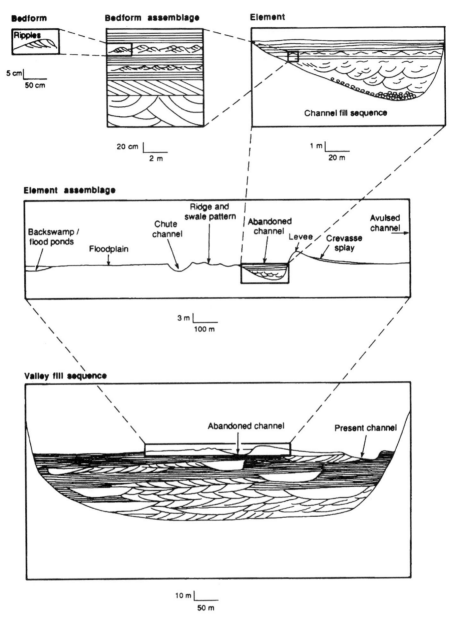

Figure 8.1 Scales of river sediment analysis. Extent of vertical exaggeration varies from figure to figure

dimensional depositional units demarcated by distinct unconformities or bounding surfaces (Table 8.1). These have been variously termed macroforms (Jackson, 1975; Crowley, 1983), architectural elements (Miall, 1985, 1988a), or morphostratigraphic units (e.g. Frye and Willman, 1960; Brierley, 1991a). For simplicity, these units are referred to in this chapter as elements.

Table 8.2 Sedimentary scales and their use in palaeoenvironmental reconstruction

Sedimentary scale	Primary interpretative significance
Grain	Nature of available sediment Reflects source-area geology Distance from source?
Bedform-scale facies	Hydrodynamic conditions at time of deposition Palaeocurrent direction Flow depth? Sedimentation rate?
Facies assemblage	History of flow events Preservation potential/degree of reworking
Element	Form/process association Geometry of preserved geomorphic unit
Element assemblage	River style Inferred depositional environment Controls on stacking arrangement of elements Predictive modelling of interconnectedness of depositional units Reservoir geometry
Basin analysis	Regional controls on alluvial regime: topography, tectonics, climate, vegetation, etc., and their changes over time

Table 8.3 Sedimentary scales and their recognition

Sedimentary scale	Field technique						
	Hand specimen	Drill/ auger hole	Average outcrop	Large vertical outcrop	Large horizontal outcrop	3-D outcrop	subsurface seismic
Grain	+ (if coarse)	+	+	+	+	+	
Bedform	If small	If small	+	+	+	+	
Bedset		If small	+	+	+	+	
Element		Maybe	Maybe	+	+	+	Maybe
Element assemblage		Maybe	Maybe	+	+	+	Maybe
Basinfill							+

When applying geomorphic principles to fluvial sedimentology, preservational controls on depositional units must also be considered. Within-system autocyclic controls on river style which determine the character of, and relationship between, channel and floodplain elements, need to be unravelled from externally induced allocyclic controls on river behaviour, such as tectonism, eustasy and climate changes which determine the longer-term preservation of basin fills on a regional (or broader) basis. Element assemblages, when preserved in the stratigraphic record, may be termed an allostratigraphic unit (NASCN, 1983; Autin 1992; Blum pers. comm. see also Chapter 12).

Elements essentially represent geomorphic units and are defined in terms of their geometry, bounding surface, and associated sedimentological characteristics. Assessment of the spatial assemblage of elements provides insight into river style, as element-scale units directly reflect river behaviour. There has been some criticism regarding the practical application of three-dimensional element-based approaches to sediment inventory (e.g. Walker, 1990; Bridge 1993), but since three-dimensional reconstructions are the desired end-point in any assessment of river style, it only seems sensible to use a geometrically based framework as a starting point for analysis. The challenge to fluvial sedimentologists is to relate the spatial assemblages of differing elements to channel morphology. This has typically been applied at the channel planform scale.

CHANNEL PLANFORM SEDIMENTOLOGY

A vast range of morphologies exists for alluvial channels (e.g. Schumm, 1985; Figure 8.2), and it is only for reasons of practical communication and academic convenience that the continuum of morphological complexity is differentiated into discrete types (Richards, 1986). River styles are generally classified on the basis of channel planform, differentiated on the basis of three criteria, namely:

(a) sinuosity, the ratio of thalweg length to valley length
(b) braided index, the number of bars or islands per meander wavelength, and
(c) lateral stability of the channel(s)

Five major channel planform types are recognized: braided, wandering, meandering, anastomosed and low-sinuosity or straight (Figure 8.3). Following Rust (1978a), distinction of planform styles is made in the first instance on the basis of the number of channels. Meandering and straight reaches have only one channel, while braided and anastomosing configurations are multichannelled. Wandering gravel-bed reaches are somewhat intermediate, and typically have two or three channels. Rust (1978a) proposed a sinuosity value of 1.5 to differentiate meandering and anastomosing channels from straight and braided channels. Once more, wandering gravel-bed reaches fall into an intermediate category. Sedimentological inventories of these planform styles are briefly summarized in the following sections.

Braided river depositional environment

In braided river reaches, flow diverges and rejoins around bars and/or islands on a scale of the order of channel width. Channel sinuosities are less than 1.5. Series of broad, shallow, rapidly shifting channels and bars, along with dry channels and elevated, frequently vegetated areas, result in several topographic levels (Williams & Rust, 1969).

Braiding occurs at points in river channels at which either local flow competence is exceeded, or the channel becomes overloaded with sediment, and deposition occurs. These conditions are determined by competence and capacity limits respectively. Once initiated, bars trap further sediment in a positive feedback manner, thereby diverting

Figure 8.2 Oblique photograph of the Channel Country, southwest Queensland, Australia, taken at flood stage on 6 April 1949. Note the amazing complexity of alluvial styles within the one photograph, with a braided configuration in the foreground, highly sinuous channel threads in mid-photograph, and sheetflood conditions in the background

flow and creating new channels. Bars form continuously in association with channel thalweg shifting and become part of the floodplain upon channel abandonment. Bar development may include lateral accretion (e.g. Bristow, 1987), although longitudinal growth is generally favoured. In general, this produces alluvial plains with a network of channels, but no clearly defined overbank terrain. In some situations, increases in discharge may lead to reoccupation of former channels.

Braided river deposits generally exhibit random interbedding of trough and tabular cross-bed sets, often with reactivation surfaces and ripple cross-laminations

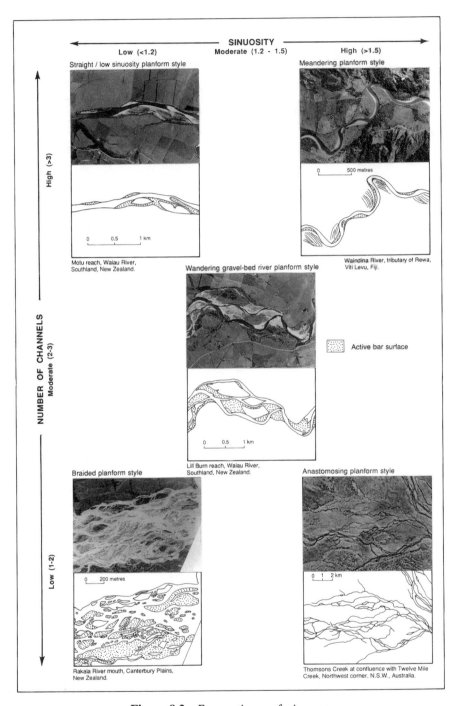

Figure 8.3   *For caption see facing page*

preserved above the sets (e.g. Doeglas, 1962; Krigstrom, 1962; Ore, 1964). Local scour units, discontinuous lags, and diffuse pebble sheets are common, whereas muds are seldom observed, accumulating only in abandoned channels. The relative proportion of sheet and channel deposits depends upon local circumstance, especially the discharge regime. Channel-fill deposits vary widely, with lenses of bedload sediment (structureless or trough cross-stratified deposits), or broad, multilateral sand belts composed of tabular, dip-oriented sands. Systematic upward change of particle size or cross-bedding characteristics is unlikely to be observed except during channel abandonment, as sequences demonstrate large facies changes over short distances.

In general, braided rivers are observed in high-energy geomorphic environments and are associated with coarse-grained alluvial systems. However, examples of muddy braided systems have been observed in arid environments in which clay aggregates act as sand-sized particles (Nanson *et al.*, 1986). Given the diversity of environmental and textural settings in which braided channels have been observed, a range of braided facies models have been developed (summarized in Miall, 1977, 1978b; Rust, 1978b).

Wandering gravel-bed river depositional environment

Wandering gravel-bed river reaches are characterized by fewer channels and smaller areas of active bar platform than in braided reaches. Generally there is one dominant channel, which is irregularly sinuous and splits around vegetated islands (Neill, 1973; Church, 1983; Desloges and Church, 1987; Morningstar, 1988). Bars grow and channels wander primarily across- rather than down-valley, and their upper surfaces are composed of roughly horizontally bedded sands with scour and fill units in old slough-channel fills. Fines are found as either thin drapes or as thicker, massive fills in back channels on the distal floodplain (Brierley & Hickin, 1991).

Wandering gravel-bed rivers typically demonstrate reach-by-reach variability in planform character. Stable channel reaches consist of gravel-based, lateral and point bars with cross-bedded sands and occasional silt drapes (Desloges & Church, 1987). Lateral channel shifting results in islands accreting onto the floodplain, with channel abandonment resulting in a slough between the former island and floodplain deposits (e.g. Schumm & Lichty, 1963; Nordseth, 1973; Morningstar, 1988). Unstable sections, or sedimentation zones (Church, 1983), also have a single dominant channel, with sporadically active mobile bed gravels and sands. These unstable reaches have medial bars and islands that are made up of massive and planar gravels,

---

Figure 8.3 Five end-members along the continuum of morphologic complexity demonstrated by alluvial rivers (after Rust, 1978a). Note how the proportion of active bar surfaces varies from style to style, reflecting greater instability in meandering, wandering gravel-bed and braided reaches. Bar type also differs among these end-member planform styles, with no active bars in the anastomosing reach, alternating lateral bank-attached bars in the low sinuosity reach, point bars in the meandering reach, dissected mid-channel and bank-attached compound bars in the wandering reach, and highly dissected mid-channel compound bars with a compendium of unit bars in the braided reach

with occasional horizontally bedded sands. Reoccupation and re-excavation of pre-existing channel segments may be more frequent than development of new channels (Gottesfeld & Gottesfeld, 1990).

Meandering river depositional environment

Meandering river reaches generally are single-channelled and have sinuosities greater than 1.5. The channel has a relatively low width–depth ratio with well-defined banks. Meandering reaches tend to be either overcompetent or underloaded. At bankfull stages, helical flow in bends carries sediment up the convex slope of point bars, while the concave bank is scoured. In this lateral accretion process, bedload materials are deposited on point and lateral bar surfaces. The characteristic asymmetrical channel form is maintained as the river migrates laterally, often producing a series of scroll bars in an alternating ridge and swale pattern (e.g. Sundborg, 1956; Nanson, 1980). As the channel migrates, within-channel deposits form the basal component of the floodplain, and are overlain by vertical accretion deposits laid down by overbank processes (Wolman & Leopold, 1957). Overbank deposits may form a range of secondary features away from the well-defined channel (Happ *et al.*, 1940; Fisk, 1944; Sundborg, 1956), reflecting clear process differentiation from the channel zone. Although the distinctly upward-fining nature of sediments associated with channel movement has been long recognized (see review in Miall, 1978a), it was not until the 1960s that the process of lateral accretion was incorporated into a model of floodplain growth by point bar development (Bernard *et al.*, 1962; Bernard & Major, 1963; Allen 1963, 1965). In the classic point bar model, dune structures rest atop imbricated channel lag deposits with indistinct bedding. These deposits grade upwards to ripple structures of finer grain size. As a result of progressive lateral and/or downvalley channel migration, these sequences successively overlie former positions of the channel thalweg, now recorded by a basal erosion surface. Vertical accretion of fine sediment subsequent to meander abandonment or channel avulsion completes the formation of a depositional cycle, with upward reduction in both particle size and sedimentary structure. The relative importance of lateral versus vertical accretion deposits in each cycle is dependent largely upon the stability of the channel position and subsidence rate (Bridge & Leeder, 1979).

This classic meandering facies model applies only to fine-grained meandering systems (see review in Jackson, 1978; Miall, 1985). The wide range of geomorphic settings in which meandering regimes are observed results in pronounced variability in their associated sedimentology. The classic model has limited relevance for coarse-grained meandering river reaches, where point bars are often modified by chute channels and chute bars (e.g. McGowen & Garner, 1970; Levey, 1978; Brierley & Hickin, 1992). These sequences characteristically exhibit little vertical fining; indeed, some of the coarsest grain sizes and largest structures may be found high up sequences in chute or chute bar deposits (McGowen & Garner, 1970). Under even coarser gravel-bed conditions, Gustavson (1978) showed upper point bar sequences to be dominated by thin gravel sheets deposited as transverse gravel bars.

Anastomosing river depositional environment

Anastomosing rivers are multiple-channel systems that have distinctively stable channels which transport a high proportion of materials as suspended load relative to bedload (Schumm, 1968). Interconnected networks of low slope, narrow and deep, straight to sinuous, channels are pinned in place by well-vegetated, fine-grained banks (D.G. Smith & Smith, 1980). Adjacent to channels are well-developed levees with occasional crevasse splays. The levees separate channel zones from wetlands composed of finer-grained sediments and vegetation. Vegetated wetlands cover more than 80% of the floodplain (D.G. Smith & Smith, 1980; D.G. Smith, 1983, 1986).

Channel fills in anastomosing planform styles fill vertically, and are typically significantly coarser-grained than surrounding deposits. These stringers are texturally well-segregated from sandy-silt levees, which act to confine channel movement. Off-channel crevasse splay sand and/or fine gravel sheets tend to be relatively thin, but extend some distance from the channel (D.G. Smith, 1983, 1986). Wetlands between channels can be differentiated into floodpond, marsh and backswamp environments composed of differing proportions of clay-rich mud, mud/organic matter and peat bog materials.

Just as braided and meandering planform styles can be found in a range of environmental settings, so can anastomosing channels. In the various examples studied by Derald Smith, the anastomosing channel style is viewed to reflect fairly rapid aggradation of channel and floodplain deposits, either within confined valley settings or in tectonically subsiding lowland plains. However, anastomosing planform styles have also been observed in arid regions characterized by limited vegetation and low sedimentation rates (e.g. Rust, 1981; Nanson *et al.*, 1988; Schumann, 1989). The continuum of river behaviour from anastomosing, distributary, anabranching and avulsive situations is clearly represented in the transition from a meandering to an anastomosing channel condition following avulsion of a part of the lower course of the South Saskatchewan River (N.D. Smith *et al.*, 1989).

Low-sinuosity river depositional environment

Flume experiments have clearly demonstrated the instability of flow–sediment coupling in straight channels, and the adoption of laterally deflected flow around alternating bars (e.g. Schumm *et al.*, 1987). Bridge *et al.* (1986) described the sedimentology of a gravel-based, low-sinuosity stream in which the channel is occupied by alternating lateral and point bars, although emergent areas form nuclei for mid-channel bars. Channel migration and within-channel sedimentation occurs by bend expansion and translation downstream and lateral growth of islands and chute cut-offs. These bars are composed of large-scale cross-stratified sands overlain by small-scale cross-stratified sands interbedded with plant debris. Deposits of lateral and point bars do not differ substantially from those of migrating braid bars (Bridge & Gabel, 1992). The floodplain itself is made up of abandoned bars and associated features (such as scrolls and chute channels), abandoned channels and narrow levees.

LIMITATIONS OF CHANNEL PLANFORM FACIES MODELLING

Of the five primary planform styles described above, most research in fluvial sedimentology focused initially on identification of characteristics of a meandering river style. In the mid-1960's, two criteria were considered diagnostic of a meandering channel planform, namely the presence of upward-fining cycles of both grain sizes and sedimentary structures, and epsilon cross-stratification (Moody-Stuart, 1966). Both of these phenomena result from changing depositional conditions on point bars as bends migrate laterally and flow depths decrease. However, these features are not necessarily diagnostic of meandering channel planform styles, as they are not observed in all meandering rivers, nor are they confined to meandering channels. Spiral flow associated with channel curvature depends upon local flow geometry and river stage, and is not unique to any particular channel planform. Lateral accretion deposits and epsilon cross-stratification have been observed in braided river depositional environments (e.g. Bluck, 1979, 1980; Ori, 1983; Allen, 1983; Bristow, 1987), while bars and islands have been described for high-sinuosity meandering rivers (e.g. Schwartz, 1978; Forbes, 1983). Indeed, none of the criteria that have been used to differentiate among sediment sequences associated with differing river channel planform styles has proven to absolutely discriminate between individual channel planform styles (Jackson, 1978; Bridge, 1985, Miall, 1985; Collinson, 1986).

Limitations of channel planform facies modelling can be summarized under three headings, namely: problems of planform definition, modelling considerations, and the lack of appreciation for geomorphic process.

Problems of planform definition

To be effective, a modelling procedure must summarize available data into physically meaningful groups which can be differentiated on the basis of discrete phenomena. A wide range of laboratory, empirical and theoretical studies have demonstrated that a continuum exists between channel patterns and their flow and sedimentary patterns (Bridge, 1985). Within this broad spectrum various key cases have been identified, but these are not necessarily discrete. Furthermore, end-member situations are defined using different criteria. For example, meandering rivers are defined primarily on the basis of their sinuosity, braided channels are distinguished by their multichannelled character, and anastomosing river systems are differentiated by their laterally stable multichannelled configurations. Beyond this, many river reaches demonstrate different planform styles at different flow stages; braids can meander, meanders locally braid, and the anabranches of an anastomosing river can be sinuous or straight. Finally, individual field situations may have unique attributes, such that they do not conform to any particular fluvial style. For example, the geometry and configurations of channels in many arid-region rivers bear little resemblance to any planform facies model described elsewhere (see Figure 8.2).

Modelling considerations

Over the past 25 years, two modern fluvial facies models have dominated as analogues in reconstructions of river style from the ancient record: models based on the meandering Mississippi River (Fisk, 1944) and the braided Brahmaputra River (Coleman, 1969). These two models, however, constitute end-member situations in the spectrum of morphological complexity demonstrated by rivers and their validity as analogues may have been somewhat overextended. Shortcomings of the limited number of sediment inventories from modern alluvial settings that can be used as analogues are illustrated by the recent profusion of reinterpretations of river styles in ancient sedimentary sequences following detailed sedimentological descriptions of anastomosing river deposits (e.g. Flores & Hanley, 1984; Johnson & Pierce, 1990; Kirschbaum & McCabe, 1992). Many of these re-analyses mark significant shifts in assessment of alluvial style, once familiarization with the anastomosing facies model became more prevalent.

In recent years there has been a profusion of fluvial facies models used to describe differing field situations. There is no magic number of models that defines the spectrum of morphological complexity demonstrated by rivers. Rather, individual models fit in particular positions along the continuum of stream power and grain size trends (Reading & Orton, 1991). Forcing data into end-member facies models may severely constrain past-environmental interpretation of river styles, thereby providing misleading insights into distribution and stacking arrangement of depositional units within a basin fill.

Perhaps the greatest problem, however, is not the manner by which facies models are constructed, but the manner in which they are used. Unless all limitations are considered, facies models may be used beyond the scope of their applicability, and subsequent interpretations may be seriously flawed. Several philosophical and methodological issues confound facies modelling procedures, such as:

(a) How many detailed descriptions of field situations are needed to define a "general summary" model?
(b) Any model is only as good as the data on which it is built, and all their associated vagaries. Just how reliable a norm are existing models? For example, Blair & McPherson (1992) argued that widely held facies concepts for the Trollheim fan facies model are inconsistent with the actual characteristics of the fan itself.
(c) As data are extracted from field descriptions and summarized into model forms, the simplicity of the ensuing models often means that they have limited real-world sense (e.g. Jackson, 1978; Collinson, 1978; Miall, 1980, 1985, 1987; Friend, 1983; Anderton, 1985; Bridge, 1985; Reading, 1987; Dott, 1988).
(d) In providing a general summary of a specific sedimentary environment the configuration of features demonstrated by the model may never actually be observed in any one field situation, and the model may be viewed cynically as a meaningless average scenario.

Relation to geomorphic process and environment of deposition

Individual channel planform types do not reflect specific geomorphic processes that occur under unique sets of circumstances; rather, they reflect fluvial adjustment to combinations of interrelated variables, in which limiting factors may impose a particular morphologic response (see review by Ferguson, 1987). Discriminating functions used to differentiate among planform types have been based on relationships among channel slope, discharge and particle size functions (Lane, 1957; Leopold & Wolman, 1957; Henderson, 1961). The failure of discriminant analysis applied in this manner may be attributed to problems of definition and interdependence of the terms used (neither discharge nor slope are independent), and the fact that functions derived in one region, with one particle size range, cannot necessarily be applied to another (Carson, 1984).

Given the lack of geomorphological distinctiveness of individual channel planform styles, it is scarcely surprising that planform–sediment correlates are far from unequivocal. Similar bedform-scale facies assemblages may be viewed independently of planform style and there appear to be no sedimentary structures that are peculiar to individual channel planform types (Bridge, 1985; Brierley, 1989). The principle of convergence, in which depositional units stack in a similar manner for different planform styles, has also been demonstrated for element assemblages that comprise the floodplains of contiguous braided, wandering and meandering reaches of a gravel-bed river (Brierley & Hickin, 1991).

In Table 8.4, geomorphic elements which make up river systems are related to the five primary channel planform styles. Although this summary table is static, and does not consider factors which control preservation of depositional units, the striking feature of Table 8.4 is the degree of overlap in the presence of the same geomorphic features in differing planform styles. There are no unique elements for

Table 8.4 Element presence/absence for primary channel planform styles (frequency of occurrence in truly alluvial reaches: + +, always; +, commonly; 0, occasionally)

	Braided	Wandering	Meandering	Anastamosing	Straight
Within-channel element					
Primary channel	+	+ +	+ +	+	+ +
Secondary channel	+ +	+ +	0	+ +	0
Avulsed channel	+	+	0	+	0
Chute cut-off	+	+	0	0	
Ox-bow	0	+		0	
Swale	0	+	+		0
Downstream accretion element	+ +	+	+		0
Lateral accretion element	+	+ +	+ +	0	0
Scroll bar	0	+	+		0
Oblique accretion		0			
Concave bank bench				0	
Channel-marginal element					
Levee	0	0	+	+	0
Crevasse splay	0	0	0	+	0
Floodbasin elements					
Floodplain	+	+	+ +	+ +	+
Backswamp		0	+	+	

particular planform styles other than the seldom-observed oblique accretion benches and concave bank benches associated with a meandering planform type. The overlap in the range of depositional sequences for differing planform styles has profound implications for the predictability of the sediment signature for differing planform reaches and hence interpretation of alluvial style from ancient depositional sequences.

ELEMENT ANALYSIS: FRAMING A "CONSTRUCTIVIST" APPROACH

Recognition of the non-diagnostic nature of facies models derived for end-member planform situations has resulted in comprehensive re-evaluation of the association between channel morphology and sedimentological attributes. For example, Miall (1985) documented the element-scale sedimentology of 12 alluvial styles, while Nanson and Croke (1992) outlined a genetic framework for floodplain classification which is differentiated into 13 suborders on the basis of stream power and grain size/cohesiveness. Once more, however, the differing alluvial styles characterized in these reviews represent either end-member or intermediary sedimentological models. As Bridge (1993) indicated, classification needs to be based on easily measurable parameters used to define mutually exclusive classes, and terms used to refer to classes should be explicit. However, the adopted framework for classification must also be open-ended, to allow for field-based descriptions which do not fall in any predetermined box. The response suggested in this chapter is to change the frame of reference away from the planform scale to the element scale, and view sediment sequences as assemblages of elements rather than as representative of any particular river style.

In recent years, numerous attempts have been made to integrate principles from fluvial geomorphology into sedimentology. Building on the work of Friend (1983) and Allen (1983), Miall (1985, 1988a) defined a series of fluvial architectural elements on the basis of their geometry, bounding surface, position within a sediment sequence, and sediment character, namely: channel, gravel bars and bedforms, sandy bedforms, foreset macroforms, lateral accretion deposits, sediment gravity flows, laminated sand sheets, and overbank fines. Differing parameters are used to define individual elements. For example, geometry, texture and depositional process are the primary factors used to differentiate among channel elements, gravel *versus* sand bedforms, and sediment gravity flows respectively. In this chapter a more consistent framework for element classification is adopted. Given the geomorphological perspective of this review, and the emphasis on modern depositional environments, elements are viewed simply as the geomorphic building blocks of a river system. Three components of river depositional systems can be differentiated, namely within-channel deposits, floodplain elements, and features deposited at the margin of the channel and floodplain (e.g. Happ et al., 1940; Allen, 1965; Task Committee, 1971; Lewin, 1978a).

These various components can themselves be differentiated into a limited number of geomorphic units or elements (Table 8.5). The stacking arrangement of these

Table 8.5   An elemental framework for sedimentological analysis of river deposits (this table is schematic, and is not intended to be all-inclusive)

Within-channel elements
Channel fill sequences
  – hierarchy of scales of channel fill may be evident
  – fixed vs. sheet channel bodies
  – range of smaller-scale channel-fill elements produced by channel abandonment e.g. meander cut-offs, swale-fills
  – flood channels that scour floodplains (back channels)
  – chute-channel fills

Components of channel fill
  – Basal lag: coarse materials that are sorted out and left as a residual accumulation in channels
  – Bars: areas of net sedimentation of size comparable in magnitude to the channels in which they occur. Used to infer lateral and/or downstream component of channel movement.
  – unit bars: simple, single-unit depositional forms; longitudinal, diagonal, lateral, point forms
  – compound bars: composed of a range of morphostratigraphic units (e.g. ridge, bar platform, chute channel elements); mid-channel, bank-attached and compound point bar forms
  – anomalous features: concave bank benches; oblique accretion benches, etc.

Channel-marginal elements
Levees
  – wedge-shaped ridges which border stream channels
  – generally scaled proportional to dimensions of their adjacent channel

Alluvial ridge
  – meander belt elevated above the floodplain surface, within levees

Crevasse-splay deposits
  – narrow to broad, localized tongues of sediment which are sinuous to lobate in plan

Sand wedge
  – wedge-shaped sheets of sand launched onto floodplains in non-levee situations

Sheet-flood deposits
  – horizontally bedded sands, common in ephemeral river styles

Floodplain elements
Floodplain
  – poorly drained, relatively featureless areas located adjacent to or between abandoned or active channels
  – typically tabular, prismatic bodies of rectangular cross-section
  – proximal–distal gradation in grain size is common

Backswamp or floodpond conditions may develop in the distal floodplain
  – may be organic-rich

---

elements within any alluvial suite is the primary indicator of alluvial style. Processes operative at the element scale not only dictate the style of flow/sediment interaction that results in particular assemblages of bedform-scale facies, but they also determine the likely manner by which deposits become incorporated into floodplains, and hence may be preserved. In this constructivist approach to facies modelling, depositional suites are viewed summarily as element assemblages rather than as similar to any particular fluvial facies model. This presents a flexible methodological framework with which to make inferences about river style. As Collinson (1978, p. 579) comments, "It is better to try to characterise a sequence not by a single distilled sequence but by some description of the variability of *observed* sequences, ideally with known spatial relationships". Rather than viewing any data set as "closely akin to model a, b or whatever", or "intermediary in character between models x and y", a constructivist approach simply views the data as an assemblage of elements, and uses knowledge of the geomorphic form–processes associations of elements and their stacking arrangement to make inferences about river style and behaviour.

The characters of the various morphological elements which make up a river system are described in subsequent sections. There is no pretence that the range of element types characterized below is exhaustive. Indeed, an open-ended approach to element inventory is favoured, such that additional elements can be added to the framework as they are recognized. In general, modern examples are used in the following discussion. In some instances, however, elements are better unravelled in the ancient record, where exposures are more extensive and longer-term histories of channel/basin fill and their preservation can be invoked. For example, channel geometry and fill character are often better characterized from ancient depositional sequences, although Galloway (1981) observed that channel-fill deposits commonly show little resemblance to well-described modern analogues. Channels can only be truly defined if the concave-up channel scour surface can be identified (Miall, 1985). Such negative characteristics may be easier to characterize in the depositional record than constructional, positive characteristics such as within-channel deposits (e.g. Best, 1987; Cowan, 1991).

Within-channel elements

Schumm (1960) indicated that width–depth ratio of channels increases with the percentage of cohesive materials in banks, and sinuosity increases with higher proportions of suspended load, other parameters being equal (Schumm & Khan, 1972). Accordingly, sequences with a large proportion of fines are generally indicative of sinuous streams, while instances with little preservation of fines are indicative of low-sinuosity situations. This framework of bedload, mixed-load and suspended-load streams has served as the primary framework for channel classification and assessment of river behaviour and metamorphosis (Schumm, 1981). A wide range of channel types and associated fill sequences have subsequently been described from the ancient depositional record (e.g. Friend *et al.*, 1979; Ramos *et al.*, 1986; Marzo *et al.*, 1988; Munoz *et al.*, 1992).

The nature of within-channel deposits is dependent on the style of channel, its behaviour through time, and the manner of its abandonment. Channels that are stable between avulsive events create ribbon bodies, whereas channels that steadily migrate within a channel belt create sheet-like bodies, whether the channel is meandering or braided (Friend, 1983). A width–thickness ratio of 15 : 1 has been suggested to separate these bodies (Friend, 1983; Cowan, 1991). Ribbon and mobile sheets stack vertically as multistorey and multilateral units respectively. Preservation of channel-fill sequences, and their incorporation into the basin fill, are contingent on the manner of channel shifting (whether by lateral accretion or avulsion), the extent of subsequent reworking, and the rate of aggradation and/or subsidence.

In general, fixed channels have relatively low width–depth ratios and well-defined channel margins. Channel banks are characteristically composed of finer-grained, more cohesive materials than the channel-bed material. Such stationary channels may be abandoned and subsequently infilled following avulsion. Style of avulsion may be significant. Gradual avulsion of a river channel may lead to preservation of an entire channel-fill sequence encased within floodplain fines (e.g. R.M.H. Smith, 1987). Conversely, rapid avulsion may simply leave an extensive depression in the landscape which subsequently infills with fine-grained materials in the form of a clay plug (e.g. Fisk, 1944, 1947). Sheet sandbodies are produced either by aggradation of large, laterally unstable, multichannel fluvial tracts, or by large-scale sheet flooding (Friend, 1983). In general, sheet sandbodies are indicative of bedload-dominated situations, and may consist of several coalesced sequences. Todd & Went (1991) describe an alternative mechanism in which fluvial sand-sheet bodies are produced by unidirectional channel combing across a valley fill.

A range of smaller-scale channel-fill elements is produced by channel abandonment, such as meander cut-offs and swale-fills. Cut-offs occur whenever a stream shortens its course, thereby locally increasing its slope. Allen (1965) differentiates two styles of meander cut-off, namely neck and chute cut-offs. Neck cut-off is the primary mechanism of meander loop abandonment. Such cut-offs occur late in the development of the loops, either by the gouging of a new channel across the narrow neck of land between two loops, or through the capture of one loop by the next upstream. Bedload sediment rapidly plugs the ends of the abandoned channel to produce an ox-bow lake. Filling is completed by sediment brought in from overbank flows, producing upward-fining particle size trends in units with well-defined morphologies. Within-channel particle size patterns may record variations in flow regime over time, however, with influxes of coarser materials under differing discharge conditions (e.g. Erskine *et al.*, 1992). Abandoned channels which form clay plugs may subsequently resist lateral migration of the channel.

Chute channels short-circuit the main body of flow in a river. Enlargement of the chute channel and plugging of the old channel proceed gradually, resulting in a chute cut-off. Because of the small angular difference between the old channel and the chute channel, the stream continues to flow through the old channel for some time, depositing bedload sediment at the upstream and downstream ends and on the floor and sides until terminal closure of the cut-off is complete (Allen, 1965). Chute-channel fills are notably straighter in outline than either meander cut-offs or swales, and commonly consist of high-energy depositional sequences (e.g. McGowen & Garner, 1970; Teisseyre, 1977; Levey, 1978; Brierley, 1991a,b; Brierley & Hickin, 1992). Given their position on the inside of bends, chute-channel fills may be preferentially preserved in the alluvial pile.

In some laterally migrating channels, swales are found in an alternating sequence with ridges on the inside of bends. Swale-fill deposits are narrow, arcuate bodies of prismatic cross-section. Their fine-grained fill thickens downstream, with bottom surfaces that are convex downward (Allen, 1965).

Coarse materials that are sorted out and left as a residual accumulation in channels may form a basal lag deposit. Typically lag deposits are found in the deeper part of stream beds, where they accumulate in lenticular fashion and in time become buried by finer-grained bar deposits. Lag deposits may take the form of erosion pavements, atop which bedload deposits are transported as thin sheets. Various styles of channel-bottom processes have been described, including diffuse gravel sheets (Hein & Walker, 1977), bed traction carpets (Todd, 1989) and hyperconcentrated flood flows (G.A. Smith, 1986). These various channel bottom processes result in basal lag sequences composed of heterogeneous deposits reflecting rapid deposition and/ or prolonged winnowing. Massive sandstone bodies, with flat, unscoured bases and lobate margins, may result from sediment gravity flows associated with bank collapse features (e.g. Jones & Rust, 1983; Rust & Jones, 1987; Turner & Monro, 1987).

The primary type of within-channel element is bars. Bars are areas of net sedimentation of size comparable in magnitude to the channels in which they occur (Allen, 1965; N.D. Smith, 1978). There are two main components in bar form. The basal feature, or platform, is made up of coarse material and is overlain by supraplatform deposits of varying forms which are subject to removal and replacement during floods (Bluck, 1976). Interpretation of bar type is often critical in elucidation of alluvial style, as bars are the key indicator of within-channel processes. Bars are readily reworked as channels shift position over their floodplains or within the channel belt. Bank-marginal features are much less likely to be reworked than mid-channel forms. The long-term preservation of bars is conditioned by factors such as the aggradational regime and the manner of channel movement.

Bars adopt many varied morphologies, ranging from simple unit bars to complex compound features. Bar character is controlled primarily by local-scale flow and grain-size characteristics. The spatial assemblage of elements on bars and their associated palaeocurrent indicators are used to infer the primary migration pattern of bars, whether lateral or downstream. Unit bars are simple features composed of one depositional style (N.D. Smith, 1974), and are found at characteristic locations such that downvalley transition in character can be discerned (Church & Jones, 1982). Bed-material character, and the competence of flow to transport it, determine formation of longitudinal bars (Leopold & Wolman, 1957). With flow oriented obliquely to the long axis of the bar, a diagonal feature is produced (Church, 1972). In highly sediment-charged sandy conditions, flow divergence results in transverse or linguoid bars (Collinson, 1970; N.D. Smith, 1974; Cant & Walker, 1978). Lateral and point bars are found at channel margins under both sandy and gravelly conditions, and record sediment accretion on the convex slopes of river bends. A range of examples of point bars has been described, reflecting differing flow and textural regimes and varying radii of curvature of bends (e.g. Sundborg, 1956; Bluck, 1971; Jackson, 1976; Levey, 1978; Lewin, 1978b; Nanson, 1980; Shepherd, 1987). Scroll bars on the inside of bends may form a distinct element in themselves, while former positions of the channel may be recorded by a series of accretionary ridges, commonly in a series of ridges and swales (e.g. Sundborg, 1956; Nanson, 1980). Outstanding examples of these features have been recorded in the rock record (e.g. R.M.H. Smith, 1987).

Most river bars are not simple unit features, but are complex, compound features made up of a mosaic of morphostratigraphic units such as bar platforms, ridges, chute channels, etc. The spatial pattern of these units determines bar sedimentology (e.g. Brierley, 1991a). Compound bars can be differentiated into three primary forms, namely mid-channel and bank-attached compound bars and compound point bars. On mid-channel compound bars, chute channels may dissect the bar surface into a chaotic pattern of remnant units. Resulting sediment sequences exhibit no within-bar trends for facies type or particle size, whether assessed down-bar, laterally or vertically. On bank-attached compound bars, the down-bar or diagonal orientation of chute channels may result in sediment sequences which are continuous down-bar, but exhibit discontinuous lateral and vertical facies and particle-size trends. Within-bar sediment sequences on compound point bars exhibit both around-the-bend and lateral trends (Brierley, 1991a). Several styles of within-channel compound bar features have been described for sand-bed channels, such as linguoid bars (Collinson, 1970; Blodgett & Stanley, 1980), macroforms (Crowley, 1983), sand flats (Cant, 1978; Cant & Walker, 1978), sand waves (Coleman, 1969) and sand sheets (N.D. Smith, 1970).

A range of within-channel forms has been characterized for laterally constrained sinuous channels, such as point dunes (Hickin, 1969), gravel counterpoint bars (S.A. Smith, 1987), and convex bar deposits (Goodwin & Steidtmann, 1981). In low to moderate sinuosity sand-bedded channels with moderately stable banks, oblique-accretion benches may form as muddy drapes which are lapped onto the relatively steep convex bank (e.g. Bluck, 1971; Nanson & Croke, 1992). In quite different circumstances, concave bank benches may form at the outside of moderately tight bends (e.g. Carey, 1969; Nanson & Page, 1983). These features, which are composed of slackwater deposits comprising thinly interbedded fine-grained sands and organic sequences, become incorporated into the floodplain as the bend assemblage migrates downstream.

Channel-marginal elements

The relationship between processes that operate in river channels and on their adjacent floodplains is conditioned by channel-marginal elements. These elements, namely levees and crevasse splays, are generally scaled in proportion to dimensions of the adjacent channel. In three-dimensional geometry levees are sinuous, ribbon-like, prismatic bodies of triangular cross-section which border channels (Allen, 1965). Levees are best developed at the concave bank of bends, where they commonly form steep high banks. Levees and splays are the coarsest top-stratum (overbank) element, and are similar lithologically because both are deposited as rhythmites from decelerating floods that overtop natural levees and crevasse channels (Farrell, 1987). When a stream overtops its banks, flow velocity is checked, so that not all of the previous load can be transported, and sediment is deposited adjacent to the banks. The coarsest debris is laid down close to the channel and the finer material further down the levee at a greater distance from the channel (e.g. Alexander & Prior, 1971). Deposition rate is greatest close to the channel and declines down-levee, giving the slope into the floodbasin.

If levees are well developed along extensive, fine-grained floodplains, it infers that the channel is relatively stable and that segregation between the channel zone and the backswamp is well defined. Levee construction, and restriction of a stream to a meander belt, leads to substantial local elevation of the floodplain surface between the levees. This is commonly referred to as an alluvial ridge. The greater the height of the levees above the general floodplain, the more likely it is that local crevassing within the active meander belt will result in some permanent change of stream course, away from the present meander belt and into the flanking basin (e.g. Brizga & Finlayson, 1990).

In most instances levees are composed of thinly interbedded flood-cycle deposits reflecting rising- and waning-stage deposits (e.g. R.M.H. Smith, 1980; Platt & Keller, 1992). Upward-coarsening sequences are common, reflecting an increase in bed shear stress related to individual flood events (e.g. Wizevich, 1992). As levees are only occasionally inundated, they typically have vegetated surfaces, and their deposits may have a significant proportion of roots and other organic matter. Unless obliterated by bioturbation, levees are commonly characterized by small-scale, cross-laminated sands. In some instances these abound with fossil vertebrates (eg. Bown & Kraus, 1987).

Crevasse-splay deposits are narrow to broad, localized tongues of sediment which are sinuous to lobate in plan. Splays are fed by a crevasse channel that cuts the channel-marginal levee. Once a crevasse is initiated, the floodwaters deepen the new course and develop a system of distributive channels on the upper slope of the levee (e.g. Coleman, 1969; O'Brien & Wells 1986; N.D. Smith *et al.*, 1989). Crevasse splays generally extend well beyond levee toes onto floodbasin deposits. Coleman (1969) identified two styles of crevasse splay along the Brahmaputra River.

(1) A single channel cuts the crest of a relatively steep levee. Once the more gentle gradient beyond the levee is reached, the channel terminates abruptly in a radiating, steep-faced, delta-like fan consisting of multiple channels. Little coarse sediment extends beyond the delta onto the adjacent basin. Grain size varies little from the edge of the main channel to the delta fan. The splay typically comprises small-scale, ripple cross-laminated fine sands, with thin silt/clay drapes that are readily disturbed by plants.

(2) Water leaves the main channel system through a multichannelled anastomosing pattern. Termination of the crevasse splay is not abrupt, and a broad zone of fine-grained material extends into the basin. Coarser sands seemingly override the finer-grained deposits of a previous crevasse unit, with little basal scouring. Small-scale cross-bedding is dominant and small cut-and-fill structures are also evident.

Since the master channels of crevasse systems often tap sediment at relatively low levels in the parent streams, bedload-derived sediments observed in crevasse splays are typically both coarser and better sorted than the levee and floodbasin deposits they overlie (e.g. Singh, 1972; Bown & Kraus, 1987). Fills of crevasse channels commonly exhibit coarsening-upward sequences, with palaeocurrent trends oblique to the main channel (Bridge, 1984). Crevasse channel sandbodies have symmetrical lenticular forms with low width–depth ratios (Ghosh, 1987). Longitudinal facies variation and diminution in grain size may be discerned with distance from the main channel. Large-scale cross-stratification and horizontal stratification may be evident, in addition to small-scale cross-laminations, but orderly vertical variation of sedimentary structures may not be apparent (e.g. Tyler & Ethridge, 1983).

Crevasse splay and levee facies are inextricably linked, as a levee feature must be present for splays to form, otherwise sheetflood facies would be found (see below). The bulk of coarser-grained levee deposits may actually accumulate as crevasse splays, while finer-grained and poorly sorted sediments represent falling-stage and sheet-type overflow. Hence, in sections parallel to the channel, levees appear to consist of a series of overlapping and interfingering lenses of coarse, well-sorted sands, capped by finer-grained materials that accumulate during falling stage (Coleman, 1969). Overbank flow in leveed rivers does not tend to result in sheet deposits, as the majority of floodwaters are funnelled into the basin via distinct flood channels. In certain situations within confined valleys, levees may constrain flows to such an extent that when floods do breach levees they induce significant scour on adjacent floodplains (e.g. Nanson & Erskine, 1988).

Channel banks are more commonly breached in non-levee situations, in which sheets of sands or coarse bedload deposits are launched onto the adjacent floodplain. These elements, referred to as sand wedges (Brierley, 1991b), are characteristically composed of trough and planar cross-bedded sequences transitional upwards to finer-grained flood cycles consisting of rippled and laminated interbeds. These deposits rest atop the proximal floodplain with a scoured basal contact. In ephemeral depositional environments, laterally extensive, horizontally bedded sand sheets are thrust onto the floodplain during flood events (e.g. Stear, 1980, 1985; Tunbridge, 1981, 1984; Sneh, 1983; Olsen, 1988, 1989; Beer & Jordan, 1989; Deluca & Eriksson, 1989). These floodplain sheets differ from splays both in terms of their morphologic expression and their lateral extent across basin fills. Splays are much less extensive features which demonstrate distinct thinning with distance from the associated channel/levee assemblage.

Floodplain elements

Floodbasins are poorly drained, relatively featureless areas, located adjacent to or between abandoned or active channels, that act as stilling basins in which suspended fines settle from overbank flows. The size, shape and position of flood basins depend on the history of the floodplain. Because they are confined by the valley margins and existing alluvial ridges, floodbasins are elongated parallel to active streams and are generally much longer than they are wide. As such, floodbasin masses are typically tabular, prismatic bodies of rectangular cross-section (Allen, 1965). In general, study of floodbasin facies has been somewhat neglected at the expense of channel facies, yet floodbasin deposits are volumetrically much more substantial (Jackson, 1978; Bridge, 1984; Platt & Keller, 1992).

Floodplains generally comprise the finest topstratum deposits, in which there is progressive gradation in grain size away from the main channel. The basic sedimentation unit on floodplains are flood-cycle bedsets produced by discrete overbank floods (e.g. Fisk, 1944, 1947; Sundborg, 1956, Wolman & Leopold, 1957;

McKee *et al.*, 1967; Carey, 1969; Coleman, 1969; Williams, 1971; Singh, 1972; Farrell, 1987; Brierley, 1991b). These units commonly fine upwards from an erosive base, reflecting scour during the rising stage of a flood, followed by deposition during waning flood velocities. Bases are less likely to be scoured with flow over vegetated surface. Increasing flow energy during the rising flood stage may result in upward-coarsening sequences at the base of the flood-cycle unit.

Proximal–distal relations on floodplains are determined in large part by the character of channel-marginal elements, in particular the capacity for high-energy flows to breach levees and launch relatively coarse-grained sediments onto proximal channel zones. The progressive decrease in grain size with distance from the channel accentuates the development of backswamp or floodpond conditions in the distal floodplain (e.g. Fisk, 1947; D.G. Smith & Smith, 1980; D.G. Smith, 1983). Under backswamp conditions with infrequent incursions of clastic sediments, organic debris typically abounds. These deposits are generally extensively bioturbated and are uniform or massive in appearance (e.g. Farrell, 1987). Sediments of ponds and marshes developed in floodbasins locally yield freshwater shells. The development and nature of peat under backswamp conditions is dependent on the distribution of clastics into the distal floodplain along with associated climatic conditions (McCabe, 1984).

Floodbasins often exhibit an inherited network of older channels. This factor, along with other controls on floodplain relief, may have pronounced impacts on velocity distributions on floodplain surfaces, and the associated deposition of materials of varying calibre. For example, if levees are not well developed, considerable energy may be exerted by flows on the floodplain surface, and bedload materials may be distributed and reworked considerable distances from the main channel(s). A range of field studies (e.g. Gretener & Strömquist, 1987; Walling & Bradley, 1989; Walling *et al.*, 1991) and mathematical modelling procedures (e.g. James, 1985; Pizzutto, 1987; Gee *et al.*, 1990; Bates *et al.*, 1992) have recently been applied to predict flood routing and distribution of sediments over floodplain surfaces.

APPLICATION OF ELEMENT ANALYSIS

Depositional systems are three-dimensional assemblages of genetically and spatially related sedimentary facies (Galloway, 1992). The form–process association implicit to an element approach enables particular patterns of element assemblages to be predicted. For example, crevasse splay and levee sediments may migrate across a peat swamp, but it is unlikely that point bar facies would directly overlie swamps (McCabe, 1984). Similarly, levee features must be observed marginal to channel features. Analysis and interpretation of element assemblages framed around such principles of spatial association provide direct insight into river style and system evolution.

The obvious starting point in assessment of alluvial style is to evaluate channel type, particularly in terms of its scale, cross-sectional geometry, lateral stability, nature of channel fill, and relationship with floodplain facies (Ethridge & Schumm,

1978). Lack of identifiable channel features in ancient depositional sequences may reflect the limited presence of channels (i.e. sheetflood conditions prevailed) or the fact that channels were not well defined, either due to the lack of textural variation between channel and floodplain facies or because margins were too shallow to allow recognition (Friend, 1983). Collinson (1978) outlined four characteristics that are used to indicate channel type from the ancient record in relation to modern-day sequences, namely the ratio of coarse member thickness to fine member thickness, palaeocurrent patterns, internal geometry of the coarse member sandbody, and the shape of the coarse member sandbody. Insights into channel-fill mechanisms can be gained by analysis of the geometry and orientation of smaller-scale elements that can be used to interpret bar type. From this, the width/depth of channels and their cross-sectional geometry can be inferred, although the relationship between thickness of a channel sandstone and depth of the formative channel is complex (e.g. Collinson, 1978; Fielding & Crane, 1987). Within- and between-channel palaeocurrent indicators can be utilized to infer the sinuosity of the channel(s), but once more this is not straightforward as the pattern of palaeocurrent indicators is often complex and its relationship to channel sinuosity may not be very clear (e.g. Collinson, 1978, 1986). These channel-fill data can then be used to interpret channel style, whether fixed and abandoned channel fills, mobile sheet sands (combed or laterally unstable), or intermediate in character. Channel-marginal conditions are indicated by the presence or absence of levee and splay elements. Interfingering with floodbasin deposits is a primary indicator of the lateral stability of the channel(s) and the flood history.

Element analysis focuses directly on mechanisms by which sediments are deposited in channels and become incorporated in the floodplain. A constructivist approach to facies modelling does not necessarily equate to, or lead towards, sedimentological anarchy, as suggested by Walker (1990, p. 779), but provides a rigorous framework with which to analyse the complexity of river deposits, thereby permitting more reliable interpretation of river depositional environments and their evolution. At present there are simply insufficient data of suitable refinement, at enough points along the spectrum of morphological complexity demonstrated by rivers, to build a representative range of models. Undue emphasis on a limited number of models detracts from the broader complexity of river styles that is physically real.

Three-dimensional reconstructions of elemental configurations in modern depositional systems provide the ideal testing ground for assessing the application of an element-based approach. After all, if the complexities of river depositional environments cannot be unravelled in modern river systems, where data can be attained with relative ease at predetermined sites, the possibilities of reliably interpreting ancient depositional environments must be conjectural to say the least. Detailed field sediment inventories of modern floodplain architecture are required across the spectrum of river styles. One option may be to employ an elemental approach to re-evaluate field sites used in derivation of existing vertical-profile facies models. Beyond this, element-scale assemblages need to be assessed for rivers in differing environmental settings, such as arid-zone rivers.

The role of antecedent controls on river behaviour needs to be better understood in terms of their control on the geometry of subsequent river deposits. A variety of intergradational fluvial sequences may occur at the same stratigraphic horizon within a given local area (cf. Jackson, 1978; Galloway, 1981). Differing segments of floodplains may be produced by floods of differing magnitude. This complexity needs to be integrated into approaches used to unravel river style. Analysis of controls on element stacking and preservation needs to be extended to evaluate mechanisms, by which alluvial assemblages become incorporated into basin fills. In particular, the roles and consequences of channel shifting and avulsion-related mechanisms, and the reworking potential of subsequent flows to scour floodplains, are critical components in determining the relative proportions of within-channel, channel-marginal and floodplain elements preserved in basin fills. It would seem sensible to concentrate research efforts within modern basin settings in which the preservation history can be reliably inferred. Differential compaction of elements and implications for assessment of their geometry (McCabe, 1984) also needs to be integrated with research into tectonic, climatic and eustatic controls on the preservation of element-scale units.

Quantitative modelling procedures have been developed to evaluate the stacking arrangement and connectivity of elements in basin fills resulting from channel-belt avulsion and co-eval basin subsidence (e.g. Allen, 1978; Leeder, 1978; Bridge & Leeder, 1979). These have subsequently been applied to differing tectonic situations (e.g. Alexander & Leeder, 1987). Such models need to be used cautiously, however. For example, models developed by Allen (1978) and Bridge & Leeder (1979) suggest that subsidence affects the geometry of sandstone bodies deposited by meandering river systems. In these models, streams avulse and cut into the underlying fine-grained floodplain facies, and it is inferred that high rates of subsidence will result in isolated ribbon sandstone bodies while low rates of subsidence will result in multistoried sheet sandstones. However, Beer & Jordan (1989) document a rapidly subsiding system in which streams do not erode and rework overbank areas. A significantly higher rate of net subsidence is accompanied by a sharp decrease in the amount of fine-grained material preserved, with no accompanying change in the manner of storey development.

Finally, it is interesting to note the application of element-based approaches in fluvial sedimentology to the resource extraction industry. In this era of enhanced hydrocarbon recovery from reservoirs in alluvial basins, analysis of the geometry of depositional units and their associations with river styles is destined to play an increasing role in resource exploitation and extraction strategies. Improved assessments of the complexity of alluvial depositional environments will greatly aid predictive modelling of alluvial morphology, especially for those interpretations based on limited data. The greater the reliability of the reconstruction of three-dimensional geometry of river depositional units in any basin fill, the greater the potential to assess and quantify reservoir compartmentalization and the interconnect-edness of depositional units, and hence successfully model reservoir heterogeneity (e.g. Tyler & Ethridge, 1983; Tyler, 1988; Miall, 1988b; Kerr & Jirik, 1990; Chapin & Mayer, 1991; Tye, 1991; Ambrose et al., 1991).

Present depositional models of alluvial architecture are too crude, and are applied at too broad a scale for the needs of reservoir exploitation, such that much of the interwell hydrocarbon resource remains untapped (Tyler & Finley, 1991).

Reconstruction of alluvial suites at the channel scale is of primary significance in economic geology, as this element scale permits elucidation of the macroscopic heterogeneity of complex reservoir bodies. Channel fills, channel-marginal elements and floodplain elements are the principal building blocks of a basin fill and the geometry, orientations, and vertical and lateral relations of these various units fundamentally control reservoir distribution and architecture. Detailed studies of element assemblages that characterize modern floodplain architecture may play a significant role in developing a better understanding of controls on the geometry and heterogeneity of reservoirs in alluvial basins.

SUMMARY

Implications and applications of a constructivist approach to facies modelling can be summarized as follows

(a) Planform−sediment correlatives are far from unequivocal and facies models applied at the planform scale may be unreliable.

(b) Form−process associations in river depositional sequences are best defined at the element scale and river styles can be viewed as assemblages of geomorphic elements. This facies modelling procedure provides a suitable methodology with which to analyse the character and evolution of river depositional systems. Rather than focusing attention on end-member facies models, the constructivist approach is open-ended and analyses individual sediment sequences at any point along the spectrum of river styles. This approach to facies modelling supports the adage outlined by Bridge (1993) wherein "a picture is worth a thousand words". The message of this review, however, is that the scalar framework used to construct our picture of river styles must integrate information from the granular scale to the catchment scale. The scale of depositional unit which best describes form−process associations in rivers, and hence provides most insight into river style and environment of deposition, is that of "geomorphic units", which equate to morphostratigraphic units or elements.

(c) There is a surprising lack of detailed data on element-scale assemblages for modern fluvial depositional systems, and floodplain architecture needs to be assessed along the spectrum of morphological complexity demonstrated by rivers, particularly larger rivers. Undue emphasis continues to be placed on end-member situations, in particular the meandering Mississippi (after Fisk, 1944) and the braided Brahmaputra (after Coleman, 1969).

(d) Controls on element presence/absence and their reworking potential need to be unravelled for differing river styles, with particular emphasis placed on the relationship between channel and floodplain elements. Preservation potential of element-scale depositional units and their incorporation into valley (and hence basin) fills needs to be evaluated in terms of both autocyclic and allocyclic controls. Autocyclic controls determine the stacking arrangement of these units, while allocyclic controls determine the make-up of element assemblages preserved in allostratigraphic units and formations. The limited range of subsiding basins probably restricts the range of river styles which are likely to be recorded in the rock record. Excellent opportunities for such work lie in detailed stratigraphic analysis of Quaternary deposits with tight dating control in areas of known environmental change.

(e) Element-scale sediment inventories from the modern record need to be manipulated in ways such that the range of one-dimensional core-log data and two-dimensional slices (outcrops) can be recognized and interpreted. Used in this way, reliable interpretation of river styles as an assemblage of element scale units may have profound significance in studies of reservoir heterogeneity.

ACKNOWLEDGEMENTS

Numerous colleagues have read various draft versions of this document, and I am grateful to them all for their thoughts. I am especially grateful to John Collinson for the thorough, critical reading of this manuscript and his numerous suggestions for both substantive and stylistic changes. Tables 8.2 and 8.3 were suggested by Ken Woolfe, who also offered much constructive comment on earlier versions of this manuscript. Figures were drafted by John Cleasby at Macquarie University. Michael Blum provided several useful points on the relationship between element-scale approaches and allostratigraphy.

REFERENCES

Alexander, J. & Leeder, M.R. 1987, Active tectonic control on alluvial architecture, In: Ethridge, F.G., Flores, R.M. & Harvey, M.D. (Eds), *Recent Developments In Fluvial Sedimentology*, Society of Economic Palaeontologists and Mineralogists Special Publication Number **39**, 243–252.

Alexander, C.S. & Prior, J.C. 1971, Holocene sedimentation rates in overbank deposits in the Black Bottom of the lower Ohio River, southern Illinois, *American Journal of Science*, **270**, 361–372.

Allen, J.R.L. 1963, The classification of cross-stratified units, with notes on their origin, *Sedimentology*, **2**, 93–114.

Allen, J.R.L, 1965, A review of the origin and characteristics of recent alluvial sediments, *Sedimentology*, **5**, 89–191.

Allen, J.R.L. 1978, Studies in fluviatile sedimentation: an exploratory quantitative model for the architecture of avulsion-controlled alluvial suites, *Sedimentary Geology*, **21**, 129–147.

Allen, J.R.L. 1983, Studies in fluviatile sedimentation: bars, bar-complexes and sandstone sheets (low sinuosity braided streams) in the Brownstones (L. Devonian), Welsh Borders, *Sedimentary Geology*, **33**, 237–293.

Ambrose, W.A., Tyler, N. & Parsley, M.J. 1991, Facies heterogeneity, pay continuity, and infill potential in barrier-island, fluvial and submarine-fan reservoirs: examples from the Texas Gulf Coast and Midland Basin, in: Miall, A.D. & Tyler, N. (Eds), *The three-dimensional facies architecture of terrigenous clastic sediments and its implications for hydrocarbon discovery and recovery*, SEPM (Society for Sedimentary Geology), Concepts in Sedimentology and Palaeontology, Volume **3**, 13–21.

Anderton, R. 1985, Clastic facies models and facies analysis, in: Blenchley, P.J. & Williams, B.P.J. (Eds), *Sedimentology: Recent Developments and Applied Aspects*, Geological Society of London, Special Publication Number **18**, 31–47.

Autin, W.J. 1992, Use of alloformations for definition of Holocene meander belts in the middle Amite River, southeastern Louisiana, *Geological Society of America Bulletin*, **104**, 233–241.

Bates, P.D., Anderson, M.G., Baird, L., Walling, D.E. & Simm, D. 1992, Modelling floodplain flows using a two-dimensional finite element model, *Earth Surface Processes and Landforms*, **17**, 575–588.

Beer, J.A. & Jordan, T.E. 1989, The effects of Neogene thrusting on deposition in the Bermejo Basin, Argentina, *Journal of Sedimentary Petrology*, **59**, 330–345.

Bernard, H.A. & Major, C.F. 1963, Recent meander belt deposits of the Brazos River: an "alluvial" sand model, *Bulletin of the American Association of Petroleum Geologists*, **47**, 350.

Bernard, H.A., LeBlanc, R.J. & Major, C.F. 1962, Recent and Pleistocene geology of southeast Texas, in: *Geology of Gulf Coast and Guidebook of Excursion*, Geol. Soc. Houston, Houston, Texas, 175–224.

Best, J.L., 1987, Flow dynamics at river channel confluences: implications for sediment transport and bed morphology, In: Ethridge, F.G., Flores, R.M. & Harvey, M.D. (Eds), *Recent Developments in Fluvial Sedimentology*, Society of Economic Palaeontologists and Mineralogists, Special Publication Number **39**, 27–35.

Blair, T.C. & McPherson, J.G. 1992, The Trollheim alluvial fan and facies model revisited, *Geological Society of America Bulletin*, **104**, 762–769.

Blodgett, R.H. & Stanley, K.O. 1980, Stratification, bedforms, and discharge relations of the Platte braided river system, Nebraska, *Journal of Sedimentary Petrology*, **50**, 139–148.

Bluck, B.J. 1971, Sedimentation in the meandering River Endrick, *Scottish Journal of Geology*, **7**, 93–138.

Bluck, B.J. 1976, Sedimentation in some Scottish rivers of low sinuosity, *Transactions of the Royal Society of Edinburgh*, **69**, 425–456.

Bluck, B.J. 1979, Structure of coarse grained braided stream alluvium, *Transactions of the Royal Society of Edinburgh*, **70**, 181–221.

Bluck, B.J. 1980, Structure, generation and preservation of upward fining braided stream cycles in the Old Red Sandstone of Scotland, *Transactions of the Royal Society of Scotland*, **71**, 29–46.

Bown, T.M. & Kraus, M.J. 1987, Integration of channel and floodplain suites, I. Development of sequence and lateral profiles of alluvial paleosols, *Journal of Sedimentary Petrology*, **57**, 587–601.

Bridge, J.S. 1984, Large-scale facies sequences in alluvial overbank environments, *Journal of Sedimentary Petrology*, **54**, 583–588.

Bridge, J.S. 1985, Paleochannel patterns inferred from alluvial deposits: a critical evaluation, *Journal of Sedimentary Petrology*, **55**, 579–589.

Bridge, J.S. 1993, Description and interpretation of fluvial deposits, *Sedimentology*, **40**, 801–820.

Bridge, J.S. & Gabel, S.L. 1992, Flow and sediment dynamics in a low sinuosity, braided river: Calamus River, Nebraska Sand Hills, *Sedimentology*, **39**, 125–142.

Bridge, J.S. & Leeder, M.R. 1979, A simulation model of alluvial stratigraphy, *Sedimentology*, **26**, 617–644.

Bridge, J.S., Smith, N.D., Trent, F., Gabel S.L. & Bernstein, P. 1986, Sedimentology and morphology of a low-sinuosity river: Calamus River, Nebraska Sand Hills, *Sedimentology*, **33**, 851–870.

Brierley, G.J. 1989, River planform facies models: The sedimentology of braided, wandering and meandering reaches of the Squamish River, British Columbia, *Sedimentary Geology*, **61**, 17–35.

Brierley, G.J. 1991a, Bar sedimentology of the Squamish River, British Columbia: Definition and application of morphostratigraphic units, *Journal of Sedimentary Petrology*, **61**, 211–225.

Brierley, G.J. 1991b, Floodplain sedimentology of the Squamish River, British Columbia, relevance of element analysis, *Sedimentology*, **38**, 735–750.

Brierley, G.J. & Hickin, E.J. 1991, Channel planform as a non-controlling factor in fluvial sedimentology: the case of the Squamish River floodplain, British Columbia, *Sedimentary Geology*, **75**, 67–83.

Brierley, G.J. & Hickin, E.J. 1992, Floodplain development based on selective preservation of sediments, Squamish River, British Columbia, *Geomorphology*, **4**, 381–391.

Brierley, G.J., Liu, K. & Crook, K.A.W. 1993, Sedimentology of coarse-grained alluvial fans in the Markham Valley, Papua New Guinea, *Sedimentary Geology*, **86**, 297–324.

Bristow, C.S. 1987, Brahmaputra River: channel migration and deposition, In: Ethridge, F.G., Flores, R.M. & Harvey, M.D. (Eds), *Recent Developments In Fluvial Sedimentology*, Society of Economic Palaeontologists and Mineralogists, Special Publication Number **39**, 63–74.

Brizga, S.G. & Finlayson, B.L. 1990, Channel avulsion and river metamorphosis: the case of the Thomson River, Victoria, Australia, *Earth Surface Processes and Landforms*, **15**, 391–404.

Cant, D.J. 1978, Development of a facies model for sandy braided river sedimentation: Comparison of the South Saskatchewan River and the Battery Point Formation, In: Miall, A.D. (Ed), *Fluvial Sedimentology*, Canadian Society of Petroleum Engineers, Memoir **5**, Calgary, 62–639.

Cant, D.J. & Walker, R.G. 1978, Fluvial processes and facies sequences in the sandy braided South Saskatchewan River, Canada, *Sedimentology*, **25**, 625–648.

Carey, W.C. 1969, Formation of flood plain lands, *Journal of the Hydraulics Division, American Society of Civil Engineers*, **95**, HY3, 981–994.

Carson, M.A. 1984, The meandering-braided river threshold: a reappraisal, *Journal of Hydrology*, **73**, 315–334.

Chapin, M.A. & Mayer, D.F. 1991, Constructing a three-dimensional rock-property model of fluvial sandstones in the Peoria field, Colorado, In: Miall, A.D. & Tyler, N. (Eds), *The three-dimensional facies architecture of terrigenous clastic sediments and its implications for hydrocarbon discovery and recovery*, SEPM (Society for Sedimentary Geology), Concepts in Sedimentology and Palaeontology, Volume **3**, 160–171.

Church, M. 1972, Baffin Island sandurs: a study of arctic fluvial processes, *Geological Survey of Canada Bulletin*, **216**, 208 pp.

Church, M. 1983, Pattern of instability in a wandering gravel bed channel, In: Collinson, J.D. & Lewin, J. (Eds), *Modern and Ancient Fluvial Systems*, International Association of Sedimentologists, Special Publication Number **6**, Blackwell, Oxford, 169–180.

Church, M. & Jones, D. 1982, Channel bars in gravel-bed rivers, In: Hey, R.D., Bathurst, J.C. & Thorne, C.R. (Eds), *Gravel-bed Rivers: Fluvial Processes, Engineering and Management*, Wiley, Chichester, 291–338.

Coleman, J.D. 1969, Brahmaputra River: Channel processes and sedimentation, *Sedimentary Geology*, **3**, 129–239.

Collinson, J.D. 1970, Bedforms of the Tana River: Norway, *Geografiska Annaler*, **52A**, 31–55.

Collinson, J.D. 1978, Vertical sequence and sand shape body in alluvial sequences, In: Miall, A.D. (Ed.), *Fluvial Sedimentology*, Canadian Society of Petroleum Geology, Memoir **5**, Calgary, 577–586.

Collinson, J.D. 1986, Alluvial sediments, In: Reading, H.G. (Ed.), *Sedimentary Environments and Facies*, Blackwell Scientific, Oxford 20–62.

Cowan, E.J. 1991, Reservoir geometry of a fluvial sheet sandstone: an architectural reinterpretation of the Jurassic Westwater Canyon Member, Morrison Formation, USA, In: Miall, A.D. & Tyler, N. (Eds), *The three-dimensional facies architecture of terrigenous clastic sediments and its implications for hydrocarbon discovery and recovery*, SEPM (Society for Sedimentary Geology), Concepts in Sedimentology and Palaeontology, Volume **3**, 80–93.

Crowley, K.D. 1983, Large-scale bed configurations (macroforms), Platte River Basin, Colorado and Nebraska: primary structures and formative processes, *Geological Society of America Bulletin*, **94**, 117–133.

DeCelles, P.G., Gray, M.B., Ridgeway, K.M., Cole, R.B., Pivnik, D.A., Pequera, N. & Srivastava, P. 1991, Controls on synorogenic alluvial-fan architecture, Beartooth Conglomerate (Palaeocene), Wyoming and Montana. *Sedimentology*, **38**, 567–590.

Deluca, J.L. & Eriksson, K.A. 1989, Controls on synchronous ephemeral- and perennial-river sedimentation in the middle sandstone member of the Triassic Chinle Formation, northeastern New Mexico, U.S.A., *Sedimentary Geology*, **61**, 155–175.

Desloges, J.R. & Church, M. 1987, Channel and floodplain facies in a wandering gravel-bed river, In: Ethridge, F.G., Flores, R.M. & Harvey, M.D. (Eds), *Recent Developments in Fluvial Sedimentology*, Society of Economic Palaeontologists and Mineralogists, Special Publication Number **39**, Tulsa, Oklahoma, 99–109.

Doeglas, D.J. 1962, The structure of sedimentary deposits of braided streams, *Sedimentology*, **1**, 167–190.

Dott, R.H. 1988, Perspectives: Something old, something new, something borrowed, something blue – hindsight and foresight of sedimentary geology, *Journal of Sedimentary Petrology*, **58**, 358–364.

Erskine, W., McFadden, C. & Bishop, P. 1992, Alluvial cutoffs as indicators of former channel conditions, *Earth Surface Processes and Landforms*, **17**, 23–37.

Ethridge, F.G. & Schumm, S.A. 1978, Reconstructing paleochannel morphologic and flow characteristics: methodology, limitations and assessment, In: Miall, A.D. (Ed.), *Fluvial Sedimentology*, Canadian Society of Petroleum Geology, Memoir **5**, Calgary, 703–721.

Farrell, K.M. 1987, Sedimentology and facies architecture of overbank deposits of the Mississippi River, False River region, Louisiana, In: Ethridge, F.G., Flores R.M. & Harvey, M.D. (Eds), *Recent Developments In Fluvial Sedimentology*, Society of Economic Palaeontologists and Mineralogists, Special Publication Number **39**, 111–120.

Ferguson, R.I. 1987, Hydraulic and sedimentary controls of channel planform, In: Richards, K., (Ed.), *River Channels: Environment and Process*, Blackwell, Oxford, 129–158.

Fielding, C.R. & Crane, R.C. 1987, An application of statistical modelling to the prediction of hydrocarbon recovery factors in fluvial reservoir sequences, In: Ethridge, F.G., Flores, R.M. & Harvey, M.D. (Eds), *Recent Developments in Fluvial Sedimentology*, Society of Economic Palaeontologists and Mineralogists Special Publication Number **39**, 321–327.

Fisk, H.N. 1944, *Geological investigation of the alluvial valley of the lower Mississippi River*, Mississippi River Commission Waterways Experiment Station, Vicksburg, Mississippi, 78 pp.

Fisk, H.N. 1947, *Fine-grained alluvial deposits and their effects on Mississippi River activity*, Mississippi River Commission Waterways Experiment Station, Vicksburg, Mississippi, 82 pp.

Flores, R.M. & Hanley, J.H. 1984, Anastamosed and associated coal-bearing fluvial deposits: Upper Tongue River Member, Palaeocene Fort Union Formation, northern Powder River basin, Wyoming, U.S.A., In: Rahmani, R.A. & Flores, R.M. (Eds), *Sedimentology of Coal and Coal-bearing Sequences*, International Association of Sedimentologists, Special Publication, **7**, 85–103.

Forbes, D.L. 1983, Morphology and sedimentology of a sinuous gravel-bed channel system: lower Babbage River, Yukon coastal plain, Canada, In: Collinson, J.D. & Lewin, J. (Eds), *Modern and Ancient Fluvial Systems*, International Association of Sedimentologists, Special Publication Number **6**, Blackwell, Oxford, 195–206.

Friend, P.F. 1983, Towards the field classification of alluvial architecture or sequence, In: Collinson, J.D. & Lewin, J. (Eds.), *Modern and Ancient Fluvial Systems*, International Association of Sedimentologists, Special Publication Number **6**, Blackwell, Oxford, 345–354.

Friend, P.F., Slater, M.J. & Williams, R.C. 1979, Vertical and lateral building of river sandstone bodies, Ebro Basin, Spain, *Journal of the Geological Society of London*, **136**, 39–46.

Frye, J.C. & Willman, H.B. 1960, Classification of the Wisconsinan Stage in the Lake Michigan glacial lobe, *Illinois State Geological Survey Circular*, **285**, 16 pp.

Galloway, W.E. 1981, Depositional architecture of Cenozoic coastal plain fluvial systems, In: Ethridge, F.G. & Flores, M.R. (Eds), *Recent and ancient nonmarine depositional environments*: *models for exploration*, Society of Economic Palaeontologists and Mineralogists Special Publication, No. **31**, 127–155.

Galloway, W.E. 1992, *Depositional systems and sequences in the exploration for sandstone reservoirs and stratigraphic traps*, Short Course Notes, 11th Australian Geological Convention, Ballarat, 17–18 January 1992.

Gee, D.M., Anderson, M.G. & Baird, L. 1990, Large-scale floodplain modelling, *Earth Surface Processes and Landforms*, **15**, 513–523.

Ghosh, S.K. 1987, Cyclicity and facies characteristics of alluvial sediments in the Upper Palaeozoic Monongahela-Dunkard Groups, central West Virginia, In: Ethridge, F.G., Flores, R.M. & Harvey, M.D. (Eds), *Recent Developments in Fluvial Sedimentology*, Society of Economic Palaeontologists and Mineralogists, Special Publication Number **39**, 229–239.

Goodwin, C.G. & Steidtmann, J.R. 1981, The convex bar: member of the alluvial channel side-bar continuum, *Journal of Sedimentary Petrology*, **51**, 129–136.

Gottesfeld, A.S. & Gottesfeld, L.M.J. 1990, Floodplain dynamics of a wandering river: dendrochronology of the Morice River, British Columbia, *Geomorphology*, **3**, 159–179.

Gretener, B. & Strömquist, L. 1987, Overbank sedimentation rates on fine grained sediments. A study of the recent deposition in the lower River Fyrisan, *Geografiska Annaler*, **69A**, 139–146.

Gustavson, T.C. 1978, Bed forms and modern stratification types of modern gravel meander lobes, Nueces River, Texas, *Sedimentology*, **25**, 401–426.

Happ, S.C., Rittenhouse, G. & Dobson, G.C. 1940, Some principles of accelerated stream and valley sedimentation, *United States Department of Agriculture Technical Bulletin*, **695**, 134 pp.

Hein, F.J. & Walker, R.G. 1977, Bar evolution and development of stratification in the gravely, braided, Kicking Horse River, British Columbia, *Canadian Journal of Earth Sciences*, **14**, 562–570.

Henderson, F.M. 1961, Stability of alluvial channels, *Journal of the Hydraulics Division, American Society of Civil Engineers*, **87**, 109–138.

Hickin, E.J. 1969, A newly identified process of point bar formation in natural streams, *American Journal of Science*, **267**, 999–1010.

Jackson, R.G. 1975, Hierarchical attributes and a unifying model of bedforms composed of cohesionless material and produced by shearing flow, *Geological Society of America Bulletin*, **86**, 1523–1533.

Jackson, R.G. 1976, Depositional model of point bars in the lower Wabash River, *Journal of Sedimentary Petrology*, **46**, 579–594.

Jackson, R.G. 1978, Preliminary evaluation of lithofacies models for meandering alluvial streams, In: Miall, A.D. (Ed), *Fluvial Sedimentology*, Canadian Society of Petroleum Geologists, Memoir **5**, Calgary, 543–576.

James, C.S. 1985, Sediment transfer to overbank sections, *Journal of Hydraulic Research*, **23**, 435–452.

Johnson, E.A. & Pierce, F.W. 1990, Variations in fluvial deposition on an alluvial plain: an example from the Tongue River Member of the Fort Union Formation (Palaeocene), southeastern Powder River Basin, Wyoming, USA, *Sedimentary Geology*, **69**, 21–36.

Jones, B.G. & Rust, B.R. 1983, Massive sandstone facies in the Hawkesbury Sandstone, a Triassic fluvial deposit near Sydney Australia, *Journal of Sedimentary Petrology*, **53**, 1249–1261.

Kerr, D.R. & Jirik, L.A. 1990, Fluvial architecture and reservoir compartmentalisation in the Oligocene Middle Frio Formation, South Texas, *Transactions, Gulf Coast Association of Geological Societies*, **XL**, 373–380.

Kirschbaum, M.A. & McCabe, P.J. 1992, Controls on the accumulation of coal and on the development of anastamosed fluvial systems in the Cretaceous Dakota Formation of southern Utah, *Sedimentology*, **39**, 581–598.

Krigstrom, A. 1962, Geomorphological studies of sandur plains and their braided rivers in Iceland, *Geografiska Annaler*, **44**, 328–346.

Lane, E.W. 1957, *A study of the shape of channels formed by natural streams flowing in erodible material*, Missouri River Division Sediments Series, **9**, United States Army Engineer Division, Missouri River, Corps of Engineers, Omaha, Nebraska, 106 pp.

Leeder, M.R. 1978, A quantitative stratigraphic model for alluvium, with special reference to channel deposit density and interconnectedness, In: Miall, A.D. (Ed), *Fluvial Sedimentology*, Canadian Society of Petroleum Geologists, Memoir 5, 587–596.

Leopold, L.B. & Wolman, M.G. 1957, *River channel patterns, braided, meandering and straight*, United States Geological Survey Professional Paper, **282-D**, 35–85.

Levey, R.A. 1978, Bed-form distribution and internal stratification of coarse-grained point bars, Upper Congaree River, South Carolina, In: Miall, A.D. (Ed.), *Fluvial Sedimentology*, Canadian Society of Petroleum Geologists, Memoir 5, 105–128.

Lewin, J. 1978a, Floodplain geomorphology, *Progress in Physical Geography*, **2**, 408–437.

Lewin, J. 1978b, Meander development and floodplain sedimentation: a case study from mid-Wales, *Geological Journal*, **13**, 25–36.

Marzo, M., Nijman, W. & Puigdefabregas, C. 1988, Architecture of the Castissent fluvial sheet sandstones, Eocene, South Pyrenees, Spain, *Sedimentology*, **35**, 719–738.

McCabe, P.J. 1984, Depositional environments of coal and coal-bearing strata, In: Rahmani, R.A. & Flores, R.M. (Eds), *Sedimentology of coal and coal-bearing sequences*, International Association of Sedimentologists, Special Publication, **7**, 13–42.

McGowen, J.H. & Garner, L.E. 1970, Physiographic features and stratification types of coarse- grained point bars: modern and ancient examples, *Sedimentology*, **14**, 77–111.

McKee, E.D., Crosby, E.J. & Berryhill, H.L. 1967, Flood deposits, Bijou Creek, Colorado, June 1965, *Journal of Sedimentary Petrology*, **37**, 829–851.

McPherson, J.G., Shanmugan, G. & Moiola, R.J. 1988, Fan deltas and braid deltas: conceptual problems, In: Nemec, W. & Steel, R.J. (Eds), *Fan Deltas: Sedimentology and Tectonic Settings*, Blackie, Glasgow, 14–22.

Miall, A.D., 1977, A review of the braided-river depositional environment, *Earth Science Reviews*, **13**, 1–62.

Miall, A.D. 1978a, Fluvial sedimentology: An historical review, In: Miall, A.D. (Ed.), *Fluvial Sedimentology*, Canadian Society of Petroleum Geologists, Memoir 5, Calgary, 1–47.

Miall, A.D. 1978b, Lithofacies types and vertical profile models in braided river deposits: A summary, In: Miall, A.D. (Ed.), *Fluvial Sedimentology*, Canadian Society of Petroleum Geologists, Memoir 5, Calgary, 597–604.

Miall, A.D. 1980, Cyclicity and the facies model concept in fluvial deposits, *Bulletin of Canadian Petroleum Geology*, **28**, 59–80.

Miall, A.D. 1985, Architectural-element analysis: A new method of facies analysis applied to fluvial deposits, *Earth Science Reviews*, **22**, 261–308.

Miall, A.D. 1987, Recent developments in the study of fluvial facies models, In: Ethridge, F.G., Flores, R.M. & Harvey, M.D. (Eds), *Recent Developments in Fluvial Sedimentology*, Society of Economic Palaeontologists and Mineralogists, Special Publication Number **39**, Tulsa, Oklahoma, 1–9.

Miall, A.D. 1988a, Architectural elements and bounding surfaces in fluvial deposits: Anatomy of the Kayenta Formation (Lower Jurassic), Southwestern Colorado, *Sedimentary Geology*, **55**, 233–262.

Miall, A.D. 1988b, Reservoir heterogeneities in fluvial sandstones; lessons from outcrop studies, *The American Association of Petroleum Geologists Bulletin*, **72**, 682–697.

Miall, A.D. 1991, Hierarchies of architectural units in terrigenous clastic rocks, and their relationship to sedimentation rate, In: Miall, A.D. & Tyler, N. (Eds), *The three-dimensional facies architecture of terrigenous clastic sediments and its implications for hydrocarbon discovery and recovery*, SEPM (Society for Sedimentary Geology), Concepts in Sedimentology and Palaeontology, Volume 3, 6–12.

Moody-Stuart, M., 1966, High- and low-sinuosity stream deposits, with examples from the Devonian of Spitsbergen, *Journal of Sedimentary Petrology*, **36**, 1102–1117.

Morningstar, O.R. 1988, *Floodplain construction and overbank deposition in a wandering reach of the Fraser River, Chilliwack, B.C.*, Unpublished MSc thesis, Simon Fraser University, 129 pp.

Munoz, A., Ramos, A., Sanchez-Moya Y. & Sopena, A. 1992, Evolving fluvial architecture during a marine transgression: Upper Bundsandstein, Triassic, central Spain, *Sedimentary Geology*, **75**, 257–281.

NACSN (North American Commission on Stratigraphic Nomenclature) 1983, North American Stratigraphic Code, *The American Association of Petroleum Geologists Bulletin*, **67**, 841–875.

Nanson, G.C. 1980, Point bar and floodplain formation of the meandering Beatton River, northeastern British Columbia, Canada, *Sedimentology*, **27**, 3–30.

Nanson, G.C. & Croke, J.C. 1992, A genetic classification of floodplains, *Geomorphology*, **4**, 459–486.

Nanson, G.C. & Erskine, W.D. 1988, Episodic changes of channels and floodplains on coastal rivers in New South Wales, In: Warner, R.F. (Ed.), *Fluvial Geomorphology in Australia*, Academic Press, Sydney, 201–221.

Nanson, G.C. & Page, K. 1983, Lateral accretion of fine-grained concave benches in meandering rivers, In: Collinson, J.D. & Lewin, J. (Eds), *Modern and Ancient Fluvial Systems*, International Association of Sedimentologists Special Publication Number **6**, Blackwell, Oxford, 133–143.

Nanson, G.C., Rust, B.R. & Taylor, G. 1986, Coexistent mud braids and anastamosing channels in an arid-zone river: Cooper's Creek, central Australia, *Geology*, **14**, 175–178.

Nanson, G.C., Young, R.W., Price, D.M. & Rust, B.R. 1988, Stratigraphy, sedimentology and late-Quaternary chronology of the channel country of western Queensland, In: Warner, R.F. (Ed.), *Fluvial Geomorphology of Australia*, Academic Press, Sydney, 151–175.

Neill, C.R. 1973, *Hydraulic and morphologic characteristics of Athabasca River near Assiniboine*, Highway and River Engineering Division Report, **REH**/73/3, Alberta Research Council, Edmonton, 23 pp.

Nordseth, K. 1973, Floodplain construction on a braided river: The islands of Koppangoyene on the River Glomma, *Norsk Geografisk Tidsskrift*, **27**, 109–126.

O'Brien, P.E. & Wells, T.A. 1986, A small, alluvial crevasse splay, *Journal of Sedimentary Petrology*, **56**, 876–879.

Olsen, H. 1988, The architecture of a sandy braided-meandering river system: an example from the Lower Triassic Solling Formation (M. Buntsandstein), in W-Germany, *Geologische Rundschau*, **77**, 797–814.

Olsen, H. 1989, Sandstone-body structures and ephemeral stream processes in the Dinosaur Canyon Member, Moenave Formation (Lower Jurassic), Utah, U.S.A., *Sedimentary Geology*, **61**, 207–221.

Ore, H.T. 1964, Some criteria for recognition of braided stream deposits, *University of Wyoming Contributions to Geology*, **3**, 1–14.

Ori, G.G. 1983, Braided to meandering channel patterns in humid-region alluvial fan deposits, River Reno, Po Plain (northern Italy), *Sedimentary Geology*, **31**, 231–248.

Pizzutto, J.E. 1987, Sediment diffusion during overbank flows, *Sedimentology*, **34**, 301–317.

Platt, N.H. & Keller, B. 1992, Distal alluvial deposits in a foreland basin setting – the Lower freshwater Molasse (Lower Miocene), Switzerland: sedimentology, architecture and palaeosols, *Sedimentology*, **39**, 545–565.

Ramos, A., Sopena, A. & Perez-Arlucea, M. 1986, Evolution of Buntsandstein fluvial sedimentation in the northwest Iberian Ranges, (central Spain), *Journal of Sedimentary Petrology*, **56**, 862–875.

Reading, H.G., 1987, Fashions and models in sedimentology: a personal perspective, *Sedimentology*, **34**, 3–9.

Reading, H.G. & Orton, G.J. 1991, Sediment calibre: a control on facies models with special reference to deep-sea depositional systems, In: Muller, D.W., McKenzie, J.A.) Weissert, H. (Eds), *Controversies in Modern Geology: Evolution of Geological Theories in Sedimentology, Earth History and Tectonics*, Academic Press, London, 85–111.

Richards, K.S. 1986, Fluvial geomorphology, *Progress in Physical Geography*, **10**, 401–420.

Rust, B.R. 1978a, A classification of alluvial channel systems, In: Miall, A.D. (Ed.), *Fluvial Sedimentology*, Canadian Society of Petroleum Geologists, Memoir **5**, Calgary, 187–198.

Rust, B.R. 1978b, Depositional models of braided alluvium, In: Miall, A.D. (Ed.), *Fluvial Sedimentology*, Canadian Society of Petroleum Geologists, Memoir **5**, Calgary, 605–625.

Rust, B.R. 1981, Sedimentation in an arid-zone anastamosing fluvial system: Cooper's Creek, central Australia, *Journal of Sedimentary Petrology*, **51**, 745–755.

Rust, B.R. & Jones, B.G. 1987, The Hawkesbury Sandstone south of Sydney, Australia, Triassic analogue for the deposit of a large braided river, *Journal of Sedimentary Petrology*, **57**, 222–233.

Schumann, R.R. 1989, Morphology of Red Creek, Wyoming, an arid-region anastamosing channel system, *Earth Surface Processes and Landforms*, **14**, 277–288.

Schumm, S.A. 1960, *The shape of alluvial channels in relation to sediment type*, United States Geological Survey Professional Paper, **352-B**, 17–30.

Schumm, S.A. 1968, Speculations concerning paleohydrologic controls of terrestrial sedimentation, *Geological Society of America Bulletin*, **79**, 1573–1588.

Schumm, S.A. 1981, *Evolution and response of the fluvial system, sedimentological implications*, Society of Economic Palaeontologists and Mineralogists, Special Publications, **31**, 19–29.

Schumm, S.A. 1985, Patterns of alluvial rivers, *Annual Review of Earth and Planetary Sciences*, **13**, 5–27.

Schumm, S.A. & Khan, H.R. 1972, Experimental study of channel patterns, *Geological Society of America Bulletin*, **83**, 1755–1770.

Schumm, S.A. & Lichty, R.W. 1963, *Channel widening and flood-plain construction along Cimarron River in southwestern Kansas*, United States Geological Survey Professional Paper, **352-D**, 71–88.

Schumm, S.A., Mosley, M.P. & Weaver, W.E. 1987, *Experimental Fluvial Geomorphology*, Wiley, Toronto, 413 pp.

Schwartz, D.E. 1978, Hydrology and current orientation analysis of a braided-to-meandering transition: the Red River in Oklahoma and Texas, In: Miall, A.D. (Ed.), *Fluvial Sedimentology*, Canadian Society of Petroleum Geologists, Memoir **5**, Calgary, 105–127.

Shepherd, R.G. 1987, Lateral accretion surfaces in ephemeral-stream point bars, Rio Puerco, New Mexico, In: Ethridge, F.G., Flores, R.M. & Harvey, M.D. (Eds), *Recent Developments in Fluvial Sedimentology*, Society of Economic Palaeontologists and Mineralogists, Special Publication Number **39**, 93–98.

Singh, I.B. 1972, On the bedding in the natural-levee and the point bar deposits of the Gomti River, Uttar Pradesh, India, *Sedimentary Geology*, **7** 309–317.

Smith, D.G. 1983, Anastamosed fluvial deposits: modern examples from western Canada, In: Collinson, J.D. & Lewin, J. (Eds), *Modern and Ancient Fluvial Systems*, International Association of Sedimentologists, Special Publication Number **6**. 155–168.

Smith, D.G. 1986, Anastamosing fluvial deposits, sedimentation rates and basin subsidence, Magdalena River, northern Colombia, South America, *Sedimentary Geology*, **46**, 177–196.

Smith, D.G. & Smith, N.D. 1980, Sedimentation in anastamosed river systems: Examples from alluvial valleys near Banff, Alberta, *Journal of Sedimentary Petrology*, **50**, 157–164.

Smith, G.A. 1986, Coarse-grained nonmarine volcaniclastic sediment: Terminology and depositional process, *Geological Society of America Bulletin*, **97**, 1—10.

Smith, N.D. 1970, The braided stream depositional environment: Comparison of the Platte River with some Silurian clastic rocks, north central Appalachians, *Geological Society of America Bulletin*, **81**, 2993–3014.

Smith, N.D. 1974, Sedimentology and bar formation in the Upper Kicking Horse River, a braided outwash stream, *Journal of Geology*, **82**, 205–223.

Smith, N.D. 1978, Some comments on terminology for bars in shallow rivers, in: Miall, A.D. (Ed.), *Fluvial Sedimentology*, Canadian Society of Petroleum Geologists, Memoir **5**, Calgary, 85–88.

Smith, N.D., Cross, T.A., Dufficy, J.P. & Clough, S.R. 1989, Anatomy of an avulsion, *Sedimentology*, **36**, 1–23.

Smith, R.M.H. 1980, The lithology, sedimentology and taphonomy of flood-plain deposits of the Lower Beaufort (Adelaide Subgroup) strata near Beaufort West, *Transactions of the Geological Society of South Africa*, **83**, 399–413.

Smith, R.M.H. 1987, Morphology and depositional history of exhumed Permian point bars in the southwestern Karoo, South Africa, *Journal of Sedimentary Petrology*, **57**, 19–29.

Smith, S.A. 1987, Gravel counterpoint bars: examples from the River Tywi, South Wales, In: Ethridge, F.G., Flores, R.M. & Harvey, M.D. (Eds), *Recent Developments in Fluvial Sedimentology*, Society of Economic Palaeontologists and Mineralogists, Special Publication Number **39**, 75–81.

Sneh, A. 1983, Desert stream sequences in the Sinai Peninsula, *Journal of Sedimentary Petrology*, **53**, 1271–1279.

Stear, W.O. 1980, Channel sandstone and bar morphology of the Beaufort Group uranium district near Beaufort West, *Transactions of the Geological Society of South Africa*, **83**, 391–398.

Stear, W.O. 1985, Comparison of the bedform distribution and dynamics of modern and ancient sandy ephemeral flood deposits in the southwestern Karoo region, South Africa, *Sedimentary Geology*, **45**, 209–230.

Sundborg, Å. 1956, The River Klarälven – A study of fluvial processes, *Geografiska Annaler*, **38A**, 127–316.

Task Committee 1971, Sediment transportation mechanics: Genetic classification of valley sediment deposits, *Journal of the Hydraulics Division, American Society of Civil Engineers*, **97**, HY1, 43–53.

Teisseyre, A.K. 1977, Meander degeneration in bed-load proximal streams: repeated chute cut-off due to bar-head gravel accretion – a hypothesis, *Geologica Sudetica*, **12**, 103–120.

Todd, S.P. 1989, Stream-driven, high density gravely traction carpets: possible deposits in the Trabeg Conglomerate Formation, SW Ireland and some theoretical considerations of their origin, *Sedimentology*, **36**, 513–530.

Todd, S.P. & Went, D.J. 1991, Lateral migration of sand-bed rivers: examples from the Devonian Glashabeg Formation, SW Ireland and the Cambrian Alderney Sandstone Formation, Channel Islands, *Sedimentology*, **38**, 997–1020.

Tunbridge, I.P. 1981, Old Red Sandstone sedimentation – An example from the Brownstones (highest Lower Old Red Sandstone) of south central Wales, *Geological Journal*, **16**, 111–124.

Tunbridge, I.P. 1984, Facies model for a sandy ephemeral stream and clay playa complex: The Middle Devonian Trentishoe Formation of North Devon, U.K., *Sedimentology*, **31**, 697–716.

Turner, B.R. & Monro, M. 1987, Channel formation and migration by mass-flow processes in the Lower Carboniferous fluviate Fell Sandstone Group, northeast England, *Sedimentology*, **34**, 1107–1122.

Tye, R.S. 1991, Fluvial-sandstone reservoirs of the Travis Peak Formation, East Texas Basin, In: A.D. Miall & Tyler, N. (Eds), *The three-dimensional facies architecture of terrigenous clastic sediments and its implications for hydrocarbon discovery and recovery*, SEPM (Society for Sedimentary Geology), Concepts in Sedimentology and Palaeontology, Volume 3, 172–188.

Tyler, N. 1988, New oil from old fields, *Geotimes*, **33**, 8–10.

Tyler, N. & Ethridge, F.G. 1983, Fluvial architecture of Jurassic uranium-bearing sandstones, Colorado Plateau, western United States, In: Collinson, J.D. & Lewin, J. (Eds), *Modern and Ancient Fluvial Systems*, International Association of Sedimentologists, Special Publication Number **6**, 533–547.

Tyler, N. & Finley, R.J. 1991, Architectural controls on the recovery of hydrocarbons from sandstone reservoirs, In: Miall, A.D. & Tyler, N. (Eds), *The three-dimensional facies architecture of terrigenous clastic sediments and its implications for hydrocarbon discovery and recovery*, SEPM (Society for Sedimentary Geology), Concepts in Sedimentology and Palaeontology, Volume 3, 1–5.

Walker, R.G. 1990, Facies modelling and sequence stratigraphy, *Journal of Sedimentary Petrology*, **60**, 777–780.

Walker, R.G. & James, N.P. 1992, (Eds), *Facies Models: Response to Sea Level Change*, Geological Association of Canada, St Johns, Newfoundland, 409 pp.

Walling, D.E. & Bradley, S.B. 1989, Rates and patterns of contemporary floodplain sedimentation: A case study of the River Culm, Devon, UK, *GeoJournal*, **19**, 53–62.

Walling, D.E., Quine. T.A. & He, Q. 1991, Investigating contemporary rates of floodplain sedimentation, In: Petts. G.E. & Carling, P.A. (Eds), *Lowland Floodplain Rivers: Geomorphological Perspectives*, Wiley, Chichester, 165–184.

Williams, G.E. 1971, Flood deposits of the sand-bed emphemeral streams of central Australia, *Sedimentology*, **17**, 1–40.

Williams, P.F. & Rust, B.R. 1969, The sedimentology of a braided river, *Journal of Sedimentary Petrology*, **39**, 649–679.

Wizevich, M.C. 1992, Sedimentology of Pennsylvanian quartzose sandstones of the Lee Formation, central Appalachian Basin: fluvial interpretation based on lateral profile analysis, *Sedimentary Geology*, **78**, 1–47.

Wolman, M.G. & Leopold, L.B. 1957, *River flood plains: some comments on their formation*, United States Geological Survey Professional Paper, **282-C**, 87–107.

9 Process Deduction from Fluvial Sedimentary Structures

SIMON P. TODD
BP Exploration, Aberdeen, UK

INTRODUCTION

The previous chapters in this volume have discussed fluvial sedimentary structures mainly from a viewpoint of the fluid processes and their resultant products. This chapter takes the opposite perspective, and discusses the advances in the deduction of depositional process from preserved sedimentary structures in the rock record. The emphasis switches from the potential range of fluid processes and products, to the consideration of potentially preserved sedimentary structures in an ancient sedimentary succession.

At smaller scales (1 mm to several metres), fluvial sedimentary structures, being commonly dependent only on fluid flow and grain population, are repeatable within many different types of river channel. This is one reason why fluvial channel facies models that rely solely on comparative analogy with modern fluvial channel facies associations are at best ambiguous. Instead, the comprehensive fluvial process deduction must rely on a full description of fluvial sedimentary structures and bounding surfaces at small and large (hundreds of metres to kilometre) scales. Therefore the scope of this chapter is broad and while the interpretation of bedform and barform structure is emphasized, briefer discussion of channel form, floodplain architecture and basin-fill style is also included.

The interpretation of mesoscopic channel and floodplain structures is discussed first. The debate centres on deduction of fluid processes from the preserved facies. Traditional interpretations are only outlined – the emphasis is rather placed on developments of particular note that have occurred mainly in the last ten to fifteen years. These issues are addressed under a series of subheadings in gravel, sand and mud categories. The interpretation is designed to remain only qualitative – quantitative interpretation of fluvial discharge regime through palaeohydraulics is left to the previous and subsequent authors in this volume. The second portion of this chapter involves the spatial and temporal association of different fluvial sedimentary structures. Spatial changes within a single linked depositional system as recorded in one depositional unit are discussed first; thereafter the controls on alluvial architecture are discussed. The evolving techniques of sequence stratigraphy

Advances in Fluvial Dynamics and Stratigraphy. Edited by P.A. Carling and M.R. Dawson.
© 1996 John Wiley & Sons Ltd.

are here advocated as one good method of relating the sedimentary building blocks
into a comprehensive fluvial stratigraphy.

MESOSCOPIC STRUCTURES IN GRAVEL

Fluvial sedimentary structures in gravel are notoriously difficult to interpret. The
principal reason for this is the continuing dearth of experimental and natural
observations of gravel depositional processes. This is understandable, because much
gravel displacement tends to occur in high-energy, infrequent events. Hence in this
area much understanding of depositional process derives from intuitive interpretation of depositional products, combined with the scantier observations of process.
Nevertheless, there has been considerable progress in the interpretation of fluvial
conglomerates, particularly from simple comparisons with Miall's (1978) braided
river facies models. Several advances are described here, including the use of grain
fabric, texture and grading to interpret flow conditions, the use of bed thickness
versus maximum particle size diagrams, the interpretation of subhorizontal gravel
couplets, the variable geometry of cross-stratification formed by the interaction of
bars, pools and confluences, and the collation of fluvial process through the
understanding of lateral and vertical associations of gravelly structures.

Grain fabric, texture and grading

Gravels and conglomerates lend themselves conveniently to qualitative analysis of
grain size, shape, sorting and orientation studies that may provide much information
about process. However, there can be considerable textural variation in alluvial
conglomerates. Mud-matrix-supported conglomerates are clearly the products of
cohesive debris flows or mudflows (Blackwelder, 1928; Bluck, 1964; Rodine &
Johnson, 1976; Nemec & Steel, 1984; Wells, 1984). Such flows of sediment move
en masse and have the rheology of Bingham plastics with a yield strength. The
larger clasts in the flow are supported principally by the strength of the cohesive
mud matrix (Hampton, 1975, 1979; Rodine & Johnson, 1976). Buoyancy often
forces the larger clasts towards the top of the flow, causing inverse grading
(Hampton, 1979). Whilst cohesive debris flows of this kind have the ability to move
on very low slopes (<1°) (Rodine & Johnson, 1976), the presence of the products
of cohesive debris flows is nonetheless strongly suggestive of deposition on the
comparatively high-gradient slopes of a proximal alluvial fan (e.g. Blackwelder,
1928; Bluck, 1964).
   Gravels with an abundant sandy matrix are more difficult to interpret. Strongly
bimodal sediment with a well sorted, flat-stratified gravel and an abundant,
moderately sorted sand matrix are commonly interpreted as sheet-flow or low-relief
braided stream deposits (e.g. Flint & Turner, 1988). In contrast, texturally less
mature gravels with polymodal gravel that are framework-supported or matrix-
supported conglomerates might be products of rapid deposition within flashy,
watery and turbulent channelized or sheet flows, of turbulent, sediment-charged
"hyperconcentrated" flows (Blissenbach, 1954; Bull, 1963, 1972; Wasson, 1977;

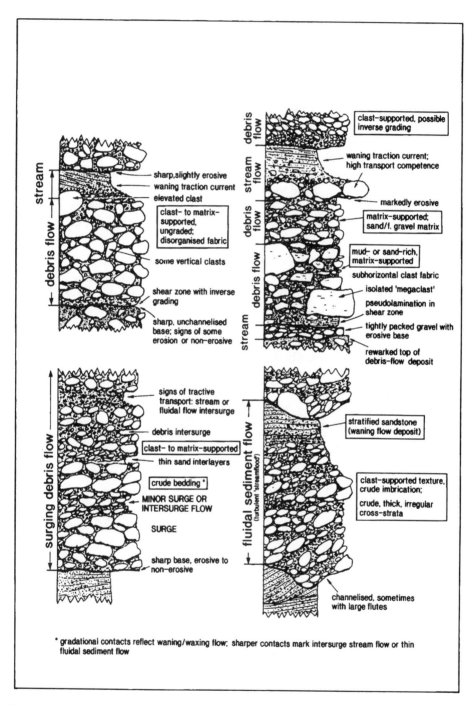

Figure 9.1 Some typical features of subaerial mass-flow deposits (from Nemec & Steel (1984, fig. 10); reproduced by permission of the Canadian Society of Petroleum Geologists)

Nemec & Muszynski, 1982; G.A. Smith, 1986; Pierson & Costa, 1987; Blair, 1987a), or of turbulent or laminar, non-cohesive mass or debris flows (Nemec & Musynski, 1982; Nemec & Steel, 1984; G.A. Smith, 1986; Waresback & Turbeville, 1990; Bøe & Sturt, 1991) (Figure 9.1). A further "end-member" is the possibility of high-density stream flows segregating into bipartite gravelly, turbulent to laminar sediment flows or traction carpets (cf. Lowe, 1982), overlain by more watery, sand-charged turbulent flow, to produce similar deposits in a channelized setting (Todd, 1989a). Because these processes are likely to form a continuous spectrum in a higher gradient fluvial environment (e.g. alluvial fan), the interpretation of the deposits must be based on careful observation of the textural and other attributes of the deposits, as well as lateral and vertical organizational relationships (Figure 9.1).

Clast orientation is considered to be a powerful tool in distinguishing mass flows (complete flow unit or lower traction carpet of bipartite flow) from watery flows (Walker, 1975; Collinson & Thompson, 1989). Disc- and blade-shaped clasts may be imbricated in two main styles.

(1) a(t) b(i) imbrication, with the long a axes of the clasts transverse to flow, is common in watery, fluidal flows in which clasts roll and saltate in a low-density bedload and fluid–solid momentum transfer is dominant (Todd, 1989a, table 1) (Figure 9.2). This type of imbrication is common in braided and meandering gravel-bed rivers in and upon gravel bars (e.g. Bluck, 1979; Hein, 1984) (Figure 9.2). The imbricated clasts may be organized in non-periodic clusters or dams which are groups of smaller clasts typically dammed or trapped either side of a larger nucleus

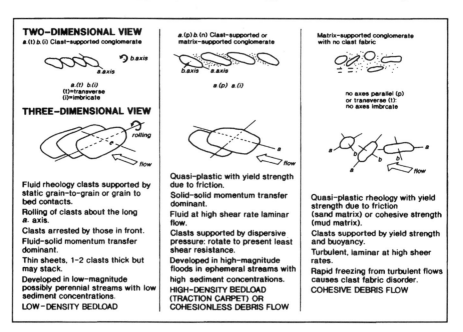

Figure 9.2 Process deduction from the orientation of disc- and blade-shaped clasts in gravels and conglomerates (modified after Collinson & Thompson (1989, fig. 7.7); reproduced by permission of Unwin Hyman Ltd)

clast (Brayshaw, 1984; Bluck, 1987), or as transverse ribs which are oriented perpendicular to flow and are usually periodic (Koster, 1978; Bluck, 1987). Transverse ribs, although seldom identified in the rock record, are powerful palaeohydraulic indicators (see Koster, 1978).

(2) $a(p)$ $b(i)$ imbrication, with the long a axes parallel to flow, forms in high-viscosity flows in which solid–solid momentum transfer is dominant and clasts are forced parallel to the shear stress by grain-to-grain collisions (Rees, 1968; Hein, 1982; Postma *et al.*, 1988; Todd, 1989a) (Figure 9.3). Imbricated fabrics or bed-parallel fabrics with the a axes parallel to flow are found in the deposits of cohesionless mass flows or high-density bedload (traction carpet) deposits in which grain-to-grain momentum transfer is dominant and a component of bed-parallel laminar shear has influenced clast fabric (Nemec & Steel, 1984; Hein, 1984; Todd, 1989a). Mass-flow gravels that have a disorganized gravel fabric, in which there may be many upright clasts, are usually interpreted as the result of rapid collapse and freezing of turbulent, non-laminar mass flows (e.g. Nemec & Steel, 1984; Wells, 1984; Bøe & Sturt, 1991) (Figure 9.1).

Grading in individual conglomerate units is often used as contributory evidence of process in the gravelly flow unit. For suspected mass-flow deposits, normal grading is taken as evidence of grain settling within fully turbulent flows. Coarse-tail inverse grading is taken as evidence of buoyancy (e.g. Wells, 1984) in muddy, cohesive mass flows or dispersive pressure in sandy, non-cohesive mass flows or high-density bedload (traction carpets) (Nemec & Steel, 1984; Todd, 1989a). Lack of grading is interpreted as the product of turbulent flows that have frozen without grain sorting in the flow. For suspected stream-deposited, low-density bedload deposits of sheet-flood, stream-flood or stream-flow conglomerates, normal grading is interpreted as the waning of flood discharge (Nemec & Musynski, 1982; Nemec & Steel, 1984; G.A. Smith, 1986; Todd, 1989a; Jolley *et al.*, 1990; Bøe & Sturt, 1991), or the construction and gradual abandonment of gravel bars (Gustavson, 1978; Hein & Walker, 1977; Hein, 1984), and inverse grading is interpreted as the product of downstream migration of gravel bars (Bluck, 1986).

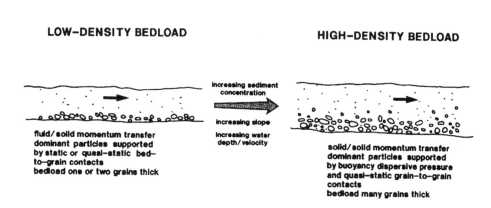

LOW–DENSITY BEDLOAD

HIGH–DENSITY BEDLOAD

increasing sediment concentration

increasing slope

increasing water depth/velocity

fluid/solid momentum transfer dominant particles supported by static or quasi-static bed-to-grain contacts bedload one or two grains thick

solid/solid momentum transfer dominant particles supported by buoyancy dispersive pressure and quasi-static grain-to-grain contacts bedload many grains thick

Figure 9.3 Transition from low-density (normal) to high-density (traction carpet) bedload

Horizontal gravel couplets from bedforms

Increasingly recognized in gravelly fluvial sediments are couplets of variably sorted and packed gravel beds, which together essentially form a normally graded, although bipartite flow unit (e.g. Steel & Thompson, 1983; Siegenthaler & Huggenberger, 1993). Siegenthaler & Huggenberger (1993), for example, describe decimetre-scale,

Figure 9.4 Gravel couplets and pool deposits from the Pleistocene Rhine gravels, Switzerland (photographs courtesy of Peter Huggenberger and Christoph Siegenthaler). The upper photograph is of pool deposits with a trough-shaped scour filled by planar cross-strata. The lower photograph illustrates gravel couplets

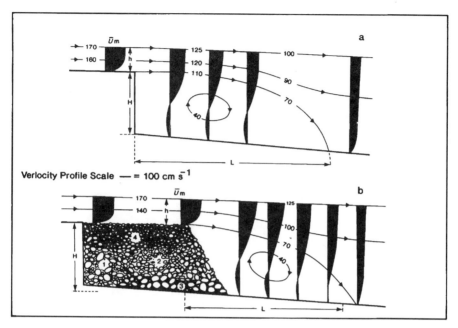

Figure 9.5 Binary gravel deposition at a bar accretion front (from Carling & Glaister (1987, fig. 2); reproduced by permission of the Geological Society of London)

fining-upward couplets from the Pleistocene Rhine gravel with a basal bimodal gravel of well sorted pebbles and cobbles with matrix of well sorted medium sand, overlain by an open framework gravel completely free of sand matrix (Figure 9.4). A mechanism for the formation of these couplets has been identified by Carling & Glaister (1987) and Carling (1990). This involves gravel bedforms with an accretion front in which grainflow is an unimportant mechanism of transport and instead particle overpassing was the dominant process (Figure 9.5). Coarser materials are concentrated at the base of the slope with a sand matrix supplied from suspension and bedload in the separated counterflow. Finer gravels accumulate in the upper part of the accretion front, but tend to be open framework because of the weaker nature of the reverse flow. Downstream movement of such bedforms would typically produce normally graded couplets with no discrete inclined stratification. However, it is also possible to envisage this process contributing to inclined cross-strata (Siegenthaler & Huggenberger, 1993) (Figures 4 and 9.6) and it thus provides an explanation for binary stratification in many gravelly deposits that may have hitherto been attributed an origin through pulsed flow or contribution from parasitic bedforms (e.g. Rust, 1984).

Bed thickness (BTh) versus maximum particle size (MPS)

Following the original suggestion by Bluck (1967), the thickness of a conglomerate unit and the size of the unit's maximum particle size (usually taken as the mean of the ten largest clasts) have often been taken as expressions of the flow's thickness and competence, respectively (e.g. Steel, 1974; Gloppen & Steel, 1981; Nemec &

Figure 9.6   Bar accretion front with low-stage sand wedge from the Devonian Glashabeg Formation, SW Ireland (for background see Todd *et al.*, 1988a,b, 1991)

Muszynski, 1982; Nemec & Steel, 1984; Todd, 1989a). When BTh–MPS data are carefully collected from an assemblage of genetically related conglomerate flow units, statistically tested linear relationships between BTh and MPS can provide powerful information on flow process and behaviour (see particularly Nemec & Steel, 1984, pp. 22–29).

Stream-lain fluvial conglomerate beds typically show no BTh–MPS relationship, because the clasts in thin (less than three clasts thick), low-density bedload sheets are generally supported by static and quasi-static grain-to-bed and grain-to-grain contacts; even if such sheets are completely mobile, the weight of the rolling and sliding clasts is supported by the bed. Conglomerates deposited *en masse*, from mass flows or high-density bedload, more often show a strong linear relationship between BTh and MPS. The processes of grain support, buoyancy, dispersive pressure and turbulence are thought to be all closely related to flow thickness (BTh). Hence, the competence of the flow (reflecting MPS) is proportionately related to BTh (Nemec & Steel, 1984). Whilst there may be problems in overestimation or underestimation of BTh, and some mass-flow deposits may not show a relationship, the method can distinguish cohesive from non-cohesive debris flows, high from low gradients, and subaerial from subaqueous flows (Nemec & Steel, 1984).

Nevertheless, care needs to be exercised in the interpretation of the relationships, particularly when derived for flow units of an origin transitional between mass and stream flow. For example, Todd (1989a) used a strong BTh–MPS relationship ($r = 0.91$) from beds in the Devonian Trabeg Conglomerate Formation in SW Ireland, interpreted to have high-density bedload deposits, to infer that there was a direct relationship between the thickness and competence of the gravelly high-density bedload or traction carpets. This proportional relationship was considered to be the result of the action of dispersive pressure, which is directly proportional to the shear strain, which in turn can be considered controlled by the thickness of the flow (Lowe, 1976, 1982, p. 286; Nemec & Steel, 1984, p. 21; Todd, 1989a). However, it may be that the relationship in the Trabeg conglomerates instead reflects an indirect relationship between sediment discharge and competence of the flows that generated the traction carpets. In this model, the BTh–MPS relationship records the proportionality between the fluid discharge and the sediment entrainment/suspension capacity through fluid turbulence of the flow. In other words, the most powerful floods were capable of entraining the coarsest and the greatest amount of sediment, and of initiating and driving the thickest and coarsest bedload sheets.

Bedform/barform structural assemblages

Gravelly bedforms/barforms can produce a large variety of interrelated structures. Flat, bed-parallel stratification can be produced by the stacking of several pulsatory, low-density bedload sheets or through downstream migration of inclined bar margins where particle overpassing rather than avalanching is the important process (see above). Low-angle cross-stratification in gravel probably forms in bars with inclined riffles, but without steep avalanching slipfaces (e.g. Ramos & Sopeña, 1983). Planar cross-stratification forms due to avalanching on slipfaces of straight-crested bars (Massari, 1983). Trough cross-stratification in gravels and pebbly sands is formed through the migration of sinuous-crested dunes (Harms *et al.*, 1975; Bluck, 1974, 1976; e.g. Middleton & Trujillo, 1984) or sediment avalanching into pools between bars (Siegenthaler & Huggenberger, 1993). The difference between troughs related to dune migration and those related to pool development may in some cases be distinguished by the lateral extent of the coset relative to the scale of the form. Dune deposits are more likely to form laterally extensive cosets in which many foresets record downstream migration of dunes, whereas bar/pool deposits are more likely to be characterized by less extensive accretion and will contain multiple reactivation surfaces (Bluck, 1974, 1976, 1979; Gustavson, 1978, McGowen & Garner, 1970; N.D. Smith, 1971, 1974).

Several students of gravelly fluvial deposits have used the relationships between different accretion or erosion surfaces to deduce bar and channel type (Fig. 9.7). Bluck (1980) was able to identify the deposits of braided stream bars of two different types in pebbly sandstones of the Scottish Old Red Sandstone. Medial bars were recognized either by large, radiating foresets or by identifying bar-tail sediments with converging dips to their foresets. Lateral bars were recognized by the growth of major sandstone sheets oblique to the direction of channel trend and over slough mudstone. Billi *et al.* (1987) recognized that Plio-Pleistocene gravel bodies isolated in an abundance of muddy overbank sediments in Italy are the product of low-sinuosity streams through identification of similar directional properties of the cross-strata and other palaeocurrent indicators such as imbrication. Coarse grained point bars were identified by similar analysis of ancient deposits by Nijman & Puidefabregas (1978).

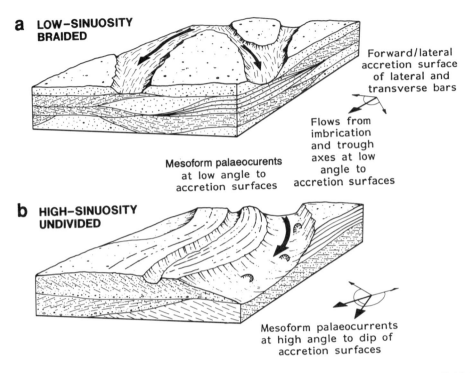

Figure 9.7 Structural attributes of low-sinuosity, braided and high-sinuosity, undivided gravel river deposits

Morrison & Hein (1987) recognized both trough and scoop-shaped cross-stratification in the Plio-Pleistocene White Channel gravels of Canada. Trough cross-strata were considered deposited from in-channel dunes, but the scoop-shaped cross-strata were interpreted to be the result of "cut and fill" in shallow scours or pools within active channels. Such scoured pools or dunes between bars probably resulted in the low-angle trough cross-strata interdigitated with planar cross-strata described by Todd & Went (1991) and concordantly or asymmetrically filled troughs in the Pleistocene Rhine gravel were also interpreted as pool deposits by Siegenthaler & Huggenberger (1993). Moreover, the latter authors suggest the pool deposits, rather than bar deposits, may be the more commonly preserved structures in other braided river deposits (e.g. Bluck, 1967; Steel & Thompson, 1983; R.M.H. Smith, 1990).

BEDFORM AND BARFORM STRUCTURES IN SAND

The understanding of the depositional processes that lead to the development of sandy bedforms and barforms in rivers was greatly advanced in the late 1960s and 1970s through the work of researchers like Allen (1970a,b, see also 1983a,b), Bridge & Jarvis (1978), Cant & Walker (1978), Coleman (1969), Collinson (1970),

Harms *et al.* (1975), Jackson (1976a,b), Jopling (1965), N.D. Smith (1971, 1972), Williams (1971) and Williams & Rust (1969), amongst others. Advances have continued in the the 1980's by way of laboratory flume studies (e.g. Saunderson & Lockett, 1983) and field observations of both modern rivers and their deposits (e.g Crowley, 1983) and ancient sediments (e.g. Allen, 1983a,b; Hazeldine, 1983). Arguably most progress has been made, however, in the study of how bedforms and barforms interact in rivers to produce discrete structural assemblages (the "elements" of Miall (1985) and Brierley (1989); see also Chapter 8). New systems of facies architecture description and interpretation (e.g. Miall, 1985) emphasize the potential hierarchial organization of fluvial bedforms in time and space, often leading to a more dynamic interpretation of channel and barform process, product and preservation.

In the discussion below, bedforms are considered as individual depositional structures developed at a spatial and temporal scale below that of the channel (mesoforms of Jackson (1975)). They are scaled to the depth of the flow and are periodic. All bedforms, whether two- or three-dimensional, large or small, are best referred to as *dunes*, but may be qualified with the appropriate descriptor (Ashley, 1990). Barforms are developed at a temporal and spatial scale of the channel.

Ripple cross-lamination

Cross-lamination of fine sands is a common expression of the migration of ripple microforms in lower-flow-regime conditions. Preservation of both stoss and lee side laminae indicates that the ripples migrated during a period of strong bed aggradation. Ripples are more common in slack flow, on upper bar surfaces, but may also be superimposed on dunes. Intimate interbedding of ripple-laminated sandstone with mudstone (flasar or wavy bedding) is a common indicator of pulsatory or even cyclic flow, reflecting periodic flood, seasonal or tidal flow.

Parallel lamination

Parallel lamination is a common phenomenon in stream-lain sandstones and is particularly diagnostic of sandy, high-energy sheetfloods (Tunbridge, 1981). The parallel laminae internally are typically only a few grains thick and are defined by subtle variations in grain size or grain fabric, particularly in sands with higher concentrations of platy grains like mica. Parallel lamination has specific implications for the regime of the formative flow. The theory of Bagnold (1966), supported by Allen & Leeder (1980), is that bedforms are created only when turbulence is free to mould the stream bed. In flows where enough grains are entrained as bedload so that a complete grain layer excludes fluid turbulence from acting on the bed, no bedforms can be produced and an *upper-flow-regime flat bed* is the stable configuration (Bagnold, 1966; Allen & Leeder, 1980).

Concentration of sand bedload is a function of several variables but is principally a function of the ratio of the applied bed shear stress to the critical threshold stress required for initial movement of the grains. Bedform stability fields, predicted from experimental data (e.g. Leeder, 1983; see also Chapters 3 and 6), show that upper-flow-regime flat-bed conditions are the stable-bed configuration in flows that exert a comparatively high shear stress on the stream bed. Such conditions produce parallel lamination in the sands (e.g. Turner, 1981; Saunderson & Lockett, 1983; Bridge & Best, 1988; Paolo *et al.*, 1989; Best & Bridge, 1992). Bed shear stress can be considered in terms of Shields' function, which increases with increasing depth or velocity. For a constant grain size, an increase in Shields' function will tend to make upper-flow-regime flat-bed conditions more stable. Thus the development of parallel lamination may be the result of deposition of the sand in greater flow depths or greater flow velocities than the flows in which the dunes formed. This interpretation is supported by the lateral transition of parallel lamination into tabular foresets commonly observed in fluvial sediments.

Alternatively, the production of parallel lamination may be promoted in shallow but fast flows (e.g. McKee *et al.*, 1967; Tunbridge, 1981). Bagnold (1973) suggested one approach to conceptualizing the production of a complete bedload layer that precludes turbulent erosion of the stream bed. Bedload transport rate may be given as:

$$i_b = e_b / \tan \alpha$$

where i_b is the bedload transport rate, e_b is an efficiency factor, and $\tan \alpha$ is a coefficient of the dynamic friction (Bagnold, 1973). The efficiency factor is inversely proportional to the depth : grain size ratio and hence shallow flows are more efficient than deep, by virtue of the greater effective velocity acting on the bed grains in shallow flows (Bagnold, 1973; Leeder, 1979, p. 74). This concept explains how upper-flow-regime flat-bed conditions and therefore parallel lamination can result from fast and shallow flows. Thus parallel-laminated sandstones may be the product of shallow flows across bar tops, comparing well with recent parallel-laminated sands deposited on the tops of the bars in the Platte River (Crowley, 1983), and parallel-laminated sandstones overlying tabular cross-strata in the rock record (e.g. Todd & Went, 1991). Alternatively, it may be the product of fast, shallow sheet floods in both modern ephemeral streams (e.g. McKee *et al.*, 1967) and ancient alluvial aprons (e.g. Tunbridge, 1981, 1984; Hubert & Hyde, 1982; Graham, 1983; Lawrence & Williams, 1987; Olsen, 1987). Discrete lenticular forms in the parallel laminae may be due to the preservation of extremely low-relief bedforms in the upper-flow-regime bed (Bridge & Best, 1988).

A third mechanism for inducing a flat bed as the stable configuration in a stream flow is by increasing sediment concentration. This has the effect of increasing density and viscosity, and decreasing turbulence. The dampening of turbulence can lead to the drastic reduction of bedform production because turbulent flow separation is no longer possible (Todd, 1989a; see also Chapter 3). The result can be flat-bed conditions, even in slower-moving flows. Abundant parallel lamination in sandstones may therefore be partially or wholly a result of high sediment concentrations in sheet or stream floods. This type of parallel lamination which reflects high sediment rates may be partially or completely disrupted by deformation associated with fluid escape leading to the preservation of structureless, massive sands.

Low-angle lamination

Low-angle cross-stratification often occurs in fluvial sandstones, typically in association with parallel lamination. It is interpreted to be the product of the migration of bedforms (low-relief dunes) that have a high wavelength/amplitude ratio. Such bedforms have been demonstrated by experiment in laboratory flumes (e.g. Saunderson & Lockett, 1983; Bridge & Best, 1988) to be formed in conditions transitional between lower and upper flow regime. The hump-back dunes of Saunderson & Lockett (1983) have no slipfaces and such bedforms can only produce leeside laminae by grain fall and grain creep across the form. These leeside laminae will be inclined at a low angle and such a process may be responsible for the low-angle stratification in fluvial sediments. Similar to the comments on parallel lamination above, the dominance of low-angle cross-stratification of this scale in a fluvial deposit may also be due to the damping of turbulence through higher sediment concentrations.

Hummocky-type stratification

High suspended loads and finer sand may also be responsible for a variety of fluvial sedimentary structures that resemble hummocky cross-stratification (Harms *et al.*, 1982). Hummocky cross-stratification (HCS) is a three-dimensional structure with an undulatory lower erosion surface with domes and basins overlain by accretion laminae that thicken into the domes or hummocks. Swaley cross-stratification is characterized by thickening into the basins or swales.

Cotter & Graham (1991) describe from a Late Devonian coastal plain succession (Toe Head Formation, SW Ireland), a series of sandstone bedsets with horizontal parallel lamination, and inclined parallel lamination in association with structures that meet the four criteria set by Harms *et al.* (1982) for HCS (Figure 9.8). The fine to very fine sand sizes and possibly the amount of suspended load were considered by Cotter & Graham (1991) as the cause of the symmetrical, low-relief bedforms that generated these structures in flows considered to have been broadly unidirectional. Similar structures have been described by Picard & High (1973), Tunbridge (1984), Stear (1985), Røe (1987) and Dam & Andreasan (1990). Cotter & Graham (1991) point out that symmetrical bedforms that generate HCS may be as much a consequence of sediment-charged unidirectional flows, as bidirectional, oscillatory flows in stormy shallow marine environments above storm wave base, which are traditionally deduced from HCS.

Rust & Gibling (1990a) also described three-dimensional forms from fluvial sediments, in this case from the top of a channel-fill sequence in the braided alluvial South Bar Formation (Carboniferous) of eastern Canada. They recognized the similarity of these forms with hummocks and swales but preferred to interpret them as preserved antidunes on the grounds of the fluvial sedimentological context and the ubiquitous, well defined current lineations parallel to unidirectional palaeoflow indicators in the same sandstone body.

Trough cross-stratification

Trough cross-stratified sandstone, which is very common in fluvial deposits, is formed by the migration of sinuous-crested dunes in a cohesionless sandy bed

Figure 9.8 Hummocky-type structures in the Devonian Toe Head Formation of SW Ireland (see Cotter & Graham (1991) for background; photograph courtesy of John Graham)

(Harms *et al.*, 1975; Jackson, 1976a; Collinson & Thompson, 1989). Dunes are often observed to ornament the bottoms of modern sandy streams, particularly in the deeper reaches in and adjacent to the thalweg (e.g. Allen, 1965a; Williams & Rust, 1969; Coleman, 1969; Collinson, 1970; Williams, 1971; Jackson, 1975, 1976a). Trough cross-stratification in ancient sandy alluvium is usually ascribed an origin by deposition from dunes (e.g. Allen, 1964, 1965a,b; 1970a).

Tabular cross-stratification

In modern sand-bed streams, tabular cross-stratification, either planar or asymptotic, is formed by the movement of repetitive, straight-crested dunes or by avalanching on the slipfaces of simple bars (e.g. Collinson, 1970; N.D. Smith, 1971, 1972; Williams, 1971; Bluck, 1976; Cant & Walker, 1978). Asymptotic foresets with tangential bases are formed where the base of the leeside slipface of the bar is washed out by flow returning upstream after separation at the bar crest (Jopling, 1965; Harms *et al.*, 1975). Such asymptotic forms imply faster and/or deeper flows (e.g. Saunderson & Lockett, 1983).

Barforms and bar structures

Fluvial channel deposits are often characterized by complex lateral and vertical changes in structural attributes or facies. Many recent studies of fluviatile sediments have recognized these complexities, described them in terms of architecture, and related them to depositional processes involving bedform interaction in turbulent stream flows. Most commonly emerging from these studies of spatial facies distribution is the ability to identify what type of barform produced the preserved sedimentary element. This type of barform identification may be extremely powerful because it may lead to a more confident deduction of channel type than the previous, simple facies models or sequences of the 1970s were able to deliver.

However, the intricate deduction of barform type and structure from the lithofacies, architecture and directional properties of fluvial sediments is to some extent hindered by the potentially confusing immense array of bars and terms used to describe these bars (e.g. Crowley, 1983, p. 117). Barform classification is difficult, but nonetheless one approach to definition is outlined below and illustrated in Figure 9.9. This approach is based on the reasoning of Jackson (1975) who considered that bedforms could be thought of as belonging to micro-, meso- and macroscopic scales, both in spatial and temporal scales (Table 9.1). Thereafter the macroforms, or bars, which are capable of greatest complexity because of their relatively large size and their comparatively long duration of formation, may be considered in terms of two main attributes, outlined below.

(1) *Bar shape, internal geometry and accretion direction* are recorded by the sedimentary structures. The most basic tier of this classification is whether the bar is *simple*, having its own slipface, or *compound* or slipfaceless with parasitic bedforms (dunes) on the accretionary margin of the bar. The directional properties of the slipface deposits in simple bars with respect to one another and to channel orientation, may yield substantial information about bar type. The directional properties of accretion surfaces and parasitic dunes in compound bars may also provide strong indications of bar type.

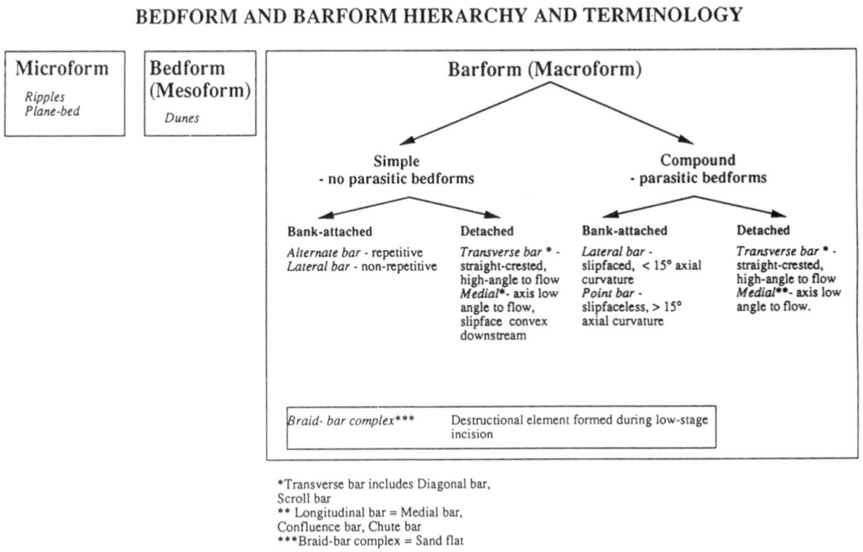

BEDFORM AND BARFORM HIERARCHY AND TERMINOLOGY

Microform	Bedform (Mesoform)	Barform (Macroform)			
Ripples Plane-bed	Dunes				
		Simple - no parasitic bedforms		Compound - parasitic bedforms	
		Bank-attached	Detached	Bank-attached	Detached
		Alternate bar - repetitive Lateral bar - non-repetitive	Transverse bar * - straight-crested, high-angle to flow Medial*- axis low angle to flow, slipface convex downstream	Lateral bar - slipfaced, < 15° axial curvature Point bar - slipfaceless, > 15° axial curvature	Transverse bar * - straight-crested, high-angle to flow Medial**- axis low angle to flow.
		Braid- bar complex***	Destructional element formed during low-stage incision		

*Transverse bar includes Diagonal bar, Scroll bar
** Longitudinal bar = Medial bar, Confluence bar, Chute bar
***Braid-bar complex = Sand flat

Figure 9.9 A classification of barforms

Table 9.1   Spatial and temporal scales of fluvial sedimentary structures

Form	Example structure	Temporal scale	Spatial scale
Microform	Flat bed Ripple	Minutes, hours (10^{-1} to 10^{0} h)	Centimetres (10^{-2} to 10^{-1} m)
Mesoform	Dune	Hours, days (10^{0} to 10^{2} h)	Centimetres to metres (10^{-1} to 10^{1} m)
Macroform	Bar	Days, weeks, years (10^{2} to 10^{6} h)	Metres (10^{1} to 10^{2} m)
Megaform	Channel Channel belt	Weeks, years (10^{4} to 10^{8} h)	Metres, kilometres (10^{2} to 10^{4} m)
	Floodplain Basin	Years, millennia (10^{6} to 10^{10} h)	Kilometres (10^{4} to 10^{5} m)

(2) *Bar position and orientation in the channel* is reflected in the architectural relationships of resultant facies and the directional properties of different sedimentary structures in the deposits. The most basic tier of this aspect of the classification is whether or not the bar is attached to the bank. Note that bank-attached bars are generally the same as the "channel forms" of Ashley (1990) and are commonly scaled to the width of the channel. Detached bars are generally the same as the "unit bars" of Ashley (1990) and are scaled to depth but are different from dunes being quasiperiodic or non-periodic. Orientation in the channel controls alignment of the slipface (if present) relative to the mean flow direction and is often recorded within the variation of directional properties of the alluvium.

Potentially one of the greatest ambiguities in the classification arises from the distinction of lateral from *point bars*. In its most basic form, any barform attached to the inner bank of a fluvial channel can be termed a "point bar" (e.g. Bridge, 1985). The basic form and position of the bar results in its position in the helicoidal flow that occurs in any stream. However, in studies of coarser-grained, low-sinuosity rivers, the bank-attached bars tend to be referred to as "lateral" or "side" bars where the former are slipfaced and the latter are slipfaceless (e.g. Collinson & Thompson, 1989, fig. 6.28). In finer-grained, higher-sinuosity rivers, the bank-attached bars tend to be referred to as "point" bars. The common theme of described *lateral bars* is the presence of a pronounced slipface that accomplishes downstream and possibly some lateral migration. Point bars commonly have no discrete slipfaces, save on parasitic chute and scroll bars, and downstream and lateral migration is instead accomplished through the movement of inclined surfaces that lie at below the angle of repose (the epsilon cross-stratification of Allen (1963)). Epsilon or lateral-accretion surfaces commonly form bounding surfaces between cross-bedded units in which the cross-stratification results from migration of parasitic bedforms such as dunes. Following this generally established usage, the term "lateral" is retained for a simple, bank-attached bar with a discrete slipface. Deposits of this origin must be clearly distinguished from point bar deposits through evidence that the tabular cross-strata were produced from avalanching on the principal slipface rather than by chute or scroll bar migration. "Point" bars are compound, bank-attached macroforms that have no slipface. Their deposits will commonly contain epsilon cross-stratification that may act as bounding surfaces to cross-bedded units. The term "side bar" is abandoned because slipfaceless bank-attached bars even in low-sinuosity rivers are point bars in this classification. "Sand flats" or "braid–bar complexes" are *medial* or bank-attached lateral or point bars formed by switching of subchannels or incision of subchannels during falling stage. The sand flat is an erosive remnant, possibly developed during several phases of growth and amalgamation of constructional/migratory bars. Some sand flats emergent at low stage will contain only structures generated by in-channel bedforms.

Several studies of ancient successions in recent years have illustrated the power of architectural description and interpretation of bar and channel deposits which allow confident bar-type and hence channel-type identification. Several of these studies are briefly described below and the interpreted barforms placed in the classification depicted in Figure 9.9. The basic tenet of bar-type identification is the geometrical relationship between bounding surfaces that record longer-term bar accretion and the cross-strata that record shorter-term flow around bedforms (Table 9.2). In all cases outlined below, it is the geometry and directional properties of the barform depositional/erosional (macroform) surfaces and their relationship to both the smaller-scale (mesoform) bedform depositional/erosional surfaces and the larger-scale channel/channel-belt (megaforms) structures that reveals most information concerning bar type and hence channel type.

Simple lateral bar structure

Lateral bars are probably preferentially formed in relatively shallow, narrow and low-sinuosity streams, for example on alluvial fans or aprons. Such a setting was appealed to by Todd (1989b) for a sandstone facies association (Facies Association A) that dominates the basal portion of the Devonian Trabeg Conglomerate Formation (see also Todd *et al.*, 1988b, 1991) (Figure 9.10). This association is composed mainly of sandstones from very fine sand grade to very coarse sand grade, the coarser grades predominant over the finer. Thin conglomerates and siltstones are also included in the association. Sandstone beds range from a few centimetres to nearly 2 m thick, but are generally less than 30 cm thick. Basal bedding planes are usually planar, with little or no relief, but scoured surfaces also occur with a relief up to 20 cm, within the short constraints of the exposure. Grain size is often variable within individual units; normal grading is most common. Scattered, outsize clasts up to very large pebble grade occur in some of the sandstone beds and intraformational clasts of mudrock and calcrete are also common. Both types of gravel-grade clast tend to be concentrated in the basal portions of sandstone beds. Tabular cross-stratification is most frequent, with asymptotic foreset profiles more common than planar ones. Cosets of tabular foresets vary in thickness from less than 0.1 m to nearly 1 m. Sandstone beds may be constituted by one or more cosets. Low-angle planar lamination also occurs in some sandstone beds, often in association with parallel lamination which is also common. Trough cross-stratification, less common than tabular, is typically composed of stacked concave-up scours and associated laminae. Individual troughs vary in height

Table 9.2 Identification of bar type from the nature and directional properties of accretion surfaces and subordinate intrasets

Bar	Accretion surface	Parasitic bedforms	Palaeocurrents	Channel type	Ancient examples
Lateral	Moderately inclined planar tabular cross-strata or low-angle planar asymptotic cross-strata. Accretion surfaces constructed from channel erosion surface.	None.	Cross-stratal dips at 30 to 60° to channel axis marked by trough axes of in-channel dunes etc.	Low-sinuosity, possible braided, high gradient often in alluvial aprons or fans.	• Trabeg Conglomerate Formation (Todd, 1989b, this chapter). • Alderney Sandstone Formation (Todd & Went, 1991). • Glashabeg Formation (Todd & Went, 1991).
Alternate	Moderately inclined planar tabular cross-strata or low-angle planar asymptotic cross-strata. Accretion surfaces constructed from channel erosion surface.	None.	In vertical successions, cross-strata may display marked bipolarity. Cross-stratal dips at 30 to 60° to channel axis marked by trough axes of in-channel dunes etc.	Low-sinuosity, possibly braided, high gradient often in alluvial aprons or fans.	Trabeg Conglomerate Formation (Todd, 1989b, this chapter).
Point	Low-angle inclined, planar accretion surfaces, possible bounding cosets of trough and/or tabular cross-strata or ripple lamination.	Trough cross-strata formed by sinuous-crested dunes. Tabular cross-strata formed by straight-crested dunes or parasitic scroll or chute bars.	Accretion bounding surfaces dip at high (>60°) angle to orientation of channel, as indicated by axes of dune-formed troughs, for example. Low-angle, planar to curviplanar or sigmoidal bounding surfaces are common. Dispersion of 90 to 120° in one sandbody common. High angle (commonly c. 90°) between parasitic cross-strata and accretion surfaces.	Low- to commonly high-sinuosity, typically undivided rivers.	• Catskill magnafacies (Bridge & Gordon, 1985; Gordon & Bridge, 1987). • Fort Union Formation (Diemer & Belt, 1991).

Medial	Moderately inclined, planar asymptotic/tabular cross-strata.	None.	Angle of repose cross-strata dip at high (>60°) angle to orientation of channel, as indicated by axes of dune-formed troughs, for example. Dispersion of 90 to 120° in one sandbody common.	Shallower, low- or high-sinuosity braided channels.	
Transverse	Moderately to steeply inclined, planar asymptotic/tabular cross-strata.	None.	Angle of repose cross-strata dip at low (<60°) angle to orientation of channel. Dispersion of 90 to 120° in one sandbody common.	Shallower, low- or high-sinuosity braided channels.	
Compound medial	Low to moderately inclined accretion surfaces bounding tabular or trough cross-strata.	Trough or tabular cross-strata formed by sinuous or straight-crested dunes on the accretion front.	Accretion bounding surfaces dip at high (>60°) angle to orientation of channel, as indicated by axes of dune-formed troughs, for example. Bounding surfaces are often shallow troughs. Dispersion of 90 to 120° in one sandbody common.	Deeper, low- or high-sinuosity braided channels.	• Brownstones Formation (Allen, 1983a). • Slea Head Formation (Todd et al., 1988b, 1991, this chapter).
Compound transverse	Low to moderately inclined accretion surfaces bounding tabular or trough cross-strata.	Trough or tabular cross-strata formed by sinuous or straight-crested dunes on the accretion front.	Accretion bounding surfaces dip at low (<60°) angle to orientation of channel. Low to moderate-angle planar to curviplanar bounding surfaces. Dispersion of 90 to 120° in one sandbody common.	Deeper, low- or high-sinuosity braided channels.	• Carboniferous, northern England (Hazeldine, 1983). Hawkesbury Formation (Rust & Jones, 1987).

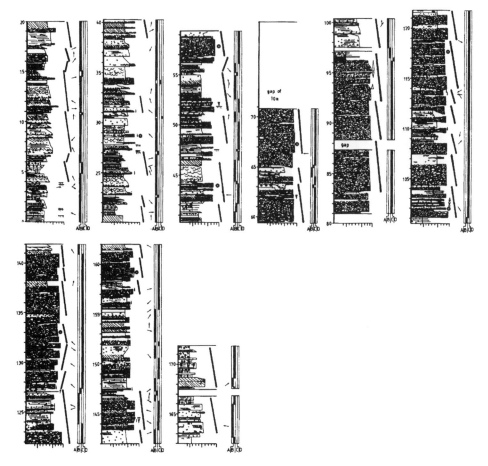

Figure 9.10   Facies log through the Devonian Trabeg Conglomerate Formation, SW Ireland
(see also Todd *et al.*, 1991)

from about 10 cm to 40 cm, with a mean of about 19 cm. The trough cosets tend to
be thinner than the neighbouring tabular cosets. Asymmetrical ripple lamination
occurs occasionally in the topmost, finer-grained portions of some sandstone beds.
Other beds are internally massive, sometimes due to overprinting of cleavage,
sometimes due to disruptive bio- and pedoturbation.

The tabular cross-stratification, either planar or asymptotic, was formed by the
movement of repetitive straight-crested dunes or repetitive or non-repetitive
transverse bars (e.g. Collinson, 1970; N.D. Smith, 1971, 1972; Williams, 1971;
Bluck, 1976; Cant & Walker, 1978). Cosets of tabular cross-stratification in these
Trabeg sandstones normally occur in isolation or in pairs and are interpreted to have
formed by the migration of bars with axes at a high angle to flow. Asymptotic
foresets with tangential bases, which predominate over planar foresets with angular
bases, were formed where the base of the leeside slipface of the bar is washed out
by flow returning upstream after separation at the bar crest (Jopling, 1965; Harms

et al., 1975). The paired sets of tabular cross-strata suggest *alternate bars* with crests aligned at an angle to the channel, and attached to one or other bank and arranged in streams so that successive bars downstream are attached to alternate banks. The pairing of cosets and some bimodal palaeocurrent data for these cross-strata support the conclusion that the tabular foresets were formed on the slip-faces of simple alternate lateral bars (Figure 9.11). Because of the small limits of the exposure, the attachment of accretion surfaces to channel erosion surfaces is seldom observed and some of the bars may well have been detached from the bank, implying simple transverse bars instead (Table 9.2).

The interdigitated stacked trough cross-stratification in the Trabeg sandstone association is formed by the migration of sinuous-crested dunes. Typically, trains of dunes form in the deeper portions of rivers adjacent to bars (e.g. Cant & Walker, 1978) and hence the trough cross-stratification of this sandstone association is interpreted to represent deposition from dunes in the deeper parts of the stream channel, between the alternate bars (Figure 9.12A). This is supported by the observation of larger tabular foresets passing laterally into smaller trough foresets (e.g. 13 m above the base of the Slaudeen log; Figure 9.10).

Parallel lamination in these sandstones is observed both overlying tabular cross-strata and as a facies which is laterally equivalent to both tabular and trough cross-strata. As discussed above, the parallel lamination in these sandstones could have formed in fast and shallow flows across the tops of the lateral or transverse bars or

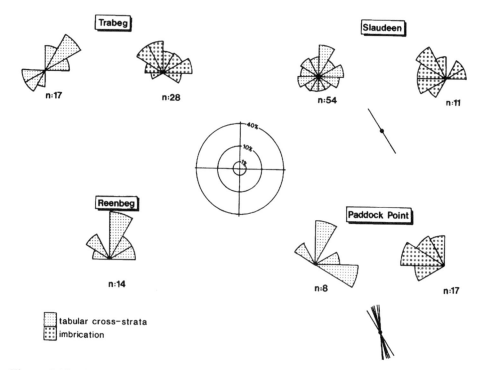

Figure 9.11 Palaeocurrent data from four different localities in the Trabeg Conglomerate Formation. Lines indicate clast lineations

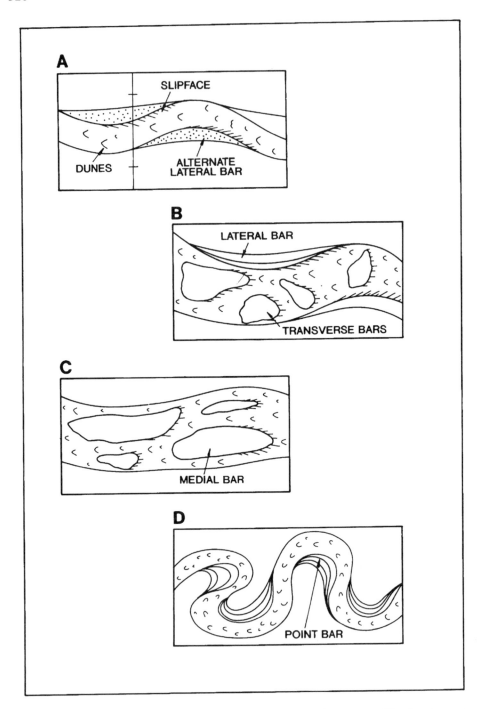

Figure 9.12 Bar types and resultant bounding surfaces and cross-stratification in sandy fluvial sediments. (A) Low sinuosity with simple, lateral bank-attached bars. (B) Low sinuosity and divided with compound lateral and medial bars. (C) Low sinuosity with compound lateral and transverse bars. (D) High sinuosity with point bars

in deeper and faster flows where the usual dunes are washed out in favour of upper-flow-regime flat beds. Thus a model emerges for these Trabeg sandstones of simple, lateral, probably bank-attached bars in low-sinuosity rivers. They are considered similar to the modern ephemeral streams described by Karcz (1972) (Figure 9.12A).

A well exposed assemblage of larger lateral bar deposits was described by Todd & Went (1991) from the Quesnard Sandstone Member of the Cambrian Alderney Sandstone Formation of the Channel Islands. Large-scale tabular cross-stratified sandstone facies contain foresets that range in height from 0.45 to 3.5 m and that range in dip between 16–25° (high angle) and 5–10° (low angle). The foreset laminae in high-angle foresets are asymptotic in profile and inversely graded at the base; the low-angle foresets are usually planar. The tabular cross-strata form cosets that extend laterally in some cases over 100 m. Most commonly, these tabular cross-strata are overlain by small-scale cross-stratified or parallel-laminated sandstone and underlain by parallel-laminated sandstone. The tabular cross-stratified sandstone facies is interbedded with trough cross-stratified sandstones which are interpreted to be the product of sinuous-crested dunes in the deeper reaches of fluvial channels.

The large-scale tabular cross-strata are interpreted as the product of outbuilding of large, probably bank-attached, and at least partly slipfaced *lateral bars*. The high angle between palaeocurrents indicated by trough axes and by the dip of the tabular cross-strata indicated that the bars built out at a relatively high angle to flow. The parallel-laminated or cross-stratified sandstones that commonly overlie the tabular sets are considered to represent bar-top deposits, while the underlying parallel-laminated sandstones are interpreted to represent bar apron deposits (Todd & Went, 1991). In total, the bar and intervening channel deposits compare favourably with modern sediments in the Platte and South Platte rivers (Crowley, 1983). Of note in this example was the preferred preservation of one direction of tabular cross-stratal dip only. This was thought to be the result of the preferential preservation of only one direction of channel migration induced by tectonic tilting, with the avalanche slipface deposits dipping sympathetically with the direction of channel combing (Todd & Went, 1991).

Tectonic tilting and concomitant aggradation was also considered to be the cause of a similar preservational bias in the directional properties of the bar deposits in the Caherquin Member of the Devonian Glashabeg Formation in SW Ireland, the second example of such phenomena reported by Todd & Went (1991). These pebbly sandstone deposits do differ, however, in interpreted bar geometry from the Quesnard Member example. The relationship of large-scale tabular and trough cross-stratal surfaces to trough-shaped bounding surfaces in the pebbly sandstone sheets was considered to reflect dispersal of sediment across the top of *lateral bars* into slough channels beside the inner bank, as reported by Bluck, (1976, 1979) from modern bar deposits in some Scottish and other low-sinuosity rivers. The main bar form was considered to be slipfaced lateral bars with a prominent inner slough channel (Todd & Went, 1991). The axes of the trough cross-strata, thought to represent most closely the axial orientation of pools, sloughs and main channels in the river, are also at a high angle to the dip of the tabular cross-strata which again are preferentially preserved in one direction. However, rather than dipping away from the inner bank as in the Quesnard example, the Caherquin tabular cross-strata were inferred to dip towards the inner bank and to dip antithetically relative to the direction of channel migration (see Todd & Went (1991) for full discussion). These two contrasting examples illustrate the value of careful examination of the relationships of cross-strata to their bounding surfaces and of the directional properties of all the hierarchial surfaces in the bar deposits.

Compound medial and transverse bar structure

Allen (1983a) described from the Devonian Brownstones Formation of Wales a complex of trough and tabular cross-stratified sandstones which he interpreted through recognition of hierarchial bounding surfaces, as the deposits of compound bars. These compound bars were considered to be *compound medial bars* in a braided river without a slipface, the leeside instead mantled by dunes. Hazeldine (1983) also described compound transverse bar deposits from the Carboniferous of England which he interpreted to have a tripartite hierarchy of tabular cross-strata. These reflected sandwaves superimposed on transverse barforms.

Excellent, extensive exposures through the Triassic Hawkesbury Sandstone in eastern Australia enabled Rust & Jones (1987) to describe large cross-stratified sandstone units including variably interbedded trough cross-strata and planar cross-strata. The planar cross-strata vary in thickness from 10 cm to 7.5 m and in plan view are straight to gently curved. Individual sets commonly extend for many tens of metres, and possibly up to a kilometre. The planar cross-strata either form concordant parallel foresets at or about the angle of repose, concordant foresets at below the angle of repose, or bounding surfaces that envelop cosets of trough cross-strata. Rust & Jones (1987) interpreted these large-scale cross-strata as the deposits on the leeside of large, slightly sinuous-crested "sandwaves" (transverse bars in the classification herein) and compared them to the largest bedforms observed in the Brahmaputra by Coleman (1969). The common accreted stacks of angle-of-repose planar cross-strata indicate that these "sandwaves" commonly migrated through avalanching on a slipface; planar cross-strata at less than the angle of repose and the "intrasets" of trough cross-strata show that the largest forms were often slipfaceless with leeside accretion achieved by grain fall and creep processes, frequently moulded into parasitic dunes by separated flows. While these depositional elements were termed "sandwaves" by Rust & Jones (1987), these forms were scaled to very large braided fluvial channels and often had parasitic bedforms; they are probably better referred to as *compound transverse bars* (Figure 9.9; Table 9.2).

Fluvial deposits dominated by trough cross-stratification occur in the Devonian Slea Head Formation in SW Ireland (Todd *et al.*, 1988b, 1991) which is dominated by thick-bedded, grey, very coarse pebbly sandstone units with abundant trough cross-stratification. Beds range in thickness from less than 1 m to over 10 m and form laterally extensive sheets (Friend *et al.*, 1979; Friend, 1983) that display little relief on their lower bounding surfaces (Figures 9.13 and 9.14). Individual beds of trough cross-stratified sandstone usually possess sharp, irregular bases with relief up to 1 m but generally less than 0.3 m. In general, the troughs have width:height ratios of about 5:1 and, on average, have heights of about 0.41 m ($n = 26$), and widths of about 1 to 2 m (Figures 9.13 and 9.14). Trough set height shows a large variation,

323

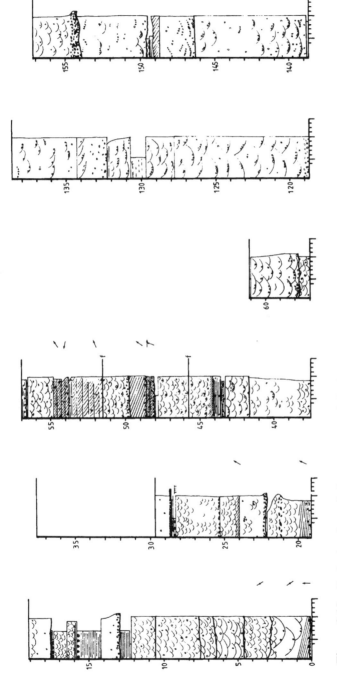

Figure 9.13 Facies log through the upper part of the Slea Head Formation, Slea Head, SW Ireland (see also Todd *et al.*, 1991)

Figure 9.14 Photographs of the facies of the Slea Head Formation. (a) Larger-scale trough cross-strata in pebbly sandstone. Compass is about 20 cm long. (b) Smaller-scale trough cross-strata. (c) Large-scale trough cross-strata in coset dipping (tectonically) at 45° to right. Tape measure is 1 m long and its top is hung on a bounding surface. (d) and (e) Mudstones with immature calcrete overlain by sharp-based sandstone bed. These immature calcretes constrast with those in the penecontemporaneous Glashabeg Formation in the same basin fill (Figure 9.15). The difference is considered to be the result of the greater rate of floodplain aggradation/alluviation in the Slea Head system relative to the Glashabeg system

however, from 0.2 to 1.0 m. Occasionally some of the undulatory relief formed by the troughs is preserved on the tops of sandstone beds and draped by a succeeding siltstone unit. Trough set height shows a large variation: troughs have width:height ratios of about 5:1 and, on average, have heights of about 0.41 m ($n = 26$), and widths of about 1 to 2 m (Figures 9.13 and 9.14). Some prominent erosion surfaces, larger than the bases of trough cosets, were recognized in the pebbly sandstone sheets. However, partly because of the nature of the exposure, none of these surfaces can be traced further than 5 m. These surfaces are mutually truncating, rather like the coset bounding surfaces of very large-scale trough cross-strata (cf. Campbell, 1976). The surfaces, as observed within the lateral limits of the outcrop, vary between being planar to undulating to concave-up trough-shaped. This trough cross-stratified sandstone facies, which dominates the deposits, is formed by the migration of sinuous-crested dunes in a cohesionless sandy bed. The low height:width ratios of the trough cosets in the Slea Head Formation suggest that the cross-strata were generated by slightly sinuous dunes with a low amplitude:wavelength aspect (Allen, 1968).

Some coarse sand (CS) to very coarse sand (VCS) sandstone beds in the Slea Head Formation also contain tabular cross-stratification, low-angle cross-stratification and horizontal parallel lamination. These facies are far less common than trough cross-stratified sandstone. Tabular cross-stratified sandstone forms beds with sharp planar bases of CS to VCS pebbly sandstone from 0.4 to 2.5 m in thickness. Cross-strata have both angular and asymptotic bases and vary in thickness from 0.2 m to 1.0 m. This facies was formed either by dunes or repetitive or non-repetitive bars. Low-angle cross-stratified sandstone forms similar beds to tabular cross-stratified sandstone, but the cross-strata are characterized by much lower angles of inclination. Low-angle cross-stratified sandstone is interpreted as the low-angle stratification that infills shallow scours or the product of migration of low-relief bedforms transitional between dunes and flat beds. Parallel-laminated sandstone also occurs in the Slea Head Formation but is more common in medium sand to very fine sand sandstones and is interpreted to represent deposition of sand in flat-bed or near-flat-bed conditions in deep flows with a high bed shear stress.

The Slea Head Formation coarse facies are distinctive in the dominance of trough cross-strata. Large-scale tabular cross-strata formed through avalanching on slipfaces of bars is rare and dunes appear to have been the dominant bedform in the river. This may be a preservational bias induced by the superior preservation potential of dunes in the thalweg relative to the positive elements of bars. However, many units of trough cross-stratified sandstone have flat or undulatory tops draped in siltstone that are not erosional bounding surfaces with the next bedset. This indicates that the preservation of the channel deposits is, in these cases, nearly complete and that slipfaced bars were not present at all. One explanation for the apparent dearth of bar deposits with tabular cross-stratification is that the Slea Head river contained *slipfaceless compound transverse bars* which were mantled in sinuous-crested dunes on the leeside and probably the stoss-side such that only trough cross-stratification was preserved. Major accretion or reactivation surfaces on these barforms may have generated the internal bounding surfaces within the pebbly sandstone sheets. Sinuous-crested transverse barforms are suggested by the troughed and curviplanar nature of some of the second-order bounding surfaces (Table 9.2). An alternative explanation is that the barforms in the Slea Head river were non-migratory, formed by preferential degradation and aggradation of the stream bed (Cant & Walker, 1978) perhaps over several flood cycles. In this case the second-order surfaces would be the result of cutting and filling of scoured subchannels (between the compound medial bars) within the fluvial tract (cf. Cant, 1978; Rust & Gibling, 1990b).

Point bar structure

Point bars and their deposits have been studied extensively in the last 25 years by fluvial sedimentologists. The point bar model of fluvial deposition formed the basis for Allen's (1965b) classic hypothesis for the generation of fining-upward cyclothems in ancient alluvium. Meandering systems often dominate coastal plains and have a high preservation potential in subsiding passive margins and foreland basins. The study of the sedimentology of modern and ancient meandering rivers continued to be intense through the 1970s (e.g. Bluck, 1971; Jackson, 1976a,b, 1978). There have been several comprehensive studies in the last ten or so years of the deposits of sandy, high-sinuosity rivers with particular reference to point bar deposits. Of particular note is the work of Bridge and his coworkers, who have interpreted both qualitatively and quantitatively the palaeodynamics of several different point bar deposits (Bridge, 1977, 1978, 1985; Bridge & Diemer, 1983; Bridge & Gordon, 1985; Gordon & Bridge, 1987). Several of these examples are described in detail below and summarized in Table 9.2.

Bridge & Gordon (1985) and Gordon & Bridge (1987) described through a series of lateral sections and facies logs the architecture of some high-sinuosity deposits in the Devonian Catskill Magnafacies of eastern USA. Large-scale trough cross-stratification is the dominant structure in the sandstone bodies, with rare large-scale planar cross-stratification. The sandbodies are multistorey with individual stories of trough cross-stratified sandstone separated by erosion surfaces which are interpreted to represent lateral-accretion surfaces. In sections approximately normal to palaeoflow, the lateral-accretion surfaces dip at about 6–10°; in sections parallel to flow, the bounding surfaces (epsilon surfaces of Allen (1970)) are generally parallel to the major basal erosion surfaces, although some gently inclined surfaces are common. The lateral-accretion beds may pass into sandstone-filled and/or siltstone-filled channels in which the relatively fine-grained beds overlie disconformably against the lateral-accretion beds. The lateral-accretion bedding in this Catskill example is considered to be the result of lateral migration of *point bars* in response to retreating cutbanks (Gordon & Bridge, 1987). Climbing lateral-accretion beds indicate net aggradation during channel migration. The trough cross-stratification is interpreted as the product of trains of sinuous-crested dunes that migrated across the lower part of the point bar surface. Major channel fills record abrupt or gradual abandonment of channel segments by cutoff or avulsion. Minor channels in the laterally accreted beds are interpreted to represent chute channels. Relatively low degrees of palaeocurrent variability suggest that the point bars formed in low-sinuosity streams, an inference supported by the palaeohydraulic analysis (sinuosity <1.25) (Bridge & Gordon, 1985;

Gordon & Bridge, 1987; see also Chapter 10). Channel and point bar geometry varied with grain size and hence generally with distance from the inferred palaeoshoreline. For example, the abundance of chute-channel fills in the coarser-grained sandbodies suggests lower sinuosity and larger degrees of braiding.

Diemer & Belt (1991) described somewhat similar deposits from the Tertiary Fort Union Formation in western USA. There, the sandstone bodies were also multistorey with lateral-accretion bedding. These deposits were similarly interpreted as point bar deposits in which most of the multistorey sandbodies probably record down valley migration of meander bends and accumulation of point bar deposits within an aggrading channel belt prior to an avulsion event. Locally thickened portions of sandstone bodies are interpreted as the product of scouring during episodic major floods. Diemer & Belt (1991) were able through palaeohydraulic analysis to discriminate the point bars and their inferred channels into three distinct size classes. Outcrop relationships suggest that the smaller channels were the tributaries and/or distributaries of the larger trunk streams represented by the larger channels.

Architecture element mapping and interpretation (Allen, 1983a; Miall, 1985) was used by Olsen (1988) to analyse quarry-face sections through the Triassic Solling Formation in Germany. The *point bar* deposits in this formation are dominated by elements of lateral-accretion bedding defined by large-scale, gently dipping, laminated sandstones. In places, these laminated sandstones are interbedded with cosets of trough cross-strata that overlie erosive bounding surfaces. These deposits were interpreted as point bar deposits in which lamination formed in the upper plane bed was the dominant structure. The trough cross-strata represent the deposits of dunes on the point bar and the inferred direction of palaeoflow slightly oblique to the strike of the low-angle epsilon beds was taken by Olsen (1988) as good evidence that the latter were formed by lateral accretion. Another form of channel element in these sandbodies is characterized by low-angle lamination with intrasets of low-angle tabular cross-stratification. The lower angle of divergence between the palaeoflow inferred from the tabular cross-strata and the strike of the low-angle bounding surfaces indicated to Olsen (1988) that these deposits originated as *medial compound bars*. This assemblage of mixed barform types led Olsen (1988) to conclude that the Solling Formation rivers had both point bars and compound medial bars, with the former dominating over the latter, and hence the rivers were braided meandering in type. Whilst the presence of point bars does not necessarily imply meandering (*sensu stricto*, sinuosity >1.5; Bridge, 1985; see above), this study still serves as a good example of deduction of bar type from the internal structure and directional properties of the resultant deposits.

The examples of point and other bar deposits described above are characterized by the availability of two-dimensional sections which allowed the structural architecture and interrelationships of the alluvium to be analysed in some detail. This is preferable by far to the attempts that characterized the 1970s to determine bar type (and hence channel type) from vertical (one-dimensional) sequence analysis (e.g. Miall, 1977, 1978; Jackson, 1978; Rust, 1978a,b). Indeed, the gathering force of two- and three-dimensional analysis to aid process deduction is one of the major advances in fluvial sedimentology of more recent times (Friend *et al.*, 1979; Friend, 1983; Allen & Williams, 1982; Allen, 1983a; Miall, 1985; see also Chapter 10).

MESOSCOPIC STRUCTURES IN MUD

Mud deposits tend to characterize the floodplain of many modern and ancient river systems, although mud can commonly form the fills of major channels. Muds and mudstones are usually considered to have deposited from suspension either in abandoned channels and bar tops (e.g Williams & Rust, 1969; Bluck, 1976) or in overbanks or muddy playas (e.g. Allen & Williams, 1982; Hubert & Hyde, 1982). However, Rust & Nanson (1989) indicate the possibility that mud may be transported in rivers as flocculated aggregates and behave much like sands grains in bedload. The recognition of such deposits in the rock record is fraught with difficulties, however, not least the destructive effects that may be exerted on mud fabric by diagenesis. The traditional interpretation of depositional structures in mudstones is covered by Collinson & Thompson (1989). It is the description and interpretation of post-depositional but early diagenetic modification of muds in soils – which include the most important advances in fluvial mud sedimentology in the last few years – that are focused on in this review. Palaeosols are now being widely recognized in the rock record and there is increasing comprehension of their power not only in indicating subaerial exposure (e.g. Wright, 1986), but also in palaeoclimatic studies (e.g. Wright & Robinson, 1988), palaeogeomorphology (e.g. Kraus & Middleton, 1987), time resolution (e.g. Bown & Kraus, 1987) and alluvial architecture analysis (e.g. Allen & Williams, 1982). Some of these aspects are discussed below in some detail, but the reader is referred to more comprehensive accounts by Wright (1986), Fenwick (1985) and Allen & Wright (1989).

Palaeosol recognition

Allen & Wright (1989) list a number of criteria that can be used to recognize palaeosols. Biogenic features include rootlets, root moulds or casts, or rhizocretions, but these have a low preservation potential and Early Palaeozoic and older soils will lack such features (Retallack, 1985). Calcretes (palaeosols characterized by calcium carbonate precipitation) are often marked by calcite precipitation around rootlets and/or burrows (Figure 9.15). Faunal features include remains of terrestrial molluscs, vertebrates and arthropods, trace fossils, and coprolites (e.g. Wright, 1986). Palaeosols are often distinctive in outcrop because of their colour. For example, reddening, resulting from the oxidation of iron, is commonly associated with pedification. Other colours include grey to white commonly associated with eluvial (leached) horizons. Darker, drab soils form in highly reducing conditions such as hydromorphic (waterlogged) or gley soils. Colour mottling, related to local changes in soil chemistry, is a typical feature of many soils (Duchaufour, 1982; Retallack, 1988).

Destratification or pedoturbation of the original alluvium can result from faunal bioturbation, root action, shrinking and swelling of expandable clays, salt structure growth, freezing and thawing, and soil gas movement (Allen & Wright, 1989). Horizonation, although essentially the opposite of destratification, may work in conjunction to produce distinctive layers within the soil profile that are marked by different texture and mineralogy. The degree of horizonation may provide some idea

Figure 9.15 Calcretes in mudstones of the Glashabeg Formation, SW Ireland. (a) Stacked soils 1 and 2. (b) Calcrete columns from calcite precipitation around roots (rhizocretions) or burrows (arrowed). These mature soils (Stage III of Maschette (1985)), are interbedded with conglomeratic channel-fill deposits. A relationship between base-level fall and gravel emplacement, terrace elevation and mature soil development is strongly suspected (see Todd *et al.*, 1991, p. 78)

of duration of soil development (Gile *et al.*, 1966; Birkeland, 1984; Maschette, 1985; Retallack, 1988) and this notion forms the basis for the use of palaeosols in stratigraphic analyses of ancient alluvium (e.g. Allen, 1974, 1986; Steel, 1974; see also next section). Granulometrics may be utilized in soil studies to decipher physical (particle size), mineralogical and chemical changes induced by pedification (e.g. Fenwick, 1985). Mineralogy and geochemistry may also be used to investigate such changes (e.g. Chatres *et al.*, 1988; Retallack, 1983, 1986; Bown & Kraus, 1987; Robinson & Wright, 1987).

Mesostructural features may also be distinctive of soils (Allen & Wright, 1989). Such structures include pseudo-anticlinal structures which are sets of parallel, often slickenslided fracture surfaces (commonly later filled by calcite) arranged in broad, gently sloping synclines and narrow, cuspate anticlines with wavelengths of 2–10 m and amplitudes of over 1 m (e.g. Blodgett, 1984; Allen, 1986; Retallack, 1986). Pseudo-anticlinal structures indicate a high content of swelling clays, a strong soil moisture regime, and locally impeded drainage (Ahmad, 1983; Allen & Wright, 1989). Other distinctive features include columnar and prismatic structures, typically a few centimetres in diameter and up to a metre long. These structures occur in a wide variety of soils but are very common in calcretes (e.g. Allen, 1986). Microstructures observed in thin section with a petrological microscope can be very distinctive because some fabrics have yet to find other sedimentary analogues (Allen & Wright, 1989). Several soil petrographic reviews give accounts of these features (e.g. Brewer, 1976; Esteban & Klappa, 1983; Fitzpatrick, 1984; Bullock *et al.*, 1985).

Implications of palaeosols

The development and preservation of a soil horizon in ancient alluvium implies that the net sediment aggradation rate in the locus where the soil developed became neutral or weakly aggradational for a time and then was buried without erosion. In floodplains this implies that the pedogenic site is either distal and/or topographically raised above the fluvial channel sources of coarser clastic sediments (Leeder, 1975; Hall, 1983; Bown & Kraus, 1987; Allen & Wright, 1989) (Figure 9.15). Bown & Kraus (1987) developed the notion of the maturity of palaeosols reflecting "proximality" of the fluvial channel through their pedofacies concept (Figure 9.16). In this model, stacked palaeosols were predicted to first become thicker and more mature upwards as the channel migrated away from the depositional site and then became thinner and less mature as the channel reapproached. In such a setting, the most mature soils grow due to sediment starvation when the channel is farthest away (see also R.M.H. Smith, 1990), but it is expected, given average fluvial aggradation and avulsion rates, that such a process cannot account for more mature soils and certainly mature calcretes (Allen & Wright, 1989). Mature soils, particularly mature calcretes, require considerable periods of stability for growth, perhaps of the order of 10^5 years for the most mature soils (e.g. Maschette, 1985) (Figure 9.17). In these cases, and probably in some cases of less mature soils, soils may have developed on fluvial terraces that are shielded from sediment input by being topographically raised above the general level of the floodplain.

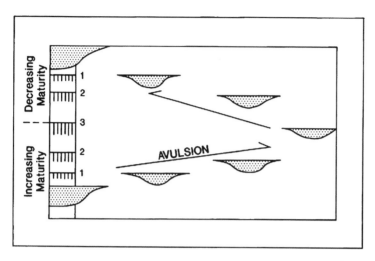

Figure 9.16 Maturity changes in a soil sequence reflecting pedofacies relationships during avulsion events. 1–3 represent increasing degrees of maturity (after Bown & Kraus (1987); from Allen & Wright (1989); reproduced by permission of J.R.L. Allen and V.P. Wright)

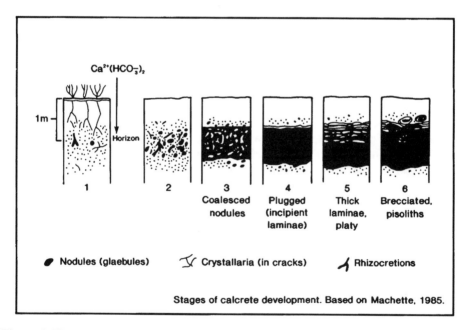

Figure 9.17 Stages of calcrete development (based on Maschette (1985); after Wright & Allen (1989); reproduced by permission of J.R.L. Allen and V.P. Wright)

CHANNEL-BELT AND FLOODPLAIN STRUCTURES

The deduction of process from short- to moderate-term fluvial sedimentary structures has been dealt with in the preceding sections of this chapter. Mesoforms such as dunes in sand and low-density bedload sheets in gravel leave relatively distinctive deposits that can be interpreted with a fair degree of ease. However, these structures are relatively short-lived, perhaps surviving one flood cycle, and can form within several component parts of the fluvial channel belt (Table 9.1). In general, fluvial mesoforms of this kind are not diagnostic of channel type. Instead the earlier discussion tried to illustrate how mesoform structures may form components or building blocks in the architecture of compound macroforms or bars (Tables 9.1 and 9.2). Macroforms are longer-lived elements of the fluvial system. While there is some latitude for overlap in the structure of these elements between channel planform types (Bridge, 1985; Brierley, 1989; Brierley & Hickin, 1991), careful analysis of the relationships of macroform bounding surfaces and the internal mesoform elements (if present) can yield sufficient data to make confident conclusions about bar type and hence channel form. The objective of this last section is to attempt to illustrate how meso- and macroscale structures both in the channel-belt and the floodplain may be linked into coherent structural assemblages of megascopic scale, the architecture of which yields much information on the nature and evolution of the alluvial basin (see also Chapters 10 and 12).

The graded stream – longitudinal architectural relationships

That rivers vary longitudinally in slope, discharge, grain size, bedforms, barforms and channel planform types is well established (e.g. Boothroyd & Ashley, 1975; Bluck, 1987). Many consider that rivers try to adopt a profile or gradient that matches best the local sediment calibre that may be transported by the shaping discharge, typically the bankfull (flood) discharge. Given sufficient time the bed roughness features of grains and bedforms will become equilibrated with local slope and discharge. This is the graded stream. Hence, longitudinal variation in bedforms and channel forms are an expression of the graded nature of the river's longitudinal profile. For example, Hein & Walker (1977) reported a longitudinal change from coarser gravel, slipfaceless longitudinal bars to finer gravel, slipfaced transverse bars. The implication for stratal assemblages is a change from horizontally stratified gravels downstream into tabular cross-stratified gravels.

Bluck (1987) demonstrated a similar relationship in other gravelly rivers showing how longitudinal clast ridges and transverse clast dams are replaced downstream by gravel sheets, bars and sandsheets with dunes (Figure 9.18). Bluck (1987) also established a logarithmic relationship between alluvial fan radius (which is proportional to drainage area (Denny, 1967)) and the rate of grain-size decline (Figure 9.19). Larger rivers are capable of dispersing a grain-size distribution along a greater channel length which is commonly graded to grain size and bedform. This relationship tends to be simplest and easiest to demonstrate on alluvial fans, but despite the complications exerted by tributaries, similar relationships of grain size, bedform, barform and channel type will also occur in non-fan graded rivers.

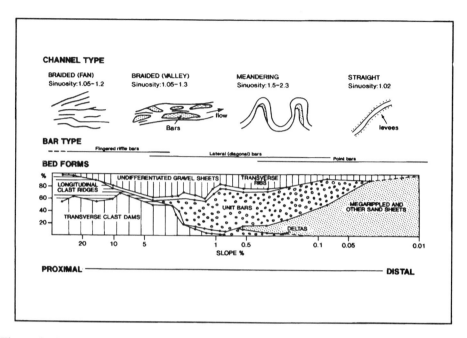

Figure 9.18 Channel type, bar type and bed forms related to channel slope of Scottish rivers and their tributaries (from Bluck (1987); reproduced by permission of the Institute of British Geographers)

In sandy streams, downstream changes from coarser-sediment, low-sinuosity, possibly braided channels on higher slopes, to finer-sediment high-sinuosity, undivided channels on lower slopes are common. In this case, upstream deposits with medial, slipfaced or slipfaceless compound bars together with slipfaced lateral bars might give way to downstream deposits with a dominance of point bars with lateral accretion bedding. Of course, such a scenario may be greatly complicated by local changes in slope (through intrabasin tectonics) or by local fluid and/or sediment discharge variation (through action of distributaries or tributaries).

Longitudinal facies changes of this kind are preserved in the rock record. Bluck (1986) described longitudinal changes in ancient river deposits in the Scottish Old Red Sandstone. He ascribed different styles of longitudinal change or "facies lineage" to different river types reflecting variable sediment and fluid discharges, sediment calibre and stream gradient — in effect different graded stream profiles. Superposition of different facies lineages may well be a record of alteration in the graded profile in response to changes in sediment supply, discharge or base level (Bluck, 1986; pers. comm. 1992). A well illustrated example of downstream variation in fluvial style has recently been documented by Hirst (1992). The Tertiary Huesca system in northern Spain was a large distributive fluvial fan which displays down fan variation in a number of attributes. The proportion of sediment deposited within channels and the thickness of sandbodies and thus palaeochannels both decrease distally; medial to distal sheet sandstones deposited by laterally unstable channels are replaced by ribbon sandstones deposited by more stable channels (Hirst, 1992; see also Ori, 1982).

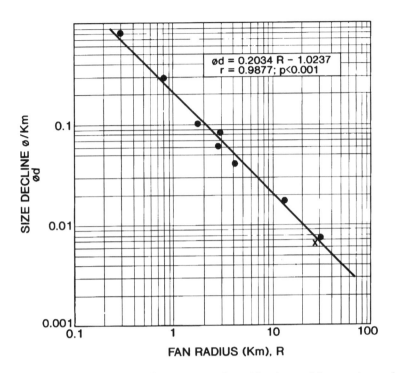

Figure 9.19   Rate of grain-size decline (expressed as phi units per kilometre) as a function of fan radius from fan head to fan toe (from Bluck (1987); reproduced by permission of the Institute of British Geographers)

Bluck's (1986) facies lineages are described from relatively well exposed horizontal sections, but commonly ancient horizontal facies relationships must be inferred from vertical facies sequences (assuming Walther's Law) where little or no horizontal control is available. An example of this was outlined by Todd (1989b) from the Devonian Trabeg Conglomerate Formation in SW Ireland. This formation is an alluvial fan deposit which contains sequences of three hierarchical scales. Four facies associations, A, B, C and D, are described from the formation. In Facies Association A, dominated by parallel-laminated and cross-stratified sandstones (see above), "typical" third-order fining-upward sequences, up to 2 m thick, are the product of channel cutting and filling, probably during flashy flooding in streams with alternate simple lateral bars. In Facies Association C, conglomerate–sandstone couplets are the product of deposition from bedload and suspension in high-magnitude floods (Todd, 1989a). Packages of Associations A and C form the greater part of the formation, but Association B – siltstone with calcretes interpreted to be abandoned channel and overbank deposits – may locally dominate parts of the succession, and Association D – mud matrix-supported conglomerates – are rare elements (see also Todd *et al.*, 1991).

The basal portion of the formation is composed of mainly sandstones of Facies Association A, with minor intervals of siltstones of Association B (Figure 9.10). Up to about 25 m above the base of the logged section, stacked fining-upward sequences, about 2 to 3 m thick, dominate the succession. Notably, the height of trough cross-strata cosets increases in a fairly systematic fashion up the formation (Figures 9.10 and 9.20). The middle portion of the succession is composed mainly of conglomerates of Facies Association C, with minor intervals of A and B (Figure 9.10). The general large-scale upward coarsening recorded in the logged section as sandstone is replaced by conglomerate and represents a local first-order sequence that ends around the mid-point of the vertical succession with the thickest and coarsest bed (Figure 9.10). There then follows a large first-order sequence that displays overall upward fining with a decrease in MPS and BTh. At the top of the Trabeg Conglomerate Formation there are several comparatively thick intervals of Facies Association B with calcretized siltstones (Figure 9.10). The top of the logged section at Trabeg records the occurrence of thick, uniform sandstones of the overlying Ballymore Formation which represent the deposits of the onlapping axial fluvial system (Todd *et al.*, 1988b, 1991).

The interbedding and gradation of one facies association into another within the Trabeg Conglomerate Formation is indicative that all the associations are the product of the same general depositional setting (and facies lineage) on a stream-dominated alluvial fan. On the basis of this fundamental assumption, it may also be concluded that Facies Association A is the distal, down-fan equivalent of Facies Association C. The implication of this is that many, if not most, of the sandstone beds of Association A were deposited as the downstream equivalents of the conglomerates of Association C, during high-magnitude, high-density stream floods. A typical fining-upward, flood-generated package in Association A comprises a thin gravelly lag member, in some cases overlying an erosional base, followed by a thicker sandstone member. These sediments were deposited in low-sinuosity rivers with

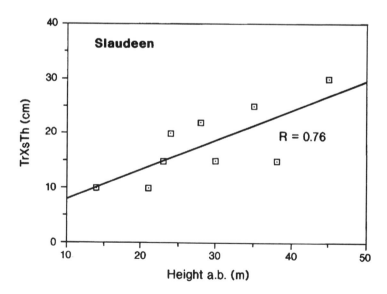

Figure 9.20 Height of trough cross-stratal thickness versus height above base in the Association A sandstones of the Trabeg Conglomerate Formation

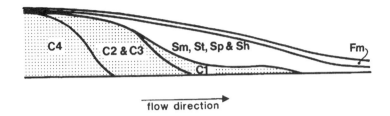

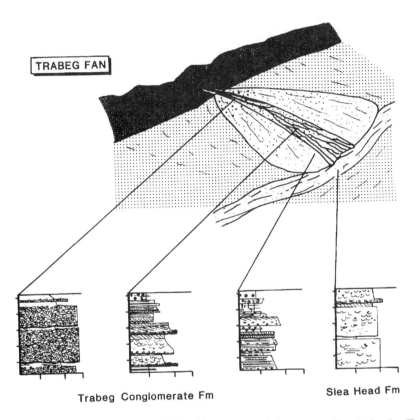

Figure 9.21 Longitudinal facies relationships generated by stream floods in the Trabeg graded distributive system. The upward increase in trough cross-stratal thickness (Fig. 9.20) is considered to record superposition of deeper and deeper channels with larger and larger dunes by a prograding distributive system

simple bank-attached alternate bars with intervening in-channel dunes (see above and Figure 9.12A). In Association C, a typical fining-upward couplet comprises a much thicker conglomerate member capped by a sandstone bed that is much reduced in thickness. The conglomerates are considered largely to be the product of high-density bedload sheets deposited at the bases of the upstream portions of high-density stream floods (Todd, 1989a). Although it is possible that the sandstone bed is reduced in thickness by erosion at the base of the succeeding conglomerate unit, there are very few examples where sandstone is completely absent and in many instances the sandstone bed is followed by siltstone. This is regarded as strong evidence that the preserved thickness of the sandstone member in the Association C conglomerate–sandstone couplets is close to the original depositional thickness of the sand bed.

Hence there is a real lateral variation between the upstream Association C fining-upward couplet, which has a thicker conglomerate member, and the Association A fining-upward sequence, which has a thicker sandstone member. By analogy with turbidite facies schemes (e.g. Walker, 1967; Hiscott & Middleton, 1979; Krause & Oldershaw, 1979; MacDonald, 1986), the reduced thickness of the sandstone member of the Association C couplets is considered to reflect by-passing of the sandy portion of the high-density stream floods past the point, or change in slope, where gravel was largely deposited. These lateral and vertical facies variations are depicted in Figure 9.21. While many of the structures may have been flood-generated, the inferred horizontal sedimentary structure change or facies lineage is considered to essentially reflect changes down the longitudinal graded profile of the river system. In this case, the system was possibly distributive, so that progradation (reflected by the initial coarsening-upward megasequence) superimposed larger channels (with larger bedforms; Figure 9.20) in successively younger beds in the vertical succession.

Lateral architectural relationships in the alluvial basin

Megascopic, lateral changes in fluvial architecture are also well documented from the modern and ancient, the latter often from application of Walther's Law in vertical section, but increasingly from two- or three-dimensional section analysis. The result of this analysis included changes from channel belt to floodplain assemblages, lateral alluvial fan to axial trunk stream transitions, and transitions into other continental facies such as lacustrine or aeolian or marginal marine such as deltaic or shoreface. Fluvial sedimentary structure of the scale of the architecture of channel or channel-belt sandbodies in an alluvial suite are increasingly common. These alluvial architectures yield information about the spatial and temporal distribution of subsidence in the alluvial basin.

Alexander & Leeder (1987), Steel (1988) and Todd (1989b) have recently discussed the controls on alluvial fans and their resulting sediment lithosomes through action on the bounding fault and deformation of the basin floor. Bridge & Leeder (1979) modelled how the relative amount and connectedness of channel-belt deposits depends on factors like aggradation rate, avulsion frequency, channel-belt width/floodplain width, and tectonic tilting. Large proportions of channel-belt deposits are developed during low aggradation rate relative to avulsion frequency rate, large channel-belt width, and areas of preferred tectonic subsidence (Bridge & Leeder, 1979; Bridge, 1985). Alexander & Leeder (1987) developed these concepts and discussed how asymmetric subsidence on a floodplain can have a pronounced effect on alluvial architecture. The essential tendency is for the river to seek the subsiding lows either through a single avulsion, a series of stepped avulsions, or by lateral migration (or "combing" (Todd & Went, 1991)). The example quoted by Leeder & Alexander (1987) of the Madison River was a case in which the meandering stream migrated towards a subsiding area adjacent to an active fault scarp by meander cut-off. In this case, point bars were preferentially preserved with their lateral-accretion surfaces dipping away from the fault scarp, i.e. antithetic to the direction of combing. It is also possible to envisage lateral accretion coupled with forward erosion of cut-banks in a more slowly sideways-moving channel belt to preferentially preserve point bars with their point bar accretion surfaces dipping towards the scarp, i.e. sympathetic to the combing direction. Todd & Went (1991) discussed similar movement of low-sinuosity, possibly braided streams and how different lateral bar types preserve different directional properties as a result of persistent, repeated channel-belt combing in asymmetrically subsiding floodplains. The avulsion or combing tendencies of rivers towards lows in the floodplain allow coarse-member channel-belt sandbodies to concentrate into these areas.

Floodplain deposits will also respond to differential tectonic subsidence in an alluvial basin. Palaeosols may be particularly sensitive to these effects. Thicker, more mature palaeosols will develop in elevated or less rapidly subsiding areas of the floodplain (interfluves). Areas of greater subsidence will receive more direct input of sediment and there will be less opportunity to develop more mature soils. Modern examples of these effects are given by Russ (1979, 1981). It is also possible that more mature palaeosols in ancient successions are due to tectonic effects. For example, Allen (1986; see also Allen & Wright, 1989) proposed that the thick, laterally extensive, mature Old Red Sandstone calcrete in Wales, the "Psamosteous" limestone, represents a tectonic event when the fluvial systems experienced a major reorganization.

Changes in fluvial architecture through time – extrabasinal controls

Extrabasinal controls on temporal changes on graded streams and their adjacent floodplains are potentially complex. They are essentially divided into two types:

(1) "upstream" controls on sediment and fluid discharge exerted by the tectonics (uplift), climate and geology (provenance) of the drainage basin; and
(2) "downstream" controls on base level which to a large extent controls the position of the graded stream profile.

Drainage-basin changes exert their influence through the amount and calibre of sediment and the fluid discharge. Clearly, any change in these attributes will change the fundamental nature of the river and its graded profile. Hence changes in uplift and erosion can cause increases and decreases in sediment supply. An increase in sediment supply may cause aggradation because the local gradient cannot cope with the new sediment discharge. Alternatively an increase in fluid discharge over sediment input might cause incision because the stream has increased capacity for transport. These changes can easily be exerted by one or a combination of the tectonic, climatic and geological factors.

Base level appears also to control the position of the toe of the graded profile (Leopold & Bull, 1979; Posamentier & Vail, 1988). Preservation of alluvial sediment is possible only where the river aggrades to maintain its graded profile. Posamentier & Vail (1988) argue that vertical changes in base level cause horizontal, longitudinal movement of the graded stream profile. Thus in a realm of rising base level, the asymptotic portion of the graded stream moves upward and backward, creating accommodation for aggradational alluviation. In contrast, base-level fall causes reorganization of the profile through the erosive backward retreat of a nick point and incision of the fluvial channel belt. Both the processes will lead to local changes in stream gradient and fluid/sediment discharge and hence will cause reorganization of grain size, bedform, barform and channel planforms that may well be recorded in vertical sequences.

In hydrologically closed lacustrine basins or in rivers far from the sea, base level is locally controlled. In rivers on coastal plains, the base level is clearly controlled by relative sea level, which in turn is influenced by the tectonics of the basin receiving the river and by eustacy (Posamentier & Vail, 1988). In the latter case particularly, it is possible to explain attributes of alluvial architecture through changes in base level (i.e. relative sea level) and hence within sequence stratigraphic concepts (Van Wagoner et al., 1988, 1990; Shanley & McCabe, 1991a,b, 1992). Combining these concepts with those of alluvial architecture and channel connectedness (Bridge & Leeder, 1979) and with pedofacies concepts (Bown & Kraus, 1987; Allen & Wright, 1989) may lead to some powerful predictions of the internal and external geometry of channel belts in alluvial sequences.

For example, during major base-level falls (Type 1 sequence boundaries associated with Lowstand Systems Tracts (Van Wagoner et al., 1988)), rivers cut into their floodplains producing initially confined incised valleys. Such incised valleys have often coarser, wider, thicker sandbodies with lower sinuosities and higher braid indices than the underlying fluvial channels because the graded profile has been shifted downwards and forwards towards the sea (Figure 9.22). Alternatively, the downward shift may be so abrupt as to cause the river system to reorganize into an altogether new profile. Such abrupt changes in the geometry of channel incision on a width and depth scale greater than that of the prevailing channels is relatively strong evidence of valley incision in an entirely continental succession, where downward shifts of alluvial facies onto marine strata are not available as documentary evidence of base-level fall. Incised valleys may be the width of just one channel belt and hence the resulting low channel belt/floodplain width ratio enhances coarse/fine member ratio, at least locally in the valley. Valley sides are likely to be terraced with soils exposed to prolonged pedification. Thus incised valley fills are likely to have relatively mature palaeosols as lateral equivalents within the facies association.

Incised valleys are produced where the relative base level (sea level) drops after having been rising. In some cases, the rate of relative base level (sea level) rise may abruptly decrease, but still remain weakly positive or aggradational. In these cases there will be no incision, but nevertheless the ratio of avulsion frequency/floodplain aggradation decreases and sharp-based, well-connected, multilateral and multistorey sandbodies may form (e.g. Shanley & McCabe, 1991a,b, 1992).

During periods of rapid base-level rise (Transgressive Systems Tracts (Van Wagoner et al., 1988)), the aggradation rate/avulsion frequency ratio (Bridge &

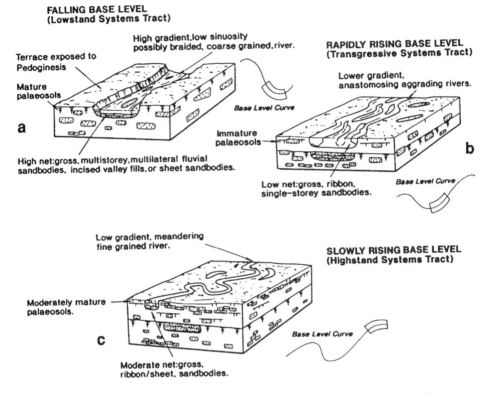

Figure 9.22   Variation in grain size, bedform type, barform structure and channel geometry and alluvial architecture relative to shifts in the graded stream profile during cyclic changes in base level

Leeder, 1979; Bridge, 1985) is high, causing channel-belt deposits to be poorly interconnected and enveloped within floodplain deposits. If climatic conditions are suitable, the Transgressive Systems Tract is often characterized by thick packages of coal. Maximum flooding surfaces in coastal plains may be represented by marine shales, but in entirely continental successions (but with a marine connection) the most rapid rate of rising relative sea level may only be reflected by higher amounts of tidal influence. This influence may extend several tens of kilometres from the coeval shoreline (Shanley & McCabe, 1992; Shanley et al., 1992) and might, for example, be recorded by rhythmic sand−mud couplets in point bar deposits (Thomas et al., 1987; D.G. Smith, 1987). Decreasing rates of base-level rise (Highstand Systems Tract) are accompanied by a decrease in the aggradation rate/avulsion frequency ratio and greater coarse/fine member ratio between channel-belt deposits (Figure 9.22). In conditions of moderate to high rates of base-level rise coupled with high rates of alluviation, only immature soils are likely to have time to develop. Thus the influence of base-level changes may be deduced by a combination of observations of channel facies lineages, channel-belt connectedness and architecture, and floodplain soil maturity.

CONCLUSIONS AND PERSPECTIVES

Since the publication of the Canadian Society of Petroleum Geologists Memoir on *Fluvial Sedimentology* edited by Andrew Miall in 1978, the sedimentology of rivers has continued to stimulate geologists and geomorphologists resulting in some major advances in the last 10 to 15 years. Laboratory and field observation of bedform/barform development in rivers has continued to pay dividends in the better understanding of process deduction from preserved sedimentary structures in ancient successions. Careful and logical deduction of deposits without direct correlation with established facies models has become more common. Particularly important is the deduction of bar type and therefore channel type from the hierarchical structural relationships of bar deposits. Such deductions will hopefully continue to be the focus of the future. Moreover, deduction of process may well be increasingly aided in the future by simulation and forward modelling in graphical computer programs (e.g Rubin, 1989). These programs can be used to mimic the development of accretion surfaces and the cosets formed by parasitic bedforms which are bounded by accretion surfaces.

As the understanding of alluvial deposits and their mineral and hydrocarbon resources becomes increasingly important, the architecture of alluvial suites will become more important. In this larger scale, computer modelling will also be important in the understanding and prediction of floodplain structure. Sequence stratigraphy, although having limitations in wholly continental successions, may well prove to be a powerful tool in the investigation of fluvial architecture. Care must be exercised, however, in the identification of bounding surfaces, particularly in wholly continental successions. For example, every channel base should not be immediately regarded as a candidate sequence boundary. Rather base-level variation should be deduced from a combination of changes in longitudinal facies associations or lineages (reflecting the graded stream profile), stacking patterns (reflecting mainly changes in aggradation rate versus avulsion rate), soil maturity and stacking patterns (also reflecting base-level changes) and provenance attributes (reflecting drainage basin and sediment supply alteration).

ACKNOWLEDGEMENTS

I thank Brian Williams for first sparking my interest in fluvial sedimentology. The flames of interest were fanned by other graduate students of Brian, namely Doug Boyd, Dave Lawrence, Colin North, Jonathan Turner, Dave Went and Liz Wild. Brian Bluck, Andrew Brayshaw, Phil Hirst, Mike Mayall, Tony Reynolds and Jonathan Turner provided constructive reviews of this article. BP Exploration Operating Company Ltd are acknowledged for permission to publish and for support in the production of diagrams.

REFERENCES

Ahmad, N. 1983. Vertisols. In: Wilding, L.P., Smeck, N.E. & Hall, G.F. (eds) *Pedogenesis and Soil Taxonomy II. The Soil Orders.* Elsevier, Amsterdam, 91–123.

Alexander, J. & Leeder, M.R. 1987. Active tectonic control on alluvial architecture. In: Ethridge, F.G., Flores, R.M. & Harvey, M.D. (eds) *Recent Developments in Fluvial Sedimentology*. Society of Economic Palaeontologists and Mineralogists, Special Publication **39**, 243–252.

Allen, J.R.L. 1963. The classification of cross stratified units, with notes on their origin. *Sedimentology*, **2**, 93–114.

Allen, J.R.L. 1964. Studies in fluviatile sedimentation: six cyclothems from the Lower Old Red Sandstone, Anglo-Welsh Basin. *Sedimentology*, **3**, 163–198.

Allen, J.R.L. 1965a. A review of the characteristics of Recent alluvial sediments. *Sedimentology*, **5**, 89–191.

Allen, J.R.L. 1965b. Fining upward cycles in alluvial successions. *Geological Journal*, **4**, 229–246.

Allen, J.R.L. 1968. *Current Ripples*. North-Holland, Amsterdam, 433 pp.

Allen, J.R.L. 1970a. Studies in fluviatile sedimentation: a comparison of fining-upwards cyclothems with special reference to coarse member composition and interpretation. *Journal of Sedimentary Petrology*, **40**, 298–323.

Allen, J.R.L. 1970b. A quantitative model of grain size and sedimentary structures in lateral deposits. *Geological Journal*, **7**, 129–146.

Allen, J.R.L. 1974. Studies in fluviatile sedimentation: implications of pedogenic carbonate units in the Lower Old Red Sandstone, Anglo-Welsh outcrop. *Geological Journal*, **9**, 181–208.

Allen, J.R.L. 1983a. Studies in fluviatile sedimentation: Bars, bar-complexes and sandstone sheets (low-sinuosity braided streams) in the Brownstones (L. Devonian), Welsh Borders. *Sedimentary Geology*, **33**, 237–293.

Allen, J.R.L. 1983b. River bedforms: progress and problems. In: Collinson, J.E. & Lewin, J. (eds) *Modern and Ancient Fluvial Systems*. Special Publication of the International Association of Sedimentologists, **6**, 19–33.

Allen, J.R.L. 1986. Pedogenic calcretes in the Old Red Sandstone facies (late Silurian – early Carboniferous) of the Anglo-Welsh area, southern Britain. In: Wright, V.P. (ed.) *Palaeosols*, Blackwell Scientific Publications, Oxford, 58–86.

Allen, J.R.L. & Leeder, M.R. 1980. Criteria for the instability of upper-stage plane beds. *Sedimentology*, **27**, 209–217.

Allen, J.R.L. & Williams, B.P.J. 1982. The architecture of an alluvial suite: Rocks between the Townshend Tuff and Picard Bay Tuff Beds (Early Devonian), southwest Wales. *Philosophical Transactions of the Royal Society of London*, **B297**, 51–89.

Allen, J.R.L. & Wright, V.P. 1989. *Palaeosols in Siliciclastic Sequences*. Postgraduate Research Institute of Sedimentology Short Course Notes No. **001**, 98 pp.

Ashley, G.M. 1990. Classification of large-scale subaqueous bedforms: a new look at an old problem. *Journal of Sedimentary Petrology*, **60**, 160–172.

Bagnold, R.A. 1966. *An approach to the sediment transport problem from general physics*. US Geological Survey Professional Paper **422-I**.

Bagnold, R.A. 1973. The nature of saltation and of "bedload" transport in water. *Proceedings of the Royal Society of London*, **A332**, 473–504.

Best, J. & Bridge, J. 1992. The morphology and dynamics of low amplitude bedwaves upon upper stage plane beds and preservation of planar laminae. *Sedimentology*, **39**, 737–752.

Billi, P., Magi, M. & Sagri, M. 1987. Coarse grained low-sinuosity river deposits: examples from Plio-Pleistocene, Valdarino Basin, Italy. In: Ethridge, F.G., Flores, R.M. & Harvey, M.D. (eds) *Recent Developments in Fluvial Sedimentology*. Society of Economic Palaeontologists and Mineralogists, Special Publication **39**, 243–252.

Birkeland, P.W. 1984. *Soils and Geomorphology*. Oxford University Press, New York, 372 pp.

Blackwelder, E. 1928. Mudflow as a geologic agent in semi-arid mountains. *Geological Society of America Bulletin*, **39**, 465–484.

Blair, T.C. 1987. Sedimentary processes, vertical stratification sequences, and geomorphology of the Roaring River alluvial fan, Rocky Mountain National Park, Colorado. *Journal of Sedimentary Petrology*, **57**, 1–18.

Blissenbach, E. 1954. Geology of alluvial fans in semi-arid regions. *Geological Society of American Bulletin*, **65**, 175–190.

Blodgett, R.H. 1984. Non-marine depositional environments and paleosol development in the Upper Triassic Dolores Formation, southwestern Colorado. In: Brew, D.C. (ed.) *Field Trip Guidebook*, Geological Society of America, Rocky Mountain section, 37th Annual Meeting, Durango, Colorado, Fort Lewis Collection, 46–92.

Bluck, B.J. 1964. Sedimentation of an alluvial fan in southern Nevada. *Journal of Sedimentary Petrology*, **34**, 395–400.

Bluck, B.J. 1967. Deposition of some upper Old Red Sandstone conglomerates in the Clyde area: a study in the significance of bedding. *Scottish Journal of Geology*, **3**, 139–167.

Bluck, B.J. 1971. Sedimentation in the meandering River Endrick. *Scottish Journal of Geology*, **7**, 93–138.

Bluck, B.J. 1974. Structure and directional properties of some valley sandur deposits in southern Iceland. *Sedimentology*, **21**, 533–554.

Bluck, B.J. 1976. Sedimentation in some Scottish rivers of low sinuosity. *Transactions of the Royal Society of Edinburgh: Earth Science*, **69**, 425–456.

Bluck, B.J. 1979. Structure of coarse grained braided alluvium. *Transactions of the Royal Society of Edinburgh*, **70**, Earth Science, 181–221.

Bluck, B.J. 1980. Structure, generation and preservation of upward fining braided stream cycles in the Old Red Sandstone of Scotland. *Transactions of the Royal Society of Edinburgh: Earth Science*, **71**, 29–46.

Bluck, B.J. 1986. Upward coarsening sedimentation units and facies lineages, Old Red Sandstone, Scotland. *Transactions of the Royal Society of Edinburgh: Earth Science*, **77**, 251–264.

Bluck, B.J. 1987. Bedforms and clast size changes in gravel-bed rivers. In: Richards, K. (ed.) *River Channels: Environment and Process*. Institute of British Geographers, Special Publication Series, **17**, 159–178.

Bøe, R. & Sturt, B.A. 1991. Textural responses to evolving mass-flows: an examples from the Devonian Asen Formation, central Norway. *Geological Magazine*, **128**, 99–109.

Boothroyd, J.C. & Ashley, G.M. 1975. Processes, bar morphology and sedimentary structures on braided outwash fans, northeastern Gulf of Alaska. In: Jopling, A.V. & McDonald, B.C. (eds.) *Glaciofluvial and Glaciolacustrine Environments*. Society of Economic Palaeontologists and Mineralogists, Special Publication **23**, 193–220.

Bown, T.M. & Kraus, M.J. 1987. Integration of channel and floodplain suites, I. Developmental sequence and lateral relations of alluvial paleosols. *Journal of Sedimentary Petrology*, **57**, 587–601.

Brayshaw, A.C. 1984. Characteristics and origins of cluster bedforms in coarse-grained alluvial channels. In: Koster, E.H. & Steel, R.J. (eds) *Sedimentology of Gravels and Conglomerates*. Canadian Society of Petroleum Geologists, Memoir **10**, 77–86.

Brewer, R 1976. *Fabric and Mineral Analysis of Soils*. Wiley, New York, 470 pp.

Bridge, J.S. 1977. Flow, bed topography, grain size and sediment structure in open channel bends: a three-dimensional model. *Earth Surface Processes*, **2**, 407–416.

Bridge, J.S. 1978. Palaeohydraulic interpretation using mathematical models of contemporary flow and sedimentation in meandering channels. In: Miall, A.D. (ed.) *Fluvial Sedimentology*. Canadian Society of Petroleum Geologists, Memoir **5**, 723–742.

Bridge, J.S. 1985. Palaeochannel patterns inferred from alluvial deposits: a critical evaluation. *Journal of Sedimentary Petrology*, **55**, 579–589.

Bridge, J.S. & Best, J.L. 1988. Flow, sediment transport and bedform dynamics over the transition from dunes to upper-stage plane beds: implications for the formation of planar laminae. *Sedimentology*, **35**, 753–763.

Bridge, J.S. & Diemer, J.A. 1983. Quantitative interpretation of an evolving ancient river system. *Sedimentology*, **30**, 599–623.

Bridge, J.S. & Gordon, E.A. 1985. Quantitative interpretation of ancient river systems in the Oneonto Formation, Catskill Magnafacies. In: Woodrow, D.L. & Sevon, W.D. (eds) *The Catskill Delta*. Geological Society of America, Special Paper **201**, 163–183.

Bridge, J.S. & Jarvis, J. 1978. Flow and sedimentary processes in the meandering river Esk, Glen Cova, Scotland. *Earth Surface Processes*, **1**, 303–336.

Bridge, J.S. & Leeder, M.R. 1979. A simulation model of alluvial stratigraphy. *Sedimentology*, **26**, 617–644.

Brierley, G.J. 1989. River planform facies models: the sedimentology of braided, wandering and meandering reaches of the Squamish River, British Columbia. *Sedimentary Geology*, **61**, 17–35.

Brierley, G.J. & Hickin, E.J. 1991. Channel planform as a non-controlling factor in fluvial sedimentology: the case of the Squamish River floodplain, British Columbia. *Sedimentary Geology*, **75**, 67–83.

Bull, W.B. 1963. Alluvial fan deposits in western Fresno County, California. *Journal of Geology*, **71**, 243–251.

Bull, W.B. 1972. Recognition of alluvial fan deposits in the stratigraphic record. In: Rigby, K.J. & Hamblin, W.K. (eds) *Recognition of Ancient Sedimentary Environments*. Society of Economic Palaeontologists & Mineralogists, Special Publication **16**, 68–83.

Bullock, P., Federoff, N., Jongerius, A., Stoops, G. & Tursina, T. 1985. *Handbook for Soil Thin Section Description*. Waine Research Publications, UK, 152 pp.

Campbell, C.V. 1976. Reservoir geometry of a fluvial sheet sandstone. *American Association of Petroleum Geologists Bulletin*, **60**, 1009–1020.

Cant, D.J. 1978. Development of a facies model for sandy braided river sedimentation: comparison of the South Saskatchewan River and the Battery Point Formation. In: Miall, A.D. (ed.) *Fluvial Sedimentology*. Canadian Society of Petroleum Geologists, Memoir **5**, 627–639.

Cant, D.J. & Walker, R.G. 1978. Fluvial processes and facies sequences in the sandy braided South Saskatchewan River, Canada. *Sedimentology*, **25**, 625–648.

Carling, P.A. 1990. Particle overpassing on depth limited gravel bars. *Sedimentology*, **37**, 345–355.

Carling, P.A. & Glaister, M.S. 1987. Rapid deposition of sand and gravel mixtures downstream of a negative step: the role of matrix-infilling and particle-overpassing in the process of bar-front accretion. *Journal of the Geological Society London*, **144**, 543–551.

Chatres, C.J., Chivas, A.R. & Walker, P.H. 1988. The effect of aeolian processes on soil development on granitic rocks in southeastern Australia. II. Oxygen-isotope, mineralogical and geochemical evidence of aeolian deposition. *Australian Journal of Soil Research*, **26**, 17–31.

Coleman, J.D. 1969. Brahmaputra River: channel processes and sedimentation. *Sedimentary Geology*, **3**, 129–239.

Collinson, J.D. 1970. Bedforms of the Tana River, Norway. *Geografiska Annaler*, **52A**, 31–56.

Collinson, J.D. & Thompson, D.B. 1989. *Sedimentary Structures* (2nd Edition). Unwin, London, 207 pp.

Cotter, E. & Graham, J.R. 1991. Coastal plain sedimentation in the late Devonian of southern Ireland; hummocky cross-stratification in fluvial deposits? *Sedimentary Geology*, **72**, 201–224.

Crowley, K.D. 1983. Large-scale bed configurations (macroforms), Platte River Basin, Colorado and Nebraska: primary structures and formative processes. *Geological Society of America Bulletin*, **94**, 117–133.

Dam, G. & Andreasan, F. 1990. High energy stream deltas: an example from the Upper Silurian Holmestrand Formation of the Oslo region, Norway. *Sedimentary Geology*, **66**, 197–225.

Denny, C.S. 1967. Fans and pediments. *American Journal of Science*, **265**, 81–105.

Diemer, J.A. & Belt, E.S. 1991. Sedimentology and palaeohydraulics of the meandering systems of the Fort Union Formation, southeastern Montana. *Sedimentary Geology*, **75**, 85–108.

Duchaufour, P. 1982. *Pedology: pedogenesis and classification*. Allen & Unwin, London.

Esteban, M & Klappa, C.F. 1983. Subaerial exposure environment. In: Scholle, P.A., Bebout, D.G.& Moore, C.H. (eds) *Carbonate Depositional Environments.* American Association of Petroleum Geologists, Memoir **33**, 1–54.

Fenwick, I. 1985. Paleosols: problems of recognition and interpretation. In: Boardman, J. (ed.) *Soils and Quaternary Landscape Evolution.* Wiley, Chichester, 3–21.

Fitzpatrick, E.A. 1984. *Morphology of Soils.* Chapman & Hall, London, 433 pp.

Flint, S. & Turner, P. 1988. Alluvial fan and fan delta sedimentation in a forearc extensional setting: the Cretaceous Coloso Basin of Northern Chile. In: Nemec, W. & Steel, R.J. (eds), *Fan Deltas: Sedimentology and Tectonic Setting.* Blackie, Glasgow, 387–399.

Friend, P.F. 1983. Towards a field classification of alluvial architecture or sequence. In: Collinson, J.E. & Lewin, J. (eds) *Modern and Ancient Fluvial Systems.* Special Publication of the International Association of Sedimentologists, **6**, 345–354.

Friend, P.F., Slater, M.J. & Williams, R.C. 1979. Vertical and lateral building of river sandstones bodies, Ebro Basin, Spain. *Journal of the Geological Society London,* **136**, 39–46.

Gile, L.H., Peterson, F.F. & Grossman, R.B. 1966. Morphological sequences of carbonate accumulation in desert soils. *Soil Science,* **101**, 347–360.

Gloppen, T.G. & Steel. R.J. 1981. The deposits, internal structure and geometry of six alluvial fan–fan delta bodies (Devonian, Norway) – a study in the significance of bedding sequence in conglomerates. In: Ethridge, F.G. & Flores, R.M. (eds) *Recent and Ancient Nonmarine Depositional Environments: Models for Exploration.* Society of Economic Palaeontologists and Mineralogists, Special Publication **31**, 49–69.

Gordon, E.A. & Bridge, J.S. 1987. Evolution of Catskill (Upper Devonian) river systems: intra- and extra-basinal controls. *Journal of Sedimentary Petrology,* **57**, 234–249.

Graham, J.R. 1983. Analysis of the Upper Devonian Munster Basin, an example of a fluvial distributary system. In: Collinson, J.E. & Lewin, J. (eds) *Modern and Ancient Fluvial Systems.* Special Publication of the International Association of Sedimentologists, **6**, 473–483.

Gustavson, T.C. 1978. Bedforms and stratification types of modern gravel meander lobes, Nueces River, Texas. *Sedimentology,* **25**, 401–426.

Hall, G.F. 1983. Pedology and Geomorphology. In: Wilding, L.P., Smeck, N.E. & Hall, G.F. (eds) *Pedogenesis and Soil Taxonomy. I. Concepts and Interactions.* Elsevier, Amsterdam, 117–140.

Hampton, M.A. 1975. Competence of fine grained debris flows. *Journal of Sedimentary Petrology,* **45**, 834–844.

Hampton, M.A. 1979. Buoyancy in debris flows. *Journal of Sedimentary Petrology,* **49**, 753–758.

Harms, J.C., Southard, J.B. & Walker, R.G. 1975. *Depositional environments as interpreted from sedimentary structures and stratification sequences.* Society of Economic Palaeontologists and Mineralogists, Short Course, **9**, Calgary, 249 pp.

Harms, J.C., Southard, J.B. & Walker, R.G. 1982. Structures and sequences in clastic rocks. Society of Economic Palaeontologists and Mineralogists, Short Course, **9**.

Hazeldine, R.S. 1983. Descending tabular cross-bedded sets and bounding surfaces from a fluvial channel in the Upper Carboniferous of northeast England. In: Collinson, J.E. & Lewin, J. (eds) *Modern and Ancient Fluvial Systems.* Special Publication of the International Association of Sedimentologists, **6**, 449–456.

Hein, F.J. 1982. Depositional mechanisms of deep-sea coarse clastic sediments, Cap Enragé Formation, Quebec. *Canadian Journal of Earth Science,* **19**, 267–287.

Hein, F.J. 1984. Deep-sea and fluvial and braided channel conglomerates: a comparison of two case studies. In: Koster, E.H. & Steel, R.J. (eds) *Sedimentology of Gravels and Conglomerates.* Canadian Society of Petroleum Geologists, Memoir **10**, 33–49.

Hein, F.J. & Walker, R.G. 1977. Bar evolution and development of stratification in the gravelly braided Kicking Horse River, British Columbia. *Canadian Journal of Earth Science,* **14**, 562–570.

Hirst, J.P.P. 1992. Variations in alluvial architecture across the Oligo-Miocene Huesca fluvial system, Ebro Basin, Spain. In: Flint, S. & Bryant, I. (eds) *Quantitative Description and Modelling of Clastic Hydrocarbon Reservoirs and Outcrop Analogues.* International Association of Sedimentologists Special Publication, **15**, 111–121.

Hiscott, R.N. & Middleton, G.V. 1979. Depositional mechanics of thick-bedded sandstones at the base of a submarine slope, Tourelle Formation (Lower Ordovician), Quebec, Canada. In: Doyle, L.J. & Pilkey, O.H. (eds) *Geology of Continental Slopes*. Society of Economic Palaeontologists and Mineralogists, Special Publication **27**, 307–326.

Hubert, J.F. & Hyde, M.G. 1982. Sheet-flow deposits of graded beds and sandstone on an alluvial sandflat-playa system: Upper Triassic Blormidon red beds, St. Mary's Bay, Nova Scotia. *Sedimentology*, **29**, 457–474.

Jackson, R.G. 1975. Hierarchial attributes and a unifying model of bedforms composed of cohesionless material produced by shearing flow. *Geological Society of America Bulletin*, **86**, 1523–1533.

Jackson, R.G. 1976a. Depositional model of point bars in the lower Wabash River. *Journal of Sedimentary Petrology*, **46**, 579–594.

Jackson, R.G. 1976b. Large-scale ripples of the Lower Wabash River. *Sedimentology*, **23**, 593–624.

Jackson, R.G. 1978. Preliminary evaluation of lithofacies models for meandering alluvial deposits. In: Miall, A.D. (ed.) *Fluvial Sedimentology*. Canadian Society of Petroleum Geologists, Memoir **5**, 543–576.

Jolley, E.J., Turner, P., Williams, G.D., Hartley, A.J. & Flint, S. 1990. Sedimentological response of an alluvial system to Neogene thrust tectonics, Atacama Desert, northern Chile. *Journal of Geological Society of London*, **147**, 769–784.

Jopling, A.V. 1965. Hydraulic factors controlling the shape of laminae in laboratory deltas. *Journal of Sedimentary Petrology*, **35**, 777–791.

Karcz, I. 1972. Sedimentary structures formed by flash floods in southern Israel. *Sedimentary Geology*, **7**, 161–182.

Koster, E.H. 1978. Transverse ribs: their characteristics, origin and palaeohydraulic significance. In: Miall, A.D. (ed.) *Fluvial Sedimentology*. Canadian Society of Petroleum Geologists, Memoir **5**, 161–186.

Kraus, M.J. & Middleton, L.T. 1987. Dissected paleotopography and base-level changes in a Triassic fluvial sequence. *Geology*, **15**, 18–21.

Krause, F.F. & Oldershaw, A.E. 1979. Submarine breccia beds – a depositional model for two-layer sediment gravity flows from the Sekwi Formation (Lower Cambrian), Mackenzie Mountains, Northwest Territories, Canada. *Canadian Journal of Earth Science*, **16**, 189–199.

Lawrence, D.A. & Williams, B.P.J. 1987. Evolution of drainage systems in response to Acadian deformation: the Devonian Battery Point Formation, Eastern Canada. In: Ethridge, F.G., Flores, R.M. & Harvey, M.D. (eds) *Recent Developments in Fluvial Sedimentology*. Society of Economic Palaeontologists and Mineralogists, Special Publication **39**, 243–252.

Leeder, M.R. 1975. Pedogenic carbonates and flood sediment accretion rates: a quantitative model for alluvial arid-zone lithofacies. *Geological Magazine*, **112**, 257–270.

Leeder, M.R. 1979. "Bedload" dynamics: grain–grain interactions in water flows. *Earth Surface Processes*, **4**, 229–240.

Leeder, M.R. 1983. On the interactions between turbulent flow, sediment transport and bed-form mechanics in channelized flows. In: Collinson, J.D & Lewin, J. (eds) *Modern and Ancient Fluvial Systems*. Special Publication of the International Association of Sedimentologists, **6**, 5–18.

Leeder, M.R. & Alexander, J. 1987. The origin and tectonic significance of asymmetric meander-belts. *Sedimentology*, **34**, 217–226.

Leopold, L.B. & Bull, W.B. 1979. Base level, aggradation and grade. *Proceedings of the American Philosophical Society*, **123**, 168–202.

Lowe, D.R. 1976. Grain flow and grain-flow deposits. *Journal of Sedimentary Petrology*, **46**, 188–199.

Lowe, D.R. 1982. Sediment gravity flows: II depositional models with special reference to deposits of high-density turbidity currents. *Journal of Sedimentary Petrology*, **52**, 279–297.

MacDonald, D.I.M. 1986. Proximal to distal sedimentological variation in a linear turbidite trough: implications for the fan model. *Sedimentology*, **33**, 243–259.

McGowan, J.H. & Garner, L.E. 1970. Physiographic features and stratification types of coarse-grained point bars: modern and ancient examples. *Sedimentology*, **14**, 77–101.

McKee, E.D., Crosby, E.J. & Berryhill, H.L. 1967. Flood deposits, Bijou Creek, June 1965. *Journal of Sedimentary Petrology*, **37**, 829–851.

Maschette, M.N. 1985. Calcic soils of the southwestern United States. In: Weide, D.L. (ed.) *Soils and Quaternary Geology of the Southwestern United States*. Geological Society of America Special Paper **203**, 1–21.

Massari, F. 1983. Tabular cross-bedding in Messinian fluvial channel conglomerates, Southern Alps, Italy. In: Collinson, J.E. & Lewin, J. (eds) *Modern and Ancient Fluvial Systems*. Special Publication of the International Association of Sedimentologists, **6**, 287–300.

Miall, A.D. 1977. A review of the braided-river depositional environment. *Earth Science Reviews*, **13**, 1–62.

Miall, A.D. 1978. Lithofacies types and vertical profile models in braided river deposits: a summary. In: Miall, A.D. (ed.) *Fluvial Sedimentology*. Canadian Society of Petroleum Geologists, Memoir **5**, 597–604.

Miall, A.D. 1985. Architectural-element analysis: a new method of facies analysis applied to fluvial deposits. In: Flores, R.M., Ethridge, F.G., Miall, A.D., Galloway, W.E. & Fouch, T.D. (eds) *Recognition of Fluvial Depositional Systems and their Resource Potential*. Society of Economic Palaeontologists & Mineralogists, Short Course **19**, 33–81.

Middleton, L.T. & Trujillo, A.P. 1984. Sedimentology and depositional setting of the Upper Proterozoic Scanlan Conglomerate, central Arizona. In: Koster, E.H. & Steel, R.J. (eds) *Sedimentology of Gravels and Conglomerates*. Canadian Society of Petroleum Geologists, Memoir **10**, 189–202.

Morrison, S.R. & Hein, F.J. 1987. Sedimentology of the White Channel gravels. Klondike area, Yukon Territory: fluvial deposits of a confined valley. In: Ethridge, F.G., Flores, R.M. & Harvey, M.D. (eds) *Recent Developments in Fluvial Sedimentology*. Society of Economic Palaeontologists and Mineralogists, Special Publication **39**, 205–216.

Nemec, W. & Muszynski, A. 1982. Volcaniclastic alluvial aprons in the Tertiary of the Sofia district (Bulgaria). *Annals of the Geological Society of Poland*, **52**, 239–303.

Nemec, W. & Steel, R.J. 1984. Alluvial and coastal conglomerates: their significant features and some comments on gravelly mass-flow deposits. In: Koster, E.H. & Steel, R.J. (eds) *Sedimentology of Gravels and Conglomerates*. Canadian Society of Petroleum Geologists, Memoir **10**, 1–31.

Nijman, W. & Puidefabregas, C. 1978. Coarse-grained point-bar structure in a molasse-type fluvial system, Eocene Castisent Sandstone Formation, south Pyrenean Basin. In: Miall, A.D. (ed.) *Fluvial Sedimentology*. Canadian Society of Petroleum Geologists, Memoir **5**, 487–510.

Olsen, H. 1987. Ancient ephemeral stream deposits: a local terminal fan model from the Bunter Sandstone Formation (L. Triassic) in the Tønder-3, -4 and -5 wells, Denmark. In: Frostick, L. & Reid, I. (eds) *Desert Sediments: Ancient and Modern*. Geological Society of London, Special Publication **35**, 69–86.

Olsen, H. 1988. The architecture of a sandy braided-meandering river system: an example from the Lower Triassic Solling Formation (M. Buntsandstein) in West Germany. *Geologisch Rundschau*, **77**, 797–814.

Olsen, H. 1989. Sandstone-body structures and ephemeral stream processes in the Dinosaur Canyon Member, Moenave Formation (Lower Jurassic), Utah, U.S.A. *Sedimentary Geology*, **61**, 207–221.

Ori, G.G. 1982. Braided to meandering channel patterns in humid-region alluvial fan deposits, River Reno, Po Plain, northern Italy. *Sedimentary Geology*, **31**, 231–248.

Paola, C., Wiele, S.M. & Reinhart, M.A. 1989. Upper-regime parallel lamination as a result of turbulent sediment transport and low amplitude bedforms. *Sedimentology*, **36**, 47–59.

Picard, M.D. & High, L.R. 1973. *Sedimentary Structures of Ephemeral Streams*. Developments in Sedimentology **17**, Elsevier, Amsterdam, 223 pp.

Pierson, T.C. & Costa, J.E. 1987. A rheological classification of subaerial sediment-water flows. In: Costa, J.E. & Wieczorek, G.F. (eds) *Debris Flows/Avalanches: Process, Sedimentology and Hazard Mitigation.* Engineering Geological Society of America Annual Review **7**, 1–12.

Posamentier, H.W. & Vail, P.R. 1988. Eustatic controls on clastic deposition II. Sequence and systems tract models. In: Wilgus, C.K., Hastings, B.S., Kendall, C.G.StC., Posamentier, H.W., Ross, C.A. & Van Wagoner, J.C. (eds) *Sea-level Changes: an Integrated Approach.* Society of Economic Palaeontologists and Mineralogists, Special Publication **42**, 125–154.

Postma, G., Nemec, W. & Kleinspehn, K. 1988. Large floating clasts in turbidites: a mechanism for their emplacement. *Sedimentary Geology*, **58**, 47–61.

Ramos, A. & Sopeña, A. 1983. Gravel bars in low-sinuosity streams (Permian and Triassic, central Spain). In: Collinson, J.E. & Lewin, J. (eds) *Modern and Ancient Fluvial Systems.* Special Publication of the International Association of Sedimentologists, **6**, 301–312.

Rees, A.I. 1968. The production of preferred orientation in a concentrated dispersion of elongated and flattened grains. *Journal of Geology*, **76**, 457–465.

Retallack, G.J. 1985. Fossil soils as grounds for interpreting the advent of plants and animals on land. *Philosophical Transactions of the Royal Society of London*, **B309**, 105–142.

Retallack, G.J. 1986. Fossil soils as grounds for interpreting long-term controls on ancient rivers. *Journal of Sedimentary Petrology*, **56**, 1–18.

Retallack, G.J. 1988. Field recognition of palaesols. In: Reinhardt, J. & Sigles, W.R. (eds) *Palaeosols and weathering through geologic time: principles and applications.* Geological Society of America, Special Paper **216**, 1–20.

Robinson, D. & Wright, V.P. 1987. Ordered illite/smectite and kaolinite/smectite as possible primary minerals in a Lower Carboniferous palaeosol sequence, South Wales? *Clay Minerals*, **22**, 109–118.

Rodine, J.D. & Johnson, A.M. 1976. The ability of debris, heavily freighted with coarse clastic materials, to flow on gentle slopes. *Sedimentology*, **23**, 213–234.

Røe, S.L. 1987. Cross strata and bedforms of probable transitional dune to upper stage plane bed origin from a Late Precambrian fluvial sandstone, northern Norway. *Sedimentology*, **34**, 89–101.

Rubin, D. 1989. *Cross Bedding, Bedforms and Palaeocurrents.* Society of Economic Palaeontologists and Mineralogists, Concepts in Sedimentology and Palaeontology, **1**, 187 pp.

Russ, D.P. 1979. Late Holocene faulting and earthquake recurrence in the Realfoot Lake area, Northwestern Tennessee. *Geological Society of America Bulletin*, **90**, 1013–1018.

Russ, D.P. 1981. *Style and significance of surface deformation in the vicinity of New Madrid, Missouri.* United States Geological Survey, Professional Paper **1236**, 95–114.

Rust, B.R. 1978a. A classification of channel patterns. In: Miall, A.D. (ed.) *Fluvial Sedimentology.* Canadian Society of Petroleum Geologists, Memoir **5**, 187–198.

Rust, B.R. 1978b. Depositional models for braided alluvium. In: Miall, A.D. (ed.) *Fluvial Sedimentology.* Canadian Society of Petroleum Geologists, Memoir **5**, 669–702.

Rust, B.R. & Gibling, M.R. 1990a. Three-dimensional antidunes as HCS mimics in a fluvial sandstone: the Pennsylvania South Bar Formation near Sydney, Nova Scotia. *Journal of Sedimentary Petrology*, **60**, 540–548.

Rust, B.R. & Gibling, M.R. 1990b. Braidplain evolution in the Pennsylvania South Bar Formation, Sydney Basin, Nova Scotia, Canada. *Journal of Sedimentary Petrology*, **60**, 59–72.

Rust, B.R. & Jones, B.G. 1987. The Hawkesbury Sandstone south of Sydney, Australia: Triassic analogue for the deposit of a large braided river. *Journal of Sedimentary Petrology*, **57**, 222–233.

Rust, B.R. & Nanson, G.C. 1989. Bedload transport of mud as pedogenic aggregates in modern and ancient rivers. *Sedimentology*, **36**, 291–306.

Saunderson, H.C. & Lockett, F.P.J. 1983. Flume experiments on bedforms and structures at the dune–plane bed transition. In: Collinson, J.E. & Lewin, J. (eds) *Modern and Ancient Fluvial Systems.* Special Publication of the International Association of Sedimentologists, **6**, 49–58.

Shanley, K.W. & McCabe, P.J. 1991a. Predicting facies architecture through sequence stratigraphy – an example from the Kaiparowits Plateau, Utah. *Geology*, **19**, 742–745.

Shanley, K.W. & McCabe, P.J. 1991b. Perspectives on the sequence stratigraphy of continental strata. American Association of Petroleum Geologists Bulletin, **78**, 544–568.

Shanley, K.W. & McCabe, P.J. 1992. Alluvial architecture in a sequence stratigraphic framework – a case history from the Upper Cretaceous of southern Utah, U.S.A. In: Flint, S. & Btyant, I. (eds) *Quantitative Description and Modelling of Clastic Hydrocarbon Reservoirs and Outcrop Analogues*. International Association of Sedimentologists, Special Publication **15**, 21–56.

Shanley, K.W., McCabe, P.J. & Hettinger, R.D. 1992. Tidal influence in Cretaceous fluvial strata from Utah, U.S.A.: a key to sequence stratigraphic interpretation. *Sedimentology*, **39**, 905–930.

Siegenthaler, C. & Huggenberger, P. 1993. Pleistocene Rhine gravel: Deposits of a braided river system with dominant pool preservation. In: Best, J.L. & Bristow, C.S. (eds) *Braided Rivers*, Geological Society Special Publication No. **75**, 147–162.

Smith, D.G. 1987. Meandering river point bar lithofacies models: modern and ancient examples compared. In: Ethridge, F.G., Flores, R.M. & Harvey, M.D. (eds) *Recent Developments in Fluvial Sedimentology*. Society of Economic Palaeontologists and Mineralogists, Special Publication **39**, 243–252.

Smith, G.A. 1986. Coarse-grained volcaniclastic sediment: Terminology for depositional process. *Geological Society of America Bulletin*, **97**, 1–10.

Smith, N.D. 1971. Transverse bars and braiding in the Lower Platte River, Nebraska. *Geological Society of America Bulletin*, **82**, 3047–3420.

Smith, N.D. 1972. Some sedimentological aspects of planar cross-stratification in a sandy braided river. *Journal of Sedimentary Petrology*, **42**, 624–643.

Smith, N.D. 1974. Sedimentology and bar formation in the upper Kicking Horse River, a braided outwash stream. *Journal of Geology*, **82**, 205–224.

Smith, R.M.H. 1990. Alluvial paleosols and pedofacies sequences in the Permian Lower Beaufort of the southwestern Karoo Basin, South Africa. *Journal of Sedimentary Petrology*, **60**, 258–276.

Stear, W.M. 1985. Comparison of the bedform distribution and dynamics of modern and ancient sandy ephemeral flood deposits in the southwestern Karoo region, South Africa. In: Collinson, J.E. & Lewin, J. (eds) *Modern and Ancient Fluvial Systems*. Special Publication of the International Association of Sedimentologists, **6**, 209–230.

Steel, R.J. 1974. New Red Sandstone floodplain and piedmont sedimentation in the Hebridean province, Scotland. *Journal of Sedimentary Petrology*, **44**, 336–357.

Steel, R.J. 1988. Coarsening-upward and skewed fan bodies: symptoms of strike-slip and transfer fault movement in sedimentary basins. In: Nemec, W. & Steel, R.J. (eds) *Fan Deltas: Sedimentology and Tectonic Settings*. Blackie, Glasgow, 75–83.

Steel, R.J. & Thompson, D.B. 1983. Structures and textures in Triassic braided stream conglomerates ("Bunter" Pebble Beds) in the Sherwood Sandstone Group. North Staffs, England. *Sedimentology*, **30**, 341–367.

Thomas, R.G., Smith, D.G., Wood, J.M., Visser, J., Calverley-Range, E.A. & Koster, E.H. 1987. Inclined heterolithic sedimentation – terminology, description, interpretation and significance. *Sedimentary Geology*, **53**, 123–179.

Todd, S.P. 1989a. Stream-driven, high-density gravelly traction carpets: possible deposits in the Trabeg Conglomerate Formation, SW Ireland and some theoretical considerations of their origin. *Sedimentology*, **36**, 513–530.

Todd, S.P. 1989b. Role of the Dingle Bay Lineament in the evolution of the Old Red Sandstone of southwest Ireland. In: Arthurton, R.S., Gutteridge, P. & Nolan, S.C. (eds) *Role of Tectonics in Devonian and Carboniferous Sedimentation in the British Isles*. Yorkshire Geological Society, Occasional Publication, **6**, 35–54.

Todd, S.P. & Went, D.J. 1991. Lateral migration of sand-bed rivers: examples from the Devonian Glashabeg Formation, SW Ireland and the Cambrian Alderney Sandstone Formation, Channel Islands. *Sedimentology*, **38**, 997–1020.

Todd, S.P., Boyd, J.D. & Dodd, C.D. 1988a. Old Red Sandstone sedimentation and basin development in the Dingle Peninsula, southwest Ireland. In: McMillan, N.J., Embry, A.F. & Glass, D.J. (eds) *The Devonian of the World*. Canadian Society of Petroleum Geologists, Memoir **14**, vol. II, 251–268.

Todd, S.P., Williams, B.P.J. & Hancock, P.L. 1988b. Lithostratigraphy and structure of the Old Red Sandstone of the northern Dingle Peninsula, Co. Kerry, southwest Ireland. *Geological Journal*, **23**, 107–120.

Todd, S.P., Boyd, J.D., Sloan, R.J. & Williams, B.P.J. 1991. *Sedimentology and Tectonic Setting of the Siluro-Devonian Rocks of the Dingle Peninsula, SW Ireland*. British Sedimentological Research Group, Cambridge, Field Guide Series, **17**, 157 pp.

Tunbridge, I.P. 1981. Sandy high-energy flood sedimentation – some criteria for recognition, with an example from the Devonian of southwest England. *Sedimentary Geology*, **28**, 79–95.

Tunbridge, I.P. 1984. Facies model for a sandy ephemeral stream and clay playa complex; the Middle Devonian Trentishoe Formation of North Devon, U.K. *Sedimentology*, **31**, 697–715.

Turner, B.R. 1981. Possible origin of low angle cross-strata and horizontal lamination in Beaufort Group sandstones of the southern Karoo Basin. *Transactions of the Geological Society of South Africa*, **84**, 193–197.

Van Wagoner, J.C., Posamentier, H.W., Mitchum, R.M., Vail, P.R., Sarg, J.F., Loutit, T.S. & Hardenbol, J. 1988. An overview of the fundamentals of sequence stratigraphy and key definitions. In: Wilgus, C.K., Hastings, B.S., Kendall, C.G.StC., Posamentier, H.W., Ross, C.A. & Van Wagoner, J.C. (eds) *Sea-level Changes: an Integrated Approach*. Society of Economic Palaeontologists and Mineralogists, Special Publication **42**, 39–45.

Van Wagoner, J.C., Mitchum, R.M., Campion, K.M. & Rahmanian, V.D. 1990. *Siliciclastic sequence stratigraphy in well logs, cores and outcrops: concepts for high-resolution correlation of time and facies*. American Association of Petroleum Geologists, Methods in Exploration Series, 55 pp.

Walker, R.G. 1967. Turbidite sedimentary structures and their relationship to proximal and distal depositional environments. *Journal of Sedimentary Petrology*, **37**, 25–43.

Walker, R.G. 1975. Generalized facies models for resedimented conglomerates of turbidite association. *Bulletin of the Geological Society of America*, **86**, 737–748.

Wasson, R.J. 1977. Last-glacial alluvial fan sedimentation in the Lower Derwent Valley, Tasmania. *Sedimentology*, **24**, 781–799.

Wells, N.A. 1984. Sheet debris flow and sheetflood conglomerates in Cretaceous cool-maritime alluvial fans, South Orkney Islands, Antarctica. In: Koster, E.H. & Steel, R.J. (eds) *Sedimentology of Gravels and Conglomerates*. Canadian Society of Petroleum Geologists, Memoir **10**, 133–146.

Williams, G.E. 1971. Flood deposits of the sand-bed ephemeral streams of central Australia. *Sedimentology*, **17**, 1–40.

Williams, P.F. & Rust, B.R. 1969. The sedimentology of a braided river. *Journal of Sedimentary Petrology*, **39**, 649–679.

Wright, V.P. 1986 (ed.). *Palaeosols: Their Recognition and Interpretation*. Blackwell, Oxford, 315 pp.

Wright, V.P. & Robinson, D. 1988. Early Carboniferous floodplain deposits from South Wales: a case study of the controls on palaeosol development. *Journal of the Geological Society of London*, **145**, 847–857.

10 Reconstructing Fluvial Channel Morphology from Sedimentary Sequences

C. BRISTOW
Department of Geology, Birkbeck College, University of London, UK

FLUVIAL CHANNEL CHANNEL MORPHOLOGY

Fluvial channel morphology is three-dimensional (Richards, 1982) but usually described in terms of planform, cross-section or long profile which are two-dimensional. In modern rivers these features can be measured directly; in contrast, measuring ancient fluvial morphology is more difficult. Charts which represent the three-dimensional form of rivers are rare, and are generally only available for larger navigable rivers. Rock outcrops occasionally reveal channel cross-sections and rarely expose palaeochannels in plan while the long profile is almost never revealed. Other sources of geological information such as boreholes give an even less complete record and cannot directly reveal any of the primary descriptors: planform, cross-section or long profile. High resolution geophysical methods may be able to map anomalies which relate to changes in lithology but they can rarely discriminate individual channels.

The interpretation of fluvial channel morphology from ancient sequences is very difficult. It requires good description, scientific reasoning and a certain amount of imagination constrained by observation. Furthermore, every modern river is different. Each is a natural system with varying slope, discharge, sediment etc. which combine in a unique form. This means that every ancient sandbody will be unique and that exceptions to every model should be expected. This chapter describes some approaches which can be used to interpret fluvial morphology from sedimentary sequences and highlights several problems and areas of uncertainty. Commonly used physical descriptors of fluvial morphology which can be measured in modern rivers are discussed first with the emphasis on a quantitative reconstruction of palaeochannel form.

QUANTITATIVE MEASUREMENTS OF MORPHOLOGY

Channel width

Channel width is usually measured perpendicular to the bank (Figure 10.1) In the ancient, width can be measured on outcrops with exceptional preservation. However,

Advances in Fluvial Dynamics and Stratigraphy. Edited by P.A. Carling and M.R. Dawson.
© 1996 John Wiley & Sons Ltd.

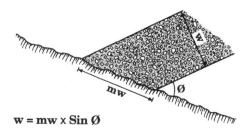

$$\mathbf{w} = \mathbf{mw} \times \mathbf{Sin}\ \varnothing$$

Figure 10.1   In modern rivers, channel width is usually measured perpendicular to the bank, while the width of a palaeochannel at outcrop depends on the orientation of the palaeochannel relative to the orientation of the cliff.

the width measured at outcrop will depend on the orientation of the palaeochannel relative to the orientation of the cliff. Reconstructing channel width, w, from an outcrop can be determined as $w =$ measured width (mw) $\times \sin \phi$ where ϕ is the acute angle between the channel edge and the cliff (Figure 10.1). The angle ϕ can be measured from the outcrop but it can also be derived from palaeocurrent measurements within the sandbody.

Channel depth and cross-sections

Another commonly used channel descriptor is depth. Depth is relatively easy to measure in a modern river but varies within a channel. Two measurements of depth are usually used, maximum depth and mean depth. Maximum depth can be determined by direct measurement, mean depth requires many measurements or a continuous profile of the channel bed. Maximum scour depth can be considered to be equivalent to preserved channel sandbody thickness although this may be reduced slightly by compaction and it is possible that there was greater scour outside of the plane of section. There are additional complications arising from partial preservation where channels are stacked and the top surface cannot be readily identified. Mean depth in modern rivers is usually calculated as cross-sectional area/channel width. In ancient channels this can sometimes be achieved given suitable outcrops and correcting for orientation. However, it should be remembered that channel depth is not the same as flow depth. Channels are rarely full to the top of their banks and flow depth varies temporally with changes in discharge.

Some rivers have stepped cross-sections which appear to correspond with different levels of discharge (McGowen & Garner, 1970; Thorne *et al.*, 1993). It has recently been suggested by Thorne *et al.* (1993) that the dominant discharge is equivalent to the top of mobile sand bars in the Brahmaputra River whereas bankfull discharge is equivalent to the level of vegetated river banks and the tops of stabilized vegetated islands or chars within the river. This observation may allow flow depth at dominant discharge to be calculated from ancient rocks where the top of sandy bar forms is taken as the dominant discharge depth (Figure 10.2). Flow depth at bankfull

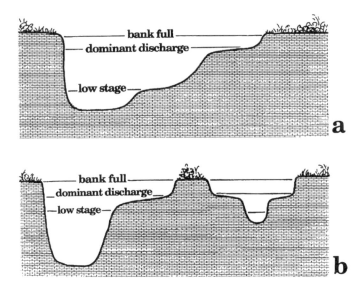

Figure 10.2   The relationship between flow stage and channel cross-section for meandering and braided rivers: (a) modified from McGowan & Garner (1970), and (b) based on the Brahmaputra River. Both sections are greatly simplified and drawn with considerable vertical exaggeration

discharge can be calculated from the top of associated in-channel fines and floodplain sediments while water depth at low flow discharge may be correlated with lower topographic levels within the river. Stepped cross-sections are occasionally preserved within the rock record (Hirst, 1991) allowing a more subtle interpretation of discharge and river regime.

Long profile

The long profile of rivers can be expressed in three ways: water surface slope, channel bed slope and valley slope. Measuring these parameters from the rock record is almost impossible so an alternative approach is required. Two methods are suggested. Basin slope, approximately equivalent to valley slope, can be estimated from the changes in thickness within a sedimentary wedge defined by two isochronous palaeosurfaces. However, as Miall (1991a) pointed out there is often a break of slope between the shelf and the fluvial systems so that the basin slope may overestimate fluvial valley slope. An alternative approach is to estimate channel slope using hydraulic equations which have been reviewed in Ethridge & Schumm (1978), while Leopold & Wolman (1957) have suggested that discharge and slope are discriminants in determining channel pattern.

Planform

Rivers are generally characterized by their appearance in planform and divided into four categories: straight, meandering, braided and anastomosed. The division of rivers into four end members based or planform is not entirely justified given that there is a continuum of channel pattern (Figure 10.3). The definitions of the channel patterns are variable and there is as yet no single descriptor of channel pattern which can be successfully fitted as a discriminant between all four channel types. Straight rivers can be seen to grade into sinuous and meandering rivers while straight and sinuous rivers can develop bars and become braided. Anastomosed rivers can have straight or sinuous anabranches, and in some cases it is possible for parts of braided rivers to become anastomosed. However, there is no better classification system available at present and the existing terminology will be followed here, although the continuum of form must be borne in mind when attempting to reconstruct channel morphology from sedimentary sequences.

Straight channels often contain alternate bars (Chang *et al.* 1971; Ikeda, 1984; Jaeggi, 1984), which are widely recognized in artificially straightened channels but rarely feature in the sedimentological literature. Bridge (1985) suggested that alternate bars are fundamental features of rivers and showed that many other bar forms may be evolved from single or multiple rows of alternate bars. However, alternate bars rarely feature in fluvial facies models and there are very few examples

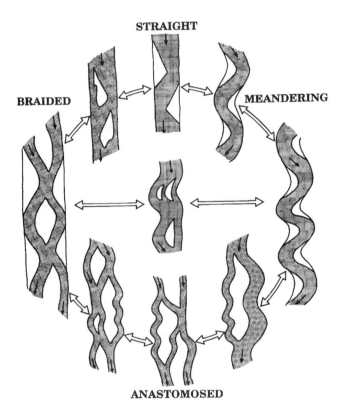

Figure 10.3 Rivers are usually described in terms of planform with reference to four end members: straight, meandering, braided and anastomosed, although there is a continuum between all four types

of alternate bars described in the rock record. The model of McCabe (1977) contains many of the features that one would expect to find on alternate bars in a straight channel.

Meandering rivers have a sinuous channel pattern where the degree of sinuosity can be described by the radius of the channel bend or by dividing the length of the river by the length of the valley through which it flows (Leopold & Wolman, 1957). Sedimentary models for meandering rivers include the classic model of Allen (1964) which presented a fining-upward point-bar deposit with associated abandoned channel deposits and floodplain sediments in an idealized block model. The broad features of this model are still useful and studies of modern meandering rivers by McGowan & Garner (1970), Bluck (1971), Jackson (1976, 1978, 1981), Nanson (1980), Arche (1983), Smith (1987) and Thomas *et al.* (1987) have provided additional details and helped to characterize the variety of meandering river deposits. Simulation models of meandering rivers (Willis, 1989) together with ancient examples which reflect the same features (Diaz-Molina, 1993; Willis, 1993a) show how well understood meandering systems have become.

A braided river is "characterized by having a number of alluvial channels with bars and islands between meeting and dividing again, and presenting from the air the intertwining effect of a braid" (Lane, 1957). The degree of braiding is characterized by a braiding index based either on the length of bars or islands in a channel reach (Brice (1964) modified by Rust (1978)) or the total length of the channels divided by the length of the widest channel in the river (Friend & Sinha, 1993). Descriptions of modern braided rivers can be found in Williams & Rust (1969), Cant & Walker (1978) and Bristow (1987, 1993), while scaled laboratory models of braided rivers are presented in Ashmore (1982, 1991). The processes of braiding and depositional models for braided river alluvium have been recently reviewed in Bridge (1993a) who also presents some new models for braided river deposition (see also Bristow & Best (1993)). The most recent models of Bridge (1993a) are based on the Calamus River described by Bridge *et al.* (1986) and Bridge & Gabel (1992). Bridge argues that bend expansion and translation seen in meandering rivers can be found in braided rivers and uses a simple downstream translation of braided channels to generate depositional morphology. Bridge (1993a) stresses the importance of examining the models from different angles to appreciate the complex internal geometry of braided river deposits. The models of Bristow & Best (1993) (Figure 10.4) are based on flume tank models of gravel-bed rivers (Ashmore, 1991) and illustrate hypothetical sections through fundamental braided river bar types. To date, there is no simulation model of braided rivers to correspond with those developed for meandering rivers, but Willis (1993b) has made a quantitative reconstruction of channel geometry and even estimated braiding index and palaeohydrology.

Anastomosing rivers are poorly defined but they are characterized by the presence of multiple channels which divide and rejoin in an independent manner. The distance between the division and rejoining of channels is greater than that in braided rivers and exceeds the wavelength of a meander or a mid-channel bar, so that the channels are seen to behave in an independent fashion. Anastomosed rivers are believed to accrete vertically and to be characterized by vertical aggradation of both channel and overbank deposits with limited lateral migration and frequent avulsion (Smith &

356

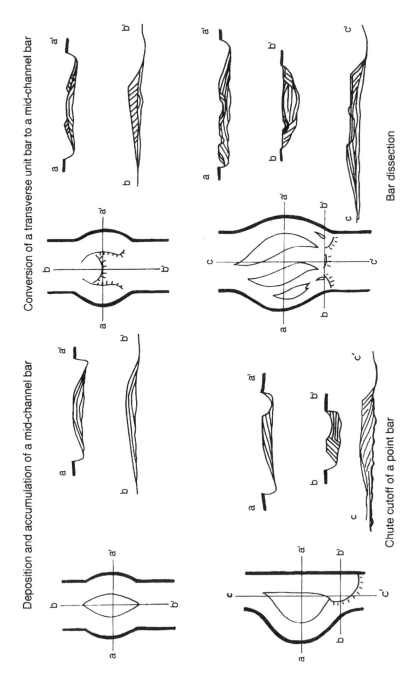

Figure 10.4 Cartoons of depositional morphology and hypothetical cross-sections based on scale models of gravel-bed braided rivers (Ashmore (1991), modified from Bristow & Best (1993)

Putnam, 1980). Recent examples from Canada have been described by Smith & Smith (1980) and examples from the Okavango Fan have been described by McCarthy *et al.* (1991, 1992) where the presence of aquatic vegetation is believed to be a significant control on channel behaviour.

TECHNIQUES: HOW TO RECONSTRUCT FLUVIAL MORPHOLOGY

A one-dimensional view: vertical sequence analysis

The idea that a fining-upward sequence would be produced by the lateral migration of a meandering channel was adopted as a model for fluvial sediments by Allen (1964). Allen (1965) provided additional explanation of how this sequence could be formed through a review of recent fluvial sediments. Miall (1977) reviewed braided river facies models and established four vertical profile models which "represent the most commonly occurring facies associations". In 1978 Miall established six vertical profile models based on modern rivers: the Trollheim, Scott, Donjek, South Saskatchewan, Platte and Bijou Creek. These models have been used for many sedimentological interpretations and because they are based on modern rivers with a known channel pattern we can be confident that they are representative of a particular river. There are problems with the models because they represent idealized sequences and cannot therefore reflect the variability present within the modern rivers, nor do they provide an infallible model because similar vertical profiles can be formed in different environments and are therefore not unique (see below). However, vertical profiles are well suited to the interpretation of borehole data such as core and logs where it is not possible to make observations of lateral facies changes or sandbody geometry. Vertical sequence analysis is also appropriate for the interpretation of sedimentary logs which are still the fundamental method of collecting and presenting field data. An additional advantage of vertical profiles is that they lend themselves to statistical analysis (Selley, 1970; Miall, 1973).

Unification and proliferation

The vertical profile models of Miall (1977) are not unique and Jackson (1978) and Bridge (1985) have pointed out the convergence of coarse-grained braided river profiles and coarse-grained meandering river models. There are similar problems with sand-bed rivers where a vertical profile though a sandy point bar in a meandering river will be very similar to a vertical profile though a braid bar with similar grain size and sedimentary structures. A vertical sequence is not unique to any given channel geometry but simply reflects the vertical arrangement of grain size and sedimentary structures. There may be a link between vertical sequence and channel pattern but this is by no means a unique solution. Another problem with the vertical profile is the potential for proliferation: with the description of each new river a new facies model would appear. Allen (1983) summed up the situation thus: "Channel behaviour and type can practically never be predicted unambiguously from vertical sequences, if only because each kind of stream is capable of generating a wide variety of local sedimentological patterns. In the search for unequivocal criteria, attention has gradually shifted toward the shape and larger scale internal geometry of fluvial sandstone bodies".

Vertical profiles do not cope well with lateral variations in facies. Some sedimentologists have tried to overcome this problem by measuring multiple vertical profiles (Bristow & Myers, 1989) but this is very time-consuming. An alternative is to sketch lateral facies changes into the logs or to make thumbnail sketches of lateral changes in the margin of the log. Allen (1983) used photomosaics of outcrops and outcrop sketches to depict the changes in sedimentary structures and facies found in the Brownstones, which led to the development of alluvial architecture analysis.

A two-dimensional view: architectural element analysis

Problems associated with vertical sequences, the lateral variation in facies and a demand from the petroleum industry for predictions of lateral variations and continuity in the subsurface led to the development of architectural analysis (Allen, 1983), where lateral profiles are used to assess the geometry of deposits and reconstruct the original depositional form. This approach became codified by Miall (1985) into architectural element analysis and in the process lost the emphasis on the use of geometry as a primary discriminant, changing the emphasis to a hierarchy of bounding surfaces and the composition of depositional elements. The use of architectural element analysis is ably demonstrated by Miall (1988b, 1993), although the mechanics of the approach have been criticized by Bridge (1993b) because the terminology used mixes texturally defined lithofacies with geometrically defined elements. Bridge (1993b) advocates the use of mutually exclusive terminology for clear description, divorcing the descriptive terms from the interpretation, and suggests geometric names for sediment body geometries. This is fine in principle but will rarely be achieved in practice since the geometries are usually complex shapes and the geologists' view of an outcrop will rarely reveal the true geometry. Another useful descriptive system has been developed by Ramos & Sopena (1983) and Ramos et al. (1986) where the descriptive facies includes grain size, sedimentary structures and bed geometry as the basic building blocks for outcrop descriptions. This is standard facies analysis but with the inclusion of geometry, which was lacking from the vertical profile models. In many respects the depositional facies of Ramos & Sopena (1983) is similar to the architectural elements described by Allen (1983) and both contain a component of description and interpretation in their classifications.

Bounding surfaces

Allen (1983) describes a hierarchical set of bedding contacts, particularly discordant erosional surfaces from the Brownstones which he claimed reflect the geometrical components and constructional patterns of sandstone bodies. In Allen's classification the zeroth-order contacts are non-erosional and concordant such as the contact between strata and laminae. First-order contacts bound such entities as individual trough cross-stratified sets or bundles of plane-bed laminae genetically associated with cross-strata. Second-order contacts bound clusters of sedimentation units of the kinds delineated by first-order contacts. These groupings Allen terms complexes which are similar to the storeys of Friend *et al* (1979). Third-order contacts divide groupings of complexes from each other and define the sandbodies themselves. These are equivalent to channel erosion or channel-scale accretion surfaces. This classification system ranks the surfaces from the smallest, zeroth, to the largest, third, in opposition to the precedent set by Brookfield (1977) in the classification of aeolian bounding surfaces. Miall (1985) adopted the bedding-contact hierarchy of Allen (1983), adding a fourth category of bounding surface to include stacked channel sandstones or groups of channels as in a palaeovalley. Subsequently Miall (1988b) has redefined the second order bedding contact into new second-, third- and fourth-order bounding surfaces thereby raising the third- and fourth-order surfaces to fifth- and sixth-order. The new categories raised by Miall include simple coset bounding surfaces as second-order surfaces, large-scale reactivation surfaces as third-order surfaces and the upper bounding surfaces of macroforms as fourth-order surfaces. Miall (1991b) attempts to translate these spatial relationships into a temporal scale which is probably unsound given the incomplete nature of the rock record.

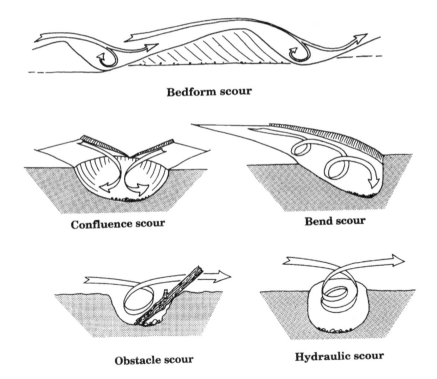

Bedform scour

Confluence scour   **Bend scour**

Obstacle scour   **Hydraulic scour**

Figure 10.5 Erosional features are common in rivers and occur over a wide range of scales. They can usually be linked to a primary process such as bedform migration, confluence scour, bend scour, obstacle scour or hydraulic scour, as shown here. Other examples related to changes in discharge and incision are shown in Figures 10.7 and 10.9 respectively

The imposition of a numerical classification system has created problems because the numbers allocated to a given surface have been changed by the principal author and because numbers are not as easy to assimilate as good clear descriptive terminology (Bridge, 1993b). Here is a suggestion for terminology which is available to replace the numbers.

Erosion surfaces

Erosion surfaces are widely recognized at a variety of scales from scour in the trough of a dune bedform, through channel erosion surfaces to palaeovalleys. In each case the surface can be linked to a primary process, bedform migration, channel migration and incision, and can be given a simple descriptive name linked to a naturally occurring geomorphic feature (Figure 10.5). This system of terminology is easy to use, providing the basis for clear communication between geologists and geographers, and can be quantified where required (Figure 10.6). There are minor problems of terminology where an erosion surface becomes an accretion surface. For example, during a flood event the water level in a river rises and the bed is scoured creating an erosion surface (Figure 10.7); during the succeeding falling stage deposition occurs and the erosion surface becomes an accretion surface. This

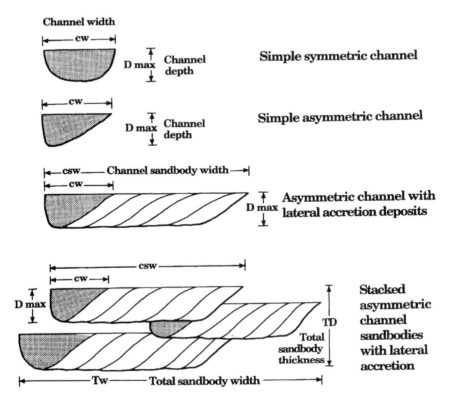

Figure 10.6 Quantification of channels, channel sandbodies and stacked channel sandstones using simple descriptive terminology and measurements

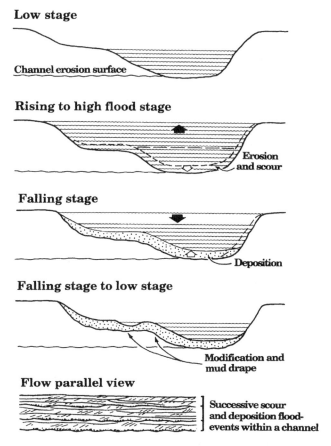

Low stage

Channel erosion surface

Rising to high flood stage

Erosion
and scour

Falling stage

Deposition

Falling stage to low stage

Modification and
mud drape

Flow parallel view

Successive scour
and deposition flood-
events within a channel

Figure 10.7 Cycle of erosion and sedimentation within a fluvial channel during a flood event

must be the case for all erosion surfaces, otherwise they would never be preserved. Another potential problem is where an internal erosion surface on a macroform passes into a channel erosion surface which locally acts as a palaeovalley erosion surface. This may sound complex but it is much easier to understand than a third-order surface passing into a fifth-order surface which locally acts as a sixth-order surface. Leaving out the numbers makes description clearer and avoids the geologist/geographer having to learn a new language.

Accretion surfaces

Accretionary surfaces in fluvial sediments are less well defined. For bedform migration we have the terms laminae, foreset, set and coset to describe the smaller-scale accretionary units of ripple, dune and plane-bed bedforms (McKee & Weir, 1953). These terms are well defined in the literature and although the original definitions are not always applied, these terms are in widespread use. This

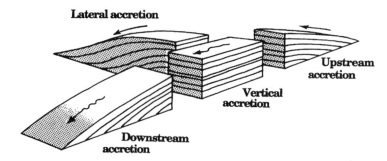

Figure 10.8 Upstream, downstream, vertical and lateral accretion units on a hypothetical bar bounded by upstream, downstream, vertical and lateral accretion surfaces respectively

terminology refers to the objects themselves and not the contacts between them. In this case the zeroth order contacts of Allen (1983) are original, although a better term could be adopted. In aeolian sediments Hunter (1977) uses the term "primary surface" which is advocated by Fryberger (1993). The larger bounding surfaces due to accretion on bars have attracted more diverse terminology: epsilon cross-stratification (Allen, 1963), lateral accretion surfaces or inclined heterolithic stratification (Thomas *et al.*, 1987) while the sediments bounded by these surfaces are known as storeys (Friend *et al.*, 1979), lateral accretion beds or inclined strata (Thomas *et al.*, 1987). In each case the surface which defines accretion onto a bar is being described so that accretion surface would seem to be an acceptable name. The term "growth surface" has recently been adopted in the description of aeolian bounding surfaces (Fryberger, 1993) but accretion surface is preferred here due to the biological connotations of a growth surface. Following this, the first-order surfaces of Allen (1983) are equivalent to bedding planes while second-order surfaces can be described as erosion or accretion surfaces within channels and third-order surfaces are equivalent to channel erosion surfaces. Examples of upstream, downstream, vertical and lateral accretion units bounded by upstream, downstream, vertical and lateral accretion surfaces are shown in Figure 10.8.

TOWARDS A THREE-DIMENSIONAL PERSPECTIVE

The integration of flow and channel pattern (Bridge 1985, 1993b) leads to the conclusion that similar morphodynamic features will be found as components of the overall fluvial system regardless of channel pattern. For instance it is suggested that braided rivers include curved channel segments which divide and rejoin around bars in a regular or repeatable pattern. The curved channel segments appear to have similar flow characteristics to curved channels in single channel meandering rivers (Bridge *et al.*, 1986). It is becoming increasingly clear that the similarities between fluvial channel types are as important as the differences. The evidence comes from both theoretical models of the origins of channel pattern where both meandering and braided channels can be evolved from straight channels with alternate bars (Bridge, 1985), and from observations of fluvial deposits. Lateral accretion deposits, once

Table 10.1 Form and process in relation to channel pattern

Fundamental morphodynamic mechanisms	Channel pattern		
	straight	meandering	braided
Channel cut and fill	yes	yes	yes
Lateral migration of channels	yes	yes	yes
Formation and migration of alternate bars	yes	yes	yes
Bend expansion and translation with point-bar formation	no	yes	yes
Meander cut-off and oxbow lake formation	no	yes	yes
Development of chute channels	yes	yes	yes
Deposition and accumulation of a central bar	yes	yes	yes
Chute cut-off of point bars	no	yes	yes
Conversion of transverse bar to mid-channel bar	yes	yes	yes
Disection of multiple bars	no	no	yes

the preserve of meandering rivers, are now widely recognized within braided river deposits. The discrimination of channel pattern will not depend on the presence or absence of such features but more on the relative abundance. For instance in an analysis of recent deposits of the Brahmaputra River exposed at low-flow stage Bristow (1987) determined that 53% was accretion to bars, 19% accretion to the river bank, 15% channel fill and 13% new medial bars. Similar quantitative studies of other rivers and simulation experiments are required to project the likely composition of fluvial sediments based on the proportions of morphological elements with respect to channel pattern.

In his flume tank experiments on simulated gravel-bed braided rivers, Ashmore (1991) indentified four braiding mechanisms: deposition and accumulation of a central bar, chute cut-off of point bars, conversion of single transverse unit bars to mid-channel braid bars, and dissection of multiple bars. Ashmore suggests that the central bar, transverse bar conversion and dissection mechanisms are all associated with flow divergence whereas chute-cut off is associated with thalweg shoaling due to a local over-supply of bedload sediment. For straight channels the fundamental morphodynamic mechanisms are channel cut and fill, lateral migration of channels and formation and migration of alternate bars. Within meandering rivers the fundamental meandering mechanisms can be summarized as bend expansion and translation with point-bar formation, meander cut-off and oxbow lake formation. The development of chute channels is secondary and probably due to discharge fluctuations. Tabulating these mechanisms and their occurrence with respect to channel pattern it is apparent that apart from dissection of multiple bars the other morphologic features are not exclusive to any channel pattern (Table 10.1).

CHANNEL HIERARCHIES

The presence of a hierarchy of channels within braided rivers was first suggested by Williams & Rust (1969), who described three orders of channel in addition to a series of levels within the river which represented active and inactive parts of the channel system. In the scheme proposed by Williams & Rust the whole river and the active channels were termed the "composite stream channel" and the "stream channel" respectively, adding two additional levels to the hierarchy. This system was modified by Bristow (1987) to a three-fold hierarchy; Bridge (1993a) suggests additional modifications to the hierarchy concept. If one accepts that the river can operate as a single entity with channels within it and that there may be different scales of channels depending on the discharge, then a three-fold hierarchy of channels is required. The first order comprises the whole river. Second-order channels are the dominant channels within the river, and third-order channels are primarily low-stage features which modify the bars deposited by the second-order channels. One implication is that ancient braided-river sandstones should also show a hierarchy of channel dimensions, whereas stacking of a single-channel river is more likely to produce similar-sized sandbodies. The presence of different magnitude channels within a single sandbody could therefore be an indication of braiding. One must be careful, however, because if the third-order channels are dependent on stage changes then the presence of several scales of channel could be due primarily to changes in stage. It has been shown above that fluctuations in stage are not necessarily associated with braiding. Therefore, a hierarchy of channel sandbodies is not an exclusive indicator of a braided river but is most likely to occur in a large braided river with a fluctuating discharge.

AGGRADATION AND PRESERVATION

Facies models for rivers are largely based on sections measured though exposed bars or river banks (Williams & Rust, 1969; McGowan & Garner, 1970; Collinson, 1970; Miall, 1977, 1978; Cant & Walker, 1978; Bristow, 1993). Given perfect instantaneous preservation these will be representative of fluvial sediments. However, lateral migration and vertical accretion within channel belts can rework the sediments and the final deposit will be modified. It is possible to conjecture that at least one bar-top sequence should be preserved after the final avulsion or migration of a channel, but in many cases the sandbody may be largely made up of the deeper channel deposits in so-called stacked channel sandstones (Friend, 1983). The deepest part of most rivers, the thalweg, is found at the outsides of channel bends and at zones of constricted or convergent flow such as confluence scours (Salter, 1993). Because of the large number of confluences within braided rivers they are believed to be an important component of braided-river alluvium (Bristow et al., 1993; Salter, 1993). Siegenthaler & Huggenberger (1993) argue that with low rates of aggradation, channels will be stacked and there will be preferential preservation of the lower topographic (deeper) parts of a river. Thus confluence scour is presented as a dominant depositional element in Rhine gravels (Siegenthaler & Huggenberger, 1993). The scour associated with channel bends should not be ignored; scour studies in the Brahmaputra River (Klaassen & Vermeer, 1988) indicate that the deepest natural scours occur at channel bends and confluences so these should both have a higher preservation potential than the shallow bar tops. Unfortunately these portions

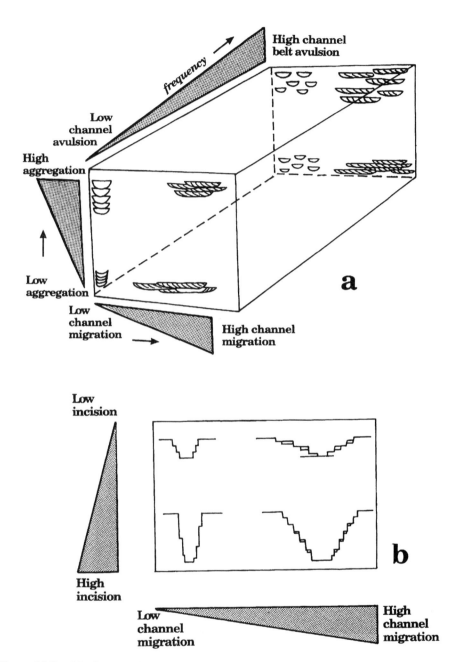

Figure 10.9 (a) Cartoon illustrating the preservation of sandbodies as a function of aggradation rate, lateral channel migration and channel-belt avulsion. The preservation of isolated channel sandbodies preserving morphology requires frequent avulsions as well as rapid aggradation. (b) For incising river systems the rate of incision and rate of migration control the number of terraces and the preservation of fluvial sediments

of modern rivers are rarely exposed and the deeper sections of facies models are sometimes conjectural. Facies models for confluence scours based on flume tank experiments and field observations have been presented by Best (1987) and Bristow *et al.* (1993) and ancient examples are suggested by Cowan (1991) and Siegenthaler & Huggenberger (1993).

Preservation is also a function of river behaviour. Meandering rivers tend to migrate across the floodplain gradually reworking older deposits and with continued slow aggradation may preserve several thalweg deposits overlain by one point bar. Aggrading braided river systems will also rework earlier deposits. In both cases, as the rate of aggradation increases there is a better chance of preserving higher topographic levels within the fluvial deposits. The relative frequency of avulsion events and rates of aggradation will affect the internal geometry as well as the external stacking pattern. A cartoon of aggradation vs. migration shows that while fines preservation increases with increased aggradation, the nature of the channel deposits should not change greatly (Figure 10.9a). The most significant changes occur with an increase in the avulsion frequency. Avulsions may be initiated by major floods or tectonic events. In alluvial architecture simulation models the rate of avulsion is random (Leeder, 1978) or follows a Weibull distribution (Bridge & Leeder, 1979) due to lack of data. Avulsion frequency appears to increase with high rates of sediment accumulation (Bridge & Leeder, 1979), since this is most likely to lead to rapid establishment of local gradients favouring avulsion. The preservation of form due to increased avulsion frequency may be associated indirectly with a high rate of accretion due to the expected increase of avulsion frequency under conditions of rapid aggradation. A similar model can be drawn for incising fluvial systems (Figure 10.9b). Here the balance between migration rate and incision clearly affects the preservation of fluvial sediments on terraces and the number of terraces generated by a degrading fluvial system. In this case terrace formation occurs when sediment supply is equal to or greater than the transport power of the stream. Under these conditions accretion occurs or an equilibrium is maintained. Erosion and incision occur when the rate of sediment supply is less than the transport power of the stream. In the degrading system it is assumed that avulsion is less likely.

CONCLUSIONS

The reconstruction of fluvial channel morphology is an engaging subject full of potential pitfalls due to the non-unique nature of rivers and the inherent lateral variation within rivers. Models should be based on modern depositional environments because here we can verify processes and confirm the relationship between planform and deposits. It is clear that while the quantitative reconstruction of palaeochannels is highly desirable, it cannot always be achieved. In many cases the potential for interpretation is limited by the data available. Facies models have some drawbacks because they are idealized and cannot therefore incorporate all of the variations found in natural systems. Every modern river is unique and therefore every ancient sandstone will be unique. The descriptive terminology developed to cope with this situation has to be very flexible. Conventional facies analysis is adequate but it is essential that geometry is included as one of the facies descriptors along with lithology, grain size and sedimentary structures. The elaborate terminology of architectural element analysis and bounding surface hierarchies is unnecessary because there is sufficient terminology already available for description and interpretation. Vertical profiles have been proved as a robust method for interpretation and are suited to the description and interpretation of vertical sections such as borehole cores. Lateral profiles are becoming increasingly popular for outcrop studies but care needs to be taken in assessing the variation of form as a function of aspect and the orientation of outcrop with respect to palaeoflow. Strictly speaking the reconstruction of morphology requires three-dimensional information which is almost never available to the sedimentologist, and a certain amount of interpolation between data points is necessary. While simple presence or absence criteria for the interpretation of channel pattern based on a single trait does not work, it is possible that proportions of different components may be significant. More quantitative studies of the area, dimensions, volumes and geometries of recent fluvial sediments are required to provide the building blocks for reconstructing ancient fluvial sequences. It should be remembered that each solution will be unique but that similarities may be found between fluvial systems that can be used as common reference points in discussion. Partial preservation of fluvial channel deposits is likely to lead to preservation of topographically lower parts of channel deposits. The deepest scours in modern rivers occur at bends or confluence scours and yet these are poorly represented in facies models or channel reconstructions. Preservation is also a function of channel behaviour, aggradation and external influences such as sediment supply, subsidence and base-level changes.

REFERENCES

Allen, J.R.L. 1964. Studies in fluviatile sedimentation: six cyclothems from the Lower Old Red Sandstone, Anglo Welsh Basin. *Sedimentology* 3, 163–198.

Allen, J.R.L., 1965. A review of the origin and character of recent alluvial sediments. *Sedimentology* 5, 89–191.

Allen, J.R.L. 1983. Studies in fluviatile sedimentation: bars, bar-complexes and sandstone sheets (low sinuosity braided streams) in the Brownstones (L. Devonian), Welsh Borders. *Sedimentary Geology* 33, 237–293.

Arche, A. 1983. Coarse-grained meander lobe deposits in the Jarama River, Madrid, Spain. In: Collinson, J.D. & Lewin, J. (Eds) *Modern and Ancient Fluvial Systems*. International Association of Sedimentologists, Special Publication 6, 345–354.

Ashmore, P.E. 1982. Laboratory modelling of gravel braided stream morphology. *Earth Surface Processes*, 7, 201–225.

Ashmore, P.E. 1991. How do gravel-bed rivers braid? *Canadian Journal of Earth Science* 28, 326–341.

Best, J.L. 1987. Flow dynamics at river channel confluences: implications for sediment transport and bed morphology. In: Ethridge, F.G., Flores, R.M. & Harvey, M.D. (Eds) *Recent Developments in Fluvial Sedimentology*. Society of Economic Paleontologists and Mineralogists, Special Publication 39, 27–35.

Bluck, B.J. 1971. Sedimentation in the meandering River Endrick. *Scottish Journal of Geology*, 7, 93–138.

Brice, J.E., 1964. *Channel patterns and terraces of the Loup River in Nebraska.* United States Geological Survey Professional Papers, **422-D**.

Bridge, J.S. 1985. Palaeochannel patterns inferred from alluvial deposits: a critical evaluation. *Journal of Sedimentary Petrology* **55**, 579–589.

Bridge, J.S. 1993a. The interaction between channel geometry, water flow, sediment transport and deposition in braided rivers. In: Best, J.L, & Bristow, C.S. (Eds) *Braided Rivers.* Geological Society, Special Publication, **75**, 13–71.

Bridge, J.S. 1993b. Description and interpretation of fluvial deposits: A critical perspective. *Sedimentology* **40**, 801–810.

Bridge, J.S. & Gabel, S.L. 1992. Flow and sediment dynamics in a low sinuosity river: Calamus River, Nebraska Sandhills. *Sedimentology* **39**, 125–142.

Bridge, J.S. & Leeder, M.R. 1979. A simulation of alluvial stratigraphy. *Sedimentology* **26**, 617–644.

Bridge, J.S., Smith, N.D., Trent, F., Gabel, S.L., & Bernstein, P. 1986. Sedimentology and morphology of a low sinuosity river: Calamus River, Nebraska Sandhills. *Sedimentology* **33**, 851–870.

Bristow, C.S. 1987. Brahmaputra River: Channel migration and deposition. In: Ethridge, F.G., Flores, R.M. & Harvey, M.D. (Eds) *Recent Developments in Fluvial Sedimentology.* Society of Economic Paleontologists and Mineralogists, Special Publication **39**, 63–74.

Bristow, C.S. 1993. Sedimentary structures exposed in bar tops in the Brahmaputra River, Bangladesh. In: Best, J.L, & Bristow, C.S. (Eds) *Braided Rivers.* Geological Society, Special Publication **75**, 277–289.

Bristow, C.S. & Best, J.L. 1993. Braided rivers: perspectives and problems. In: Best, J.L. & Bristow, C.S. (Eds) *Braided Rivers.* Geological Society, Special Publication **75**, 1–11.

Bristow, C.S. & Myers, K. 1989. Detailed sedimentology and gamma-ray log characteristics of a Namurian deltaic succession 1: Sedimentology and facies analysis. In: Whately, M.K.G. & Pickering, K.T. (Eds) *Deltas: Sites and Traps for Fossil Fuels.* Geological Society, Special Publication **41**, 75–80.

Bristow, C.S., Best, J.L. & Roy, A.G. 1993. Morphology and facies models of channel confluences. In: Puigdefabregas & Tomas (Eds) International Association of Sedimentologists, Special Publication **17**, 91–100.

Brookfield, M.E. 1977. The origin of bounding surfaces in ancient aeolian sandstones. *Sedimentology* **24**, 303–332.

Cant, D.J. & Walker, R.G. 1978. Fluvial processes and facies sequences in the sandy braided South Saskatchewan River, Canada. *Sedimentology*, **25**, 625–648.

Chang, H.Y., Simons, D.B. & Woolhiser, D.A. 1971. Flume experiments on alternate bar formation. *Journal of the Waterways, Harbours and Coastal Engineering Division, ASCE*, **79**, 155–165.

Collison, J.D. 1970. Bedforms in the Tana River, Norway. *Geografiska Annaler* **52**, 31–56.

Cowan, E.J. 1991. The large scale architecture of the fluvial Westwater Canyon Member, Morrison Formation (Jurassic), San Juan Basin, New Mexico. In: Miall, A.D. & Tyler, N. (Eds) *The three-dimensional facies architecture of terrigenous clastic sediments, and its implications for hydrocarbon discovery and recovery.* Society of Economic Paleontologists and Mineralogists Concepts and Models Series, 80–93.

Diaz-Molina, M. 1993, Geometry and lateral accretion patterns in meander loops: examples from the Upper Oligocene-Lower Miocene, Loranca Basin, Spain. In: Marzo, M. & Puigdefabregas, C. (Eds) *Alluvial Sedimentation.* International Association of Sedimentologists, Special Publication **17**, p. 115–131.

Ethridge, F. & Schumm, S.A. 1978. Reconstructing paleochannel morphologic and flow characteristics: methodology, limitations and assessment. In: Miall, A.D. (Ed.) *Fluvial Sedimentology:* Canadian Society of Petroleum Geology, Memoir **5**, 703–721.

Friend, P.F. 1983. Towards the field classification of alluvial architecture or sequence. In: Collinson, J.D. & Lewin, J. (Eds), *Modern and Ancient Fluvial Systems.* International Association of Sedimentologists, Special Publication **6**, 345–354.

Friend, P.F. & Sinha, R. 1993. Braiding and meandering parameters, In: Best, J.L. & Bristow, C.S. (Eds) *Braided Rivers*, Geological Society, Special Publication **75**, 105–111.

Friend, P.F., Slater, M.J. & Williams, R.C. 1979. Vertical and lateral building of river sandstone bodies, Ebro Basin Spain. *Journal of the Geological Society of London* **136**, 39–46.

Fryberger, S.G. 1993. A review of aeolian bounding surfaces, with examples from the Permian Minnelusa Formation, USA. In: North C.P. & Prosser, D.J. (Eds) *Characterisation of Fluvial and Aeolian Reservoirs*. Geological Society, Special Publication **73**, 167–197.

Hunter, R.E. 1977. Basic types of stratification in small aeolian dunes. *Sedimentology* **24**, 361–387.

Hirst, J.P.P. 1991. Variations in alluvial architecture across the Oligo-Miocene Huesca fluvial system, Ebro Basin, Spain. In: Miall, A.D. & Tyler, N. (Eds) *The three-dimensional facies architecture of terrigenous clastic sediments, and its implications for hydrocarbon discovery and recovery*. Society of Economic Paleontologists and Mineralogists, Concepts and Models Series, 6–12.

Ikeda, S. 1984. Prediction of alternate bar wavelength and height. *Journal of Hydraulic Engineering Division ASCE* **110**, 371–386.

Jackson, R.G. II. 1976. Depositional model of Point Bars in the Lower Wabash River. *Journal of Sedimentary Petrology* **46**, 579–594.

Jackson, R.G. II. 1978. Preliminary evaluation of lithofacies models for meandering alluvial streams. In: Miall, A.D. (Ed.) *Fluvial Sedimentology*. Canadian Society of Petroleum Geologists, Memoir **5**, 543–576.

Jackson, R.G. II. 1981. Sedimentology of muddy fine-grained channel deposits in meandering streams of the American Middle West. *Journal of Sedimentary Petrology* **51**, 1169–1192.

Jaeggi, M.N.R. 1984. Formation and effects of alternate bars. *Journal of Hydraulic Engineering Division ASCE*, **110**, 142–156.

Klaassen, G.J. & Vermeer, K. 1988. Confluence scour in large braided rivers with fine bed material. *International Conference on Fluvial Hydraulics*, Budapest.

Lane, E.W. 1957. *A study of the shape of channels formed by natural streams flowing in erodible material*. MRD Sediment series **9**, United States Army Engineering Division, Missouri River, Corps Engineers, Omaha, Nebraska.

Leeder, M.R 1978. A quantitative stratigraphic model for alluvium, with special reference to channel deposit density and interconnectedness. In: Miall, A.D. (Ed.) *Fluvial Sedimentology*, Canadian Society of Petroleum Geologists Memoir **5**, 587–596.

Leopold, L.B. & Wolman, M.G. 1957. *River and channel patterns: braided, meandering and straight*. United States Geological Survey Professional Paper **282**, 39–85.

McCabe, P.J. 1977. Deep distributary channels and giant bedforms in the Upper Carboniferous of the Central Pennines, northern England. *Sedimentology* **24**, 271–290.

McCarthy, T.S., Stanistreet, I.G., and Cairncross, B. 1991. The sedimentary dynamics of active fluvial channels on the Okavango Fan, Botswana. *Sedimentology* **38**, 471–487.

McCarthy, T.S., Ellery, W.N. & Stanistreet, I.G. 1992. Avulsion mechanisms on the Okavango fan, Botswana: the control of a fluvial system by vegetation. *Sedimentology* **39**, 779–795.

McGowan, J.H. & Garner, L.E. 1970. Physiographic features and stratification types of coarse-grained point bars: modern and ancient examples. *Sedimentology* **14**, 77–111.

McKee, E.D. & Weir, G.W. 1953. Terminology of stratification and cross-stratification. *Bulletin of the Geological Society of America*, **64**, 381–390.

Miall, A.D., 1973. Markov chain analysis applied to an ancient alluvial plain succession. *Sedimentology* **20**, 347–364.

Miall, A.D. 1977. A review of the braided river depositional environment. *Earth Science Reviews* **13**, 1–62.

Miall, A.D. 1978. Lithofacies types and vertical profile models in braided river deposits: a summary. In: Miall, A.D. (Ed.) *Fluvial Sedimentology*. Canadian Society of Petroleum Geologists, Memoir **5**, 597–604.

Miall, A.D. 1985. Architectural element analysis: a new method of facies analysis applied to fluvial deposits. *Earth Science Reviews* **22**, 261–308.

Miall, A.D. 1988a. Facies architecture in clastic sedimentary basins. In: Kleinspehn, K.L. & Paola, C. (Eds) *Frontiers in Sedimentary Geology: New Perspectives in Basin Analysis.* Springer Verlag, New York, 67–81.

Miall, A.D. 1988b. Architectural elements and bounding surfaces in fluvial deposits: anatomy of the Kayenta Formation (Lower Jurassic), Southwest Colorado. *Sedimentary Geology* **55**, 233–262.

Miall, A.D. 1991a. Stratigraphic sequences and their chronostratigraphic correlation. *Journal of Sedimentary Petrology* **61**, 497–505.

Miall, A.D. 1991b. Hierarchies of architectural units in clastic rocks, and their relationship to sedimentation rate. In: Miall, A.D. & Tyler, N. (Eds) *The three-dimensional facies architecture of terrigenous clastic sediments, and its implications for hydrocarbon discovery and recovery.* Society of Economic Paleontologists and Mineralogists, Concepts and Models Series, 6–12.

Miall, A.D., 1993. Architecture of the Upper Mesaverde Group. In: Best, J.L, & Bristow, C.S. (Eds) *Braided Rivers.* Geological Society, Special Publication **75**, 305–332.

Nanson, G.C. 1980. Point bar and floodplain formation on the meandering Beatton River, British Colombia, Canada. *Sedimentology* **27**, 3–29.

Ramos, A. & Sopena, A. 1983. Gravel bars in low sinuosity streams (Permian and Triassic, central Spain). In: Collinson, J.D. & Lewin, J. (Eds) *Modern and Ancient Fluvial Systems.* International Association of Sedimentologists, Special Publication **6**, 301–313.

Ramos, A., Sopena, A., and Perez-Arlucea, M. 1986. Evolution of Buntsandstein fluvial sedimentation in the northwest Iberian ranges (Central Spain). *Journal of Sedimentary Petrology* **56**, 862–875.

Richards, K. 1982. *Rivers: form and process in alluvial channels.* Menthuen, London, 358 pp.

Rust, B.R. 1978. A classification of alluvial channel systems. In: Miall, A.D. (Ed.) *Fluvial Sedimentology.* Canadian Society of Petroleum Geologists, Memoir **5**, 187–198.

Salter, T. 1993. Fluvial scour and incision: models for their influence on the development of realistic reservoir geometries, In: North, C.P. & Posser, D.J. (Eds) *Characterisation of fluvial and aeolian reservoirs,* Geological Society of London Special Publication **73**, 33–52.

Selley, R.C. 1970. Studies of sequences in sediments using a simple mathematical device. *Geological Society of London Quarterly Journal* **125**, 557–581.

Siegenthaler, C. & Huggenberger, P. 1993. Pleistocene Rhine gravel: deposits of a braided river system with dominant pool preservation. In: Best, J.L, & Bristow, C.S. (Eds) *Braided Rivers.* Geological Society, Special Publication **75**, 147–162.

Smith, D.G. 1987. Meandering river point bar lithofacies, models: modern and ancient example, compared. In: Ethridge, F.G., Flores, R.M. & Harvey, M.D. (Eds) *Recent Developments in Fluvial Sedimentology.* Society of Economic Paleontologists and Mineralogists, Special Publication **39**, 83–91.

Smith, D.G. & Putnam, P.E. 1980, Anastomosed river deposits: modern and ancient examples in Alberta, Canada. *Canadian Journal of Earth Science* **17**, 1396–1406.

Smith, D.G. & Smith, N.D. 1980, Sedimentation in anastomosed river systems: examples from alluvial valleys near Banff, Alberta. *Journal of Sedimentary Petrology* **50**, 157–164.

Thomas, R.G., Smith, D.G., Wood, J.M., Visser, J., Calverly-range, E.A. & Koster, E. 1987. Inclined heterolithic stratification – terminology, description, interpretation and significance. *Sedimentary Geology* **53**, 123–179.

Thorne, C.R., Russell, A.P.G. & Alam, M.K. 1993. Planform pattern and channel evolution of the Brahmaputra River, Bangladesh. In: Best, J.L. & Bristow, C.S. (Eds) *Braided Rivers.* Geological Society, Special Publication **75**, 257–276.

Williams, P.F. & Rust, B.R. 1969. The sedimentology of a braided river. *Journal of Sedimentary Petrology* **39**, 649–679.

Willis, B.J. 1989. Palaeochannel reconstructions from point bar deposits: a three-dimensional perspective. *Sedimentology* **36**, 757–766.

Willis, B.J. 1993a. Interpretation of bedding geometry within ancient point-bar deposits. In: Marzo, M. & Puigdefabregas, C. (Eds) *Alluvial Sedimentation.* International Association of Sedimentologists, Special Publication **17**, 115–131.

Willis, B.J. 1993b. Ancient river systems in the Himalayan foredeep, Chingi Village area, northern Pakistan. *Sedimentary Geology* **88**, 1–76.

11 Depositional Controls on Primary Permeability and Porosity at the Bedform Scale in Fluvial Reservoir Sandstones

A.C. BRAYSHAW[1], G.W. DAVIES[2] AND P.W.M. CORBETT[3]
[1] *BP Exploration (Alaska) Inc., Anchorage, Alaska, USA*
[2] *BP Exploration Operating Company Ltd, Sunbury-on-Thames, UK*
[3] *Department of Petroleum Engineering, Heriot-Watt University, Edinburgh, UK*

INTRODUCTION

For rocks to be oil or gas reservoirs they require two essential characteristics: the capacity to store fluid, called porosity, and the connectivity of pore space allowing fluid flow through the rock, permeability. Primary porosity comprises the intergranular spaces between grains before modification of the pore system by cementation, dissolution of soluble grains or fracturing. Permeability is a measure of a rock's specific flow capacity and is related to the continuity of pore space.

The heterogeneous nature of sand reservoirs gives rise to considerable variation in the magnitude and pattern of porosity and permeability. In fluvial reservoirs, this heterogeneity is very marked at a number of scales. Small-scale heterogeneities exist at the metre and submetre scale because of the high preservation potential of bedforms. Variability at a larger scale is determined by sandbody continuity and interconnectivity. In this chapter, we focus on the smaller-scale variability that is caused largely by depositionally inherited textural variations. The larger-scale heterogeneities are discussed in the following chapter and the various petroleum reservoir aspects have been addressed in several recent reviews (Martin, 1993; Swanson, 1993). Swanson (1993) stresses the importance of relating processes to reservoir facies for improved reservoir description.

The effects of small-scale permeability contrasts in fluvial reservoirs have been shown in the last decade to have significance for recovery efficiency (Kortekaas, 1985; Lasseter *et al.*, 1986; Weber, 1986; van der Graaf & Ealey, 1989; Hartkamp-Bakker, 1991; Høimyr *et al.*, 1993; Ringrose *et al.*, 1993). It is only in recent years, however, that the small-scale effects have been shown to also affect waterflood breakthrough times, even in sands where connectivity issues at larger scales are the dominant factor (Jones *et al.*, 1993). A review of the current understanding of the process controls on small scale reservoir properties is very timely.

Advances in Fluvial Dynamics and Stratigraphy. Edited by P.A. Carling and M.R Dawson.
© 1996 John Wiley & Sons Ltd.

PERMEABILITY DESCRIPTION IN FLUVIAL RESERVOIRS

A traditional method for the measurement of porosity and permeability in reservoir sandstones is by laboratory tests on core plugs. The dimensions of core plugs (i.e. prepared rock cylinders 0.025 m in diameter and length) are very close to the length scale of variation in petrophysical properties caused by lamination. As such, core plugs are an inappropriate volume to determine either lamina properties or the effective (or average) bedform properties. The grain-size variability and early diagenetic modifications (e.g. pedogenic reworking and rhizocretions) often present in fluvial sandstones give rise to very scattered porosity–permeability plots. These plots often show that permeability may vary by several orders on magnitude for a given porosity. From such plots (Figure 11.1) it is not possible to predict permeability from traditional density log transforms (i.e. predictors that rely on simple empirical correlations with porosity). It is also difficult to know how to average these small-scale measurements for comparison with larger-scale measures (e.g. well tests).

The traditional 0.25 m spacing between core plugs is also inappropriate for characterizing heterogeneous reservoir elements that occur at the metre scale (i.e. bedforms). This is illustrated by the core plug data from two fluvial intervals (Figure 11.2) which appear similar, in terms of variability, at first glance. Comparison between the two intervals with the benefit of probe permeability data at centimetre spacing (Figure 11.3, see p. 383) shows additional detail and just how inadequate plug data are in fluvial reservoirs. In this case, the differences in heterogeneity are

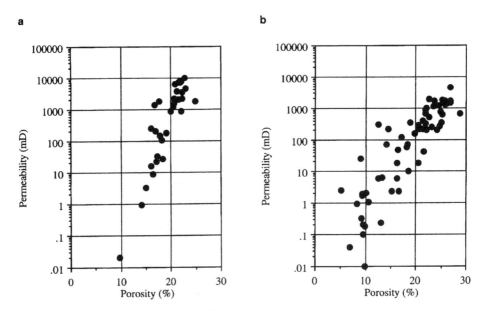

Figure 11.1 Porosity versus permeability for core plugs from two fluvial reservoirs: (a) Lower Jurassic, Norwegian North Sea (data courtesy of Statoil); (b) Triassic, southern England (courtesy of British Petroleum). Note the wide scatter of permeability for given porosites.

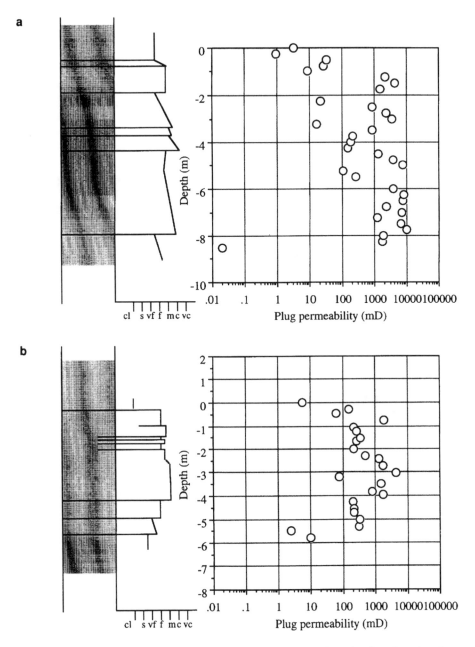

Figure 11.2 Profiles of the core plug permeability data and grain size data for the two fluvial reservoir intervals included in Figure 11.1: (a) Lower Jurassic, Norwegian North Sea; (b) Triassic, southern England. Note that permeability is correlated in each case with the grain size profile

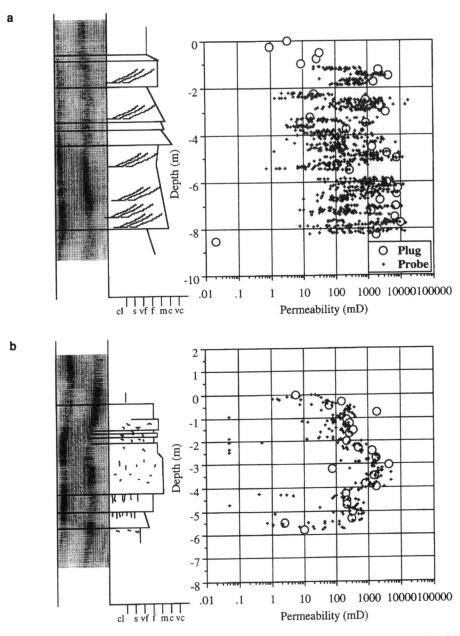

Figure 11.3 Permeability profiles from probe permeameter compared to the core plug data given in Figure 11.2: (a) Lower Jurassic, Norwegian North Sea; (b) Triassic, Southern England. Note the dramatically different degree of scatter suggesting quite small-scale petrophysical characteristics of these intervals. The petrophysical measurements are responding to the well developed cross-lamination in (a) and to a more massive sand in (b). The core plug data do not pick this character up and therefore estimates of effective permeability from core plugs in fluvial sandstones may be misleading because of the potential for small-scale grain size variations

due to the preservation of bedforms with strongly contrasting cross-bedding, (Figure 11.3a), compared with the relatively homogenized sandstone (Figure 11.3b).

The probe permeameter is a very effective device for characterizing the lamina-scale variability commonly seen in fluvial reservoirs (Corbett & Jensen, 1993) and provides an invaluable supplement to traditional sedimentological core description. The small-scale variability is clearly related to the primary depositional structure (Brendsdal & Halvorsen, 1993; Hartkamp-Bakker, 1993). The probe permeameter provides an ideal quantitative descriptor of fluvial reservoirs and a powerful link between petrophysics and the depositional process.

PRIMARY CONTROLS ON POROSITY AND PERMEABILITY

A prerequisite to unravelling the effects of bedding, lamination and sedimentary structures on porosity and permeability in fluvial sandstones is understanding the fundamental pore-scale controls on porosity and permeability. Unravelling these controls in clastic reservoirs has been the subject of a long history of research, involving progressively more sophisticated technology. Of significance is that, regardless of depositional environment, the essential factors determining porosity and permeability at a pore scale are the same.

In order to isolate these controls, early studies concentrated on using relatively simple sandstones such as Berea in the USA or Fontainbleu in Europe (e.g. Graton & Fraser, 1935; Fraser, 1935; Krumbein & Monk, 1943; Wyllie & Spangler, 1952). Both of these sandstones have negligible variations in sorting, texture or grain size. Later studies, using slightly more complex sandstones, investigated the relationship between porosity–permeability data and sediment properties (e.g. Hutchinson *et al.*, 1961; Hewitt & Morgan, 1965; Polasek & Hutchinson, 1967). The results of these studies show that for relatively simple, clean sandstones, the texture of a sediment (grain size and sorting) and fabric (packing and orientation) exert the main controls on porosity and permeability. Where sandstones are more argillaceous, then clay matrix and the proportion of ductile grains begin to have a progressively pronounced effect on porosity and permeability. However, the effect of clay matrix on permeability is more pronounced than on porosity.

Texture (grain size, sorting)

For clean sandstones, the most important textural characteristics of a sediment that control primary porosity and permeability are grain size, sorting, sphericity and shape. Of these, however, only the grain size and sorting are of major importance. For permeability, grain size is overwhelmingly more important than any other factor: permeability typically increases with increasing grain size by a factor of eight between each grain-size class. The reason is that there is a concomitant increase in pore throat size with grain size, a principle which has been demonstrated both experimentally (Beard & Weyl, 1973) and theoretically from sphere grain pack studies (Bryant *et al.*, 1993a). Permeability shows a systematic relationship with sorting, such that as sorting becomes poorer, permeability falls (Figure 11.4). So, in fluvial environments, as in any others, factors which determine the grain size of laminae and bedding will have the primary influence on permeability.

For porosity, grain size plays a less clear role: grain packing experiments show that primary porosity is theoretically independent of grain size, which is typically the case for unconsolidated sands. However, finer grain sizes sometimes exhibit higher porosities than coarser ones, because they are more loosely packed (Gaither, 1953).

In contrast to grain size, sorting does have a measurable effect on porosity: well sorted sands are more porous that poorly sorted ones, as demonstrated by Beard & Weyl (1973), amongst others. The difficulty in isolating discrete controls is illustrated by the complex interrelationship between grain size, sorting, porosity and permeability. This interrelationship is shown in Figure 11.4, established largely from the experimental work of Pryor, (1973) and Beard & Weyl (1973) for both consolidated and unconsolidated sandstones. The effects of grain size and sorting on porosity and permeability are now so well understood that they can be modelled successfully for artificial sands – for example in Bryant et al.'s (1993a,b) sphere pack model.

Fabric (packing, orientation)

In natural sedimentary rocks, the ultimate effect of grain size and sorting on reservoir porosity and permeability is complex, because of the myriad small-scale variations in these two properties that occur at the lamination scale. These variations are implicitly linked with the detailed arrangement of grains, that is, their packing and orientation. These parameters together are known as the fabric of the sediment.

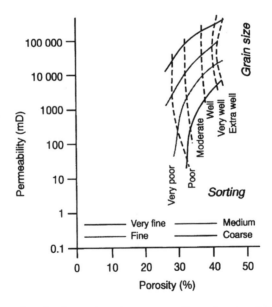

Figure 11.4   Effect of grain size and sorting on porosity and permeability established from the experimental work of Pryor (1973) and Beard & Weyl (1973)

The relationship of fabric to porosity and permeability is difficult to quantify precisely. Grain packing essentially determines the geometry of pores and their interconnectedness. Packing tightness depends partly on the extent of jostling and collisions between grains as they settle on a bed, and partly on the shape, sorting and roundness of grains as they nestle together during post-depositional compaction. Grains in fluvial environments are normally deposited with a statistically preferred long-axis orientation perpendicular to flow. Packing cannot be easily studied in artificial sediments because they would have to be too complex to capture the full variability of real sediments.

Much effort has been expended in trying to determine the initial packing of sands from fluvial depositional environments as a guide to evaluating the role of later compaction and cementation of reservoir sands. Packing of natural fluvial sands has been studied by Pryor (1973) and Atkins & McBride (1992). Atkins & McBride (1992) devised a quantitative index to assess the packing of sediment which measured the number of point contacts between grains, known as the contact index. When compared with other depositional environments, such as shoreline or aeolian, Atkins & McBride (1992) found that fluvial sands had the greatest ranges in contact index, and accordingly depositional packing.

POROSITY AND PERMEABILITY TRENDS IN FLUVIAL SANDBODIES

The interrelationships between porosity and permeability and sediment properties can only be used predictively if they can be built into scaled-up models of sandbody heterogeneity. Whilst the direct effect of textural and fabric variations can be sharply defined at a lamination scale (Kortekaas, 1985) the precise effects of these parameters are harder to resolve at the larger sandbody scale.

The building blocks of fluvial sandbodies are bedforms, with each preserved sedimentary structure resulting from systematic variations in grain packing, orientation and size. The most widely preserved and intensively studied bedform is cross-bedding. Detailed measurements of porosity and permeability in cross-bedding show that both parameters are higher in foresets than elsewhere in the bedform. This reflects the fact that grains avalanche down the lee-side of dunes resulting in poor packing, hence grains here have the lowest contact index. Porosity and permeability are lower in bottomsets because grains are characterized by a mixture of those rolling down the lee-face of the dune and finer material transported downward in suspension by eddies in the wake of the bedform. A commonly observed phenomenon is that foresets may retain reasonable permeability whilst the bottomsets may become relatively impermeable, and constitute fluid flow barriers and reduce the effective permeability (Roach & Thompson, 1959; Hartkamp-Bakker, 1993). Hartkamp-Bakker (1993) on a study of Permian and Triassic fluvial reservoirs showed that the coarse-grained foreset laminae are of consistently higher permeability, with the bottomset permeability generally, but not always, lower than that of the fine-grained foresets. In this study however, with the benefit of thin section analysis, no general rule for the relationships between grain size and permeability could be developed. This study suggests that, whilst permeability contrasts in cross-bedded sandstones are to be expected, detailed systematic measurement and analysis on a reservoir-specific basis is required.

The development of laboratory probe permeameters (Halvorsen & Hurst, 1990; Jones, 1992) has significantly improved the small-scale petrophysical description of cores. Lamina and bottomset permeabilities can be determined by probe studies on core (Figure 11.3, and Brendsdal & Halvorsen, 1993) but are not systematically measured with core plugs (Figure 11.2).

At the larger sandbody scale, the development of portable probe permeameters (Eipe & Weber, 1971; Daltaban et al., 1989) has allowed collection of a wealth of permeability data from extensive outcrops. Because of their marked heterogeneity, fluvial reservoirs have received considerable attention in the search for suitable analogues to fluvial reservoirs. Closely spaced transects through fluvial sandbodies provide invaluable data on spatial porosity and permeability distribution. These show systematic differences in permeability with direction that have important implications for fluid flow. Permeability grids measured in the field tend to be at relatively coarse spacings of 0.01–0.25 m or larger (Dreyer et al., 1990; Hartkamp-Bakker, 1993) and at these spacings it is difficult to determine the spatial correlation associated with either laminae or bedforms. Statistical measures (correlograms and semivariograms) based on too coarse a sampling scheme show the spatial correlation of permeability in fluvial sandstones to be random (Dreyer et al., 1990) or even fractal (Lake, 1993). More closely spaced probe data such as are measured on slabbed cores (0.002–0.005 m) often show a deterministic structure at the lamina and bedform scale. This has been shown in a laminated shoreface example by Corbett & Jensen (1993) but these measurements are equally applicable to fluvial reservoirs.

Measurements made on characteristic fining-upward fluvial sandbodies at a coarse scale repeatedly show a broad decrease in porosity and permeability with diminishing grain size (e.g. Hewitt & Morgan, 1965; Reineck & Singh, 1980). However, there is rarely a smooth relationship. and measurements are typified by extreme local fluctuations as illustrated by the recently available probe data.

Measuring local anisotropy caused by grain fabric as opposed to lamina contrast is problematical. The scale of lamina contrasts means that preparation of samples small enough to eliminate lamina effects result in samples that are difficult to measure without encountering boundary effects. This occurs whether the flow is confined (cubic Hassler cell type devices) or unconfined (probe permeameters). Brendsdal & Halvorsen (1993) have investigated anisotropy in small-scale (0.025 m) cubed samples and conclude that fabric anisotropies of 0.54 (i.e. vertical permeability 54% of horizontal permeability) and 0.78 could be detected under different boundary conditions, for cell and probe respectively. When the interlamina order of magnitude differences are present, however, fabric-related anisotropy (caused by preferred orientation of non-spherical grains) is much less significant.

Repeated measurements show that the direction of maximum permeability in fluvial sandbodies is characteristically parallel to the axis of the small troughs. The minimum permeability is at right angles to the trough axis, and can vary by up to 85–90% of maximum value (Hewitt & Morgan, 1965). The anisotropy could be more significant when the bounding surface permeabilities are properly measured and taken into account. The dip direction of fluvially cross-stratified sandstones is generally thought to be the direction of maximum permeability. It is commonly observed that the direction of maximum permeability tends to be parallel to the direction of the depositing currents (Figure 11.5), although, as will be discussed later, this depends on the nature of the dominant bedform structure.

Anisotropy in sediments is scale-dependent. In the absence of significant fabric anisotropy, permeability within laminae will tend to be isotropic because the sediments are relatively well sorted. As laminae of varying permeability are combined the anisotropy will increase ($k_v < k_h$). As the bedforms are taken into account, particularly if the bottomset permeabilities are low, the sediment can be significantly anisotropic with vertical transmissability severely reduced. Anisotropy measured from adjacent core plugs provides a very limited measure of lamina properties and should not be taken to represent average reservoir characteristics.

Weber *et al.* (1972) and Hewitt & Morgan (1965) presented a series of formulae for calculating directional permeability in festoon cross-bed sets based on a study of

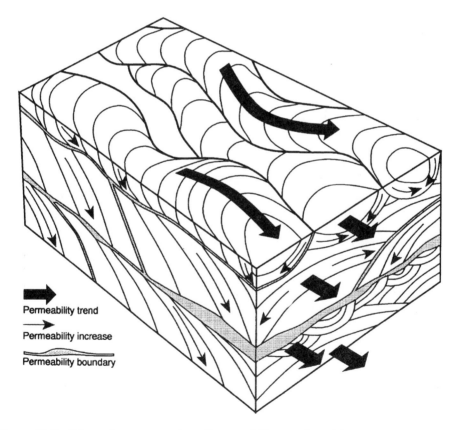

Figure 11.5 Schematic representation of permeability trends and barriers in cross-bedded fluvial sandstone. Small arrows depict direction of permeability increases (redrawn from Pryor, 1973).

fluviatile festoon cross-bed sets in a series of recent and ancient outcrops. These formulae can be useful for estimating the effective single-phase flow in sediments dominated by foresets and bottomsets.

At a sandbody scale, there is typically a systematic decrease in permeability in fluvial bars both downstream and bankwards (Pryor, 1973). This is attributed to variations in grain size and the proportion of detrital clay – both being related to changes in depositional energy. Other data demonstrate that in fluvial sandbodies permeability is generally highest in the central two-thirds of a given channel sandbody, and diminishes to the sandbody margins (Pryor, 1973).

In summary, much has been learned in the last few years about the permeability distributions in fluvial reservoir sandstones at a small scale. There is still much to do to extend this understanding to larger scales.

WATER/OIL DISPLACEMENT CHARACTERISTICS IN FLUVIAL RESERVOIRS

The internal permeability distribution of a fluvial sandbody will clearly influence the fluid flow through that sandbody, and ultimately the efficiency with which it can be swept of hydrocarbons (Corey & Rathjens, 1956; Kortekaas, 1985). The ultimate goal of any study on the spatial variability of porosity and permeability in fluvial sandbodies is therefore aimed at being able to predict water and oil displacement characteristics. In oil fields under development by waterflooding, this knowledge has a significant impact on the location of injection wells and the optimum perforation interval to ensure maximum sweep and greatest recovery.

We have reviewed the problems associated with characterizing fluvial sediments by probe or plug measurements. These measurements are single-phase (gas) permeabilities and can be scaled up or averaged by a variety of techniques – statistical, empirical or numerical. Measuring the effective permeability for multiphase flow (oil/water in the waterflood of an oil reservoir) is more complex. We will consider experimental data on fluid flow through geological media before considering the modelling and scale-up of the flooding process.

DISPLACEMENT CHARACTERISTICS IN A PHYSICALLY STRUCTURED FLUVIAL SANDSTONE

Laboratory floods

There are few published examples of well controlled quantitative two- or three-dimensional displacements in well characterized geological media. Due to theoretical considerations, traditional oil industry core floods tend to be conducted on homogeneous samples which are, not surprisingly, difficult to find in many fluvial reservoirs. There has been increasing interest in floods in heterogeneous rocks in the laboratory in recent years. These experiments are, however, extremely difficult to carry out and are correspondingly rare. We describe one such group of experiments.

383

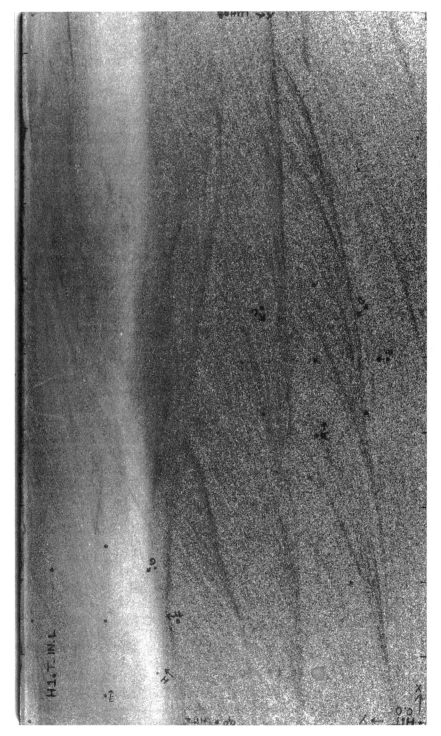

Figure 11.6 Cross-bedded Triassic fluvial sandstone ($0.28 \times 0.48 \times 0.01$ m) used for the fluid flow experiment

384

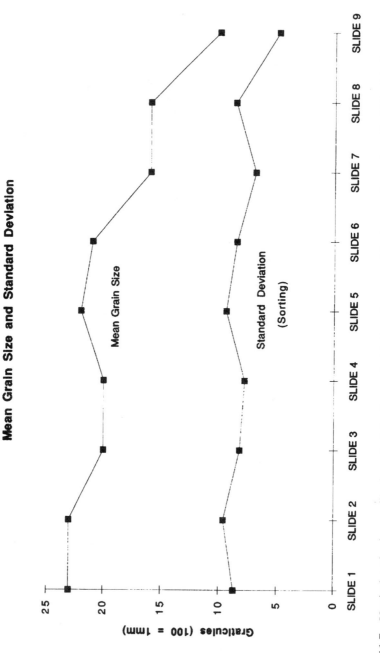

Figure 11.7 Vertical grain size profile through slab of cross-bedded sandstone used in the experimental study. Slide 9 is from the upper part of the slab in Figure 11.6, slide 1 from the lower part

A series of fluid flow displacement experiments were carried out on a sample of Triassic cross-bedded fluvial sandstone, from Hollington in the East Midlands, UK. A 1 cm slab of the rock, measuring 28 × 48 cm, was extracted for flow tests (Figure 11.6). The slab was cut perpendicular to interpreted flow direction, and exhibited three trough cross-bedded sets: two lower coarse-grained sets and a single upper finer-grained set. Each set showed decimetre-scale bedding, well developed festoon or trough sets. Individual foresets were picked out by lamina-scale variations in grain size and discrete laminations of segregated heavy minerals. Detailed grain-size analysis through the slab indicated that each set fined upwards. The average grain size of the overall slab also fined upwards, with the higher cross-bed set being overall finer than the lower (Figure 11.7).

The flow properties of the slab were characterized using a probe permeameter. The probe permeameter was used to measure the areal distribution of permeability. In samples containing complex heterogeneities, however, interpretation of probe data is far from straightforward. This is expecially the case where the heterogeneities in the rock are on a scale equal to or less than the tube diameter. The gas flow from the probe tip is unconfined. The area of investigation is therefore usually a hemispherical region of rock just below the tip of the instrument (Goggin, 1988; Daltaban et al., 1989; Hartkamp-Bakker, 1993; Corbett, 1993). The presence of small-scale heterogeneities in the region of flow will distort this flow regime making any interpretation of permeabilities based on comparisons of flow in homogeneous samples somewhat unreliable.

Flow rate is proportional to permeability, therefore flow rate maps (of measurements taken at constant pressure) can be considered to be equivalent to permeability maps. Flow rate measurements were collected at nodes of a closely spaced 0.01 m by 0.004 m grid. The recorded flow-rate value at each node is represented by the top right-hand corner of each pixel shown in Figures 11.8 and 11.9.

At a purely qualitative level, the striking similarity between the physical characteristics observed on the slab (Figure 11.6) and the flow-rate map (Figures 11.8 and 11.9) provides strong evidence of a link between depositional fabric and fluid flow, with variations in permeability mirroring accurately individual laminae and small-scale bedding variations. The highest flow rates occur along foresets, and the lowest along bottomsets. This reflects the better sorting of foreset laminae and lower packing (contact index), resulting from grains avalanching down the lee-side of the bedform. The lower permeabilities and concomitant flow rates recorded in the bottomsets result from the fact that sediments here are an admixture of grains rolling down the lee-face of the ripple and finer material transported downward in suspension by the eddies created downstream of the ripple crest. Similar permeability contrasts between foresets and bottomsets are widely reported from experimental studies of fluvial sandstones. It is clear, however, that the degree of permeability contrast between foreset and bottomset laminae will depend upon a combination of stream velocity, turbulence and available grain sizes. If velocity is relatively high and turbulence low, foresets will be formed continuously from material avalanching down the lee-face, with little contrast between lamina thickness, grain size or sorting. If, on the other hand, turbulence and velocity fluctuate more, the thickness, grain size and sorting of individual laminae will vary. In this case, the permeability of the fine laminae may be

FLOWRATE MAP AT 100 MBAR

FLOWRATE
(cc/min)

1400
1250
1100
950
800
650
500
350
200
0.00

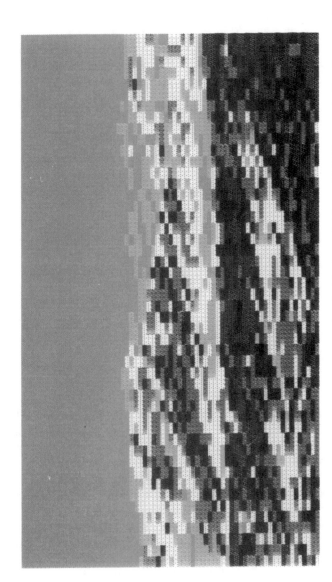

Figure 11.8   Flow-rate map of cross-bedded rock slab shown in Figure 11.7. Flow rate is directly proportional to permeability

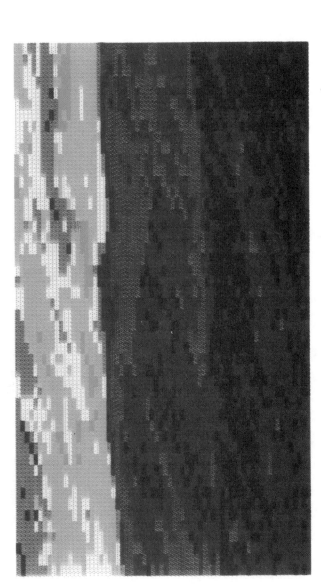

Figure 11.9 Flow-rate map of cross-bedded rock slab shown in Figure 11.7 with an expanded scale to show the variability in the upper cross-bed set. Flow rate is directly proportional to permeability

further reduced by inclusion of micas and clays. The contrasts observed in this slab are consistent with a depositional model of fluctuating currents.

Following characterization with the probe, the slab was sealed in epoxy resin and holes were drilled at discrete intervals around the perimeter to facilitate the injection of fluid. The entire pore space of the rock was then filled with water. Oil was then injected into the slab, gradually displacing the water. Finally, to simulate a waterflood, water was again injected into the slab, displacing the resident oil. In order to observe and measure each stage of displacement, a series of quantitative radiographs (X-rays) were taken to permit visualization of fluid movement.

A saturation map from the first oil displacement (oil displacing water) is shown in Figure 11.10. In this experiment, oil is injected through two holes on the left side of the slab, the water being produced from two identical holes on the opposite side. This flow simulation shows clearly that the oil displacement picks out the structure of the cross-bedding as it moves through the sandstone. Laminations characterized by lower permeability, such as the bottomsets and occasional finer-grained foresets, tend to retain water more readily than the higher-permeability laminations. These are characterized by higher residual water saturations, which largely fail to become oil-saturated. The important control of grain size over permeability is

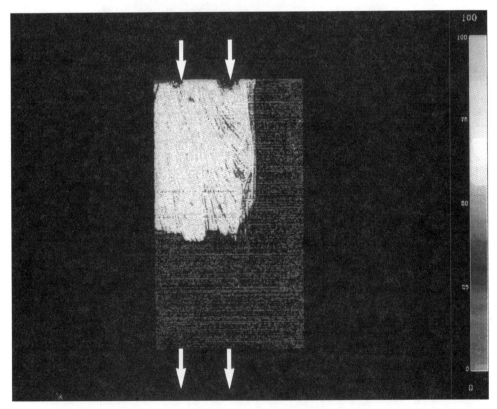

Figure 11.10 Oil displacing water in cross-bedded fluvial sandstone. Oil is injected and water displaced at discrete locations indicated by arrows. In the lowermost cross-bed set note that oil is more advanced in the foresets than in the bottomsets. Refer to Figure 11.7 for dimensions

demonstrated by the upper cross-bedding set showing greatest advancement of the flood front.

The waterflood of the same sample – water displacing resident oil – mirrors the most widespread method of increasing oil recovery from reservoirs. In the case of the experimental study, a waterflood perpendicular to the depositional trend was investigated. Water was injected from a single hole at the top of the sample and produced from a similar hole at the bottom of the slab. Unlike in the previous example, where injected fluid (oil) preferentially flows into regions of high permeability, here the reverse is the case, with water initially flowing preferentially into regions of low permeability because of the influence of capillary pressure.

The results of the waterflood experiment are shown in Figure 11.11. This example demonstrates quite clearly how the distribution of oil and water is controlled by capillary forces, the distribution of which is largely determined by primary depositional fabric. Clearly visible are areas of high oil saturation. These are the coarser-grained laminae characterized by higher permeabilities. The areas of high capillary pressure (low permeability) are picked out preferentially by high water saturation. The influence of individual laminae can be readily observed. The most

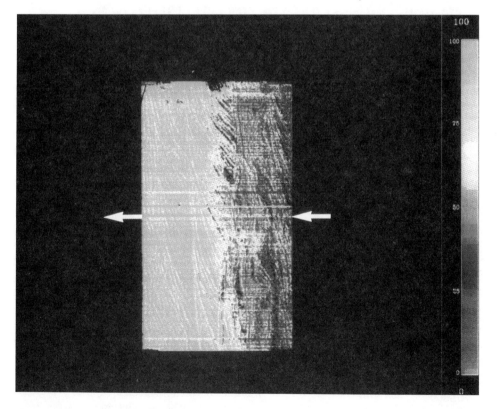

Figure 11.11 Water displacing oil in cross-bedded rock slab. Water is injected and oil produced at discrete locations indicated by arrows. Note fine definition of rock laminations by saturation distribution. This reflects small-scale capillary behaviour, itself determined by primary depositional fabric. Refer to Figure 11.7 for dimensions

detailed example of the influence of primary depositional fabric on flow characteristics can be seen in the waterflood of the same sample.

Field experiments

Outside of producing fields, which are hardly experiments, there have been few examples of large-scale multiphase flow experiments in any oil reservoirs. There have been single-phase experiments in well characterized outcrops – most notably the experiments in the Quaternary Rhine River channels (Weber *et al.*, 1972) and the "Gypsy" field project in Oklahoma (Doyle *et al.*, 1992).

The permeabilities in the Holocene sands (Weber *et al.*, 1972) were so high (tens of darcies) that capillary forces were non-existent and the flow anisotropy that might have been expected from the sedimentary structure was undetectable. In the Gypsy project (Doyle *et al.*, 1992) there was deliberate emphasis on the larger scales of fluvial heterogeneity, with no attempt to characterize or model the bedform-scale heterogeneities. These studies illustrate how difficult it is to conduct large-scale experiments in "well characterized" outcrops for realistic subsurface field conditions and processes.

RESERVOIR ENGINEERING ASPECTS OF FLUID FLOW IN FLUVIAL RESERVOIRS

Given the length scales associated with laminae and bedforms (a metre or less) the physical processes that dominate under waterflooding are capillary forces (Ringrose *et al.*, 1993). In fluvial reservoirs, with high lamina contrasts and the low flooding rates expected in the interwell volumes, this is particularly evident and well demonstrated by the flooding experiment in the cross-bedded sandstone.

Numerical modelling using computer simulation is one way in which the effects of two-phase flow can be determined. Numerical experiments based on good data and modelling at the appropriate scales can be more cost-effective than artefact-prone experiments. The experiments are necessary, however, to confirm the numerical findings. Numerical simulation is often necessary to understand the experimental data.

Kortekaas (1985) conducted numerical simulations on theoretical festoon cross-bedding sets modelled from a fluvial sandstone. By observing the measured variations in the direction and magnitude of a fluvially cross-bedded sandstone he constructed a simplified mathematical model of a small part of the cross-bedded sandstone. In his model, the foreset laminae where given a higher but variable permeability, and the bottomsets were given a lower permeability. The actual values for permeability and dimensions were based on probe permeameter measurements on cores from various cross-bedded reservoir zones. The foreset laminae were given lengths of 4 m and set heights of 0.8 m. The permeabilities of the foresets were 50–250 mD. In contrast, the bottomsets had heights of 0.02 m and permeabilities of 5–10 mD. Simulations were carried out assuming a water-wet sandstone. They were designed so that direction-dependent relative permeability and capillary pressure curves could be calculated. Kortekaas's small-scale simulations of water/oil displacements in cross-sections suggested that, in real rocks of similar character, considerable amount of movable oil would initially be left behind in the higher-permeability foreset laminae. However, where flow was parallel to foreset laminae, the displacement efficiency was considerably improved.

Kortekaas's work has been extended with the benefit of additional data (particularly from the probe permeameter) and more powerful computers to the point where the effects of cross-bedding are now well appreciated (Kasap & Lake, 1990; Hartkamp-Bakker, 1991; Ringrose *et al.*, 1992; Corbett *et al.*, 1993). However, it is clear that such mathematical models have a number of shortcomings: (i) the compromise between geological refinement and manageability; (ii) inadequate multiphase input data at the appropriate scale; and (iii) detailed understanding of bedform geometries, relationships, internal permeability patterns and stacking relationships.

There is no doubt that the contrasting small-scale characteristics of the two sands illustrated in Figure 11.3 will have different capillary-trapped volumes of oil following a waterflood. How representative of the reservoir as a whole these small intervals may be, and whether the effects will be "significant" at the field scale are questions that only careful geological analysis and geologically based modelling can infer. How we go about answering these questions is the subject of much petroleum industry interest at the present time (Jones *et al.*, 1993).

In the near future it will be possible to determine effective flow properties at the lamina and bedset scale. A start has been made using the "geopseudo" approach (Corbett *et al.*, 1992). The next key ingredient is to incorporate more realistic bedform geometries, accompanied with the development of more focused geostatistical tools and appropriate data collection strategies.

CONCLUSIONS

Achieving optimum oil or gas recoveries from reservoirs requires a comprehensive understanding of how fluids will flow in sedimentary rocks. Such fluid flow is dependent upon the relative magnitude and direction of permeability, and the presence and geometry of zones of effectively zero porosity and permeability which act as barriers to flow. The efficiency of the recovery is also dependent on the small-scale heterogeneity and the ordering of the permeability structure.

The level of porosity and permeability heterogeneity varies between sandstone reservoirs of differing depositional environment. Fluvial reservoirs are widely recognized as being characterized by great complexity because of the high preservation potential of bedforms combined with complex geometry of sandbodies. Absolute permeability varies widely as a function of different diagenetic and burial histories. The small-scale heterogeneity is, however, largely controlled by the characteristics of the bedforms which are inherited from the depositional process. The heterogeneity and ordering are functions of the depositionally derived sedimentological structure and therefore functions of the process. The process, therefore, has a significant control on the recovery potential of reservoirs. Petroleum engineers could therefore usefully learn something about depositional processes. To help their education, the geoscience community needs to focus on what the critical issues are. We hope this review will stimulate a move towards a quantitative description of reservoir sandstones.

The development of the probe permeameter in recent years and it increased usage by the petroleum industry is providing new insights for the relationship between petrophysics and primary depositional process in these complex systems. These data, coupled with better understanding of fluvial processes, will lead to improved understanding and prediction of the performance of fluvial petroleum reservoirs. Fluvial reservoirs contain a significant proportion of the world's oil reserves (Martin, 1993) so the rewards are there to be reaped.

REFERENCES

Atkins, J.E. & McBride, E.F. 1992. Porosity and packing of Holocene river, dune and beach sands. *AAPG Bulletin*, **76**, 339–355.

Beard, D.C. & Weyl, P.K. 1973. Influence of texture on porosity and permeability of unconsolidated sand. *AAPG Bulletin* **57**, 349–369.

Brendsdal, A. & Halvorsen, C. 1993. Quantification of permeability variations across thin laminae in cross-bedded sandstone. In: Worthington, P.F. (ed.), *Advances in Core Evaluation Accuracy and Prediction III*. Gordon and Breach, London, 25–42.

Bryant, S.L., Cade, C.A. & Mellor, D.W. 1993a. Physically representative network models of transport in porous media. *American Institute Chemical Engineering Journal*, **39**(3), 387–396.

Bryant, S.L., Cade, C.A. & Mellor, D.W. 1993b. Permeability prediction from geologic models. *AAPG Bulletin*, **77**, 1338–1350.

Corbett, P.W.M. 1993. *Reservoir characterisation of a laminated sediment, Rannoch Formation, North Sea*. PhD Thesis, Heriot-Watt University, Edinburgh.

Corbett, P.W.M. & Jensen, J.L. 1993. Quantification of variability in laminated sediments: a role for the probe permeameter in improved reservoir characterisation. In: North, C.P. & Prosser, D.J. (eds), *Characterization of Fluvial and Aeolian Reservoirs*, Geological Society, Special Publication, **73**, 433–442.

Corbett, P.W.M., Ringrose, P.S., Jensen, J.L. & Sorbie, K.S. 1992. Laminated clastic reservoirs: The interplay of capillary pressure and sedimentary architecture. **SPE 24699**, *Proceedings of the 1992 SPE Annual Technical Conference and Exhibition*, 4–7 October, 365–376.

Corey, A.T. & Rathjens, C.H. 1956. Effect of stratification on relative permeability. *Journal of Petroleum Technology*, December, 69–71.

Daltaban, T.S., Lewis, J.J.M. & Archer, J.S. 1989. Field minipermeameter measurements – their collection and interpretation. *Proceedings of 5th European Symposium on Improved Oil Recovery*, Budapest, 25–27 April.

Doyle, J.D., O'Meara, D.J. Jr. & Witterholt, E.J. 1992. The "Gypsy" Field Research Program in Integrated Reservoir Characterization. **SPE 24710**, *Proceedings of the 1992 SPE Annual Technical Conference and Exhibition*. 4–7 October, 493–502.

Dreyer, T., Scheie, A. & Walderhaug, O. 1990. Minipermeameter-based study of permeability trends in channel sand bodies. *AAPG Bulletin*, **74**, 359–374.

Eijpe, R. & Weber, K.J. 1971. Mini-permeameters for consolidated rock and unconsolidated sand. *AAPG Bulletin*, **55**, 307–309.

Fraser, H.J. 1935. Experimental study of porosity and permeability of clastic sediments. *Journal of Geology*, **43**, 910–1010.

Gaither, A. 1953. A study of porosity and grain relationships in experimental sands. *Journal of Sedimentary Petrology*, **23**, 180–195.

Goggin, D.J. 1988. *Geologically-sensible modeling of the spatial distribution of permeability in eolian deposits: Page Sandstone (Jurassic), Northern Arizona*. PhD Thesis, University of Austin, Austin, Texas.

Graton, L.C. & Fraser, H.J. 1935. Systematic packing of spheres with particular relation to porosity and permeability. *Journal of Geology*, **43**, 785–909.

Halvorsen, C. & Hurst, A. 1990. Principles practise and applications of laboratory minipermeametry. In: Worthington, P.F. (ed.), *Advances in Core Evaluation Accuracy and Prediction*. Gordon and Breach, London, 521–549.

Hartkamp-Bakker, C.A. 1991, Capillary oil entrapment in cross-bedded sedimentary structures of fluvial sandstone reservoirs. *SPE 22761, 66th Annual Technical Conference and Exhibition of the Society of Petroleum Engineers*, Dallas, Texas, 6–9 October.

Hartkamp-Bakker, C.A. 1993. *Permeability heterogeneity in cross bedded sandstones. Impact on water/oil displacement in fluvial reservoirs*. PhD Thesis, TU Delft, Krips repro meppel, Meppel, Netherlands. 294 pp.

Hewitt, C.H. & Morgan, J.T. 1965. The Fry in situ combustion test – reservoir characteristics. *Petroleum Transactions*, AIME, **234**, 337–343.

Høimyr, Ø, Kleppe, A. & Nystuen, J.P. 1993. Effects of heterogeneities in a braided stream channel sandbody on the simulation of oil recovery: a case study from the Lower Jurassic Statfjord Formation, Snorre Field, North Sea. In: Ashton, M. (ed.), *Advances in Reservoir Geology*. Geological Society, Special Publication, **69**, 105–134.

Hutchinson, C.A., Dodge, C.F. & Polasek, T.L. 1961. Identification, classification and prediction of reservoir non-uniformities affecting production operations. *Journal of Petroleum Technology*, **13**, 223–230.

Jones, A., Doyle, J., Jacobsen, T. & Kjønsvik, D. 1993. Which sub-seismic heterogeneities influence waterflood performance? A case study of a low net-to-gross fluvial reservoir. Paper presented at *7th European IOR Symposium*, Moscow, Russia, 27–29 October.

Jones, S. 1992. The Profile Permeameter: A new fast accurate minipermeameter. **SPE 24757**, *Proceedings of the 1992 SPE Annual Technical Conference and Exhibition*, 4–7 October, 973–984.

Kasap, E. & Lake, L.W. 1990. Calculating the effective permeability tensor of a gridblock. *Society of Petroleum Engineers Formation Evaluation*, **June**, 192–200.

Kortekaas, T.F.M. 1985. Water/oil displacement characteristics in crossbedded reservoir zones. *Society of Petroleum Engineers Journal*, **25**, 917–926.

Krumbein, W.C. & Monk, G.D. 1943. Permeability as a function of the size parameters of unconsolidated sand. *Petroleum Transactions*, AIME, **151**, 153–163.

Lake, L.W. 1993. In P.W.M. Corbett & P. Forbes, Characterisation and modelling of lateral heterogeneities in reservoirs, workshop report. *First Break*, **10**, 427–429.

Lasseter, T.J., Waggoner, J.R. & Lake, L.W. 1986. Reservoir heterogeneities and their influence on ultimate recovery. In: Lake, L.W. & Carroll, H.B.J. (eds), *Reservoir Characterization*, Academic Press, Orlando, 545–560.

Martin, J.H. 1993. Braided fluvial hydrocarbon reservoirs: The Petroleum Engineer's perspective. In: Best, J.L. & Bristow, C.S. (eds), *Braided Rivers*. Geological Society, Special Publication, **75**, 333–367.

Polasek, T.L. & Hutchinson, C.A. 1967. Characterization of non-uniformities within a sandstone reservoir from a fluid mechanics standpoint. *7th World Petroleum Congress Proceedings*, Mexico DF, **2**, 397–407.

Pryor, W.A. 1973. Permeability–porosity patterns and variations in some Holocene sand bodies. *AAPG Bulletin*, **57**, 162–189.

Reineck, H.E. & Singh, I.B. 1980. *Depositional Sedimentary Environments*. 2nd ed., Springer-Verlag, Berlin, 549 pp.

Ringrose, P.S., Sorbie, K.S., Corbett, P.W.M. & Jensen, J.L. 1992. Immiscible flow behaviour in laminated and cross-bedded sandstones. *Journal of Petroleum Science and Engineering*, **9**, 103–124.

Roach, C.H. & Thompson, M.E. 1959. *Sedimentary structures and localization and oxidation of core at the Peanut Mine, Montrose County, Colorado*. USGS Professional Paper **320**, 197–202.

Swanson, D.C. 1993. The importance of fluvial processes and related reservoir deposits. *Journal of Petroleum Technology*, April, 368–377.

Van de Graaf, W.J.E. & Ealey, P.J. 1989. Geological modelling for simulation studies. *AAPG Bulletin*, **73**, 1436–1444.

Weber, K.J., 1986. How heterogeneity affects oil recovery. In: Lake, L.W. & Carroll, H.B. (eds), *Reservoir Characterization*. Academic Press, Orlando, 487–544.

Weber, K.J. Eijpe, R., Leinjnse, D. & Moens, C. 1972. Permeability distribution in a holocene distributary channel-fill near Leerdam (The Netherlands). *Geologie en Mijnbouw*, **51**, 53–62.

Wyllie, M.R.J. & Spangler, M.B. 1952. Application of electrical resistivity measurements to problems of fluid flow in porous media. *AAPG Bulletin*, **36**, 359–403.

12 The Prediction and Modelling of Subsurface Fluvial Stratigraphy

COLIN P. NORTH

Department of Geology and Petroleum Geology, University of Aberdeen, UK

THE NATURE OF THE PROBLEM

Objectives

Sediments deposited in fluvial environments may be of great economic and industrial importance because of the fluids or minerals they contain. Fluvial sandstones form major groundwater aquifers and hydrocarbon reservoirs. They may be the locus of placer deposits such as copper, tin, gold and diamonds, and provide the site for precipitation of minerals such as uranium and vanadium. Most of the world's coal deposits formed under the influence of fluvial processes. In addition to direct economic importance, alluvial deposits can be extremely useful to give both local and regional information for plate-tectonic, basin-scale reconstructions and palaeoclimate analysis, as river systems are very sensitive to extrabasinal controls such as tectonism, climate change and sea-level variation.

A knowledge of the spatial distribution of both channel and floodplain sediments is crucial to economic exploitation of the resources. In most cases, however, the deposits of interest are partly or wholly buried beneath the surface, so a forecast has to be made of the expected stratigraphy and character of the deposits before the resource can be extracted. Information has to be gathered to make this prediction, usually an iterative cycle of data collection and refinement of the prognosis. The current understanding of the subsurface geology is expressed as a model, which is constructed by the combination of direct observations and indirect estimates. A model is considered to have predictive value if it makes precise statements about unsampled volumes of the subsurface.

The purpose of this chapter is to review the arsenal of processes and procedures now available for predicting and modelling subsurface fluvial geology. Firstly it is necessary to reconsider what it is that makes subsurface prediction difficult, and the nature of the data that may be collected from the subsurface, for these set limits on what is achievable. It is also necessary to summarize the requirements for models that may be produced, since the success of the prediction is going to be measured against the expectations. Before individual methods are considered in detail, there is

Advances in Fluvial Dynamics and Stratigraphy. Edited by P.A. Carling and M.R. Dawson.
© 1996 John Wiley & Sons Ltd.

an overview and discussion of the main approaches taken to subsurface modelling, as the underlying philosophy also imposes limits on the strength of each method. This discussion also provides an opportunity to clarify some of the terminology used in this area. The remainder of the chapter reviews the individual methods presently in use. In practice, it is common to employ several methods concurrently.

From an economic viewpoint, the main requirement is to provide engineers, be they hydrologic, petroleum or civil in discipline, with as much quantitative detail as possible on lithology distribution and internal sediment-body character, so that they can use this to plan and manage effectively the extraction or mining of the resource. Basin analysis and sedimentological research traditionally has produced large-scale, regional geologic models. These are primarily relevant to exploration because they focus on location and external geometry of sediment bodies, but they underplay the importance of the detailed sediment framework and variation in character. However, information at smaller scales is also essential; for example, it has been estimated that geological heterogeneity traps up to 40% of movable oil reserves in fluvial reservoirs (Tyler & Finley, 1991). Failure to provide such detail severely inhibits the application of new and sophisticated technologies to improving oil recovery efficiencies, which are lamentably low in far too many cases, efficiencies less than 25% being common (Tyler & Finley, 1991). Coal seams may be split by crevasse-splay sands, or washed out completely by channels, so the ability to predict sand-body trend may save considerable costs during coal mining operations (e.g. Horne *et al.*, 1978). Many of the prevailing depositional models are too descriptive and imprecise to satisfy these demands, and do not provide the quantitative component at the 100–1000 m scale critical to the engineering of resource extraction.

From a practical standpoint, there are two fundamental objectives for subsurface prediction. The first is to estimate the spatial distribution of rock type (sand, mud) and the detailed petrophysical properties of each lithology. The second is to estimate the uncertainty in this spatial distribution. This information can then be used to optimize exploitation, such as minimizing the number of wells needed in an oil field, or routing shafts and roadways in underground mines. Most crucially, this information can be used at the project planning stage to assess the effect on the economics of the geologic risk involved.

For fluvial sediments, subsurface prediction has to be tackled at two different spatial scales:

(a) At the largest scale, it is the overall fluvial architecture that is the first concern. An understanding of the architecture of the various types of channel fill is critical since most of the coarse, porous beds that host hydrocarbons and mineral deposits were formed within channels. When exploring in the subsurface, the problem is then generally one of determining the exact location of the sandstones, particularly in sequences produced by high-sinuosity rivers dominated by suspended load, where the channel belts are preserved as thin ribbons of sand encased in considerable volumes of floodplain mudrock (e.g. the Manville Group of western Canada (Putnam, 1983)).

However, it is important not to become excessively preoccupied with the channel sandstones. The floodplain sediment is often volumetrically more significant than the channel fill. Therefore, it will be encountered in boreholes more often than the channels fills, so we need to learn to read the runes in this material if we are to make the most of the available data. The floodplain is not always a passive player in the fluvial system as it may be highly important as a control on channel location; for example, because of fine grain size, plant growth and soil formation, the floodplain may be less easy to erode and so will stabilize channel position. The floodplain may contribute directly to the economic value of a fluvial succession. For example, it is the site for plant growth and thus ultimately coal formation. In this case the emphasis in subsurface mapping is reversed: now it is the channel sandstones that are the nuisance.

(b) Though an understanding of fluvial architecture is essential, it is not the only factor that must be considered. Once the broad sediment distribution has been ascertained, then the variability of mineralogy and texture within the deposits has to be mapped and quantified at scales one or more orders of magnitude smaller. These variations have a strong effect on the flow of fluids (water, oil, gas) through the rock. Such flow contributes to diagenetic alteration of the sediments. Diagenetic fabric generally follows trends in depositional fabric, so prediction of the diagenesis in the subsurface itself requires knowledge of the depositional system. Such diagenesis may have added value to the sequence, for example by concentrating ores, or it may have caused the deposition of minerals blocking pore throats, in which case it is detrimental. Variations in fluid flow characteristics must be taken into account when planning hydrocarbon or groundwater extraction.

Problems fluvial deposits present

The paramount difficulty presented by fluvial deposits is the degree and wide range of complexity possible in overall architecture and internal heterogeneity (e.g. Ambrose *et al.*, 1991). It is the high rate of change of lithology and petrophysical properties that creates the largest problem for prediction, and requires closely spaced sampling for any accuracy to be achieved. Indeed, for such sequences it can be remarkably demanding just to correlate major units from one borehole to another to establish the gross morphology of the system, let alone establish the more subtle detail. Fluvial hydrocarbon reservoirs are renowned for internal anisotropy, and for possessing sporadic permeability barriers that can only be detected after many wells have been drilled. For example, it is known the continuity of the main reservoir sands generally will be best parallel to palaeoslope. But crevasse-splay sands, which constitute significant oil reserves in many fields, do not honour this rule and are extremely difficult to locate and delineate. Grain size may vary from large boulders to clay over lateral distances as short as a few tens of metres, and vertical distances of just a few centimetres. Diagenesis may commence immediately sediment is deposited, particularly on the floodplain, as a result of subaerial exposure and due to soil processes and compaction, so increasing the variability of the deposits. Furthermore, the normal character of fluvial deposits may be obscured by reworking and modification by aeolian processes, and the interaction with other environments, such as those of glacial, lacustrine, deltaic and marine realms.

The planform pattern of a fluvial system can be complex and vary downstream. Whilst it is typically distributary on fan surfaces in its upper reaches, and becomes contributory in lower reaches, it will commonly become distributary again in the lowest reaches such as on terminal fans and in deltas. But the planform pattern is not reliably diagnostic of setting as a change to distributary and back to contributory can also occur in middle reaches as a result of tectonically induced gradient changes or valley confinements, such as occur along low-gradient arid-region rivers in central Australia (e.g. Cooper Creek, east and west of Innamincka, South Australia).

Schumm (1977) has shown a strong empirical link between mode of sediment transport and cross-sectional geometry of the channel. Thus there appears to be an important genetic divide between systems dominated by bedload transport, suspended-load transport and mixed-load transport. The dominant transport mode usually imparts the primary character to grain-size distribution and sedimentary structures. At first sight, therefore, this link appears to offer some hope to those attempting to describe subsurface deposits, as, if this link truly holds, study of a small sample such as obtained from a borehole ought to reveal the likely channel geometry. However, transport modes show a continuous spectrum, and any given river system may display contemporaneously any or all of these modes of transport, and the dominant process in any given channel reach may change with time. Furthermore, this relationship can break down, and it is possible for a river to display any geometry (braided, meandering, straight) with any mode of transport (bedload, suspended-load, mixed-load) if the conditions are suitable (see also discussion by Brierley in Chapter 8). The geometry of a single channel reconstructed from outcrop of ancient alluvium does not necessarily indicate the channel pattern within the overall channel belt (Bridge, 1985). It is clear the geometry is the result of the interactions between discharge, sediment supply, bank material and valley slope (see discussion by Knighton & Nanson (1993) on the factors which induce anastomosing channel patterns), but as yet we do not fully understand the links and the role of other factors.

Contemporary rivers in the same basin may possess greatly differing discharge characteristics, sediment load and channel geometry. Consequently, each fluvial axis displays differing facies components and internal sedimentary features. Physical proximity to an understood example is no guide for subsurface geologic prediction; compare, for example, the mud-rich Brazos River in Texas with its sand-rich neighbour the Colorado River (Texas) (Galloway, 1981).

Expectations and accuracy

Situations requiring the subsurface prediction of stratigraphy are varied, and each makes different demands of the geologist, but a common feature is the requirement for quantification of the geometry and petrophysical properties of the rocks. Ideally, a numerical representation (model) of the geology is the desired product of the exercise, since this can be used directly by engineers. It is also the safest way for geologists to communicate the forecast, since they have taken the responsibility for analysing and preparing the data into the required form, rather than letting a non-geologist extract the information needed. Just as each situation imposes different demands, so each has a particular accuracy specification. A common thread here is the need to estimate the magnitude of the errors, for this factor, the geological risk, can then be incorporated into the economic assessment, which is one of the prime objectives in any resource assessment.

In the past, geological descriptions have tended to be too qualitative and broad-brush to be of direct use in engineering analysis. Sedimentary facies models are intrinsically deficient in this respect, as discussed below, and there has been an over-reliance on this approach. They may be adequate at the exploration phase of a project, where identification of the depositional realm is the objective of facies analysis. But they are woefully inadequate for the modern demands of resource management, where extraction efficiencies have to be many-fold better than in the past.

The last decade has seen a slow but progressive improvement in the quantity and detail of information incorporated in geological descriptions, particularly as a result of the increasingly common publication of lateral profiles to accompany the more traditional vertical profiles (as strongly advocated by Miall (1985)). Now, however, geologists may be accused of being too detailed in their descriptions of the subsurface. This accusation usually comes from the engineers tasked with resource extraction, who find they cannot incorporate all the detail into their extraction plans. This is particularly true in connection with petroleum reserves, because the knowledge of the physics of flow of multiphase fluids (water, oil and gas) in rocks is as yet too imperfect to allow for an analytical solution to fluid flow through a complex suite of rocks, and computing power is inadequate to solve the problem numerically. Geologists feel obliged to describe every detail of heterogeneity, yet such detail may be made irrelevant by the type of extraction method being employed – the static geologic picture is only the starting point, the dynamic picture is what is needed.

The important message here is that the geoscientists and engineers, the suppliers and users, respectively, of the information, should work together on each project. The geological emphasis needed is often determined by the economic and engineering parameters of the project. So in a hydrocarbon reservoir analysis, for example, while the geologist may be fretting over the sinuosity of the ancient river, the engineer may be much more concerned by the impact on channel-sand permeability and porosity of the variations in diagenesis.

Description or prediction?

The task of the geologist is to produce a full and quantitative description of the sediment distribution and character in the subsurface. As will have been deduced from earlier chapters in this book (Chapters 7 to 10), reconstructing the fluvial environment can be difficult even with well-exposed outcrops. When trying to describe sequences in the subsurface, the problems are compounded by lack of access to the rocks. Reflection seismic investigation from the surface is plagued by the problems of attenuation by the rock of the acoustic signal, particularly the higher frequencies, and by noise swamping the signal. At depths of more than a kilometre, spatial resolution is rarely better than 20 m, and generally worse than 30 m. Traditionally, seismic is used for mapping structure, and it is unsuitable for detailed mapping of the sediments except in special circumstances (e.g. Weber, 1993).

Boreholes provide the only direct samples of the subsurface, by means of core material and downhole geophysical logging techniques. Typical core diameters are 10 cm (4") or less. Wireline logging methods have distances of investigation of up to one metre from the borehole, and only achieve greater penetrations with a considerable loss of spatial resolution. Therefore, even when the number of boreholes is large, the proportion of the subsurface actually sampled for description is still extremely small. When the study area is onshore, drilling costs are relatively cheap, and boreholes may be closely spaced. Some of the older oil fields in Texas, Kansas and California have wells 70 m or less apart (e.g. El Dorado, Kansas, started in 1915 (Tillman & Jordan, 1987)). Such close spacing may also be necessary when extracting heavy oils by means of steam-flooding (e.g. Tia Juana, Venezuela (Kruit, 1987)) or *in situ* combustion. Appraisal of coal deposits for opencast extraction is customarily done by means of a rectangular grid of boreholes spaced 50 m apart. Yet even if every one of these boreholes were cored and logged, which they may be in the case of shallow coal appraisal holes, the quantity of rock directly sampled by core amounts to just one-thousandth of one per cent; the proportion imaged by downhole logging methods is better, but is still only one-tenth of one per cent. Table 12.1 shows the area that would be sampled by 10 cm diameter core and wireline logs for borehole spacings commonly employed in hydrocarbon extraction, assuming all holes are cored and logged. Note that the area sampled is an inverse quadratic function of hole spacing, so as well spacing is doubled, the area sampled reduces by a factor of four. The examples given are for fully developed mature fields: during exploration and appraisal stages, especially offshore where drilling costs are very high, boreholes are usually several kilometres apart.

Given the extremely small proportion of the subsurface that can be directly sampled and described (Table 12.1), attempting to map and quantify the subsurface is dominantly an exercise in *prediction* not description. The geologist has to take the

Table 12.1 Area sampled by core (10 cm diameter) and downhole wireline logging for a range of well spacings typical of highly developed oil or gas fields

Distance between wells (m)	No. wells per km^2	Equivalent acre spacing	Area sampled by core (%)	Area sampled by logs (%)	Field example	Reference
50	400	0.62	0.001257	0.1257	Opencast coal appraisal	
70	204	1.21	0.000641	0.0641	El Dorado, Kansas	Tillman & Jordan (1987)
100	100	2.47	0.000314	0.0314	South Belridge, Ca., USA	Miller *et al.* (1990)
200	25	10	0.000079	0.0079	Tia Juana, Venezuela	Kruit (1987)
					Daqing, China	Qiu *et al.* (1987)
400	6	40	0.000020	0.0020	Peco, Alberta, Canada	Gardiner *et al.* (1990)
					Little Creek, Miss., USA	Werren *et al.* (1990)
500	4	62	0.000013	0.0013	Brent, UKCS	Struijk & Green (1991)
					Prudhoe Bay, Alaska	Atkinson *et al.* (1990)
1000	1	247	0.000003	0.0003	Forties, UKCS	Wills (1991)
					Leman, UKCS	Hillier & Williams (1991)
1500	0.44	556	0.000001	0.0001	Auk, UKCS	Trewin & Bramwell (1991)
2000	0.25	989	0.0000008	0.00008	West Sole, UKCS	Winter & King (1991)

limited information available and produce representations of the most probable distribution and character of the sediments – this is an exercise in *modelling*. The resultant picture of the subsurface is more than 99.9% prediction, and less than 0.1% description.

If prediction is such an important part of the task, what are the factors that are crucial for it to be successful? Although a complex issue, two main points stand out. First, it is vital to identify correctly the environment of deposition of the geologic components. Correct identification comes about through detailed knowledge of the processes operating in each environment, and the products they can generate. At the heart of this is facies recognition, and there have been, and still are, far too many cases of misidentification of fluvial deposits.

The second crucial factor in subsurface prediction is adequate general knowledge of the geologic components identified. Detailed information is needed on the way processes operate within the environment, the nature and scale of variability within each depositional subsystem, the way facies do or do not correlate, and ultimately the impact of the fabrics on fluid flow. For example, although it is known that gold is likely to be deposited within a river channel, mining experience has shown that in-channel processes result in preferred sites of concentration (Nami & Ashworth, 1993). Most knowledge of this type comes from studies of outcrops and modern environments, and from laboratory experiments.

Given the extremely small amount of data available from the subsurface, the best way we can make use of analogue outcrop and modern environment studies is to improve our understanding of the processes operating in the depositional setting, since this knowledge has predictive power, whereas descriptions of sediments do not.

Subsurface datasets – characteristics and limitations

The significance of the limitations inherent in the data from the subsurface is that most of the methods applied at the surface either are of limited value (e.g. because of lack of data), or cannot be applied at all (e.g. because of lack of resolution in the data). Though conventional sedimentological analysis is still of use, and has led to many successes, it is these limitations that have resulted in the application to the prediction of subsurface geology of new technologies such as statistical modelling and process-based forward modelling.

Reflection seismology is used to get an image of the subsurface over a wide area, and it is the geographical extent and density of sampling of these data that is of such great value (e.g. see Kearey & Brooks (1991) for an introduction and explanation). The method is dependent on variations in acoustic impedance of the rocks, which in turn is a function of many variables such as the density, porosity and fluid content. Such variations have to be of relatively large magnitude before it is possible to detect their presence; for example, a mud to sand transition will influence the seismic signal, but subtle changes in sand grain size or internal organization will be transparent. The method is also very dependent for success on the energy magnitude of the seismic source for depth of penetration and distance that can be imaged, and on the frequency of the original signal for the spatial resolution that can be achieved. A particular problem here is that rocks attenuate high frequencies much more rapidly than low frequencies, so increased depth of penetration is at the expense of resolution. As a rule of thumb, the thickness of any unit that can be resolved by reflection seismology is approximately one-quarter the wavelength of the seismic signal reaching that unit. If seismic reflectors are closer together than a quarter wavelength, the signal from each tends to add and cannot be separately discriminated (Sheriff, 1985). When achieving depths of investigation of 2 to 4 km, the depth to many hydrocarbon prospects, the frequency of the reflected seismic energy arriving back at the surface is typically less than 100 Hz, and often less than 50 Hz. The minimum sandstone bed thickness that can be resolved under these conditions is then of the order of 15 to 25 m (see Table 12.2).

The acquisition of three-dimensional (3D) instead of two-dimensional (2D) seismic cannot overcome these difficulties. The physical basis of both survey methods is the same, the fundamental difference between the two being increased density of sampling: traverse lines are about 25 m apart for a 3D survey instead of 100 or 200 m apart for a 2D survey. The reduced distance between samples allows corrections to be made for the dip of reflectors in any direction rather than just dip parallel to the traverse line. The much-enlarged matrix of samples can also lead to the recognition of features not readily apparent in a 2D survey, but 3D surveys cannot improve the resolution or depth of investigation.

Seismic exploration is of use at the large scale in mapping out the extent of the major stratigraphical packages, and in picking out gross morphologies. It may, for example, be possible to show that a sequence formed on an alluvial plain, but this is usually by inference and context, not direct measurement. The overall morphology of large-scale sedimentary features can usually be mapped. In this respect, seismic stratigraphy has a large role to play (Payton, 1977), and is an important starting point for determining sequence stratigraphy. As an example of the efficacy of seismic

Table 12.2 Effective spatial resolution that can be obtained with surface-run seismic exploration for moderate to deep targets. The resolution depends on velocity of sound in the rock (expressed here as interval transit time Δt, which is the inverse of velocity), and the dominant frequency reaching the target, which is governed by the attenuation by overlying layers. Approximate best resolution is given for two typical arrival frequencies, and for a range of likely alluvial lithologies after compaction due to moderate burial

	Sandstone (compacted)	Mudstone (compacted)	Mudstone (less compacted)	Limestone
Arrival frequency 100 Hz				
$\Delta t(\mu s\ ft^{-1})$	55	60	170	48
Resultant wavelength (m)	55	51	18	64
Best resolution (m)	14	13	4	16
Arrival frequency 50 Hz				
$\Delta t(\mu s\ ft^{-1})$	55	60	170	48
Resultant wavelength (m)	111	102	36	127
Best resolution (m)	28	25	9	32

surveying, individual lobes on submarine fans can be mapped with 3D seismic, though not usually with 2D alone. It is only in very favourable circumstances that seismic can be used to fill in the details of fluvial stratigraphy (e.g. Ruijtenberg *et al.*, 1990), though in these cases a 3D survey will be better than a 2D survey. Methods have been devised to compute porosity directly from seismic data (Doyen, 1988; Doyen *et al.*, 1989), but although accuracies to within 4% of the bulk rock porosity have been achieved, the spatial resolution is still too poor for detailed subsurface modelling.

The restricted quantity of subsurface information is certainly the most significant limitation on the prediction of fluvial stratigraphy in the subsurface. There are also other limitations, prominent amongst which is the fundamental problem of the ambiguity inherent in each and every observation of the subsurface.

Borehole core is not immune from this problem. Though undoubtedly the most useful single type of data for the sedimentologist and stratigrapher, and often the only way to assess many features, such as bioturbation, the small diameter of such core (5–15 cm) makes it difficult to identify sedimentary structures unequivocally. For example, it is not easy to separate planar cross-stratification from large-scale trough cross-bedding, yet the difference in interpretation is important. In this example, the type of criteria that may distinguish trough from planar foresets are: repeated truncations at high angles through several decimetres of core, thickening of foreset laminae in one direction, and high frequency of random variations in grain size over short vertical distances (Weimar & Tillman, 1980). Lateral facies transitions and many large-scale features, such as lateral accretion surfaces and low-angle truncations, cannot be recognized directly at all from core. Lateral accretion may be inferred from the vertical stacking of small-scale fining-up packages, and upward transition from dune to ripple structures, but these criteria are not unequivocal or unique to lateral accretion sediment bodies. The main difficulty is lack of information on the geometry and context of the feature under observation. In the extreme is the classic problem of the correct identification of fluvial sediments from deposits of other environments. It is because of these difficulties that core workshops are so important and common, and published core photos and descriptions, such as the many texts from the American Association of Petroleum Geologists and SEPM, are so useful.

Despite the deficiencies, borehole core has some advantages over outcrops. In particular, mudrock is generally much better sampled in core, as are the contacts between units. Delicate structures and trace fossils may be better preserved. Core is invaluable for calibration of wireline logging tools and for assessment of petrophysical properties. In theory, coring offers the opportunity to sample much longer and more complete stratigraphic sections than typically exposed at the surface. In practice, however, the high costs of offshore coring (due to the cost of rig time) mean usually just a small percentage of holes are cored, then only over short (hopefully representative?) sections. This limitation is combined with drilling problems, which may result in core damage or loss, and operator error, which means typically that the top or bottom of the core, just where things are getting interesting, is missing because coring started too late or stopped too soon.

Downhole logging, using geophysical remote-sensing instruments lowered on cables or attached to the drill string, supplies much additional information on the subsurface. In conjunction with sandstone thickness mapping, this has long been the primary route to determining subsurface stratigraphy (e.g. Harms, 1966). Whilst downhole logging provides invaluable information to the geologist (see summary in Cant (1992) or Rider (1991)), particularly at the medium to large scale, it must be appreciated that there are basic limitations that can prevent the elucidation of fluvial stratigraphy in the subsurface other than in a broad-brush fashion. Each instrument responds to a particular physical property of the rocks or fluids they contain, such as electrical conductivity or natural radioactivity. The prime use of these methods is to determine lithology and pore fluid characteristics. Yet one measurement type alone cannot unambiguously identify lithology, so several types of instrument are used together. Even then, the interpretation may be open to doubt; for example it is difficult to differentiate between bioclastic carbonates and sandstones heavily cemented with diagenetic calcite. Furthermore, there are no instruments that, downhole, can directly measure textural properties of the rock such as grain size and sorting, aspects normally considered vital when reconstructing fluvial morphology at outcrop. Changes in grain size of sandstones are normally deduced because of the common association of increasing clay content, and hence increasing natural radioactivity, with decreasing size of sand grain: alternatively, there is a relationship between porosity and grain size that can be used. But great care has to be taken to rule out alternative explanations, such as the presence in the sandstone of radioactive heavy mineral grains, like monazite.

Another weakness of downhole logs for detailed sedimentological interpretation is that the spatial resolution of many of the logging instruments may be too poor to discriminate the level of detail desired. This is due to the innate physics of the logging tools (see Rider (1991) for a more complete account). For example, there is an inverse exponential relationship between distance and radiation energy reaching the detector of the natural gamma-radiation (GR) tool; whilst 50% of the signal comes from within a radius of approximately 18 cm, there is a progressively decreasing contribution from up to a metre away from the detector. The consequence of this is that thin beds cannot be fully resolved with the GR tool, and may not be detected at all. For example, clay drapes on ripple lamination will not be imaged, and fluctuations in clay/sand ratio and grain size will be averaged out and be indistinguishable from more homogeneous sequences. Such detail can be crucial in facies recognition and in discerning the type of fluvial system.

The downhole dipmeter tool, which can measure the orientation of dipping surfaces in sedimentary structures, has the potential to provide considerable palaeocurrent detail which can be of great use to identify fluvial types. Prior to about 1984, the tools available were of low resolution, dip values being determined typically about every 50 cm in the along-hole direction; such low frequency of sampling is often inadequate for the current purpose, the main use being for structural analysis. Since the mid 1980s, high-resolution tools have been in use which, under favourable circumstances, can yield dip information every 10–20 cm. This can be most useful, but sadly the information is often not available because the dipmeter is not routinely used for cost reasons. In addition, the resolution of dipmeter tools suitable for use with oil-based drilling mud is not as good. Interpretation of the dip information is not always straightforward, due to noise and stray dip values; the application of filters and comparative study with outcrops is improving the situation (e.g. Herweijer *et al.*, 1990; Cameron, 1992; Cameron *et al.*, 1993; Williams & Soek, 1993).

A recent advance has been the development of acoustic borehole imaging tools, and extremely high-resolution resistivity tools such as the Formation Microscanner (FMS), that give an image of the entire borehole wall. This source of data can be of great value in identifying sedimentary structures and palaeocurrents, and hence fluvial morphology (e.g. Harker *et al.*, 1990; Luthi, 1990), but data analysis requires substantial effort, the identification of resistivity features is subjective (Bourke, 1992), and the downhole running costs are usually considerably more than for the dipmeter tool, so such measurements are as yet rarely available.

Recognition of fluvial facies in core and with downhole geophysical logs

The first and perhaps most important step in prediction and modelling of fluvial sequences in the subsurface is to ascertain that the rock sequence in question actually is of fluvial origin. This may seem a trite and obvious point, but it is not always easy to establish. It is of paramount importance when one gets to the modelling stages since the outcome hinges on the assumptions put in about fluvial style, and controls the choice of analogue data.

Sedimentological studies have relied heavily on vertical profiles, particularly up until the mid-1980s. The premise for this has been Walther's Law of Facies (1894) which states that "... it is a basic statement of far-reaching significance that only those facies and facies areas can be superimposed, without a break, that can be observed beside each other at the present time" (translation by Blatt *et al.*, 1972). This is usually taken to indicate that "facies occurring in a conformable vertical sequence were formed in laterally adjacent environments and that facies in vertical contact must be the product of geographically neighbouring environments" (Reading, 1986). Borehole analysis depends very heavily on this premise.

Walther stressed that the law only applies to successions without major breaks, a point often ignored by geologists who have failed to describe the type of contact between facies. If contacts are sharp, even when erosion cannot be demonstrated, the facies may have been formed in depositional environments which were widely separated.

Identification of both fluvial origin and style requires the assembly of a large body of evidence, much of which may be circumstantial or non-unique in origin, and which is highly dependent on having plentiful core material. One needs to bring to bear all available knowledge of fluvial systems and their deposits as established from outcrops and modern rivers, i.e. a full sedimentologic analysis as reviewed in earlier chapters of this book and elsewhere (e.g. Collinson, 1986). Given the limitations of subsurface data, effort tends to focus on recognition of processes from sedimentary structures, and identification of bedforms (Chapter 9). The procedure is not unlike that followed at outcrop (Chapter 10), but is always exacerbated by restricted width of core, and general paucity of information. The simple presence of certain types of sedimentary structure, such as asymmetric current ripples, may be ambiguous evidence, unless present in large quantities, as almost all primary structures that can be detected in subsurface data may be produced in any depositional environment. The presence of hummocky cross-stratification (HCS), for example, has been used to prove that a sequence is not fluvial but shallow marine, yet there is evidence that antidunes may produce an HCS-like structure in fluvial deposits (Rust & Gibling, 1990).

Useful diagnostic features include: palaeocurrent information (from dipmeter data or foreset measurements from oriented core) indicating essentially unidirectional flow; freshwater body and trace fossil occurrences (e.g. Koster, 1987), and either relatively low levels of bioturbation or low levels of species diversity where bioturbation intensity is high; inclusion of specifically terrigenous material such as plant remains (maybe as coal) and microflora (spores, pollen), either interbedded with sands or included as clasts and grains; evidence of frequent subaerial exposure such as desiccation cracks, palaeosols, wind reworking and inclusion of thin aeolian sand-sheets; and context, such as association with aeolian or lacustrine sequences. However, none of these criteria alone is absolute, as it is possible for these phenomena to occur in lacustrine and shallow marine environments, though usually to a lesser degree than in fluvial environments. A more circumstantial approach is to look for similarity in vertical profile to sequences known to be fluvial in origin (the facies model approach), though this carries other problems as discussed below. Much reliance is often placed on textural characteristics such as grain-size profiles along the borehole and degree of sorting. For example, a fining-upward grain-size profile is considered by many to be diagnostic of fluvial channel deposits, representing point-bar deposition or waning flow (e.g. Visher, 1965). However, fining-up profiles can form in other environments, such as submarine fans, and the absence of fining-up profiles cannot be taken as proof that a deposit is not fluvial in origin, as many modern river systems do not leave behind such sequences, and many ancient fluvial deposits lack this feature (e.g. Dreyer et al., 1990).

Recognition of fluvial facies with downhole geophysical logs alone is fraught with problems of non-uniqueness, and it is essential in each study area to calibrate the logs with detailed analysis of core material. Environment determination is based heavily on lithology identification, and facies interpretation usually relies on log shape, because of the common linkage between log response and grain size in sandstone bodies. Schemes to describe the geometry of log shape have been proposed, based on shape (trend), curve characteristics, and the nature of lower and upper contacts (Figure 12.1), but though of value in communicating and summarizing log information, these are subjective and are certainly not diagnostic. Such schemes are most commonly applied to the natural gamma-ray (GR) log, though they were originally devised for the self-potential (SP) electric log. Many workers have gone on to document the log shapes found or expected for a wide variety of depositional environments. They depend on the close link between GR response and clay content frequently observed, but this is not always reliable owing to the presence of diagenetic cements or heavy minerals (see Rider (1990) for an in-depth discussion). In any case, grain-size trends themselves are not diagnostic. For example, fining-upward profiles may be produced by a fluvial point bar or a transgressive marine bar.

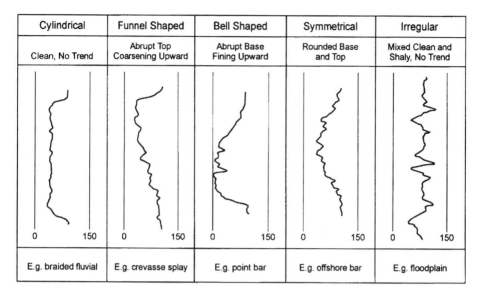

Cylindrical	Funnel Shaped	Bell Shaped	Symmetrical	Irregular
Clean, No Trend	Abrupt Top Coarsening Upward	Abrupt Base Fining Upward	Rounded Base and Top	Mixed Clean and Shaly, No Trend
0 150	0 150	0 150	0 150	0 150
E.g. braided fluvial	E.g. crevasse splay	E.g. point bar	E.g. offshore bar	E.g. floodplain

Figure 12.1   Idealized wireline log shapes used in facies analysis of the subsurface. When paired with another log such as resistivity or sonic, the patterns are mirrored, resulting in the bell and funnel shapes. Such log shapes are not diagnostic of environment but are useful descriptors

Wireline logs, can be of great assistance in interpreting the subsurface, but it is strongly recommended that a suite of logs be used rather than relying on a single log such as the GR (see chapter 12 of Rider (1991) for a suggested procedure), and in all cases logs must be cross-calibrated to core material *from the same location*. A supplementary technique is to obtain geophysical log signatures, such as GR, from analogue outcrops (e.g. Myers & Bristow 1989; Slatt *et al.*, 1992).

Once fluvial origin is established, then ideally one would like to identify the style of the fluvial system, such as low or high sinuosity, and the morphology of the channels, but the diagnostic features are usually not detectable with subsurface data. Having dismissed many traditional sedimentological criteria used with outcrop information, Bridge (1985) has concluded that there are four factors useful for distinguishing channel pattern:

(1) proportion of channel fills relative to lateral-accretion deposits, which increases with degree of braiding;
(2) mean grain size of channel fills relative to lateral-accretion deposits, which decreases with increasing sinuosity;
(3) palaeocurrent variance, which may increase with sinuosity;
(4) bankfull discharge, slope, and width/depth of palaeochannels throughout the channel belt, which must be quantitatively reconstructed from channel-bar and channel-fill deposits.

Unfortunately, the first two criteria are usually impossible to apply in the subsurface, since the low dip angle of lateral accretion surfaces makes them hard to identify in boreholes; even establishing a horizontal reference surface is difficult. The third criterion is regularly applied, as palaeocurrent data in the form of dipmeter logs may be available; caution is still required, as these data may be difficult to interpret owing to ambiguities in data collection, such as lack of knowledge of the structure producing the dipping surface. The fourth criterion may be difficult to apply owing to insufficient information about lateral extent of sediment bodies – channel depth, in the form of preserved sediment thickness, is not difficult to assess, but channel width is much harder to determine.

In summary, determining whether a given rock sequence in the subsurface is of fluvial origin can be done by assessment of a spectrum of data, and is not particularly difficult for the careful practitioner. But determining the details of the fluvial system, its style and architecture, is not so easy and requires a large dataset; even then, the interpretation is often vague ("it's probably high sinuosity"), and imprecise.

PHILOSOPHY

Models and modelling

A *hypothesis* is a suggested explanation of observed facts or phenomena, adopted as a basis from which to deduce what further critical facts are needed to test its validity. Geologists have to use the method of multiple working hypotheses because they deal with incomplete data and because several processes may have contributed to the final product. *Models* are idealized and simplified representations of reality set up to aid our understanding of complex natural phenomena and processes. The creation of a model, the modelling process, is based on analysis of available data and erection of a set of working hypotheses. There are many different types of model which vary according to the use to which each is to be put, and which may quite deliberately set out to represent just one or a few attributes of the real world. *Experimental models* are constructed to investigate the factors responsible for particular features, for example when sand transport is modelled in a flume under controlled conditions. *Visual models* are pictorial representations of working hypotheses to aid in the appreciation of the link between process and product, and the relationships within and between environments. *Mathematical models* simulate complex processes either through the equations that describe the processes (analytical models) or through a matrix of numbers which describe in detail the process products at many points in space and time (numerical models).

As well as being descriptive and of interpretive value, models also have predictive power as they make a statement about places and times for which measurements have not been made. But it is vital to remember that in nature there are no models – every place and time is unique, and although two environments may possess many similarities, they are never exactly the same. Models constructed from a particular dataset will be totally valid only for that one dataset.

Prediction approaches

There are several different fundamental ways of approaching subsurface prediction, each of which follows a particular philosophy. It is worth briefly reviewing these here as terms are used differently by different authors, leading to confusion and misunderstanding. Each philosophical approach to modelling has inherent assumptions and weaknesses. It is therefore important to appreciate the underlying philosophy of a chosen modelling approach since it will be limited by the characteristics of that philosophy. Failure to appreciate this is one of the main causes of disappointment with sophisticated predictive modelling – no matter how good the data, how expensive the computer, or how fancy the graphics and colours, there are inbuilt limits to the power of each approach. Greater insight into the geology may come from trying more than one technique. Later sections of this chapter will illustrate many of the approaches possible, but, in essence, the main choices to be made are between:

(1) inverse or forward modelling?
(2) process based or product-based modelling?
(3) deterministic or stochastic modelling?

The first choice to be made is crucial. The traditional approach to interpreting a given body of data is to work backwards from specific observations to the causative events and thus gain a general picture of the geological evolution of the study area. This approach is one of data inversion, and the exercise is *inverse modelling* (sometimes referred to as reverse modelling). A simple example is that of contouring, such as the construction of lines of equal sandstone thickness (isopach mapping). There are two weaknesses with this approach. Firstly, it takes no account of phenomena that were not sampled, and is thus very sensitive to data density. Secondly, the interpretation is ambiguous and non-unique – more than one set of processes and events can produce the same outcome, and there is no way to distinguish between them. For example, similar fining-upward grain-size profiles can be achieved by a variety of very different fluvial processes. Furthermore, the rock record is such that there is often more missing than preserved. Phases of erosion or non-deposition may be hard or impossible to detect, so leading to additional ambiguities in the interpretation. When faced with several possible interpretations, geologists apply a principle, known as Ockham's Razor, that one should invoke the simplest possible explanation for the observed facts and avoid unnecessary complexities. Such a simplification may, however, just be naïve. Despite the problems, inverse modelling is still the most widely followed approach as people feel most comfortable with it.

An alternative approach is that of *forward modelling*. General characteristics are extracted initially from the dataset (e.g. "the sequence is fluvial, probably bed-load, low-sinuosity, arid-region"). Then a description of the geology (the model) is built up using a set of rules appropriate for those characteristics. In effect, we are trying to simulate what we believe went on, and redeposit the sequence. The problem of this approach is that the resultant model may not match the measured observations at the sampling locations, since our knowledge of natural processes is imperfect and the rules will be incomplete. *Model validation* becomes a major part of the exercise, and is not easy.

The second decision reflects our confidence in understanding geological processes. Ideally, modelling should be *process-based*, that is, follow strictly the physical processes we believe to have operated. However, our knowledge of these processes is incomplete, especially when considering an entire depositional environment such as the river valley, rather than a small segment of it such as a channel reach. So an alternative is *product-based modelling*, where the rules used to build the model reflect observations made elsewhere (modern environments, outcrop, experience of other subsurface areas) and no attempt is made to obey physical laws. Such modelling is usually based on a statistical analysis of observations, and perhaps could be classed as *statistical modelling*. Note that the term statistic means "a summary value calculated from a group of data, often (but not necessarily) as an estimation of some population parameter" (Kendall & Buckland, 1982).

The final question to be addressed is whether the modelling will be deterministic or stochastic (see Figure 12.2 for additional explanation). A deterministic process is one where all events are strictly determined by preceding events. *Deterministic modelling* (e.g. Figure 12.2B,C) is thus governed by strict rules, and has a single outcome that can be predicted in advance with certainty. For example, most contouring techniques are deterministic in nature. So a contour plan of, say, sandstone thickness is a deterministic model of that sandstone.

A stochastic process is one where events are correlated to some degree and show some trends, but which contains a random element. It is important to remember that the term "random" does not mean unknown, as colloquially it is sometimes used, but means independent and uncertain, and following a statistical probability which may be known in advance. The rules used in *stochastic modelling* (e.g. Figure 12.2D–F) include the requirement to honour the statistical probabilities, but there will be an infinite number of ways to achieve this, and hence an infinite number of possible outcomes (called realizations).

Consider a real-life example. The manager of the local branch of a bank needs to know how many staff to employ to provide the counter service to the customers. This requires a balance to be made between, on the one hand, the cost of employing and training staff and, on the other hand, the flow of customers into the branch and the variability in types of enquiry. The manager doesn't want staff at the counter if there is nothing for them to do, but equally there is a need to keep queues short so customers have the minimum wait before being served. The usual approach to this problem is for the manager to commission a survey of the counter activity, usually over a period of a week or month. This shows the typical number of customers expected, and gives some idea of maximum load on the counter staff. Trends may be identified, such as more customers come in on Fridays than other days, the busiest part of each day is around lunch time and just before closing, and there are more enquiries about getting loans (a time-consuming type of enquiry) towards the end of each month than at the beginning of the month. On the basis of this information, the manager can draw up a staff rota (*the model*) to cope with the expected demands.

If the model is based solely on the main trends in customer traffic, perhaps by using an average of the number of customers each day to decide how many staff to have on duty, then that is a deterministic model. But the actual flow of customers (*the process*) is usually much more erratic than expected and, as everyone who has visited a bank knows from personal experience, there are often times when the queues are unacceptably long. The reason is that the process is stochastic, and there are random

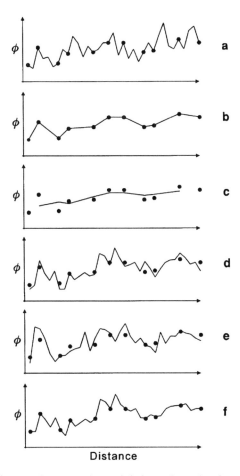

Figure 12.2 The difference between deterministic and stochastic modelling, and between conditioned and unconditioned modelling. Consider a rock property such as porosity (ϕ) that varies with distance as shown by the solid line in (A). We have only limited samples (solid circles) to establish this, so we need to make a prediction (a model) of the way that rock property varies between the measured values. (B) shows a simple linear interpolation between the samples. This assumes the actual variation is smooth; it fails to predict the real variability, and suppresses the numerical range. (C) shows a three-point moving average between the samples, a statistical technique which is useful to detect overall trends (ϕ increases upwards to the right). Both the procedures in (B) and (C) are deterministic, with a single possible outcome. Procedure (B) honours the data, and is a conditioned model, but procedure (C) is unconditioned and does not honour the sample values. (D) is a stochastic model calculated by adding random noise to the linear interpolation of (B). For this example, the noise was produced using a random-number generator, where it was arranged that the random numbers should be selected from a Normal (Gaussian) distribution with mean of zero and a standard deviation the same as that of the measured values. This approach preserves the overall trends seen in the measured values, but makes the variability more "realistic" (compare to (A)). But repeated runs of this model produce different results, and a second run is shown in (E). Neither (D) nor (E) are conditioned to the data. Example (F) is conditioned – it is the same model as in (D) but now forced to pass through the sample points. The procedure (F) is followed to create fractal models: instead of generating random noise with a Normal distribution, the noise is generated with a fractal distribution using procedures such as those in Voss (1985)

fluctuations in customer flow that are not allowed for in the model. In our banking example, a cause of fluctuation might be a sudden heavy rain shower that leads people to decide now rather than later might be the best time to transact bank business. The manager needs to modify the model to allow for unexpected surges in customer numbers. The practical way to do this, in cases such as the banking example, is to build redundancy into the model, that is have extra staff on hand who can help out when needed. By producing a stochastic model of customer numbers with time, based on the observed flow of customers in the survey (the statistical probabilities), the manager could assess the probability of the number of staff on duty being insufficient at any moment. The result might be a decision to provide staffing to ensure customers have a maximum wait, say, of 5 minutes for 90% of the time.

The choice we make, between deterministic or stochastic modelling, reflects our confidence in understanding of natural systems; stochastic modelling is usually selected because of uncertainties about the physical processes. Note that while stochastic modelling is a variety of statistical modelling, statistical modelling does not have to be stochastic and include a random component. For example, moving-average contouring is statistical, in that we calculate a statistic (the mean in this case) from a subset of the data and use that to represent the unknown value, but it is not stochastic – there is only ever one possible outcome of this method.

One issue remains in choosing an approach to predictive modelling, namely whether the procedure honours the data. That is, at places where measurements have been made, does the model have the same value as the measurement? By their very nature, inverse modelling methods usually do honour the data. On the contrary, forward modelling methods usually do not, and steps have to be taken to *condition* the model so that data values are respected (Figure 12.2F). A common solution to this is iteratively to compare model and data, adjust the starting parameters of the modelling procedure, then rerun the modelling procedure. In this way, the data and model hopefully converge. A difficulty with this solution is that natural systems frequently display non-linearity and chaotic behaviour, such that very small changes in starting value may produce very large differences in end result (e.g. Lorenz, 1969; May, 1976); it may therefore be difficult to find convergence.

In practice, subsurface prediction normally incorporates two or more of the above modelling philosophies in hybrid methodologies. For example, process-based forward modelling requires parameters to be set initially, which is done by preliminary interpretation of the data, itself an inversion step. Furthermore, it is usual to compare the results of two or more approaches (as distinct from several runs of the same approach) to seek convergence. This will be discussed further below when each method is described in more detail.

CONCEPTUAL MODELS AND DESCRIPTIVE STUDIES

Conceptual methods

The formation of conceptual and qualitative explanations is the traditional, and usually first-applied, approach to interpreting natural phenomena. Even when it is intended to construct numerical or analytical models, it is useful to go through this stage to ensure the correct rule-set is brought to bear (i.e. make sure one is in the right "ball-park"). The danger is that this is so inherent in geological science that the method may be applied subconsciously to cover deficiencies in the quantitative methods, or inadequacies in data collection. For example, how many times have you heard the comment about contours on a map that "they don't look right"? It is not that the contours do not correctly reflect the data, but that they do not fit the pre-conceived ideas of the mapper; this may be due to insufficient data, or the application of the wrong conceptual model. It is because of this danger that space is devoted here to discussion of conceptual methods and models.

Dangers of "classification mentality"

The desire to simplify Nature by pigeon-holing complex problems is understandable, and imparts a sense of security. If the classification scheme is based on the key factor(s) that create differences, then the classification may aid comprehension. But too often the classification precedes interpretation, and is a device of convenience. Furthermore, it tends to imply that natural systems occupy discrete states when in reality there is a spectrum. Examples in the literature on fluvial systems are numerous, such as the common division between bed-load, suspended-load and mixed-load rivers (e.g. Galloway & Hobday, 1983; Flores *et al.*, 1985), which still dominates the thinking of practising hydrocarbon geologists (Figure 12.3B). Usually a conceptual model is constructed for each component in the classification.

On the positive side, these classifications have been useful in resource exploitation by highlighting the variation in heterogeneity with scale (indeed scale is commonly the basis of the classification), and have been a guide to engineers unfamiliar with the complexities of "real" geology. However, classification does not in itself help one determine subsurface geology. In particular, the classifications published so far are purely descriptive and empirical, and have no predictive power. For instance, the recent scheme from Weber & van Geuns (1990) is founded on a tripartite division of hydrocarbon reservoir types into layercake, jigsaw-puzzle or labyrinth (Figure 12.3A). They emphasize the product not the processes that created them and, with their focus solely on sand-body connectivity, fall into the trap of mixing several processes into a single category.

Facies models

A *facies* is a body of rock with specified characteristics. When the definition is based on the physical and chemical characteristics, strictly the term *lithofacies* should be used; if the considerations are the fauna and flora, the term *biofacies* is more appropriate. Rocks may be grouped also as *seismic facies*, according to their seismic properties such as reflector continuity, amplitude and velocity, or as *log facies*, according to their influence on downhole radioactivity, sonic and electric logs. Whatever the characteristics employed, a facies should ideally be defined objectively on observable, preferably measurable, features.

Whilst it is expedient that the definition of a facies be descriptive, and not require interpretation of the origins of the rock, it is now accepted good practice that where

414

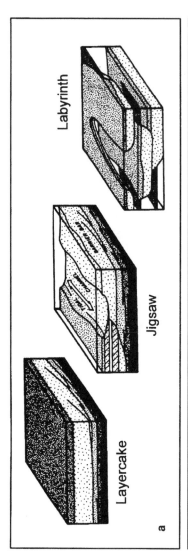

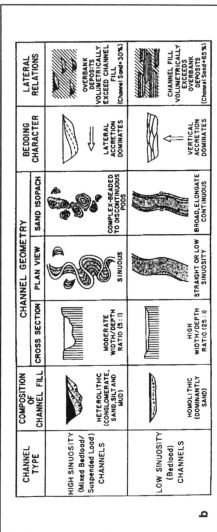

Figure 12.3 Examples of classification schemes in common use. (A) Framework for constructing reservoir simulation models, proposed by Weber & Van Geuns (1990). The authors' purpose was to illustrate to reservoir engineers the likely internal complexity of a hydrocarbon reservoir, but it is now being cited as a way to subdivide (classify) reservoir types for predictive modelling. Reproduced from Weber & Van Geuns (1990), "Framework for constructing clastic reservoir simulation models", *Journal of Petroleum Technology*, by permission of SPE. (B) Classification of fluvial channel characteristics, modified by Davies *et al.* (1993) from a scheme by Galloway (1981). This scheme separates high-sinuosity from low-sinuosity systems, and implies that bedload-dominated rivers are always low-sinuosity (see comments in channel type column). Reproduced from Davies *et al.* (1993) by permission of the Geological Society

possible a facies should be defined in such a way as to reflect a particular physical process or small group of processes (though note a movement to return to the use of grain size in the definition, such as in Reading & Orton (1991, their figure 6.1)). For example, in a sandstone sequence, one might select out a trough cross-bedded facies and a current-ripple cross-laminated facies; the former is the product of dune bedforms created by unidirectional currents, the latter is the product of much smaller bedforms which, whilst also formed by unidirectional currents, are produced by quite different energy conditions.

Miall (1977, 1978) has erected a classification of fluvial facies which has been used by many authors since. Taken at face value, this is an example where classification is misleading as it makes what is in reality a continuous spectrum appear discrete. Bridge (1993a) discusses this problem, and shows how the Miall scheme suffers from non-uniqueness of definition. However, it is important to appreciate that Miall intended one of the main values of the classification to be that "use of the code should aid in standardizing lithologic descriptions and ... facilitate comparisons between different fluvial sequences" (Miall, 1978 p. 599). Bridge (1993a) has objected to the proliferation in the literature of acronyms (lithofacies codes) of the type initiated by Miall. This aspect is one of personal preference and should not diminish the original objective of Miall to standardize descriptions so as to improve communication and allow comparison. The message here is as usual, think carefully about what you are doing, and keep an open mind – if the Miall scheme works on a particular sequence (and before trying it see the full details in Miall (1978) and read the discussion by Bridge (1993a)), perhaps with personal adaptations, then that is fine. But don't distort the facts just to make them fit the scheme.

Facies are the small-scale building blocks. *Facies associations* are groups of facies that occur together and are considered to be genetically related. The order of the facies in the association may be fixed, or vary spatially and temporally, and one or more facies may be missing at any place, but it is the high frequency of occurrence together that justifies the grouping. In fluvial environments, for example, a sequence of gravel facies and thick cross-bedded sand facies may be identifiable as a *fluvial-channel facies association*.

When the term *depositional environment* is used, it is in a physiographic sense to refer to a part of the Earth's surface that can be distinguished from adjacent parts because of variation in the totality of all the conditions, physical, chemical and organic, that influence the surface. Each environment is characterized by a particular balance between the physical processes operating, and the energy involved in each. There is thus a clear connection between the type of environment and the nature of the sediment. An environment is defined by its grouping of processes, so when attempting to determine the environment of deposition of a sedimentary rock sequence, one does so by determining the patterns of association of facies.

A *facies model* is a summary or type model which attempts to show which facies and facies associations occur in a particular environment, and the relationships between them. According to Walker (1992), a facies model is created by a "distillation process", boiling out from the local details of many modern and ancient examples that which is the "pure essence" of each environment. Geologists generally subscribe to the view that there are a relatively limited number of basic types of sedimentary environment, which in turn give rise to a relatively limited number of associations of lithologies, faunas and floras. Given the hypothesis that there are a limited number of possible environments, then there are a limited number of facies models needed to describe and illustrate all depositional environments. This requires acceptance of the philosophy that there is system and order in Nature.

Walker (1984b, 1992) argues that

> the generality embodied in a facies model, as opposed to a summary of one particular example, enables the facies model to assume four main functions:
> 1) it must act as a *norm*, for purposes of comparison;
> 2) it must act as a *framework* and guide future observations;
> 3) it must act as a *predictor* in new geological situations;
> 4) it must act as an integrated basis for *interpretation* for the system that it represents.

Proponents of this approach argue that the great value of facies models is in throwing out confusing detail in order to see the patterns and signature of each environment. Such models provide a framework so that what initially appeared random now appears comprehensible. They provide a norm against which new examples can be compared so as to identify important new features, and allow one to question whether they are of local or systematic importance. The models have merit in identifying the diagnostic features of each environment (this is largely the way Walker (1992) uses the term *predictor*), and being of predictive value in suggesting the expected overall spatial arrangement of sediment. When modified to take account of dynamic factors such as changing base level or climate, the models also predict the temporal evolution of deposits in an environment.

As an example of the power of facies models as the basis for interpretation, Walker (1992) cites the Bouma (1962) sequence for turbidites (Figure 12.4A). Prior to introduction of the model, each turbidite bed was interpreted individually and in isolation. Bouma generalized the internal structure for hundreds of turbidite flows, so providing the basis for understanding waning flow and deposition from turbidity currents. The same principle applies for other models of other situations such as fluvial point bars in meander bends (Figure 12.4C) (Allen, 1963, 1965, 1970; Bridge, 1975, 1977; Jackson, 1978) which have improved understanding and increased awareness of deposition by lateral accretion.

From the 1950s to the early 1980s, sedimentologists expended much effort on producing facies models for each and every environment and subenvironment. A sedimentological paper was not reputable if it did not describe a new model, and the creation of the models seemed to be the end in itself. Unfortunately, many of the new models contravened Walker's criteria, and were merely summaries of specific cases rather than a condensation of common characteristics. One difficulty is in knowing which features are general and which are local detail, when just a few examples have been studied.

New recognition of variations in fluvial style, and the need for additional models, can be seen by noting that Visher's 1972 review dealt only with meandering rivers, Miall in 1977 had four types, yet Schumm in 1981 illustrated 14 types, this flying in the face of previous studies of air photos (Galay *et al.*, 1973; Mollard, 1973) which had demonstrated an almost continuous spectrum of fluvial styles from low to high sinuosity, and single to multiple channels.

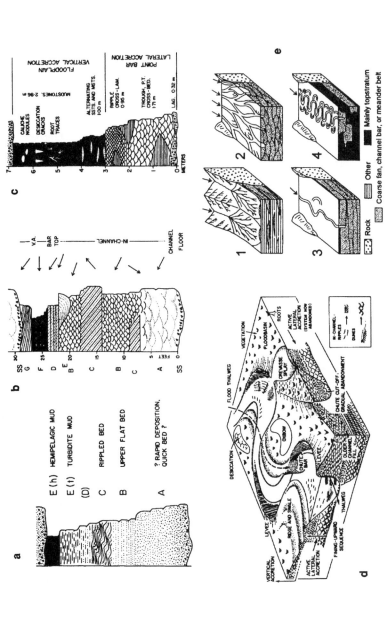

Figure 12.4 Examples of facies models. (A) The Bouma (1962) turbidite sequence representing the deposits of waning flow. Note the absence of scale in this model. Reproduced from Walker (1984a), in *Facies models*, by permission of the Geological Association of Canada. (B) Cant & Walker (1976) summary sequence for the Devonian Battery Point Sandstone, Quebec. This was developed by Markov analysis of lithofacies observed at outcrop. Arrows show palaeoflow directions, letters indicate facies. VA stands for vertical accretion. This is commonly cited as a typical braided river model. (C) Model for lateral and vertical accretion deposits of meandering rivers, from Allen (1970), synthesized from a variety of outcrop examples. (D) Block model showing morphological elements of a meandering river system, from Walker & Cant (1984). (B) to (D) reproduced from Walker & Cant (1984), in *Facies models*, by permission of the Geological Association of Canada (E) Hypothetical models produced by Allen (1965) to illustrate textural and geometrical characteristics of common alluvial facies: (1) piedmont formed of alluvial fans; (2) braided stream; (3) low-sinuosity stream; (4) strongly meandering stream. Note: Allen shows that low-sinuosity streams do not have to be braided – compare to Figure 12.3B. Reproduced from Allen (1965), "A review of the origin and characteristics of Recent alluvial sediments", *Sedimentology*, by permission of Blackwell Science Ltd

The products of this endeavour are epitomized by the two editions of the book called *Facies Models* edited by Walker (1979, 1984a), but no self-respecting sedimentology text book would be without its full set of facies models. Individual models are generally presented in one or both of two forms. The first is as an idealized vertical sequence (e.g. the Battery Point model for braided-stream deposition of Cant & Walker (1976)) (Figure 12.4B), which shows grain-size variation, sedimentary structures, and often an idealized downhole log response. The second form is as a block diagram of the environment (e.g. Figure 12.4D), showing the three-dimensional distribution of facies and indicating the processes operating at various points in the environment.

There are also two classes of facies model, though these are end-members of a continuous spectrum. The first, in accordance with Walker's rules, is a presentation of facts and observations. The model for the Battery Point Sandstone of Cant & Walker (1976) (Figure 12.4B) is clearly in this class. The second is when the model is used primarily to present interpretation, and where the author is trying to communicate ideas. Such would be the models of Allen (1965) (Figure 12.4E) illustrating the broad differences to be expected in the facies of low- and high-sinuosity rivers, models which it is clearly understood are hypothetical. The problem is that not all authors have understood and kept the distinction, and have presented models in which the two aspects are inextricably intertwined. This is not helpful for those who wish to use the models, and ultimately undermines the value of the original work.

Attempts to condense detail and summarize interpretations can be taken too far, and there is a danger that the models are fundamentally misleading. Expectations may sometimes exceed reality. Blair & McPherson (1992) have shown recently that the Trollheim-type vertical sequence published by Miall (1978) as an alluvial-fan braided gravel river model (Figure 12.5A) contains elements such as fining-up profiles and cross-bedded sand that are not actually present in the type section, the Trollheim fan, Owens Valley, California (Figure 12.5B). Indeed, this fan is formed entirely of debris-flow deposits, and fluvial sediments are absent.

There is also a danger that the models become the accepted wisdom, and govern later interpretations. For example, Walker & Cant (1984), in discussing the Cant & Walker (1976) summary vertical sequence for the Battery Point Sandstone, which is interpreted to be the result of a low-sinuosity fluvial system, note that fine-grained vertical accretion deposits are very thin compared with the meandering norm. The first point to note here is that Walker & Cant state emphatically that the Battery Point sequence they present is not a [facies] model, but is only a summary of a local example that could be redistilled with local examples from other areas to produce a general facies model. However, though it is not a facies model according to the criteria of these authors, it is nonetheless a model of sorts in that it is an idealized and simplified representation of reality, and it has been used by countless other workers as though it were a general facies model. For instance, Collinson (1986) reproduces the vertical profile with a caption that includes the words "facies model". The second point to note is that others have taken a low proportion of fine-grained accretion deposits to be a diagnostic feature of low-sinuosity systems, and have interpreted their own data in this light. Yet, as discussed by Bentham *et al.* (1993),

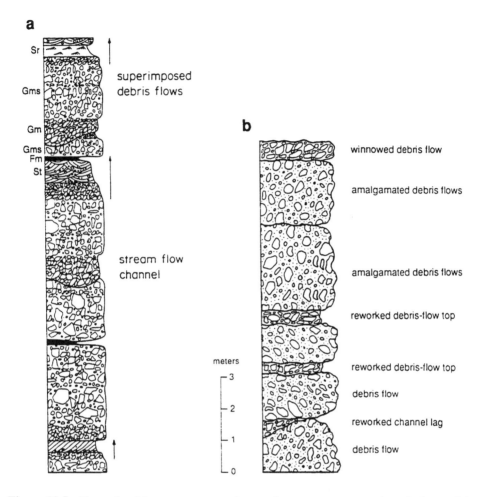

Figure 12.5 Example of how attempts to characterize sequences to produce facies models can introduce errors. (A) is Miall's (1978) schematic vertical section for the Trollheim fan, including debris flow (Gms), channel gravel (Gm), cross-bedded and ripple-bedded sand (St, Sr), and overbank clay (Fm) deposits. Arrows denote fining-up cycles. (B) Idealized sequence compiled by Blair & McPherson (1992) from their own fieldwork. They see the sequence as dominated by vertically amalgamated debris flows, with minor clast-supported gravel occurring as reworked mantles of the debris flows, or as localized channel-lag insets. Note the absence in this version of cross-bedded sands and clays Reproduced from Blair & McPherson (1992) by permission of the Geological Society of America

most of the facies models in the literature have been constructed from fluvial deposits produced in laterally confined and regionally degrading environments, and the paucity of vertical accretion deposits probably reflects these factors, not the river sinuosity.

There should be no doubt that conceptual facies models have been, and still are, a most useful tool to sedimentary geologists and those concerned with reconstructing past environments from information in the rock record, particularly when trying to interpret subsurface data. For example, analysis of core material typically starts with identification of individual facies (lithofacies usually), and these are interpreted by comparing the facies and facies associations described to the expected facies in published models. The models are an aid to understanding the context of core material, and assist determination of lithology from wireline logs, because they predict the local patterns of sediment variation. They are an aid to identification of the depositional environment because they suggest diagnostic criteria. They are an aid with well-to-well correlation, because they predict the relationships between facies and the probable spatial extent. Most importantly, they are a valuable visualization tool, which helps the geologist preserve her/his sanity when dealing with inadequate datasets and when faced with apparent chaos. Johnson & Stewart (1985) show how these aspects of facies models have been important in the search for hydrocarbons beneath the North Sea.

The use of conceptual facies models is an example of qualitative forward modelling, and so carries with it the weaknesses inherent in such a philosophy, as outlined above. As a conceptual and qualitative model however, it is easy to condition it to the data simply by adapting the model at the sample locations (is this cheating?). But there are problems with using this method to model the subsurface, some practical, and some philosophical.

From a practical stance, the biggest flaw is the qualitative nature of facies models, and the resultant imprecision. Such models cannot be used directly to construct numerical representations of the type required for simulating the effects on fluids of depleting an oil or gas reservoir, or for modelling the passage of radioactive contaminants around a waste-disposal site. Most published models are devoid of numerical information, and many even lack scale information, or merely show the range (usually wide) of length scales over which the features might be expected to develop. Valuable information on the natural variability within an environment is absent because it has been distilled out by the very process of creating the model. Information on spatial variation in facies and palaeocurrents is usually absent. yet is fundamental to full understanding of fluvial systems. We cannot expect a single vertical facies sequence to be representative of a single channel bar, much less a class of channel pattern (Bridge, 1985). Detailed studies of porosity and permeability variation in sandstones (e.g. Lasseter et al., 1986; Corbett & Jensen, 1992) have highlighted that, for hydrocarbon extraction, it is the variability that is often more important than the norm. Until fairly recently, raw data were rarely published, which is one reason for the resurgence of outcrop studies, to go and collect it again (e.g. papers in Miall & Tyler, 1991).

Another thorny problem is what is the right scale for creation of a facies model? The smaller and more distinct the environment, the easier it is to characterize it and construct a meaningful model. Thus the point-bar model has stood well. But the larger the area considered, the more features it includes, so the harder it is to find points in common and instead everything seems to be important. This explains the plethora of models, each only slightly different.

A weakness of facies models is that they inherently include both autogenic and allogenic factors, without distinguishing between them or explicitly identifying them. Perhaps another confusion comes in because of a failure to take adequate account of the *rates* of processes – facies models hardly ever attempt to embrace the time dimension. Major differences in preserved sediment character and distribution can be attributed to the balance between deposition and erosion rate in the alluvial setting (e.g. Bridge & Leeder, 1979), or the balance between the allogenic factors such as sea-level change, tectonic activity and sediment supply (e.g. Sloss, 1963; Harbaugh & Bonham-Carter, 1970). Should we create a separate facies model for every permutation of these factors?

Furthermore, are facies models a valid approach to tackling natural systems? The school of thought behind adoption of facies models is that there are a relatively limited number of environments. Others take the view, however, that there is an infinite number of environments, each of which grades slowly one into another such that it is impossible to identify main characteristics (Galloway & Hobday, 1983; Anderton, 1985; Miall, 1985). Anderton (1985) is contented that facies models have value as a framework and predictor, but disagrees on the norm and interpretation basis. The problem here may be that of scale, mentioned above: there are a limited number of distinct processes, but a very large number of ways these processes can be arranged and interact. Or it may be that every environment is unique and there are few steady states that can be described as "typical". The alluvial realm, for example, should perhaps be regarded as a continuous spectrum, where fluvial facies models are snap-shots in time and space of the range of possibilities (e.g. Reading & Orton, 1991, their figure 6.1). Brierley (1991, see also Chapter 8) has proposed the concept of morphostratigraphic element assemblages as a way to handle the spectrum of variation in river systems. Caution certainly is needed, as models can lead to a false sense of security.

Architectural element analysis and lateral profiling

Vertical profiles have been the basic tool used in constructing fluvial facies models, and have been regarded as of diagnostic potential in their own right (e.g. Cant & Walker, 1976; Miall, 1977; Rust, 1978). However, there has been an increasing recognition that such profiles have been less than successful in predicting lateral facies relationships or channel migration and stacking patterns. This might be expected from a theoretical consideration alone (Collinson, 1978). It has now been demonstrated (e.g. Allen, 1983; Ramos & Sopeña, 1983; Miall, 1985) that full understanding of fluvial deposits cannot be expected without considering the lateral geometry of the facies as well as the vertical profile.

Architectural elements are components of a depositional system equivalent in size to, or smaller than, a channel fill, and larger than an individual facies unit (Allen, 1983; Miall, 1985). They are characterized by a distinctive facies assemblage (facies association), internal geometry, external form and (in some but not all instances) vertical profile. Implicit in their definition is the three-dimensional morphology, and they can be recognized only by a combination of vertical and lateral profiling of rock sequences. Architectural elements have been used by some workers since the 1960s to describe and interpret sedimentary sequences (e.g. Horne *et al.*, 1978), but it was not until the 1980s that attempts were made to formalize the procedures and to define the terminology.

Miall (1985, 1988a) has expanded and synthesized earlier work on fluvial systems, notably that of Allen (1983) on the Devonian braided stream deposits in the Welsh Borders, to develop a scheme with eight basic types of architecture element (Table 12.3, Figure 12.6). The elements are proposed as constants in the fluvial environment, whereas the ways to combine the elements are infinitely variable. It should be noted, though, that the elements defined by Miall can vary internally, and it is this aspect that has proved confusing to some workers, and has lead to objections from others (e.g. Bridge, 1993a) that this is one reason the scheme is unworkable. Miall (1995) acknowledges this point, but repeats the warning given in his 1985 paper on this aspect. The alternative scheme suggested by Bridge (1993a) has its own deficiencies (Miall, 1995) and still does not handle the problems of gradational contacts and transitions between packages. In practice, it is always necessary to find some pragmatic scheme for subdivision of a rock sequence, in order to try to find patterns (i.e. the primary building blocks) in the complexities of natural systems so as to attempt interpretation. Provided the basis of the scheme is clearly explained, it does not matter so much which scheme is employed.

Table 12.3 Architectural elements of Miall (1985, 1992), with critical comments of Bridge (1993a). See Figure 12.7 for explanation of third and fourth order surfaces

Element symbol	Element	Geometry and relationships	Critical comments (modified from Bridge (1993a))
CH	Channels	Finger, lens or sheet; concave-up erosional base; internal third-order erosion surfaces	A composite unit – all other elements can occur in channels
GB	Gravel bars and bedforms	Lens, blanket, tabular; commonly interbedded with SB	Not unique element as can occur as part of elements CH, LA and DA
SB	Sandy bedforms	Lens, sheet, blanket, wedge; occur as channel fills, crevasse splays, minor bars	Same as GB; crevasse splays and channel fills cannot be represented by a single unit
DA	Downstream-accretion macroforms	Lens on flat or channelled base; convex-up third order surfaces, fourth order upper bounding surface	Ambiguous definition, as most channel bar forms result from a combination of downstream and lateral accretion. Function of way element sectioned
LA	Lateral accretion macroforms	Wedge, sheet, lobe; internal third order surfaces	As DA
SG	Sediment gravity flows	Lobe, sheet, typically interbedded with GB	A lithofacies (Miall's facies code Gms) not an element
LS	Laminated sand sheet	Sheet, blanket	A lithofacies not an element; not unique, as can occur as part of elements LA, DA, CH
OF	Overbank fines	Blankets; commonly interbedded with SB; may fill abandoned channels	"Overbank" facies (Miall's codes Fm & Fl) are not always overbank deposits, and can occur as part of CH, LA, DA

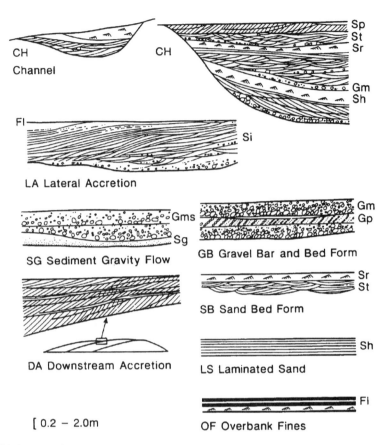

Figure 12.6 The eight basic architectural elements in fluvial deposits, as proposed by Miall (1985, 1992). No vertical exaggeration. Note the variable scale. Reproduced from Miall (1992), in *Facies models: response to sea level change*, by permission of the Geological Association of Canada

Miall (1988a, 1991a, 1992) also emphasizes the hierarchy of scales of depositional units, and the *bounding discontinuities* enclosing the architectural elements. He suggests that fluvial depositional units can be subdivided into ten natural groups based on their physical size, sedimentation rate, and the time scale represented by each unit (Table 12.4, Figure 12.7). The bounding discontinuities include the contacts between beds and sedimentary structures, channel scours, the base and top of stratigraphic units, and the surfaces which define the major allogenic subdivisions of alluvial successions, such as regional unconformities.

Miall advocates lateral profiling and *architecture element analysis* as a useful tool for field studies, facilitating field description and leading to more comprehensive interpretations. It has proved to be a useful descriptive framework (Miall, 1995) as shown by the many papers which have embraced the methodology (e.g. several outcrop studies described in Miall & Tyler (1991)). The approach is valuable in forcing consideration of lateral and vertical extent, and the relationships between the

Table 12.4 Hierarchy of depositional units in alluvial deposits, from Miall (1991a). See Figure 12.7 for explanation of bounding surface terminology

Group	Time scale of process (years)	Example of processes	Instantaneous sedimentation rate (m ka^{-1})	Fluvial depositional units	Rank and character of bounding surfaces
1	10^{-6}	Burst-sweep cycle		Lamina	Zeroth-order, lamination surface
2	10^{-5}–10^{-4}	Bedform migration	10^5	Ripple (microform)	First-order, set bounding surface
3	10^{-3}	Bedform migration	10^5	Diurnal dune increment, reactivation surface	First-order, set bounding surface
4	10^{-2}–10^{-1}	Bedform migration	10^4	Dune (mesoform)	Second-order, coset bounding surface
5	10^{-0}–10^1	Seasonal events, 10-year flood	10^2–10^3	Macroform growth increment	Third-order, dipping 5–20° in accretion direction
6	10^2–10^3	100-year flood, bar migration	10^2–10^3	Macroform (point bar, levee, splay)	Fourth-order, convex-up macroform
7	10^3–10^4	Long term geomorphic processes	10^0–10^1	Channel	Fifth-order, flat to concave-up channel base
8	10^4–10^5	Fifth-order (Milankovitch) cycles	10^1	Channel belt sequence	Sixth-order, flat, regionally extensive
9	10^5–10^6	Fourth-order (Milankovitch) cycles	10^{-1}–10^{-2}	Depositional system, alluvial fan, sequence	Seventh-order, sequence boundary; flat, regionally extensive
10	10^6–10^7	Third-order cycles, tectonic and eustatic processes	10^{-1}–10^{-2}	Basin-fill complex	Eighth-order, regional disconformity

components. Miall (1993, 1994a) has shown with some detailed examples how analysis in two-dimensional outcrop of downstream and laterally accreting elements allows reconstruction of the three-dimensional fluvial architecture.

On the minus side, the technique is not always easy to apply even at outcrop. It can be difficult to trace out surfaces, which often become indistinct and gradational. Actual fluvial rocks are not as strictly hierarchical as suggested; for example channels (bounded by fifth-order surfaces) are commonly filled not by macroforms such as bars (bounded by fourth- or third-order surfaces) but by the deposits of dunes (bounded by second-order surfaces). It is not clear that Miall's eight elements (Table 12.3) are any advance on the components already used in the literature (e.g. Friend, 1983) such as mesoforms and macroforms, or dunes, bars (qualified as gravel or sandy) and channels, though a particular strength is that it does emphasize

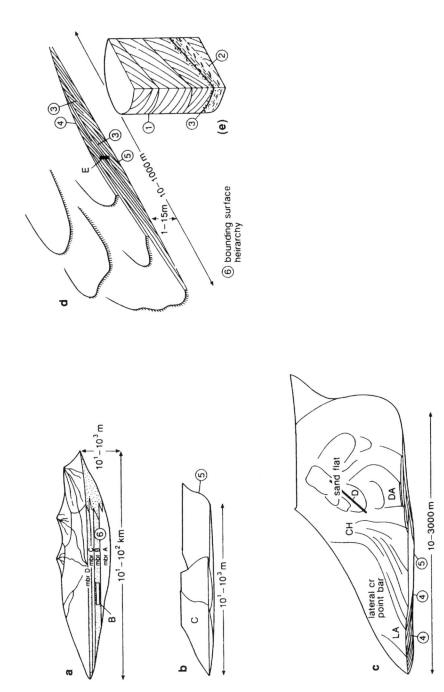

Figure 12.7 The scales of depositional elements in a fluvial system, as proposed by Miall (1988b). This also illustrates the bounding-surface hierarchy listed in Table 12.4. A circled number indicates the rank of the bounding surface. In (c), CH is a channel element (see Table 12.3), LA is a lateral accretion element, and DA is downstream accretion. In (d), the sand flat is shown as building up from migrating "sand waves". From Miall (1988b); reprinted by permission of AAPG

the distinction of lateral and downstream accreting forms. Several workers have found the list of elements too restrictive or inappropriate (e.g. DeCelles *et al.*, 1991; Fielding *et al.*, 1993), and have been forced to define their own set.

By demonstrating a correspondence between his architectural element hierarchy and the scales of heterogeneity recognized by petroleum geologists as influencing fluid flow in the subsurface (e.g. Tyler, 1988; Figure 12.8), Miall (1988b) also advocates the adoption of this approach for subsurface reservoir description and modelling. It is difficult to see how this can be used for subsurface prediction, and is a puzzling recommendation, since Miall himself recognizes that, by definition, architecture element analysis cannot be accomplished on one-dimensional datasets from the subsurface, such as borehole core, since it requires three-dimensional observation of the elements (Miall, 1985).

In his 1988 paper advocating use of this method for reservoir heterogeneity studies, Miall (1988b, p. 694) advises reservoir modellers to "... use the bounding surface hierarchy to examine each component separately in its correct scale context". He gives guidelines on how bounding surfaces of various orders, and thus the architectural elements, may be recognized in subsurface data. However, Miall himself says (1988b, p. 686) "... third- and fourth-order surfaces may be very difficult to distinguish in individual wells". Later he notes "Fourth-, fifth- and sixth-order surfaces may appear very similar to third-order surfaces in core". He admits (p. 686) that "Even in excellent outcrop, the correct classification of bounding surfaces is not always easy". The only way to identify features of this scale is with closely spaced wells, but the inter-well spacings required are in the range 50–500 m, which as already discussed is not often achieved, and never at the field appraisal stage when much of the modelling needs to be done.

There may yet be an important use for architecture element analysis in subsurface prediction of stratigraphy. As Miall (1985) points out, a strength of the approach is that it is purely descriptive, and frees the sedimentologist from rigid adherence to any preconceived (facies) model. Many of the quantitative methods of subsurface modelling, described below, require a supply of information gathered from outcrop analogues. As advocated by Miall himself (e.g. Miall, 1994a), this scheme or something like it may be of considerable value as a standardized descriptive framework for the collection of outcrop data.

Systems tracts and sea-level change

Systems tracts are defined as "a linkage of contemporaneous depositional systems" (Posamentier *et al.*, 1988). Contemporaneity may be established biostratigraphically, but commonly the systems tracts are identified on the basis of their bounding discontinuities. Many discontinuities form, directly or indirectly, as a result of fluctuations in relative sea level, so immediately we can discriminate three main systems tracts – highstand, lowstand and transgressive. Although too large to be modelled in the same way as individual depositional environments, the concept of systems tracts is of significance because it allows prediction to be made from one depositional environment to another. If we can identify the type of systems tract and recognize an environment at one point within it, then potentially we are able to

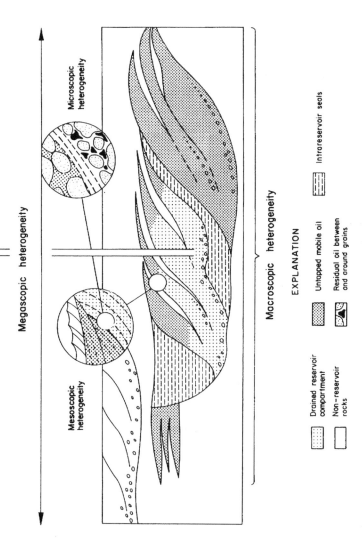

Figure 12.8 Illustration of the scales of heterogeneity occurring in a hypothetical meander-belt reservoir sandstone, from Tyler & Finley (1991) after Tyler (1988). This also illustrates the relation of heterogeneity to swept and uncontacted reservoir components. Compare this to the scales selected by Miall for his architectural elements (Figures 12.6 and 12.7, Tables 12.3 and 12.4). Reproduced by permission of SEPM

predict the distribution of the other environments within that tract. For example, the recognition of a shelf-edge delta in a lowstand tract leads to the prospect of a related deep-sea submarine fan.

There can be no doubt that fluvial stratigraphy is intimately linked to changes in base level of the river system. The base level may be a regional one such as sea level, or a more local one such as an intermontane lake. Arguably, even local base levels are genetically connected, ultimately, to sea level. Therefore, modelling concepts based on sea-level change ought to have as much application in the continental realm, and to alluvial stratigraphy, as they do to the marine realms for which they were originally developed. They certainly apply to many marine-fringing alluvial plains and related fluvio-deltaic environments, which form probably the majority by volume of the sequences that one is trying to predict in the subsurface. The difficulty, however, is that fluvial systems take a long time to reach equilibrium with their base level, so even when that base is sea level, which keeps changing, the fluvial environment will always be out of step and out of phase. In addition, due to the numerous internal differences between river systems, even adjacent rivers may respond differently to the same base-level rise.

Many workers now believe that it is the global or eustatic changes in sea level that have been the major influence on the boundaries of depositional systems and preservation in the geological record (e.g. Haq *et al.*, 1988). If this is true, then knowledge of the variation of global sea level over geological time can be used to make a first-pass prediction on the spatial and temporal distribution of depositional environments. There are two schemes for approaching stratigraphy based on bounding discontinuities and unconformities: allostratigraphy and sequence stratigraphy.

Allostratigraphy

According to the North American Stratigraphic Code (NACSN, 1983), "an allostratigraphic unit is a mappable stratiform body of sedimentary rock that is defined and identified on the basis of its bounding discontinuities". Walker (1992) believes the definition should be extended to read "… bounding discontinuities and their correlative conformities". Bounding discontinuities can be erosional or non-erosional (conformable), and can occur at all scales from the subtle scour within a channel to angular unconformities with erosional relief of hundreds of metres. Bounding discontinuities highlight important changes in depositional conditions. Thus they define allostratigraphic units in which depositional conditions were either fairly constant, or were progressively changing but without breaks (e.g. fining-upward sequences). For this reason, allostratigraphic units are more natural subdivisions of the geological record for interpretive purposes than conventional lithostratigraphic units (Figure 12.9). The scheme in turn emphasizes the processes external to the depositional system that initiate and terminate sedimentologically related facies.

Allostratigraphy is a purely descriptive concept, but unlike conventional lithostratigraphy it does force consideration to be given to processes, especially allogenic ones. Although it has little predictive power, it is mentioned here because it

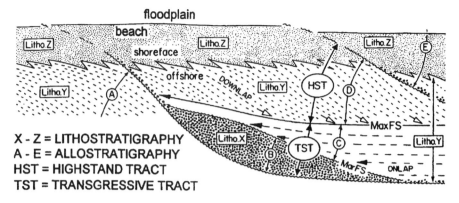

Figure 12.9 Comparison by Walker (1992) of allostratigraphic description with conventional lithostratigraphy and sequence stratigraphy. Lithostratigraphic units are X (conglomerate), Y (shales) and Z (sandstones). Allostratigraphic units are A to E. Sequence divisions are HST (high-stand systems tract) and TST (transgressive systems tract). MaxFS is maximum flooding surface, MarFS is marine flooding surface, which, with the surfaces marked with small zig-zags, are bounding disconformities. Large serrations represent gradational facies change. Reproduced from Walker (1992), in *Facies models: response to sea level change*, by permission of the Geological Association of Canada

is in effect an extension of Miall's architecture element analysis, and is the general case of sequence stratigraphy, which is described next. Martinsen *et al.* (1993) have shown that in some cases, such as where there are major changes in sediment source direction and regional tectonism, allostratigraphy is preferable to a sequence stratigraphic description.

Sequence stratigraphy

Sequence stratigraphy is a method of interpreting stratigraphic data that was formulated through the study of seismic sections (seismic stratigraphy), and later applied to outcrop and well data. The method was developed by researchers at the Exxon Production and Research Company in Houston, led by Peter Vail who had been influenced as a graduate student by his supervisor Larry Sloss. Sloss had spent many years investigating major unconformity-bounded sequences on the N. American craton, and linking transgressive–regressive cycles to orogenic events (Sloss, 1979, 1988). Although in use in the 1960s, it was not until 1977 that the Exxon school's labours had a proper public airing, with an AAPG Memoir on seismic stratigraphy (Payton, 1977). This work also showed how major seismic sequences could be related to eustatic sea-level changes, and included a chart of sea-level change for the Mesozoic to the present – the so-called "Vail curves" – derived from wide-ranging regional studies of coastal onlap (Vail *et al.*, 1977). The technique has been refined by improving the sea-level curves, culminating in the so-called "Haq curve" (Haq *et al.*, 1987). A point often overlooked is that much of the early work in this area was reliant on the use of seismic data, as is much of the current application of the technique for hydrocarbon exploration today, and so depends on assumptions about the resolution and interpretation of these data (Thorne, 1992).

By incorporating lithological data and facies successions, seismic stratigraphy has given rise to the more geologically oriented sequence stratigraphy (see Wilson (1992) for an excellent summary, and Wilgus *et al.* (1988) for a full account). The fundamental unit is the sequence, which is "a stratigraphic unit composed of a relatively conformable succession of genetically related strata bounded at its top and base by unconformities and their correlative conformities" (Mitchum *et al.*, 1977). An immediate problem is that the terms "genetically related" and "relative conformities" are undefined, and this has been the cause of much controversy (Walker, 1990). However, by defining packages of strata bounded by unconformities, sequence stratigraphy emphasizes the external controls on sedimentation, and encourages a chronostratigraphic rather than lithostratigraphic analysis. Note that, by this definition, sequence boundaries can occur *within* lithostratigraphic units.

Fundamental to the evolution of stratigraphic architecture is the concept of *accommodation space*, which is described by Jervey (1988) as "the space made available for potential sediment accumulation" where "... in order for sediments to be preserved, there must be space available below base level (the level above which erosion will occur)". Schumm (1977) has shown that river systems can be divided broadly into three zones, an upstream sector dominated by erosion, a midstream sector dominated by sediment transport, any storage being temporary, and a downstream sector dominated by deposition. Whether deposition occurs above sea level on alluvial plains, or below sea level (non-fluvial) on the continental shelf, depends on the balance between the rates of sediment supply, tectonic subsidence, and eustatic sea-level change. The balance between these factors controls the size of the accommodation space in which sediment accumulates; if there is no space available, sediment ultimately will be transported further away. What matters is the *relative* sea-level height, and the rate and direction of any change in it. If sediment supply remains constant, it is the rate at which relative sea level rises – and so creates new accommodation space – that determines the extent to which sediment aggrades or progrades. A slow rise (or stillstand) favours progradation (insufficient space), a rapid rise favours aggradation and the geometry and architecture of the resultant strata will be different. The Vail school believe that it is eustatic sea-level changes that are dominant in modulating the rate at which accommodation space is created or removed. For reasons as discussed shortly, others disagree, but this is a point of detail in the method, and should not be regarded as a defect in the basic concepts of sequence stratigraphy.

The Exxon researchers have produced a series of hypothetical models that predict the succession of related depositional systems (*depositional systems tracts*) associated with the different parts of a cycle of eustatic rise and fall (Posamentier & Vail, 1988) (Figure 12.10). In part conceptual, in part based on forward mathematical modelling of the processes (Jervey, 1988), these models make explicit predictions about how the fluvial systems will behave, and the nature, albeit broadbrush, of the sediment that ought to be preserved. According to these models, significant fluvial deposition is expected during early sea-level rise (transgressive tract), and during highstands and the early part of sea-level fall (highstand tract).

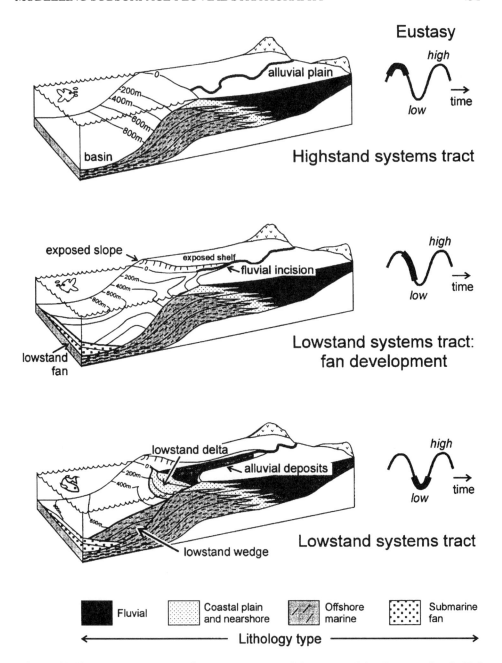

Figure 12.10 Three examples of the sequence models proposed by Posamentier & Vail (1988) to show hypothetically the change in depositional environments caused by eustatic sea-level change. These show the impact on fluvial systems caused by a eustatic fall in sea level. This first triggers fluvial incision on the exposed marine shelf, then leads to fluvial aggradation within the incised valley. The exact sequence of events depends on factors not shown in these models – for full details, see discussion in Posamentier *et al.* (1988b) and Posamentier & Vail (1988). Reproduced by permission of SEPM

Furthermore, fluvial deposition will be confined to incised valleys during sea-level rise, but will become "widespread" as the highstand is reached and passed (Posamentier & Vail, 1988). In this way, it is considered theoretically possible to predict fluvial stratigraphy in the subsurface from regional mapping of depositional environments (such as from seismic surveys) and using a global sea-level curve (e.g. the Haq *et al.* (1988) curve) to find the rate of sea-level change at the time the sediment is believed to have been deposited.

Building on earlier work at the Texas Bureau of Economic Geology, Galloway (1989a,b) has devised an alternative scheme of sequence stratigraphy. The Exxon school pick the boundaries between depositional sequences on the basis of unconformities, since these are readily apparent on seismic sections. Furthermore, their definition of an unconformity is a surface "along which there is evidence of subaerial erosional truncation ... or subaerial exposure" (Posamentier *et al.*, 1988). In contrast, Galloway believes genetic sequences are naturally divided by transgression events, so sequences should be bounded by maximum-flooding surfaces (Figure 12.11 shows the difference between these definitions). It is argued that these

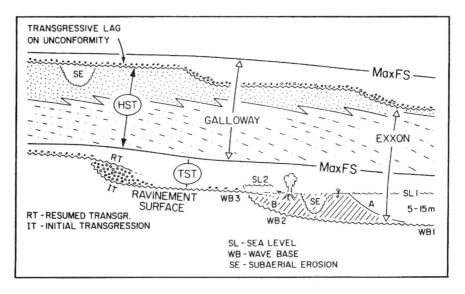

Figure 12.11 Diagrammatic comparison by Walker (1992) between the Exxon sequence boundaries and the Galloway genetic stratigraphic sequence boundaries (maximum flooding surfaces, MaxFS). The transgression erosion surfaces are shown stepped, implying alternations of stillstand or very slow transgression, and more rapid transgression. During very slow transgression (SL1 to SL2), the shoreface erodes landward (A to B lower right). removing the diagonally ruled section. Note that, as a result, all evidence of subaerial erosion (trees, roots, shallow incised channels) is removed. During rapid transgression, wave base rises from WB2 to WB3, and the top of the former beach is eroded (below letters SL2). This gives rise to the concept of separate erosion surfaces associated with initial transgression and resumed transgression. Only those channels incised more than 5–15 m (SE, top left) can be preserved below ravinement surfaces. This depth approximates the depth from sea level to fair-weather erosive wave base. Reproduced from Walker (1992), in *Facies models: response to sea level change*, by permission of the Geological Association of Canada

surfaces are more widespread, are easier to detect than unconformities, and there is no necessity to try and prove subaerial erosion or exposure. Galloway then develops a series of depositional models related to relative sea-level change and subsidence that are broadly similar to those of the Exxon school, but which differ in internal detail. The Exxon scheme as originally proposed was intimately linked to eustatic sea-level changes. Galloway's scheme recognizes the importance of other controls on depositional architecture, and is thus more appealing to fluvial stratigraphers. There has been considerable argument as to whether the Galloway approach or the Exxon approach is "correct". Both have advantages and disadvantages, which relate partly to whether you are working with seismic data or rocks at outcrop. The Galloway approach is more lithofacies oriented. Though both approaches claim their sequences are *genetic*, they are only genetic in the sense that they develop during one complete cycle of relative sea-level fluctuation. They are not sedimentologically genetic, because most sedimentological parameters change when an unconformity or maximum-flooding surface is crossed (Walker, 1990, 1992). With the aid of a regime-based numerical simulation model, and comparisons to modern sedimentation, Thorne & Swift (1991) have compared the Exxon and Galloway models for depositional sequences. They conclude that both approaches have their merits, and suggest refinements to both. But they come out in favour of the Galloway direction because it places greater emphasis on sediment input to a basin, and less reliance on eustatic changes.

In practice, it has proved much more difficult to test sequence stratigraphy predictions in alluvial strata than in marine sequences, even at outcrop (Walker, 1990). In thin successions of alluvial strata where incision has occurred into underlying marine strata, recognition of regionally significant sequence boundaries related to changes in stratigraphic base level is comparatively straightforward. The approach has been successfully applied to shelf and delta deposits at the surface and in the subsurface (e.g. Van Wagoner *et al.*, 1990), and as early as 1964 allowed Exxon to predict the occurrence of submarine fans in the North Sea Tertiary before any drilling had been done. But studies on fluvial deposits have been frustrated by abrupt lateral facies changes, typically poor biostratigraphic resolution and limited age-dating, numerous internal erosion surfaces, the absence of through-going marker horizons, the limited extent of outcrops on which to gain experience, and the virtual impossibility of distinguishing alluvial from marine sediments on seismic. The emphasis in the definitions of sequence stratigraphy elements on marine flooding surfaces is bound to work against its use on non-marine rocks.

One outstanding example of sequence stratigraphic analysis of an alluvial package has been carried out on the Upper Cretaceous rocks exposed in the Kaiparowits Plateau of S. Utah, over distances up to 100 km (Shanley & McCabe, 1991, 1993). Superb exposures in the cliffs here allow surfaces to be traced from marine strata into the alluvial sequence. Coals provide good correlation markers. In addition to abrupt change in facies development across the sequence boundary unconformities, these authors have documented a fundamental change in fluvial architecture. Below the sequence boundary, the pattern is of multistorey sand bodies that contain complete fining-upward units in the uppermost storey; channels are isolated in fine-grained floodbasin facies. Above the boundary, channels are laterally amalgamated, coarse-grained and pebbly, fining-upward storeys are incomplete, and the system is one of lower sinuosity. Such an upward change in style was anticipated by the simulation modelling of Bridge & Leeder (1979), as discussed in a later part of this chapter.

By synthesizing their observations within the sequence stratigraphic framework, Shanley & McCabe (1993) have produced a predictive model (Figure 12.12) that can account for the observed change from amalgamated channels filling incised valleys during lowstands, through to isolated high-sinuosity channels at highstands, including a phase during transgression of tidally influenced fluvial deposits. They emphasize that the key to this approach, and the prime difference from previous attempts to explain the architectural changes, is the recognition of regionally extensive, time-significant unconformities (the sequence boundaries). They advocate the use of the method, and their model, for subsurface studies of coastal-plain fluvial aquifers and petroleum reservoirs, and suggest that it explains the occurrence of low-sand-content, isolated, high-sinuosity channels observed in such diverse examples as the Triassic Ivishak Sandstone of Prudhoe Bay in Alaska (Atkinson et al., 1990), the Middle Jurassic Ness Formation of the Brent Group in the northern North Sea (Livera, 1989), and Pleistocene deposits of the South Belridge oil field in the San Joaquin Valley, California (Miller et al., 1990).

Aside from the practical difficulties of applying the technique to fluvial strata, even at outcrop, as outlined earlier, the application of the technique to subsurface prediction is dogged by concerns over the inherent theoretical basis of the hypothetical models themselves, and the construction and validity of the global sea-level curves. There is increasing realization that the geomorphic response of river systems to intrinsic thresholds and external triggers is complex (Wescott, 1993). The significance of these debates is that different assumptions about the geomorphic response change the time in the cycle of sea-level rise and fall that fluvial aggradation or incision is expected, and the predicted style of the river system.

From the fluvial standpoint, one of the key unanswered questions is the exact nature and amount that a change in sea level (base level) will influence the river system upstream. Part of the debate has been caused by differing definitions of the term "base level" (Schumm, 1993), and disagreements (e.g. Miall, 1991b) on the point to which the fluvial system adjusts (the bayline of Posamentier & Vail (1988)). Posamentier & Vail (1988) have built their model using the concept of river equilibrium profile, and predict aggradation or incision as this profile moves basinwards or downwards. Miall (1991b), Schumm (1993) and Wescott (1993) all highlight that there are fundamental flaws in the models that have been constructed because of errors in the reasoning over the equilibrium profile. For example, the presumption has been made that after an incision or aggradation event, a river attempts to restore its gradient, along its *whole* length, to that before the event occurred. The reasoning was that because the river discharge and sediment load were constant, the river had to return to the same gradient.

It can be shown (e.g. Schumm, 1993) that it is not necessary for the entire river to adjust, a fact supported by observations on modern rivers, which show that the distance upstream affected by incision is only a fraction of the whole river (e.g. Autin et al. (1991) on the Mississippi River; Blum (1992) and Blum et al. (1994) on

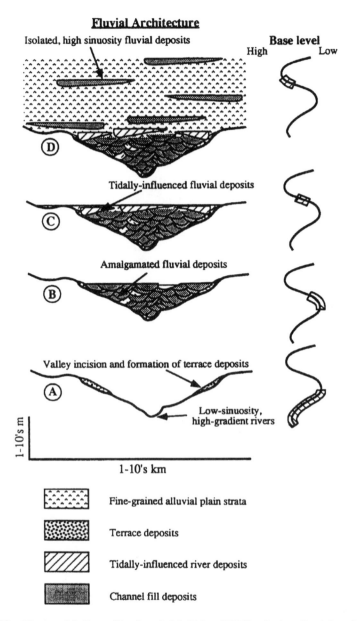

Figure 12.12 The model from Shanley & McCabe (1993) relating fluvial architecture to base-level change in a series of transverse sections. (A) Slow rates of base-level rise leading to base-level fall produce fluvial incision and alluvial terraces. (B) During reduced rates of base-level fall, and slow base-level rise, valleys initially aggrade with amalgamated fluvial deposits. (C) Increased rates of base-level rise produce tidally influenced fluvial deposits. (D) Reduced rates of base-level rise that are approximately balanced by rates of sedimentation result in isolated meander belt sandstones that "float" in a matrix of alluvial plain deposits. Reproduced from Shanley & McCabe (1993) "Alluvial architecture in a sequence stratigraphic framework: a case history from the Upper Cretaceous of southern Utah, USA", in *The geological modelling of hydrocarbon reservoirs and outcrop analogues*, by permission of Blackwell Science Ltd

the Texas Colorado River). An aspect usually overlooked during debates about this issue is that it is often difficult, especially with subsurface data alone, to tell the difference in an alluvial suite between a surface formed by incision or formed by the intrinsic scour that occurs in all rivers (Salter, 1993). It is highly likely that valley incision will be accompanied by the formation of terraces during periods of degradation (e.g. Blum *et al.*, 1994). The internal stratigraphy of these terrace deposits is complex and makes recognition of a single sequence boundary difficult using data from the subsurface alone. Incision or aggradation occurring in response to base-level change may modify the sediment load, so altering the gradient. Initial reasoning has been two-dimensional, and there has been a failure to distinguish between valley gradient and channel gradient. Rivers can adjust to base-level change by adjusting channel pattern, shape, length and roughness. For example, Schumm (1993) shows that by increasing its sinuosity, and thus increasing channel length, a river can adjust to a degree of base-level fall without changing valley gradient. The river system takes a long time to adjust to changes at the river mouth, a time that may far exceed the length of a eustatic cycle, so the river may never fully adjust to base-level change.

Base-level change is not the only way to achieve valley incision or aggradation. The relative role of tectonics and climate change may be hard to unravel, especially as they may be interdependent and are not fully understood. A methodological problem exists in studying such complex interactions, namely that of convergence, where different processes can produce similar results (Schumm, 1991). Incision can be triggered by base-level fall, by climate change (e.g. Hall, 1990), or by tectonic uplift (Sloss, 1991). Depth of incision does not have to equal, or even approach, the level of base-level fall. Indeed, deep levels of incision are much more likely to be the product of tectonic uplift than base-level fall (Schumm, 1993). Much work is still required in this area. Flume-based experimental studies such as those of Koss *et al.* (1994) can be most useful in this respect. A consensus view is emerging that near the river mouth eustatic effects are likely to dominate, but tectonic and climatic effects become more important in the upstream direction. Shanley & McCabe (1993) speculate that the current sequence models, based primarily on eustatic controls, are probably relevant within 100–150 km of the coast, because many major river systems, such as the Gironde in France, show tidal influence this far inland. In their Upper Cretaceous example, they recognized tidal influence up to 65 km from the shore. Inland of this, the influence of eustatic controls diminishes, or may be non-existent. Figure 12.13 is taken from Shanley & McCabe (1994), who report the deliberations of a working group at the 1991 NUNA conference on high-resolution sequence stratigraphy. It attempts to express qualitatively the relative balance between the external factors controlling fluvial systems, but there was no agreement at the conference on the appropriate horizontal scale.

Martinsen *et al.* (1993) have shown that the sequence stratigraphy models, developed largely from extensional passive continental margins, do not predict correctly the development of compressional foreland basins, and neither the Exxon nor the Galloway model can easily be applied even for description.

The other major concerns with sequence stratigraphy relate to the so-called global sea-level change curve (Haq *et al.*, 1987). Most of those applying this technique

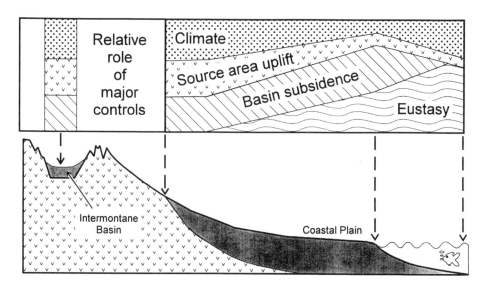

Figure 12.13 Schematic graph showing the way the relative influence of allogenic controls on fluvial stratigraphy changes with distance from the sea. From Shanley & McCabe (1994); reprinted by permission of AAPG

attempt to correlate major unconformities to this curve, and many claim a successful match (e.g. Shanley & McCabe (1993) discussed above). But there is still doubt about the accuracy and resolution of the curve (e.g. Miall, 1986, 1991b) partly because much of the data on which it is based have yet to be published. An example of the concerns is that the length of many of the claimed eustatic cycles is much shorter than the potential errors in dating techniques (Miall, 1991b). Furthermore, there are inconsistencies within and between publications on the ages of specific events. It is difficult to see how others are going to match their observed sequences to this curve except fortuitously. The curves are as yet not independently tried and tested.

Perhaps the most crucial concern is whether the sea-level curve is, or could ever be, truly global. Miall (1986, 1991b) has shown that the basis of the curve is flawed because several of the type areas used to determine sea-level change events are tectonically unstable, a point which has recently been demonstrated convincingly for the Jurassic of the North Sea by Underhill & Partington (1993). Tectonic controls will operate only over localized areas. The curve has been heavily criticized for being based almost exclusively on passive margin types of sequence, which are inherently tectonically less active. Watts & Thorne (1984) have shown that Exxon-type coastal onlap curves can be produced without eustatic changes solely by considering crustal deformation caused by thermal contraction and sediment loading. Cloetingh (1988, 1991) and Cloetingh & Kooi (1992) have demonstrated that intraplate stresses can cause widespread relative sea-level changes of the type and size required, but such changes *could not* be global in effect. The fact that an apparently global chart has been constructed suggests that its basis should be closely examined (Miall, 1991b). Glacial control is the driving force usually postulated for global sea-level changes of the size and frequency given in the Haq curve, since the only other suggested control, changes in ocean volume caused by changes in size of plate spreading ridges, operates at much too slow a rate. The presence in the curve for the Mesozoic of many eustatic cycles is therefore an enigma, since this is a time when the polar caps were believed to have been ice-free (Frakes, 1979). The enigma would be solved if the controls were tectonic.

Sequence stratigraphic models based on depositional systems and regimes potentially overcome the complaints about facies models not allowing for natural variability. However, it is clear that the first-generation sequence stratigraphic models require much refinement, though they have already been of considerable value (Posamentier & Weimar, 1993). Many examples are coming to light of sedimentation patterns that do not fit the models even in the marine realm (e.g. Walker, 1990; Kolla & Perlmutter, 1993). It is becoming more widely appreciated that what matters most is not the direction of relative sea-level change (transgression or regression) but the rate of that change especially as compared to the rate of sediment supply (see Figure 12.14). It should be noted that many of the early models were constructed using the assumption that sediment supply was constant, and that depositional systems are simply translated downdip during lowstands. Studies of the Quaternary and Holocene show this assumption to be overly simplistic. Another oft-ignored aspect of the early models is that they did not take into account isostatic adjustment of the shelf due to relative sea-level lowering, which it is known will result in reduced gradients on the shelf. Shanley & McCabe (1994) review in detail current progress on applying sequence stratigraphy to continental strata, and anyone considering trying this for themselves is strongly urged to read their comments before proceeding.

Study of these situations leads to successive improvements to the predictive models. The need for such refinement should not be surprising as the sequences are examples of forward modelling, and will inevitably therefore suffer from mismatches with real data. Some of the value has been in forcing a closer look at the driving forces behind basin-scale sediment transport and deposition. As emphasized by Shanley & McCabe (1994), when well practised, sequence stratigraphy attempts to explain the formation of sequences and sequence boundaries through an understanding of all controls on sedimentation. The approach can be of value even if the relative sea-level curve is not global or dominated by eustacy, certainly has application to lacustrine strata (see examples discussed in Shanley & McCabe (1994)), and must be considered as of substantial relevance to investigations of alluvial sequences.

Floodplain deposits and the use of palaeosols

Studies of subsurface geology have traditionally paid scant regard to the fine-grained material of the floodplain. It is potentially of value as a seal to an aquifer or hydrocarbon reservoir, but otherwise it is uneconomic, and not usually the subject of modelling. Such units can be of value in well-to-well correlations, as they generally extend laterally over much greater distances than the channel sands. The outstanding

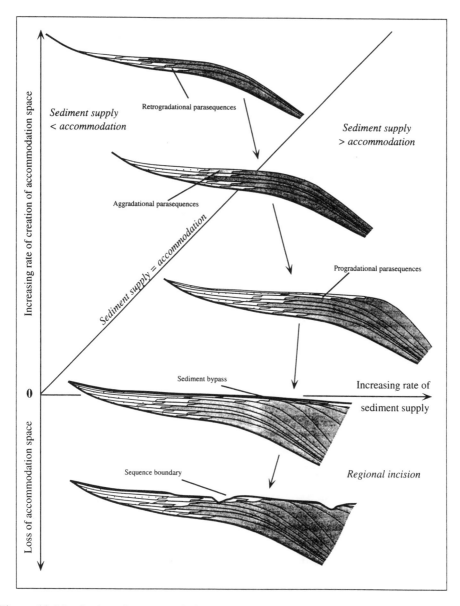

Figure 12.14 A plot of accommodation space against sediment supply illustrates the way the interplay between these factors affects the resulting stratigraphic stacking pattern. From Shanley & McCabe (1994); reprinted by permission of AAPG

exception is in the search for coal resources, where floodplain deposits have been the target of exploration, but even here, studies have usually been unsystematic. For examples of good analysis, see Ethridge *et al.* (1981) and the references therein. These authors sought to explain the occurrence in the fluvial Eocene Lower Wasatch Formation, in Wyoming, USA, of thick laterally extensive coals that have a low content of clastic material (average ash content 6%). This requires the peats to have accreted faster than the surrounding alluvium, or for the peats to have been protected from ingress of river deposits.

Recognition of the role of mudrocks as barriers and baffles to fluid flow has seen effort put into gathering data on the spatial dimensions of these facies, but the deposits themselves have been considered not worth detailed investigation. This is a gross oversight, as the floodplain deposits, and palaeosols contained therein, provide a more continuous record than the channel deposits of the types and rates of processes which operated during development of an alluvial system. However, it is not surprising that there has been this lack of attention, as prior to about 1975 the literature on floodplain sequences was sketchy, and records of pre-Quaternary soils were rare. Some pioneering work had been carried out on recent Mississippi River floodplain sequences (e.g. Coleman, 1966; Krinitsky & Smith, 1969), particularly on behalf of the US Army Corps of Engineers, but is not in widely available literature so has been overlooked by many workers. Major advances in this respect have come from Allen (1974), Leeder (1975), Bridge (1984), Fielding (1986), Farrell (1987), Wright (1989, 1990), Retallack (1986, 1990), Bown & Kraus (1987), Smith *et al.* (1989), and Dubiel (1991, 1992). Guccione (1993) has shown how grain-size distribution can be diagnostic of position on the floodplain and hence reveal the distance to the source channel.

It is only over the last decade or so that most fluvial sedimentologists have accepted that soils are a normal part of alluvial systems, and thus palaeosols should be regarded as expected, not exceptional, features. Indeed, the absence of palaeosols in a sequence in the rock record should be a cause for concern and trigger further study, as it reflects a particular combination of autogenic and allogenic controls.

Palaeosol development is controlled by topography and drainage, substrate, climate and time. Within fluvial systems, palaeosol maturity is also controlled by avulsion and terracing. It has been recognized that, to a first approximation, the presence of preserved palaeosols attests to slow rates of floodplain aggradation. The degree of maturity of the soil is inversely correlated to aggradation rate, and attempts have been made to model this quantitatively (Leeder, 1975). The implications of this model have been used by many authors since. However, as pointed out by Wright (1990), knowledge of the rates of formation of pedogenic carbonate horizons is scanty, and based on a few studies from the Quaternary of the SW USA. Until the calibrations are improved, assessment of aggradation rate in this way can only give general indications.

On the other hand, the degree of maturity of a palaeosol may be of value in identifying the position on the floodplain, relative to the channels, that the soil formed. This leads to a prediction of the approximate distance to the channel, a feature that would be most useful in studies of the subsurface. Systematic lateral changes in palaeosol maturity within floodplain deposits were recognized by Bown & Kraus (1987) in the Eocene Willwood Formation, Wyoming, USA. They introduced the term *pedofacies*, to mean "adjacent bodies of sedimentary rock that differ in their ancient soil properties because of distance from areas of relatively high sediment accumulation". In addition, they devised the *pedofacies model* which predicts that soil maturity on a floodplain increases progressively with increasing distance from active channels as a consequence of decreasing frequency of flooding and decreasing sediment accumulation rates (Figure 12.15A). They advocated the use of this model to aid reconstruction of alluvial sequences (Kraus & Bown, 1988).

The Kraus & Bown pedofacies model has predictive power in that it can be used to estimate, from an observed palaeosol sequence in floodplain deposits, the approximate distance to the related channel deposits (Kraus & Bown, 1993). The lateral variations in palaeosol maturity are expected to result in similar vertical variations, producing *pedofacies sequences*. In the Willwood Formation, simple sequences are about 3–7 m thick, where floodplain mudstones are enclosed between crevasse-splay events. Biostratigraphic zoning suggests such a sequence represents 10 000 to 20 000 years of deposition (Kraus, 1987). Compound pedofacies sequences, tens of metres thick, of stacked palaeosol profiles in floodplain deposits have been observed, bounded above and below by channel sandstones. Each pedofacies cycle in the compound sequence differs in maturity from the others. The changes in maturity reflect the progression of channel migration and avulsion. At its simplest, as a channel migrates away from a point, then progresses back again, the

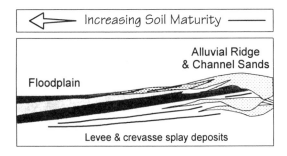

a

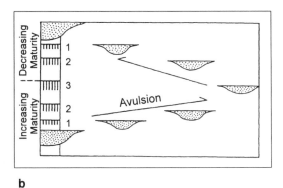

b

Figure 12.15 (A) Pedofacies association model of Bown & Kraus (1987), reproduced by permission of SEPM. Soil maturity increases with increasing distance from channel. (B) Summary of the way palaeosol maturity can be used to indicate proximity to channel sand bodies. As the channel avulses away from a site, the soil increases in maturity from 1 to 3. As the channel returns, at the same site one sees stacked soils of progressively decreasing maturity. This requires the succession to be overall aggradational

maturity of the contemporaneous soils gradually increases to a maximum, then decreases (Figure 12.15B). Deviations from this gradual change reflect the style of channel migration, and may include sudden jumps due to avulsion or meander-loop cut-off. But, at this scale, changes are attributed dominantly to autogenic factors.

In megasequences comprising hundreds of metres of sediment, and embracing several compound pedofacies sequences, the degree of maturity at the peak of maturity in each cycle may vary up-section. Change at this scale has been interpreted by Kraus (1987) to be the result of allogenic factors. In the case of the Willwood Formation, where peak maturity increases upwards, it is believed to be due to decreasing tectonic activity and subsidence rates, and thus decreasing aggradation rate.

The procedure recommended by Kraus & Bown (1993) is in many ways an oversimplification of reality. There are many factors that can affect soil maturity other than distance from an active channel. For example, the simple model suggests that the most mature calcretes, the thickest and most cemented, will be at the greatest distance laterally from channels. But the reverse relationship has been found to occur if the river is transporting significant amounts of carbonate. In this case, the best developed calcrete profiles are on bar tops and levees, and by comparison the floodplain soils appear incipient and immature. This shows how a single factor alone should not be used to define soil maturity (see discussion in Bown & Kraus, 1987). Parent material will also strongly influence soil type. Coarse-grained material is typically better-drained whereas fine-grained material may frequently be waterlogged. Variations in topography across the floodplain will also affect soil maturity, producing a catena sequence, and overprint the expected lateral variations. In a study of the Moors Cliff Formation in the Upper Silurian Old Red Sandstone of SW Wales, Love (1993) found that topographic relief of as little as 0.5 m was sufficient to cause major pedofacies variation. However, variations due to this cause occur over distances of 20 to 50 m, and in the Old Red Sandstone it was still possible to see a Kraus & Bown type pedofacies profile over scales of 500 m to 3 km. Topographic influences may not always be disadvantageous: Atkinson (1986) has used the presence of a catena to show that tectonic activity was occurring syndepositionally in the Eocene of northern Spain.

Lateral changes in soil maturity of the type in the Kraus & Bown pedofacies model are probably most relevant to periods of aggradation in alluvial settings. At other times, the fluvial channel belt dissects its own floodplain. Successive periods of incision and infilling produce a complex series of river terraces. The highest terraces are isolated longest from the depositional influence of the river, and develop the most mature soils. Successively younger (but not necessarily always successively lower) terraces have less mature soils (Wright, 1989) (Figure 12.16). This concept of soil *chronosequences* has been known to soil scientists for a long time, but has only recently been taken up by those working on palaeosols in the rock record.

Recognition of chronosequences in the subsurface is extremely difficult, and few have been documented even at outcrop (e.g. Allen & Williams, 1982; Platt & Keller, 1992; Love, 1993; Marriott & Wright, 1993). A potential complication is the effect of variations in substrate. However, there is an aspect to chronosequences that is

a

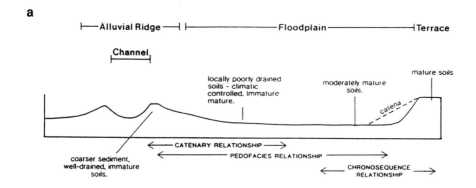

b

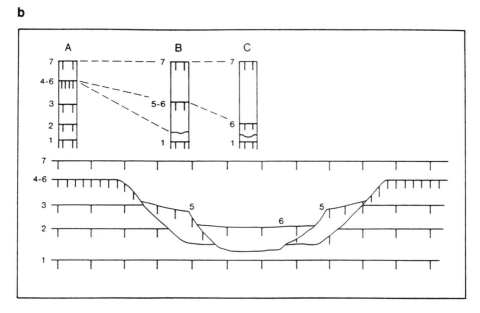

Figure 12.16 (A) Summary of where each of the possible soil relationships might occur in an alluvial setting. Catena relationships occur because of topographical variation. Pedofacies relationships relate to distance from channel. Chronosequences reflect a succession of terracing events. (B) Schematic of a chronosequence relationship created by simple phases of incision and infilling. Numbers 1–7 are geomorphic surfaces of progressively younger age. Density of short vertical lines is a guide to length of pedogenesis. A–C are cross-sections. From Wright (1989), reproduced by permission of the Postgraduate Research Institute for Sedimentology, University of Reading

often overlooked yet has considerable implications for attempting to discern in the subsurface periods of aggradation from periods of incision. Within fluvial systems, river channels are inherently unstable: the channel usually makes itself topographically higher than the floodplain, and this leads to periodic avulsion. Thus no area of the floodplain remains isolated from channel processes for long periods of time, preventing very mature soils from forming. The presence of very mature soils can therefore be taken as good evidence for a phase of river terracing (Leeder, 1975).

The Kraus & Bown pedofacies model has much attraction for those attempting to predict subsurface stratigraphy. However, there can be difficulties in identifying the type and maturity of soils with borehole data alone, even with core and geochemical profiles. In a study of the Upper Triassic Lunde and Upper Triassic to Lower Jurassic Statfjord Formations in the Snorre oil field of the northern North Sea, Love (1993) was able to identify palaeosols in core, and used them to assist reconstruction of the fluvial environment. However, the amount of core available is relatively small, which hampers significantly the power of the technique. The message for management is that apparent economies at the time the boreholes were drilled, in minimizing core cutting, have proved false economies as new technology has become available which with more core could have greatly improved understanding of reservoir complexity.

For the pedofacies model to be widely applicable to subsurface prediction, it must be possible to recognize soil profiles from their wireline log signatures alone. This is far from easy, and much more work is needed. But some success has been claimed for the Lower Cretaceous Travis Peak Formation in Texas (Williams *et al.*, 1993), by simultaneously considering as many petrophysical properties as possible.

Analogous proximal–distal spatial variation of palaeosol profile has been documented from other settings, such as Quaternary and Miocene arid-type alluvial fans (Wright & Alonso Zarza, 1990). However, the Kraus & Bown (1993) pedofacies model may only relate to thick aggrading systems such as the Willwood Formation. Furthermore, the profiles of Bown & Kraus (1987) are overlapping and composite, whereas many ancient palaeosol sequences are characterized by discrete (compound) mature profiles separated by non-pedified or lightly pedified intervals (Wright, 1989). Such sequences require episodic deposition, and will not result from the gradualistic model of Bown & Kraus (1987). It is likely that the pedofacies concept will undergo significant refinement in the near future, and may prove not to apply in the same way in all settings. Yet it offers promise that has attracted the attention of the oil industry.

DIRECT NUMERICAL MODELLING FROM OBSERVATIONS

Palaeohydrology

The rise of the relatively new discipline of palaeohydrology (Schumm, 1965, 1977; Gregory, 1983) has led to the establishment of empirical relationships between hydrologic and sediment characteristics of alluvial channels and channel morphology. The early drive for this was the efforts of geomorphologists to establish the links between channel morphology and hydrology, and the desire of hydrologists to predict discharge characteristics of ungauged streams from stream cross-sectional data. Sedimentologists then saw the potential value to the reconstruction of palaeochannels given incomplete information (such as from the subsurface). Though it is still finding application, and is often used in analysis of outcrops (e.g. see Chapter 10), it is an area where the science has stagnated.

Palaeohydrologic analysis of ancient river systems is based on some important assumptions (Maizels, 1990). The basic premise is that the individual elements of a fluvial system represent a direct response to environmental conditions, and are in equilibrium with these conditions. Following this, it is an assumption that a change in environmental conditions will be of sufficient magnitude or appropriate form to produce a significant and identifiable change in the fluvial system. If these assumptions hold, then it is theoretically possible to relate identifiable changes in the fluvial deposits to changes in hydraulic, hydrologic and environmental conditions, and through them to the expected channel morphology and architecture. There are, though, a range of problems in carrying out this procedure. The first is that the most valuable evidence of change may not get preserved in the rock record. Next is the problem again of convergence, whereby different processes and causes produce similar effects (Schumm, 1991, p. 58). Working back from the effects (the sediments) may not lead to a unique solution. The final group of difficulties relates to the reliability of the established relationships, which in turn depends on the quality, quantity and applicability of the available database on Holocene river systems.

Many empirical relationships have been established linking together factors such as discharge, slope, channel length, drainage area, silt–clay proportion, and sinuosity (Ethridge & Schumm, 1978; Williams, 1984). Some have been of value in elucidating and charting changes in the controlling parameters of palaeochannels exposed at the surface (e.g. Maizels, 1988, 1990). From the viewpoint of trying to establish the stratigraphy of fluvial deposits in the subsurface, the relationships of greatest interest are those that link the thickness of (preserved) channel-fill sediment, which can be measured from borehole data, to parameters of shape and lateral dimension, such as channel width, channel-belt width, sinuosity or meander-bend curvature. With such information, it is possible to quantify resources, such as oil-in-place, and plan extraction strategy. This approach seems to have been, and still be, most popular in North America, where it usefully puts limits on the likely dimensions of hydrocarbon reservoirs (e.g. Cornish, 1984; Lorenz et al., 1985; Davies et al., 1993).

Using the presence of lateral accretion structures as a diagnostic, point-bar sequences have been used at outcrop to estimate directly channel width (e.g. Moody-Stuart, 1966; Leeder, 1973; Elliot, 1976), by applying Allen's (1965) rule that point bars extend two-thirds of the distance across the channel. Of course, this is an option not available for the subsurface, and a more indirect route has to be taken. The most commonly adopted procedure involves the following series of calculations (see Figure 12.17 for explanation of the components involved):

(a) channel (bankfull) depth (h) from measured sediment thickness (t);
(b) channel (bankfull) width (W_c) from calculated channel depth (h);
(c) channel-belt width (W_m) (meander-belt amplitude) from channel width (W_c);
(d) meander length (L_m) from channel width (W_c).

This procedure takes no account of avulsion – the channel-belt here is that of a slowly evolving, essentially single channel – nor any overlap of sand bodies due to lateral migration or avulsion. The role of levee sands is usually not explicitly made clear in the literature, and crevasse-splay sand bodies are ignored completely, even though they may form significant hydrocarbon pools. One reason for this is that

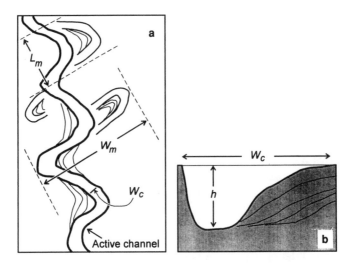

Figure 12.17   Schematic diagram of a river system identifying the main components used in palaeohydrologic analysis. (A) Plan view showing the channel (bankfull) width, W_c, channel-belt width (meander-belt amplitude), W_m, and meander length, L_m. (B) Cross-section through active channel, showing channel (bankfull) depth, h, and channel (bankfull) width, W_c

there is almost no published information on splay dimensions, and how these do or do not relate to the channel system from which they originate, though attempts are now being made to address this deficiency (e.g. Mjøs *et al.*, 1993).

Channel depth from sand thickness

Two factors are important: the proportion of the channel-fill that has been preserved, and the reduction in thickness brought about by burial compaction. It is assumed in this calculation that the thickness is the maximum for the channel, but when working with subsurface data it must be remembered that the observed thickness is probably less than the maximum since it is unlikely that the borehole has penetrated the greatest thickness of the channel fill. Sediments are compacted during burial, so the sand-body thickness used in this step should first be corrected to its original value. Many authors either ignore this factor (e.g. Davies *et al.* (1993), whose example is buried >2 km) or use the rule of thumb from Ethridge & Schumm (1978) that sands compact up to 10% of their thickness (e.g. Lorenz *et al.*, 1985). Recent advances in geohistory analysis for basin analysis have led to improved compaction modelling, and the correction ought to be calculated from a maximum burial analysis using procedures such as those of Sclater & Christie (1980).

The usual diagnostic of channel-fill packages, given confidence in the overall system being alluvial, is the presence of fining-up grain-size profiles. For temperate climate rivers, the thickness of a full profile approximates channel depth (Allen, 1966), but great care has to be taken in identifying a representative thickness (Lorenz *et al.*, 1985). The upper parts may be eroded, and sequences may be stacked. The presence of a fine-grained cap to the cycle may not be a safeguard, as such a package may represent a chute channel over a point bar rather than the main channel. Fining-up profiles may not always be present, as discussed in an earlier section.

Other workers have tried to use the thickness of sets of cross-bedding to put limits on the channel depth. This is normally considered to be a fruitless exercise because usually only a small proportion of each set is preserved. The proportion is itself a function of the rate of deposition (Rubin & Hunter, 1982), which is almost impossible to determine directly from the rock record. However, Paola & Borgman (1991) have recently shown from a theoretical study that the rate of deposition is only relevant for the migration of bedforms of uniform height. The mean set thickness for random topography can approach the mean height of the topography (bedform) even in the absence of deposition. The implications of this have yet to find application in the prediction of fluvial stratigraphy, though the potential is surely there.

To overcome some of the possible discrepancies, Fielding & Crane (1987) have attempted to arrive at a thickness : depth relationship compiled from published literature (Figure 12.18A). There is much spread in the data and all types of channel planform have been combined. Fielding & Crane found the median relationship from the data predicts channel depth h can be found from sand-body thickness t using

$$h = 0.55t \qquad (12.1)$$

It is interesting to note that, taken at face value, this relationship implies that a channel is about one-half as deep as the thickness of sediment it deposits, which ought to be physically impossible. But analysis that led to the empirical relationship of eqn 12.1 also included many examples of stacked channel-fill sands, and so the relationship is a rule of thumb that can be used when it is difficult to ascertain the amount of stacking that has occurred. This relationship has become popular, yet few using it acknowledge the diversity of data on which it is based (and Fielding & Crane do not quote the sources, so one does not even know the ages of the examples). Matters would be improved if confidence intervals had also been supplied by Fielding & Crane.

Channel width from channel depth

One route to the width of the preserved sand body (e.g. Lorenz *et al.*, 1985) is via channel width W_c using the empirical relationship with channel depth h compiled by Leeder (1973) that

$$W_c = 6.8h^{1.54} \qquad (12.2)$$

for modern rivers with sinuosity greater than 1.7. Leeder (1973) discusses at length the deficiencies of this approach, and shows that the relationship is invalid for rivers of lower sinuosity. Yet most of those using this equation do not first check that their palaeochannels have sinuosity greater than 1.7, though admittedly this is almost impossible to do in the subsurface (see discussion below).

Channel-belt width

This is the dimension usually required by resource geologists modelling the subsurface, as it is the expected width of the sand bodies. One route to channel-belt width W_m is from channel width W_c. Lorenz *et al.* (1985), in their investigation of methods to derive reservoir width from vertical borehole data alone,

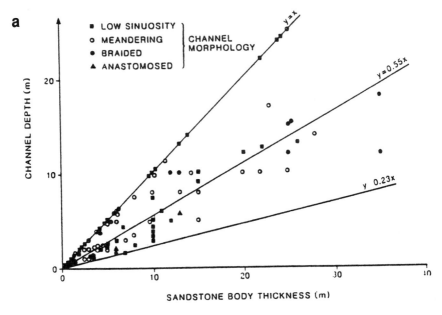

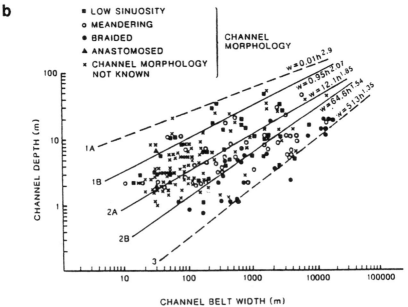

Figure 12.18	*For caption see facing page*

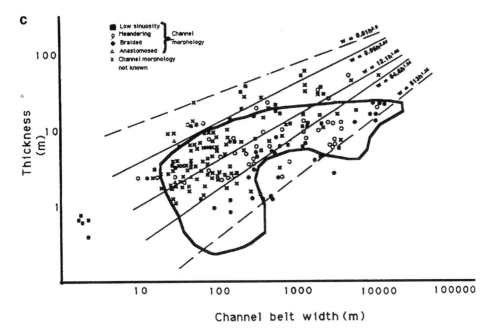

Figure 12.18 Empirical palaeohydrologic relations used in modelling. (A) Channel depth from sandstone body thickness from Fielding & Crane (1987). The two outer lines enclose the limited data available from published sources. The median line is taken to be reasonably representative. (B) Channel depth from channel-belt width for modern and ancient fluvial deposits, from Fielding & Crane (1987). The correlation lines are for different fluvial conditions: 1A is incised, straight and non-migrating channels: 1B is non-migrating, non-incised channels; 2A is geometric mean of all channel types; 2B is channels with fully developed meandering profiles; 3 is laterally unrestricted (braided) fluvial systems. (A) and (B) are reproduced by permission of SEPM. (C) Channel depth from channel-belt width, as in (B), but with simple meandering stream channels separated out. Reproduced from Martin *et al.* (1988), "Reservoir modelling of low-sinuosity channel sands: a network approach", SPE Technical Paper 18364, presented at *SPE European Petroleum Conference*, London, 17–19 October 1988, by permission of SPE.

decided to combine the data of Leopold & Wolman (1960) with those of Carlston (1965), supposedly to improve the accuracy of the technique, and arrived at the relationship

$$W_m = 7.44W_c^{1.01} \qquad (12.3)$$

However, they noted that the Carlston data seem to form a separate population, and consistently had larger channel-belt widths for a given channel width. Surely this calls into question the validity of combining these datasets (see discussion below on the database)?

The more direct, though not necessarily more accurate, route to channel-belt width W_m is from the estimate of channel depth h (e.g. as used by Davies *et al.*, 1993). Collinson (1978) has combined the relationships from Carlston (1965), linking channel width and channel-belt width to discharge, with that of Leeder (1973), linking channel width to channel depth (eqn 12.2 above), to produce the relationship

$$W_{\mathrm{m}} = 64.6 h^{1.54} \qquad (12.4)$$

This implicitly contains the assumption that sinuosity is greater than 1.7, so the provisos discussed in the preceding subsection also apply here.

An alternative approach is to compile published data on channel depth and channel-belt width to derive directly the required relationship. This has been done by Fielding & Crane (1987) (Figure 12.18B), again combining into the one plot many sources of data, for many types of river, this time including ancient as well as modern examples. The best-fit line for all the collected data gives

$$W_{\mathrm{m}} = 12.1 h^{1.85} \qquad (12.5)$$

Yet again, others are taking this relationship without considering its origin and thus the limitations. For a channel of 2 m depth, the Fielding & Crane relationship produces a channel-belt width of 43 m, whereas the Collinson (1978) relationship of eqn 12.4 predicts a width of 188 m, over four times larger. Though eqn 12.5 is for all river planforms and sinuosities, it is used by Davies et al. (1993) specifically just for the sand bodies identified as of low sinuosity (they use eqn 12.4 for high-sinuosity deposits). Their conclusions on sand-body dimensions are found to be not inconsistent with the pressure test interpretations, but such tests depend on a much simplified model of the geology (Davies et al., 1993), and there are probably more sources of error in interpreting these tests than there are in the palaeohydrological relationships.

A better approach, as advocated by Bryant & Flint (1993), is to reduce the scatter on the Fielding & Crane (1987) plot by critically grouping data by channel planform type (e.g. Martin et al., 1988) (Figure 12.18C), or, perhaps better still, grouping by systems tract after a sequence stratigraphic analysis has been undertaken. Tests of this latter approach have yet to be published. Care must also be taken to consider the tectonic setting, as tectonically influenced gradient changes, either parallel to flow or perpendicular, can influence channel-belt width by tipping the balance from incision and terracing, to aggradation or sinuosity changes (Leeder, 1993).

Meander length

Rivers of high sinuosity may leave behind sand bodies that, in the downstream direction, thicken and thin significantly, or are isolated entirely from each other (Figure 12.3B). This aspect is often ignored by reservoir geologists constructing subsurface models of reservoirs. The spacing and size of such detached elements is a function of the meander wavelength L_{m}. This may be approximated from the channel width W_{c} using the Leopold & Wolman (1960) relationship

$$L_{\mathrm{m}} = 10.9 W_{\mathrm{c}}^{1.01} \qquad (12.6)$$

or from channel depth h using the relationships rearranged by Collinson (1978) from the Leeder (1973) equation (eqn 12.2)

$$L_{\mathrm{m}} = 74.1 h^{1.54} \qquad (12.7)$$

However, many workers assume downstream dimension is approximately the same as that in the transverse direction, which is not unreasonable as the difference between eqns 12.7 and 12.4 is less than 15%.

Sinuosity and palaeocurrent analysis

Some of the difficulties of recognizing channel planform from vertical profiles have already been discussed (see also Collinson, 1978; Bridge, 1985). Yet it must be emphasized that the palaeohydrologic relationships currently available and in use in industry are only valid for restricted ranges of channel type, so it is imperative that some assessment of sinuosity is made in advance. Recognition of features such as the proportion of channel-fill relative to lateral accretion deposits, and the nature of the fine-grained channel-fill facies at the top of fully preserved channel sequences, can normally only be done with core material. Experience has shown this cannot be done with any certainty using wireline logs alone (e.g. Lorenz *et al.*, 1985; Davies *et al.*, 1993).

Some workers claim that channel type can be identified from an analysis of palaeocurrent information, which in the subsurface may be obtained with the dipmeter tool or oriented core. Certainly this is an important aspect (Bridge, 1985), but procedures such as that advocated by Le Roux (1992) suffer from calibration problems, and have limited reliability (Olsen, 1993). Computer-based models of the internal structure left by the migration of compound bedforms (Rubin, 1987) can provide an insight, but are not in themselves wholly diagnostic. Computer simulations of the expected flow in individual river bends (Bridge, 1982; Willis, 1989, 1993), and thus expected sedimentary structures, grain-size profiles and palaeocurrents, based on combined theoretical and empirical relationships, have more promise. These are of value in interpreting outcrops: by comparison of the observed channel section with a selection of computer-generated cross-sections, it is possible to estimate the parameters of the river that created the preserved sequence. It is doubtful if enough data would be available from the subsurface to try the same procedure, but the models may give an insight into some general characteristics of the ancient river.

The palaeohydrologic database

This approach to the prediction of fluvial stratigraphy is essentially a statistical one. It is therefore highly dependent on the samples collected and analysed to derive relationships. A majority of the database was accumulated prior to 1980, and almost all the relationships now in use are equally old. As reviewed by Ethridge & Schumm (1978), a very large proportion of the samples are from Holocene systems in the semi-arid to subhumid Great Plains region of the USA, rivers which range in elevation from about 800 to 1500 m above sea level, and are far removed from the coast. With few exceptions, the derived relations have not been tested on modern coastal plain streams which have a greater preservation potential, or in other climatic zones. In addition, the database is actually rather small, and relates almost entirely to moderate or small rivers (Leeder, 1973; Ethridge & Schumm, 1978). The empirical relationships that have been found are really only reliable for single-channel, meandering streams (Leeder, 1973; Ethridge & Schumm, 1978).

As pointed out by Williams (1984), there are several sources of error with the application of the published relationships, aside from the derivation issues above. First is the practice of applying an equation to a situation for which it was not derived, for example the use of Schumm's relationships derived from the semi-arid Great Plains region of the USA. Second is confusion as to the definition of a variable, such as use of maximum channel depth instead of mean channel depth. Other problems include inadvertent switching of variables when copying another author's equations, and misunderstanding about the units of measure. Finally, a serious error is that of rearranging components of equations derived by linear least-squares regression in order to derive a new equation. Thus the regression equation $Y = a_1 X_1^{b_1}$ should not be combined with $Y = a_2 X_2^{b_2}$, so that $a_1 X_1^{b_1} = a_2 X_2^{b_2}$, presumably enabling X_2 to be determined from a knowledge of X_1. This is invalid where both equations have the same dependent variable. Use of this technique introduces potentially large additional error (possibly as much as several hundred per cent (Williams, 1984)). Yet many of the relationships described in the preceding sections have been arrived at in this way.

Factors which have not been properly accounted for in these relationships are climatic and tectonic setting, and aggradation rate. Advances in understanding of fluvial systems in the last two decades show that these may be the most important controls on fluvial style and scale (e.g. Leeder, 1978, 1993; Schumm, 1993; Wescott, 1993; Alexander, 1993), but they are usually ignored in the mad hunt for the sand-body size. It has been argued (Bryant & Flint, 1993) that because the timespan for the creation of channel-fill sequences is relatively short (they quote 10^1–10^2 years), channel-fill geometries may be related more closely to autogenic processes, and it is the stacking of channel sand bodies that is controlled by allogenic processes such as climate and tectonism. This division into auto- and allogenic parameters is, however, an oversimplification. It has been shown that channel style changes with gradient, and sediment load and type (e.g. Patton & Schumm, 1981; Schumm, 1993). Incision, scour and avulsion, and the preserved sand thickness, vary with aggradation rate (Allen, 1978; Leeder, 1978; Bridge, 1985; Salter, 1993). In reality the fluvial system responds in a complex way to a wide range of factors so is rarely in equilibrium with any of them.

The deficiencies of the Holocene database have led to the use of older examples studied at outcrop. The addition into the database of ancient examples may decrease the statistical variance, and allow the fitting of regression lines with better correlation coefficients, but it does not improve the reliability of the relationships. It can be difficult to obtain reliable data at outcrop (see review by Alexander, 1993). No outcrops are truly three-dimensional, and the rock record is incomplete, so it can be difficult to be sure about channel style. Graphs such as those of Fielding & Crane (1987) (Figure 12.18A,B) blur the differences between channel style and allogenic factors such as climate and tectonic setting. The latter authors do not give the sources used to compile the relationships, so the person using them cannot select out a relevant subset. Additional factors become relevant as progressively older rock record cases are considered, namely that vegetation types and quantities were very different in the geological past. Therefore, the relationships between discharge and climatic setting change (Schumm, 1968), and the whole exercise becomes a farce.

Perhaps a better solution than the use of ancient examples would be studies leading to improved understanding of the processes and their products. Loratory experiments (e.g. Ashworth *et al.*, 1993) and computer simulation studies (e.g. Bridge & Mackey, 1993b; Maizels, 1993b) offer the most hope. Much more understanding of modern rivers is still needed.

Concluding comment

Given all the uncertainties about the application of palaeohydrologic analysis to the subsurface, perhaps the best we should expect from this approach is to set some limits on the likely distribution of channel sediments. This is the current recommendation of geologists (e.g. Lorenz *et al.*, 1985; Davies *et al.*, 1993), though reservoir engineers find the unwillingness to be more precise rather frustrating. It is therefore distressing to see, in a recent paper with reservoir engineers as intended audience (Swanson, 1993), statements such as "Enough empirical data are available about streams and stream variables that deterministic predictions can be made about fluvial reservoirs ...". Swanson presents the relationships between channel dimensions and discharge as though they are cut and dried – he quotes the above equations, and gives nomograms, without once indicating the large inherent errors.

Interpolation methods and surface modelling

It was by the construction of contour maps of sandstone thickness (isopach mapping), using data from boreholes, that some of the earliest successful predictions in the subsurface of hydrocarbon-bearing fluvial reservoirs were made (e.g. Charles, 1941; Nanz, 1954; Stokes, 1961; Berg, 1968; Cornish, 1984). Such mapping, based on detailed manual correlations between wells drawn as 2D cross-sections, still forms the cornerstone of much modelling of the subsurface even today (e.g. Tye, 1991). Whether performed manually or with the aid of computers, the mechanics of this procedure involves the interpolation of measured values, such as thickness of sand, between sample locations. The chosen method of interpolation produces a value for the mapped parameter at unsampled locations, in other words a prediction of the subsurface geology.

Contour maps have been drawn by hand since 1744. Contouring at its most fundamental is a purely deterministic operation, with only a single outcome possible for a chosen procedure, so it can reflect the data input but not take account of unsampled facts. The advantage offered by hand-contouring, one that geologists commonly forget they are exploiting, is that the manual construction of contours incorporates not only the measured values but can be aided by the use of conceptual models of the type described above. Indeed, it has been stated that this process (so-called "interpretive contouring") is an essential part of mapping (Dahlberg, 1975). In essence, the geologist is imposing a preconceived picture of the geology onto the model to make up for inadequacies in the available subsurface data. This fact has been recognized and demonstrated for decades (e.g. Rettger, 1929; Handley, 1954), yet even today many geologists forget it when constructing maps by hand, and do not appreciate that it is the prime reason for hand-drawn contours appearing "more right" than computer-generated contours.

The exact procedure adopted for manual contouring, with its implied assumptions about the way the mapped parameter varies spatially, imposes an inherent bias on the resultant predictive model (Tearpock, 1992). For example, "mechanical" contouring (Tearpock & Bischke, 1991), using ten-point proportional dividers, adjusts the spacing of contours assuming a linear variation between sample points (Figure 12.19A). Incorporating no additional geological interpretation, this is the most conservative procedure. But it produces an unnaturally "smooth" geological surface, and is oversimplistic in areas of sparse data control and rapidly changing values. Other procedures, such as equal-spaced contouring or parallel contouring (see Tearpock & Bischke, 1991), may be less conservative, but can introduce structure into the map that is purely artificial (Figure 12.19B,C). Of course, full "interpretive contouring", where the geologist modifies the contour patterns to create a "best fit" to the perceived geological structure, produces the most aesthetically pleasing results (Figure 12.19D), but is open to abuse, and is responsible for some spectacular failures in the search for mineral resources (T.A. Jones, 1989).

The major advance of the past decade has been the widespread acceptance of computer-based contouring methods (Iglehart, 1992), which partly owes its happening to the dramatic increase in availability of cheap computer power. This acceptance was slow in coming, even though research in the late 1960s and early 1970s showed the advantages in objectivity and precision conferred by the use of computer methods (Dahlberg, 1975). In the early days, the software capable of producing aesthetically acceptable contouring was very expensive, and (by today's standards) very slow. The rapid development of personal and desktop computing has produced an explosion in availability of computer-based contouring packages (see list in Wagner & Busbey, 1992). But perhaps the main reason for slow recognition was a dissatisfaction with the results from computer packages – the maps produced did not appear to be geologically acceptable. Some problems arise because of poor quality control of the data fed to the computer package. Geological review of the data being used is subconsciously undertaken during manual contouring, but it has to be consciously initiated in advance when using automated methods. However, the main cause of dissatisfaction is a failure to appreciate that hand-contoured maps include a substantive amount of geological guesswork as well as hard data.

Computer-based contouring cannot readily incorporate geological interpretation. Up until the early 1980s, the usual solution to this was to add "dummy" data-points into the input, to force the contours to fit the desired pattern (Figure 12.19E). This is messy and fiddly, so software vendors produced interactive editing options that allow the geologist to modify the results of the otherwise objective procedures. An alternative is to use a "substitution surface" constructed from geological experience, not hard data, to guide the details of the modelling procedure (Kushnir, 1992), though in principle this is just a variation on the dummy-control-points procedure.

In principle, computer-based contouring should take the hard work out of modelling, and make the process totally objective. However, there are several aspects of computer-based contouring that can strongly influence the resultant

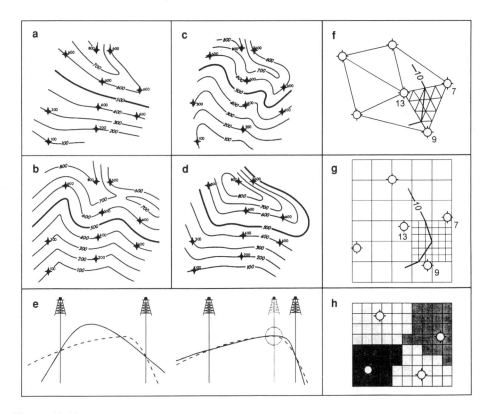

Figure 12.19 Interpolation (contouring) aspects and problems. (A) to (D) show four ways to contour manually the same samples, illustrating the impact of the choice of interpolation method on the model (the prediction) created. (A) Mechanical contouring (e.g. ten-point divider). (B) Equal-spaced contouring. (C) Parallel method. (D) An interpretive contouring. Computer-based algorithms can vary just as greatly as the manual methods. (E) Example of adding dummy control points to constrain the interpolation. On the left, the dashed line shows the real geometry of a lithological boundary. But the interpolation based on just two sample poin s (solid line) fails to come close. On the right, the addition of an extra sample point, chosen near the crest of the structure (dashed line), forces the interpolation (solid line) to be more correct. (F) Triangulation method of contouring, which guarantees that contours honour data points. The successive subdivision of the primary triangles (such as on right) allows smoother contours to be constructed. (G) Use of intermediate grid to construct contours. In the first stage values are interpolated at the intersections of all the grid lines, by a distance-weighted average of neighbouring samples. The contours are then constructed by reference to the gridded values alone, so contours may not honour sample values. Subdivision of the grid (as at right) is done to produce smoother contours. (H) Example of nearest-neighbour interpolation used to model discrete variables such as lithology types

geological prediction, many of which are often not fully appreciated by geologists using these packages. These include (for detailed review see Davis (1986), T.A. Jones *et al.* (1986), Tearpock & Bischke (1991), or see brief overview by Banks (1991)): selection of neighbouring data-points to use in calculation; the type and parameters of the mathematical interpolation function; whether contouring is "direct" such as with triangulation (Figure 12.19F), or "indirect", involving the calculation of an intermediate regular grid (the gridding stage) from which the contours are ultimately traced (Figure 12.19G); and refinement procedures to improve the aesthetics of the contours. Numerous algorithms are now available, each of which is best suited to a limited range of problems and data distributions, many of which are designed to be more "geological" (e.g. N.L. Jones & Nelson, 1992; Haecker, 1992; Davis & Sampson, 1992; Hamilton & Jones, 1992). The use of intermediate grids allows the easy calculation of areas and volumes, and comparisons to be made between multiple geological surfaces (e.g. McEachran, 1992), but may result in contours that pass the wrong side of sample points.

Most of the interpolation methods are intended for use with continuously varying parameters, such as porosity. There are few methods available for mapping out the distribution of discrete parameters such as lithology, the nearest-neighbour algorithm being the only one generally applied (Figure 12.19H).

The ability to handle non-vertical faults successfully is a feature that has only recently come to maturity in computer contouring systems. Even now, some packages cheat by considering faults solely as vertical and with constant displacement. Part of the difficulty is having sufficient information to define adequately each fault surface, for without these data fault handling can only be approximated; this is not a problem that can be laid at the door of the software vendors. Improvements in fault handling are now generally available in the major systems (e.g. Banks & Sukkar, 1992; Belcher & Hoffman, 1992). Correct handling of reverse faults and recumbent structures is still difficult because of the overlap that occurs in the surface (at any one map location, a formation boundary ends up being described by more than one elevation value), but significant progress in this direction has been reported by Mayoraz *et al.* (1992) who have been attempting to use the Earth Vision® software from Dynamic Graphics Inc. to model the highly complex thrust-fold sequences of the Swiss Alps.

The outcome of all these developments is that, although the results of computer-based contouring are strictly objective and deterministic and, most importantly, are totally reproducible whoever carries out the contouring, the geological picture is overprinted with features of a purely mathematical origin, and which change as the modelling algorithm is changed (compare Figures 12.19A–D). It is not possible to agree with Bowdon's (1992) assertion that "the main virtue of computers is objectivity – after that, it's downhill". It is crucial that the geologist takes the time to understand the processes followed in the computer, in order to avoid the many pitfalls that lurk there (e.g. Krum & Jones, 1992; Jones & Krum, 1992; Yarus, 1992).

Three-dimensional modelling and visualization

The focus at present in numerical modelling of subsurface geology is on creating truly 3D representations (T.A. Jones & Leonard, 1990). It is extremely difficult to construct 3D models by hand, but computers make it very easy. It is essential to distinguish clearly between 3D *modelling* and 3D *visualization*. For some time now, seismic data have been displayed in 3D form, but this is simply the visualization of a

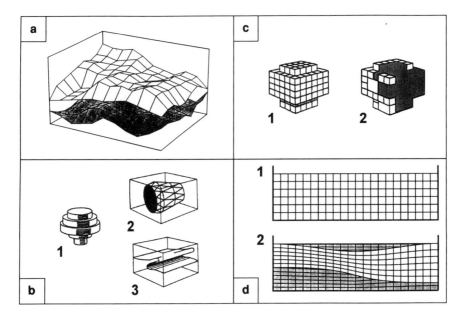

Figure 12.20 Three-dimensional (3D) modelling and visualization examples. (A) Stacking of multiple two-dimensional surfaces, each of which is calculated independently, and with no modelling of the intervening volume. (B) Surface representation of 3D volume: (1) constructive solid geometry, (2) boundary representation, (3) non-uniform recursive B-splines (NURBS). (C) Volume representation of 3D volume: (1) simple cuboids (voxels), (2) octrees (the shaded area). (D) Cross-sections illustrating choice of framework for volume representation. (1) The simple approach is to organize voxels of constant volume into a rectilinear matrix parallel to a major correlatable horizon. (2) A better approach is first to subdivide the subsurface into stratigraphic units, using previously mapped surfaces such as those in (A), and then make the voxels follow stratigraphic layering in some fashion. In this case, voxels can vary in volume, as happens in the middle layer of this example. This is stratigraphic geocellular modelling (Denver & Phillips, 1990)

3D matrix of numbers – little attempt is made to predict (model) any values between measured sample locations. 3D visualization must be a vital component now for all numerical modelling of geology in the subsurface (Lee *et al.*, 1990). A picture is undoubtedly worth a thousand words, and can yield insights into the data not otherwise achievable (e.g. Zeitlin, 1992; Pflug & Harbaugh, 1992).

The first attempts at predictive 3D representation were achieved by stacking multiple 2D surfaces (such as formation boundaries) created by contouring and mapping packages such as those described above (Figure 12.20A). But this is only multiple 2D modelling, as each surface takes account of data entirely in two dimensions, and no values are assigned to the volumes between the surfaces. So the resultant 3D representation is created in order to visualize the model, the modelling is not actually performed in 3D (see overview by Fried & Leonard, 1990).

There are two main approaches to true 3D modelling: surface representation, and volume representation (C.B. Jones, 1989; Fried & Leonard, 1990). In surface representation (Figure 12.20B), geological objects are described by calculating the envelope that surrounds them, and there are a variety of mathematical ways of doing this (e.g. Fisher & Wales, 1990; Paradis & Belcher, 1990). In volume representation (Figure 12.20C), the subsurface is first discretized into a large number of small blocks (usually referred to as "voxels", or volume elements, because of the analogy with the term "pixel", or picture element, used in 2D). Such blocks are usually cuboid in shape, for ease of definition, but they do not have to be. Geological attributes, such as lithology or porosity, are assigned to each block. At its simplest, the blocks are arranged in a single 3D matrix, usually with one surface parallel to the horizontal or present-day sea level (Figure 12.20D). A more geologically reasonable way is to start the whole process by first splitting the subsurface into a series of geologically meaningful layers with a stratigraphic framework defined by a series of 2D surfaces, and then subdivide each layer into voxels, now allowing the voxel shape to follow the shape of the stratigraphic boundaries (T.A. Jones, 1988a,b; Johnson & Jones, 1988; Denver & Phillips, 1990; Krum & Johnson, 1993) (Figure 12.20D). However, in such a scheme, the interpolation process is performed solely within a single layer, so although a complete volume is modelled, the prediction process is still not truly 3D.

The use of a stratigraphic framework is a major improvement, but it can be difficult to define the framework adequately in areas of complex geology, especially highly faulted terrains, and it requires great quantities of information. This is an aspect where availability of high-resolution 3D seismic data is of great assistance.

Nearly all these schemes rely on interpolation procedures and algorithms that are simple extensions of those used for standard contouring in 2D, so all the reservations and pitfalls outlined in the previous section apply equally to 3D modelling. For example, Krum & Johnson (1993) describe how it is necessary to add pseudo-wells as dummy control points to constrain interpolation in their 3D modelling. However, the application of a stratigraphic framework allows the subsurface to be divided into segments within which each parameter can be modelled continuously, but between which there can be discontinuities. This is much closer to real life, where, for example, porosity varies over a small range within a channel sandstone, but changes suddenly to values in a much lower range at the contact with floodplain muds.

One of the difficulties with a 3D model is how to interrogate it, as now one can be overwhelmed with information. Rotation of the model, serial sectioning through it, and statistical analysis are all desirable, but not all available. It is here that the latest advances in computer graphics hardware are important (Flynn, 1990), but as yet the cost of the fastest genuinely interactive equipment, which is similar to that used for computer animation in movie-making, puts it still out of the reach of most aspiring users.

Improved interpolation methods – geostatistics

To build a picture of the geology of the subsurface, a set of information (the sample) is collected, and, from analysis of the characteristics of this sample (i.e. the calculation of statistics), we make estimates about the entire subsurface (the population). Using analysis of samples as estimators of population parameters (e.g. taking the mean porosity from a few boreholes as a way of estimating the average porosity of an entire aquifer) is the realm of classical statistical methods. However, traditional statistical theory cannot handle parameters that are spatially correlated, so routine statistical analysis of subsurface data makes no use of the spatial information usually present in Earth science datasets. In the 1950s and 1960s, several people researching in applied statistics addressed this problem, but perhaps the most important contribution came from Matheron with his Theory of Regionalized Variables (see Matheron (1963) for list of references).

A truly random variable can take on any of its possible values independently of where it is spatially located. A regionalized variable is a random variable that has properties part-way between a true random variable and a totally deterministic one. Many natural phenomena are best described by this type of variable, especially parameters that have geographic distributions, such as the elevation of the ground, or the number of salmon in a particular river. Unlike random variables, regionalized variables have continuity from point to point, but the changes in the variable are so complex that they cannot be described by any tractable deterministic function. Consider the example of ground elevation. As anyone who has gone walking in the hills will know (and rely on), ground elevation generally does not fluctuate through all possible values over short distances; instead, in the span of a few paces the elevation changes slowly, but with a general trend (such as up or down) so that over large distances the elevation may change a great deal. Exceptions to this may be present, which are discontinuities in the variation (colloquially known as cliffs). It is not possible to derive an analytical description given by an equation that can represent the shape of the ground surface, especially over long distances.

Methods to deal with regionalized variables are particularly useful for analysing geologic phenomena, and the procedures used are collectively known as *geostatistics*. Note that many workers prefer to embrace with this term all statistical procedures that can be applied to geology, but it perhaps ought to be reserved for those phenomena involving spatially correlated parameters. Much of the development effort was carried out in South Africa and France by geologists working in the mining industry, who were concerned with estimating the distribution of precious metals (e.g. see summary of principles in Matheron (1963) and Journel (1986)). However, the methods have been extended successfully into many other areas of geological endeavour.

There are two main stages in geostatistics (Matheron, 1963): the first is description and data analysis, where the objective is to understand the nature of the geological parameters, and see how they vary spatially; the second is estimation, where the knowledge obtained from the data analysis is applied to predicting the geology in unsampled areas. Various kinds of estimate may be required: estimation of an average value over a large area, of an unknown value at a particular location, or an average over small areas. A very important part of geostatistics, and the major advantage over other methods, is that integral to the techniques is an assessment of the uncertainty of the various estimates.

The first stage, data analysis, is extremely important, yet often does not receive the attention it deserves. It is not surprising, therefore, that the estimates obtained are later found to be erroneous. One of the key tools (but not the only one) for analysing spatial variation is the *semi-variogram* (often just called the variogram), which is used to assess the degree of spatial correlation in the data. In essence, what is calculated is the average of the difference in value of all pairs of data points a given distance apart. If this statistic, the semi-variance, is small, it suggests that the data pairs are closely related to each other. As the distance between data pairs increases, it is common to find the differences increase, which suggests that there is now weaker correlation between the data pairs. The semi-variogram of the sample data (called the experimental variogram) is a plot of the semi-variance against the distance between data pairs (the lag distance) (Figure 12.21A). The variogram is then inspected to see if a general picture of spatial distribution can be determined. It is common to find that beyond a certain distance, called the range, the semi-variance no longer increases but fluctuates close to the total population variance (which appears as a flat region on the plot called the sill). The range thus defines the size of a region within which points are related. Detection of this can alone be a valuable

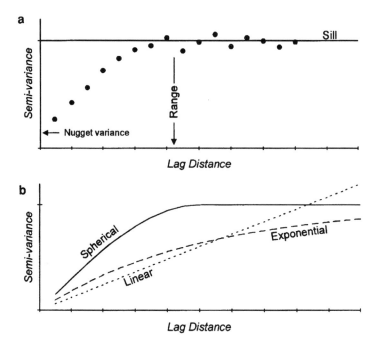

Figure 12.21 The semi-variogram for investigating spatial correlation in a data series. (A) The experimental semi-variogram, calculated by finding the difference in value between all data pairs at progressively greater distances apart (the lag distance). The range is the distance beyond which there is no significant spatial correlation between values, the point where the sill is reached. The nugget variance is a measure of the amount of variation that occurs at distances less than the inter-sample distance. If there were no spatial correlation, the nugget variance would be the same as the sample variance, and the semi-variogram would plot as a horizontal line (at the level of the sill line on this graph). (B) Examples of theoretical models of spatial variation. The linear model implies that there is always some spatial relationship between values, but the effect gets linearly weaker with increasing distance. The exponential is similar, but the effect diminishes exponentially with distance. The spherical model means there is correlation up to a point (the range distance), but beyond that there is no correlation

gain, but it is usual to go further and see if the experimental variogram can be matched to a theoretical model of spatial variation (Figure 12.21B).

If the form of spatial variation is known, or can be deduced from a semi-variogram, then geostatistics supplies a procedure to estimate the value of a parameter at any unsampled location. This procedure is called *kriging*, named after D.G. Krige, a South African mining engineer and pioneer in the application of statistical techniques to mine evaluation. Kriging is a linear least-squares estimation technique that is a variety of weighted-moving-average interpolation (see Davis (1986) for a simple summary of the procedure). The information from the semi-variogram is used to find an optimal set of weights to use in the averaging. Since the semi-variogram is a function of distance, the weights change according to the geographic arrangement of the samples. There are many subtle varieties of kriging which vary in different assumptions about the population from which the samples are drawn, and different ways to handle regional trends in values. All, though, lead to a calculation of the error in the estimate, that is the difference between the estimated value at a location and the actual value. They all inherently attempt to be unbiased by ending up with zero mean error, and aim to minimize the variance of the errors, so producing the "best" estimation possible.

Kriging can be used to make contour maps. It has been known for a long time that kriging offers many advantages over other interpolation procedures (e.g. see Matheron, 1967), not least that the calculated surface will always honour the data at the sample locations, an aspect not guaranteed with many algorithms in use, including those based on calculation of intermediate grids. But it has taken a long time for these advantages to be appreciated and acted upon in most areas of geological modelling. The petroleum industry, for example, has only recently taken the recommendations to heart, as evidenced by the fact that the early versions of one of the major commercial mapping packages, much used by petroleum geologists, had, in the late 1970s, a kriging interpolation option but it was removed during rewriting for later versions because it was not much used, and its presence was not considered important by the user community.

A problem in the past with geostatistics is that the literature on it has tended to be rather opaque to most geologists, being written by and in the language of statisticians and mathematicians. The standard text on the subject (Journel & Huijbregts, 1978) is far from easy to read. Most geologists are uncomfortable with the level of mathematics included in these works, and so have not appreciated the scope of these methods. A readable brief overview can be found in Davis (1986). Accessible and practical guides to geostatistics are thankfully now beginning to appear (e.g. Hohn, 1988; Isaaks & Srivastava, 1989; Brooker, 1991), and low-priced software packages are becoming available which implement the techniques in a relatively friendly fashion. For example, the GEO-EAS package is available free (Englund & Sparks, 1988; Nathanail & Rosenbaum, 1992), and the GSLIB library is the price of a book (Deutsch & Journel, 1992). A large number of geostatistical tools and procedures are now available. Refinements and variations are continually being announced, but the disturbing aspect to this productivity on the part of the statisticians is that the practical application of many of the theoretical advances remains obscure (Royle, 1992), and the geologist wishing to use the techniques finds it difficult to establish the relative merits of each modification.

Geostatistics tools can aid deterministic 2D and 3D modelling of the types outlined above, by giving a better understanding of the data distribution and spatial relationships, and by providing alternative predictive calculation methods (e.g. Isaaks & Srivastava, 1989; Olea, 1992; Bourgine, 1992; Nobre & Sykes, 1992). These tools are now also becoming widely used with stochastic modelling of fluvial sequences, which is discussed below. An important recent development has been the advancement of non-parametric geostatistics, where it is not necessary to assume the form of the underlying distribution. For example, it is well known that fluid flow through rocks is dominantly controlled not by the average permeability of the rock but by the extreme values of permeability – a thin high-permeability layer in a sandstone more than makes up for the rest of the sandstone being of generally low permeability (Journel & Alabert, 1988). Conventional statistics cannot cope with this feature, but indicator geostatistics can do a much better job (Journel, 1983; Journel & Alabert, 1990). This technique is also more relevant for predicting lithology type as it can handle discrete variables as well as continuous ones.

The fluvial and fluvio-deltaic Middle Jurassic Scalby Formation exposed in the cliffs of Yorkshire, NE England, has been submitted to a detailed geostatistical analysis in an attempt to develop and demonstrate the use of geostatistics for subsurface modelling (Ravenne *et al.*, 1989; Matheron *et al.*, 1987; Galli *et al.*, 1990). As well as extremely detailed study of the cliff sections, 36 boreholes were drilled and logged behind the cliff-edge and used to construct a predictive model (see Eschard *et al.* (1991) for a detailed account of the site geology). The model built with the borehole data makes a prediction of what is expected at the location of the present cliffs, which can be verified with the outcrop data.

It is crucial to remember that geostatistics, as with all statistics, is neither completely automatable nor purely objective. There is no accepted universal algorithm for determining a variogram model, cross-validation is no guarantee that an estimation procedure will actually produce good estimates of unsampled values, and kriging need not be the most appropriate estimation method. The most important decisions of any geostatistical study are taken early in the exploratory data analysis, and it is this stage that deserves the most attention. In a recent experiment conducted by the US Environmental Protection Agency, 12 independent reputable geostatisti-cians were given the same sample dataset and asked to perform the same straightforward block estimation. The 12 results were widely different due to widely different data analysis conclusions, variogram models, choices of kriging type, and sample search strategy (Journel, in foreword to Isaaks & Srivastava, 1989).

Fractals

Fractals are a form of descriptive geometry. Mandelbrot introduced the word "fractal" as a generic term for objects that are self-similar over a wide range of length scales (Mandelbrot, 1977). However, the term is not fully defined (even Mandelbrot has published two different attempts), and so although there is a large body of literature on the properties of fractals, there is no consensus as to the minimum set of properties before an entity may be classified as "fractal". For example, it is often taken that if a phenomenon displays a power law distribution then that phenomenon is proved to be "fractal"; from this, it is assumed that the phenomenon exhibits all the other properties of fractals including self-similarity at all (or many) length scales. But this line of reasoning has only weak foundation in theory and documented case studies.

Numerous geological objects have been shown to possess fractal character, from the scale of individual pore spaces (Wong & Lin, 1988) up to entire river systems (Goodchild & Mark, 1987; Harden, 1990) and coastlines (Mandelbrot, 1967). Snow (1989) has suggested that fractal dimension is a better descriptor of river sinuosity than the conventional sinuosity index. It has been said that fractals "capture the texture of reality". This has made the application of fractal geostatistics to subsurface geological data an appealing way of predicting the spatial distribution of hydrocarbon reservoir properties (e.g. Hewett, 1986; Emanuel et al., 1989; Hewett & Behrens, 1990; Crane & Tubman, 1990; Di Julio, 1993).

From the point of view of subsurface geological description, the significance of the fractals concept is that if rocks are truly fractal, then the variability at the small scale, say that of a 4 inch core, faithfully reflects the variability at much larger scales, say that of inter-borehole spacing. This concept is extended (usually without substantiation) to mean that variability measured in the vertical direction accurately reflects heterogeneity in the horizontal direction. In reality, most sequences of sediment are anisotropic, varying much more rapidly in the direction perpendicular to bedding than parallel to bedding. For the purposes of hydrocarbon reservoir modelling, noise is added to the commonly employed smooth interpolation functions so as to generate a more "realistic" picture of the unsampled geology (in a similar fashion to Figure 12.2F). The noise is generated using a fractal dimension derived from the borehole data or analogue outcrop data. This is a stochastic approach that is more appropriate for continuous variables such as permeability. The method has also been applied to the construction of contour maps (Rongey, 1992).

There are several areas of concern with regard to the application of fractal statistics for geological description. It is still not fully established that rocks are wholly fractal in character. Intuitively, one can see that a particular fractal process may be just one of the many sedimentary, diagenetic and tectonic processes that have produced the rock in the subsurface. So fractal character may be swamped by other non-fractal effects. Also, fractal processes of different dimension are difficult to separate out, so we tend to see an average result that is of lesser use (Burrough, 1983).

In such a modelling approach, the fundamental parameter that has to be measured from the geological data is the fractal dimension. The biggest problem is that our ability to measure the fractal character is extremely limited, especially in small datasets. A factor not previously recognized in the subsurface modelling literature is that the errors on the determination of fractal dimension from borehole data are so large that it makes the result rather absurd. Of the possible ways to determine fractal dimension, the rescaled-range (R/S) method of Hurst et al. (1965) has achieved some popularity (e.g. Crane & Tubman, 1990; Aasum et al., 1991; Goggin et al., 1992). The robustness of the R/S statistic (Mandelbrot & Wallis, 1969) makes the technique a more stable one than other methods such as spectral and variogram analysis. Yet recent work suggests that the R/S statistic estimated from borehole data can be extremely biased (North & Halliwell, 1994). Furthermore, the effect on which this statistic depends is more the effect of the mixing of phenomena of different scales than the infinite memory implied by the fractal concept (Mesa & Poveda, 1993).

SIMULATING FLUVIAL STRATIGRAPHY

All of the preceding numerical techniques have been reverse modelling, taking existing data and trying to derive the causative processes. An alternative approach to understanding natural systems is to use forward modelling methods. With these, we try and recreate, or simulate, the sequence of events that led to the products we now observe. Simulations fall into two main types: physical simulations, where we create an experimental model of the physical processes, such as in flumes or sedimentation tanks: or mathematical simulations, where we build an analytical or numerical model of the physical processes. The difficulty with physical simulations is in creating realistically scaled representations. Although we can scale length, it is not easy to scale proportionally factors such as viscosity and density, nor reduce gravitational forces. Mathematical simulations do not suffer from this difficulty, and are thus attractive. However, they suffer instead from the simplifications that usually have to be adopted to make the model computable. See Harbaugh & Bonham-Carter (1970) for a detailed review of these problems. Two types of mathematical models have been used to predict fluvial stratigraphy. The first is based on statistical analysis of fluvial products, and in this category most effort has gone into the use of transition analysis statistics. The second type is process-based, simulating the water and sediment transport and deposition in a fluvial environment.

Synthetic stratigraphies from transition analysis (Markov methods)

It has long been appreciated by observation that each sedimentary depositional event is dependent spatially and temporally, to a greater or lesser extent, on previous and adjacent depositional events. This can also be deduced from a first-principles consideration of the physical processes operating: it is not possible for a system to switch from one physical state to just any other state, but only to one of a limited range of related and progressive states. For example in the alluvial system, a river cannot switch from in-channel upper-flow-regime conditions to overbank deposition without passing through a succession of conditions relating to reducing discharge and channel abandonment. We may not always see this progression presented in deposits, but we know that it must have happened. In other words, the nature of deposition at any place is dependent on the events preceding it, and there is an inherent memory in the system. There is also a degree of independence in the system, as there is often a range of possibilities and there is a degree of randomness as to the actual outcome.

This mixture of dependence and independence in geological processes can be described using Markov theory. A Markov process is one in which the occurrence of a state is dependent partly on previous states, and partly on random fluctuations. A Markov chain is a sequence of events governed by Markov processes, where the probability for each of the events is dependent only on the events preceding it. A first-order chain assumes the memory in the sequence extends back just a single event; in a second-order chain, the present state is dependent to some degree on the two preceding states; and so on for *n*th-order chains. The probability of the system being in a given state at a particular time may be deduced from the knowledge of the immediately preceding states (Harbaugh & Bonham-Carter, 1970). A vertical profile through a rock sequence, such as obtained with a borehole or logging an outcrop, can be considered as a one-dimensional Markov chain, with each lithofacies representing a different state of the system. It has been common practice in geological applications to model stratigraphic sequences as first-order Markov chains, because this is the simplest type, and because it is extremely difficult to prove that a longer length chain is valid, due to the amount of data required and the inherent incompleteness of the rock record.

The properties of a one-dimensional Markov chain are described by its transition probability matrix, which expresses the probability of one state of the system being followed by each of the other possible states (Figure 12.22A). It is calculated from observations by counting the number of times a particular state is followed by each of the other states (the transition frequency, e.g. Figure 12.22D) (see Davis (1986) for details). The procedure can be extended into two and three dimensions, which are referred to as Markov fields (Figures 12.22C,D). The descriptive powers of Markov chains and Markov fields make them desirable as ways of analysing process-based simulations of the type described later in this review.

In the 1960s and 1970s there were numerous studies of outcrops, many of them composed of fluvial sequences, where a Markov chain model was applied (e.g. Zeller, 1964; Schwarzacher, 1964, 1969; Potter & Blakely, 1967, 1968; Gingerich, 1969; Allen, 1970; Miall, 1973; Hattori, 1976). These studies used this technique for two reasons. The first is because the transition probability matrix acts as a concise, quantitative summary of geological observations, and as a result was often a stepping stone to the construction of a facies model (as was the case with the Battery Point fluvial model, Figure 12.4B). The second was to test for the presence of cyclicity in lithofacies sequences, such as fining-up cycles (e.g. Allen, 1970). It was hoped that this approach would lead to greater understanding of the types and distribution of processes operating in a depositional environment. The method is still in use as a way of recording relationship data from outcrops and densely drilled fields (e.g. Mijnssen *et al.*, 1993).

Quite early on, it was appreciated by some that this technique has predictive as well as descriptive power. Potter & Blakely (1967) showed how it is possible to build a stochastic simulator for fluvial lithofacies that is driven by random sampling of a Markov transition probability matrix (Figure 12.22B). This matrix can be calculated from borehole data, or taken from analogous outcrop sections. These authors saw even then that "generation of synthetic stratigraphic sections ... offers the possibility of a better understanding of reservoir inhomogeneities". A few specialists have extended the applications into two and three dimensions (e.g. Lin & Harbaugh, 1984; Farmer, 1989). But until recently the predictive aspect has not been put into general practice by geologists modelling the subsurface, though many have argued that it should be given serious consideration (e.g. Dubrule, 1989; Haldorsen & Damsleth, 1990; Budding *et al.*, 1992). The surge of interest in the last few years

466

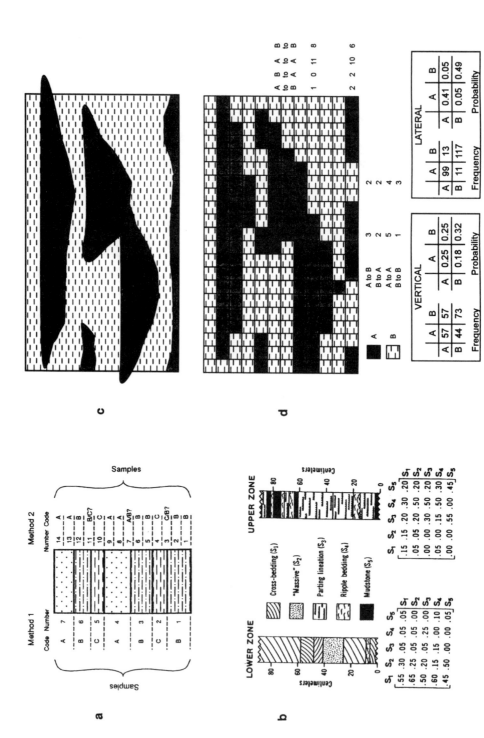

Figure 12.22 *For caption see facing page*

in stochastic modelling has seen renewed interest in using Markov methods either in isolation or in combination with other techniques. Tetzlaff (1991) combines it with a simplified fluvial process-based simulator. Fält *et al.* (1991) and Tyler *et al.* (1993) report the implementation of Markov fields, in a software package called MOHERES, to simulate the distribution of facies in North Sea fluvio-deltaic reservoirs. Their facies subdivisions are very broad, however, with categories such as low-stand channel, high-stand channel, crevasse, lower shoreface sand, offshore and coal; this is hardly a precise prediction of fluvial stratigraphy. MacDonald & Halland (1993) have adapted it to predict details of sandstone–shale distribution within a geological framework constructed from random sampling of shale continuity statistics. They applied it to the Lower Jurassic Statfjord Formation fluvial sequences in the northern North Sea.

Pang (1993) carried out a review and test of Markov methods for predictive modelling of fluvial stratigraphies. The approach has been shown to be adaptable and flexible for a variety of geological data, and it is easy to ensure that the models honour known data. But there are some important weaknesses, aside from the general problem with all stochastic simulators of multiple outcomes. Deriving reliable transition probability matrices is not easy given the usual paucity of data from the subsurface, and usually forces one to subdivide the sequence into fewer types than one would wish. Even in studies of modern fluvial environments it has been found difficult to collect enough data to build predictive models (Dawson & Bryant 1987). It is interesting to note in this context that the solution adopted by Fält *et al.* (1991) and Tyler *et al.* (1993) in the MOHERES package is to derive the transition probabilities from statistical analysis of well-to-well correlation cross-sections that have been compiled manually. In this way, they impose on the truly statistical prediction their own conceptual models for the depositional environment. Perhaps the greatest drawback when applying the approach directly to a subsurface

---

Figure 12.22 Markov transition analysis for simulation of fluvial stratigraphy. (A) A lithological sequence is analysed to derive the frequency with which each lithology is superseded by each other lithology. This can be done in two ways. In method 1, no account is taken of thickness; a simple count of the transitions is made (e.g. bed B to bed A occurs twice), and divided by the total number of transitions (six) to arrive at the probability (0.33). Thickness data should be analysed separately. In method 2, sampling is performed in increments of equal thickness, which has the advantage of building into the sampling the fact that some beds are thicker than others. However, there can be ambiguities in the analysis, such as when the sample spans two different facies. (B) Fluvial stratigraphy simulated by Potter & Blakely (1967) by random sampling of the transition probabilities (beneath each profile) derived from core and outcrop studies. Thickness was assigned by random sampling of thickness probability distribution functions. Reproduced by permission of SPE. (C) Transition statistics can be obtained from outcrop studies, or correlation panels constructed by geologists. Consider this hypothetical case of two rock types. (D) A grid is superimposed over the geology and each element of the grid assigned a lithology. The choice of grid element size determines the accuracy with which complex and small shapes are captured. It is now possible to find two-dimensional transition frequencies by counting vertically bottom to top, and laterally left to right. From these frequencies (beneath grid), probabilities can be found by dividing each count by the total number of transitions (240 laterally, and 231 vertically)

situation is the shortage of data on the lateral geological variation, most of the boreholes giving an essentially vertical view of the geology. Although the Markov theory can be applied in three dimensions, without hard data in all directions the theory is of no use. One solution to this difficulty is to use instead, or in addition, data derived from outcrop analogues (Figure 12.22C,D) – the drawbacks of this way out are discussed in a later section. Yet another problem is that the generated synthetic sequences reflect very strongly the choices made for the initial state of the system, and without care it is extremely easy to bias the modelling without meaning to do so.

Process-based prediction of fluvial stratigraphy

No-one would disagree with the assertion that the best way to simulate natural systems is to simulate the actual processes operating therein. Over the past three decades, there have been a variety of attempts to achieve this, but the earlier ones were hampered by inadequate computing power. Different approaches have been taken, and models appropriate to different scales have resulted.

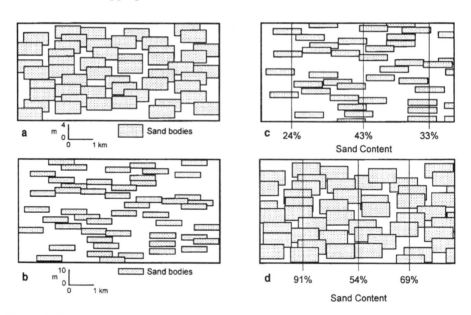

Figure 12.23 Allen's (1978) process-based exploratory model of alluvial architecture. (A) and (B) are examples from Allen (1978) showing the increase in proportion of overbank fines, and decrease in channel interconnectedness, caused solely by increasing subsidence rate from 1.59×10^{-4} m per year in (A) to 7.65×10^{-4} m per year in (B). Reproduced from Allen (1978), "Studies in fluviatile sedimentation: an exploratory quantitative model for the architecture of avulsion-controlled alluvial sites", *Sedimentary Geology*, by permission of Elsevier Science. Note the difference in vertical scale between the two diagrams. (C) and (D) show an analysis of Allen's models presented by Van de Graaff & Ealey (1989). This illustrates how the interconnectedness of sandbodies changes as the percentage of sand in a vertical section is altered (which is itself a function of subsidence rate). From Van de Graaff & Ealey (1989); reprinted by permission of AAPG

Building on earlier semi-quantitative work (Allen, 1974), Allen (1978) presented an exploratory model for alluvial suites formed in coastal plains (Figure 12.23A,B). Steady subsidence (R) is taken as the factor permitting an alluvial suite to build up. Avulsion, which is assumed to be the process causing channels to alter position bodily, is treated as a random process, and was simulated stochastically with the aid of dice or random-number tables. Allen felt these approximations made the model valid for relatively short time-periods, of the order of 10^5–10^6 years. The simulation considers just a two-dimensional section normal to fluvial transport direction, and treats the system as a simple binary one of sand and non-sand (i.e. usually mud). The period of avulsion (P) was kept constant in the order of 10^3 years, and only channel position after an avulsion is selected randomly. Channels are approximated as rectilinear bodies whose width approximates the likely meander-belt width, and whose thickness is the sum of instantaneous channel depth plus the thickness $(R \times P)$ due to subsidence during an avulsion period. Compaction is ignored except for the effects of buried sand bodies on surface relief; differential compaction of mud relative to sand is approximated simply by considering the sand/ non-sand ratio beneath a given point, with a ratio of 1 producing the maximum surface relief.

By iteratively recalculating the model after varying just the rate of subsidence, Allen showed that the character of an alluvial suite is a function solely of subsidence rate. Coarsening-upward alluvial suites can form as a result of decreasing subsidence rate alone, and do not require externally controlled changes in stream pattern such as caused by source area uplift or climate change. Of great interest to hydrocarbon geologists and hydrogeologists was the additional finding that the degree of interconnectedness and overlap of sand bodies is a non-linear function of sand-body content: when sand-body content is less than 50%, the degree of connectedness is very small, but increases rapidly as sand-body content rises (Table 12.5). This result has been used by Qiu *et al.* (1987) when attempting to assess the degree of lateral connectivity in fluvial oil and gas reservoirs in China. Allen's

Table 12.5 An analysis by Allen (1978) of his models showing how the number of sand-bodies intersected by a vertical profile is highly variable even for a single set of conditions, as well as when subsidence rate is varied. Experiments 1 to 6 represent increasing subsidence rate. The table shows for each experiment the number of vertical profiles intersecting a given number of sand bodies

Experiment	Number of sand-body intersections in vertical profile										Total number of intersections
	1	2	3	4	5	6	7	8	9	10	
1	0	0	0	3	5	8	6	4	1	1	178
2	0	0	0	3	4	5	10	3	1	2	185
3	0	0	0	0	4	9	10	2	3	0	187
4	0	0	2	2	3	5	6	8	1	1	184
5	0	0	0	3	6	4	11	1	3	0	182
6	0	0	0	4	7	8	1	1	5	2	179

(1978) work is still regularly quoted (e.g. Van de Graaff & Ealey (1989) and see Figures 12.23C,D) to support predictions of large-scale subsurface fluvial stratigraphy.

An interesting observation by Allen coming from this study was that sand bodies are not uniformly distributed throughout the sequence, but tend to cluster (Figures 12.23A,B). Therefore, it will not be possible to estimate reliably the sand content of an alluvial suite with a single vertical profile such as a borehole.

Leeder (1978) independently developed a conceptually similar quantitative model, and found similar results to Allen (1978), in particular the tendency for channels to cluster. But this new model was then implemented on a computer, and made much more sophisticated by allowing many more factors to vary randomly (Bridge & Leeder, 1979; Bridge, 1979). The model is again a two-dimensional section normal

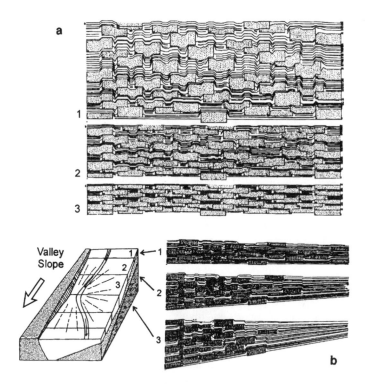

Figure 12.24 Examples of use of the Bridge & Leeder (1979) process-based model of alluvial stratigraphy. Channel sands are stippled, channel geometry is simplified. (A) shows the effects as mean channel-belt aggradation rate is reduced from (1) 0.02 m per year, through (2) 0.01 m per year, to (3) 0.005 m per year. Channel sands become more interconnected as aggradation rate decreases. Reproduced from Bridge & Leeder (1979). "A simulation model of alluvial stratigraphy", *Sedimentology*, by permission of Blackwell Science Ltd. (B) shows the way the model was utilized by Alexander & Leeder (1987) to explore alluvial architecture changes downstream in a half-graben. Variations in fault displacement and associated subsidence, from minimum (zone 1) to maximum (zone 3) fault displacement, result in changes in channel position and stacking patterns. Reproduced by permission of SEPM

to valley slope (Figure 12.24A). Floodplain fine-grained sediment is explicitly compacted as it becomes progressively buried, though the model does not compact the channel sands because its use is restricted to conditions pertaining to the top 500 m of sediment where compaction of sand is negligible (Perrier & Quiblier, 1974; Baldwin & Butler, 1985). This compaction modelling is important: after an avulsion, the channel is moved to the part of the floodplain with the lowest topographic elevation, which in turn will be a function of the sand/mud ratio under the floodplain. Tectonic tilting is simulated as well, which influences the floodplain topography, and thus the position to which channels migrate. Avulsion rate and frequency of tectonic tilting are stochastically modelled by random selection from user-supplied probability distributions.

The Bridge & Leeder (1979) model has been used by many workers to gain an understanding of channel density and interconnectedness, and has been seminal in work on the effect of synsedimentary faulting on fluvial architecture. For example, it was simulations of expected half-graben tectonics made with this model (Figure 12.24B), combined with field studies of Quaternary fluvial style in active half-grabens in Montana, that led to a better understanding of the link between asymmetric meander belts and tectonics (Leeder & Alexander, 1987; Alexander & Leeder, 1987). It can also be used to generate values to control stochastic modelling (Clemetsen et al., 1990).

The Bridge & Leeder model has been made much more detailed, adapted to three dimensions, and extensively revised to allow for many more variables and to provide much greater analysis of the outcomes (Bridge & Mackey, 1993a,b; Mackey & Bridge, 1992) (Figure 12.25). The objective is that as many parameters as possible should be computed within the system rather than be imposed by the user. The model is still fundamentally avulsion-controlled. Ideally, avulsion rate should be determined from input discharge data, but this has not yet been achievable (J.S. Bridge, pers. comm. 1993) and it is still an input parameter. The revised model is now more appropriate to subsurface modelling as it includes sand compaction for sequences which exceed 500 m thickness. It also more clearly illustrates the complexity in alluvial systems arising from the interactions between the primary controls. Given realistic values for channel-belt width to floodplain width ratio, channel-belt aggradation rate, mean avulsion period, and overbank sedimentation rate, then sandstone-body widths will increase due to any or all of either (i) an increase in bankfull depth or channel-belt width/floodplain width, or (ii) a decrease in overbank aggradation rate, channel-belt aggradation rate, or avulsion period.

More realistic results will be obtained from such process-based models when they are developed in three dimensions, so that for example the avulsion point is allowed to vary up or down valley. This development is well advanced, and preliminary results are being presented at conferences (e.g. Bridge, 1993b). At present, there can be no allowance for external factors such as base-level or climate change. However, the existing models are already being considered by some oil companies to carry out sensitivity and uncertainty studies. Though it is difficult to condition such a stochastic forward model to borehole data, this technique can give valuable insight into likely sand-body densities and arrangements when regional controls can be deduced independently from other geological studies.

472

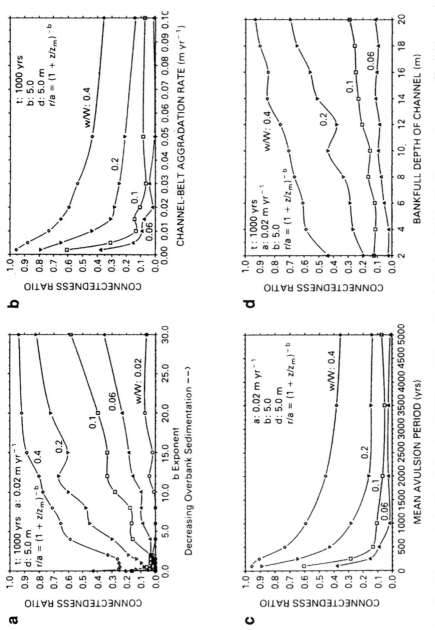

Figure 12.25 Revisions to the Bridge & Leeder (1979) model by Bridge & Mackey (1993a,b) allow much greater flexibility, and improved analysis of the outcomes. These graphs show how the model predicts that sand-body connectedness ratio will be affected as parameters in the model are varied singly in turn. (A) Variation in overbank sedimentation rate (*b* exponent). (B) Channel-belt aggradation rate. (C) Mean avulsion period. (D) Bankfull depth of channel. In each case, the connectedness ratio is computed for a variety of channel-belt width (*w*) to floodplain width (*W*) ratios. Reproduced from Bridge & Mackey (1993b), "A revised alluvial stratigraphy model" in *Alluvial sedimentation*, by permission of Blackwell Science Ltd

Harbaugh and his co-workers have taken a different approach to simulating the physical processes of sedimentation, as a result of various experiments in the 1960s and 1970s (e.g. Harbaugh & Bonham-Carter, 1970). In a series of stages, they have built a deterministic dynamic model (SEDSIM) that simulates the movement of fluid and sediment (Tetzlaff & Harbaugh, 1989). Fluid (water in the alluvial setting) is modelled by a numerical solution of the Navier–Stokes flow equations. Transport, erosion and deposition of sediment, of up to four grain-size classes, are then modelled by the inclusion of a variety of modified standard semi-empirical relationships. The system allows for multiple fluid and sediment inputs to the simulation. Advanced three-dimensional graphics are used to visualize and display the results (Lee *et al.*, 1990). Fast and high-powered computers are essential to run this simulation. The model is totally deterministic, but many of the inbuilt functions are non-linear and display apparently random behaviour (chaos) because of this. Extremely small changes to the starting conditions can lead to very large and unpredictable changes in outcome. Although work began with fluvial systems in mind, the model has been adapted to simulate deltas and turbidity-current submarine fans.

SEDSIM is a valuable tool to investigate alluvial processes, complementing experimental studies in flumes and sand tables. As with all forward models, conditioning of SEDSIM to real data is difficult and has to be achieved by iterative adjustment of the starting parameters after visual comparison between model and data. This was the approach taken when the model was used to model braided fluvial deposits in the Prudhoe Bay oil field of Alaska (Scott, 1986). Some impressive results have been obtained in this way, but Schumm's (1991) warning about convergence must be heeded yet again – a match between the model and the data means the model is valid but does not mean the model is right. Care is also needed not to be dazzled by the computer graphics of the apparently realistic images. Just because it *looks* real, doesn't mean it *is* real. SEDSIM (as published in Tetzlaff & Harbaugh (1989)) is not immune to problems with boundary conditions and initial settings of its many parameters. Much fine-tuning is still needed as it is only as reliable as the inbuilt functions describing sediment transport and fluid flow; much work is still being done by others elsewhere to improve these functions. Hydraulic engineers would not be happy with the chosen sediment transport criterion, which assumes a linear relation between critical shear stress and size of particle entrained. No account is taken during flow events of changes in bed condition such as bed armouring, and subsequent effects on transport rates (e.g. Willetts *et al.*, 1987). The use in the model of Manning's bed roughness coefficient "n" is regarded only as an empirical trick and needs to be replaced by more physically sound functions (J.K. Maizels, pers. comm. 1991). There is in the model an assumption of conservation of mass, but this is not necessarily guaranteed in nature within the boundaries of the zone being modelled. The model is usually run over time periods of the order of 10^4–10^5 years, yet no allowance is made for changing flow regime during sediment accumulation. The model must allow for the changes in eustatic sea level, tectonic subsidence and climate, and consequent changes in flood discharge and sediment supply, which are known to occur in this time-scale. Given the imprecision in many of the functions inherent in SEDSIM, it may be that useful investigative studies, to

474

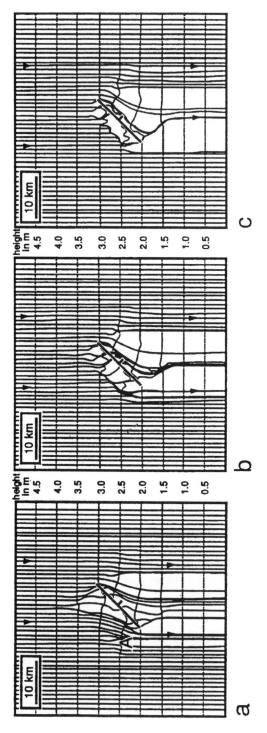

Figure 12.26 Examples of flow models produced by Weston & Alexander (1993). Here, fluid flow lines have been calculated to show the effects on flow of the deformation of the land surface by a normal fault oblique to the flow direction, downthrowing 0.5 m in the downstream direction, and with fault plane dip angle of (a) 90°, (b) 60° and (c) 45°. Reproduced from Weston & Alexander (1993). "Computer modelling of flow lines over deformed surfaces: the implications for prediction of alluvial facies distribution" in *Alluvial sedimentation*, by permission of Blackwell Science Ltd

understand primary controls, can be performed equally well with much simpler, and less expensive, software. Such a study has even been done recently by Tetzlaff (1991), one of the main originators of SEDSIM.

The SEDSIM approach of attempting to model every known process can suffer from too much complexity, with resultant instability of the models, and confusion of those attempting to use it. The model can be too sensitive to the starting parameters, which in any case are usually known only poorly. Weston & Alexander (1993), building on earlier studies by Alexander (1986), have produced a simple three-dimensional model of fluid flow over topographic surfaces to investigate the effect of synsedimentary tectonics on fluvial facies distribution (Figure 12.26). It is assumed that alluvial channels will tend to flow directly downslope and that when avulsion occurs the new channel position will be controlled by the floodplain gradient, which in turn is tectonically controlled. Their method is not strictly process-based, but works on the principle that over some long period of time much of the detail and variation in fluvial systems can be ignored as it becomes "averaged out", a principle also followed inherently by those models of allogenic controls described below. The results of Weston & Alexander (1993) suggest that the importance has been overstressed of axial channel systems, and channel deposit accumulations in the hangingwall immediately adjacent to faults, because of reliance on two-dimensional architecture simulations. Henriquez et al. (1990) have proposed that this technique be integrated with other modelling techniques in hybrid approaches to subsurface modelling.

Webb (1994) has produced a computer code to simulate the three-dimensional distribution of sedimentary units in a network of braided channels. This random-walk approach is based on geomorphological concepts melded with estimates of localized flow energy (as encompassed in the Froude number).

The SEDSIM solution reveals a great deal about the internal, autogenic controls in a fluvial system. Other workers have concentrated on understanding the external, allogenic controls. It is possible to reduce the equations of flow and sediment transport in two dimensions to a linear diffusion equation. Coupled with a sediment-partitioning procedure for calculating downstream grain-size changes, this approach has been used to study the expected grain-size distribution in an aggrading alluvial basin (Paola et al., 1992; Heller & Paola, 1992), in order to try and resolve questions such as whether periods of tectonic quiescence produce coarsening-upward or fining-upward signatures (Figure 12.27). It was found that four basic independent variables govern the model system: subsidence (rate and distribution), sediment flux, water flux, and grain-size distribution of the sediment supply (coarse material does not travel as far into the basin as fine material). These variables produce dramatically different responses at different time-scales; for example, the subsidence becomes unimportant at short time-scales. The long-term average water discharge plays a fundamental role, through the diffusivity term. The difficulty in applying such a model to real basins is in getting realistic values for the discharge.

Maizels (1993a,b) has produced an alternative computer simulation of allogenic controls. Variations in sedimentology and aggradational history are simulated from unit power relations of flows of varying sediment concentration, and from consideration of accommodation space changes due to relative sea-level change. The

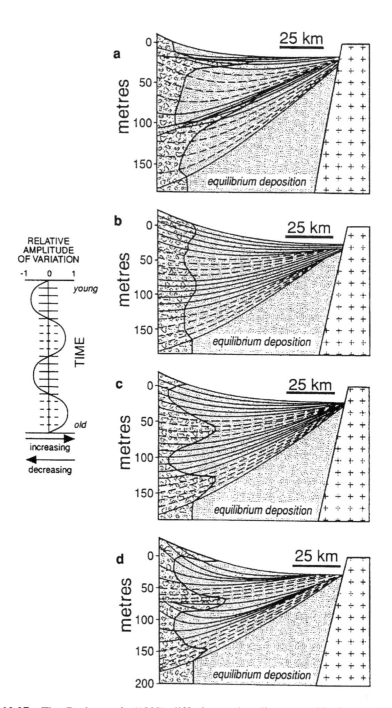

Figure 12.27 The Paola *et al.* (1992) diffusion and sediment-partitioning model of an aggrading alluvial basin. Cross-sections show the effects on a hypothetical basin of a rapid variation in (A) sediment flux, (B) subsidence, (C) gravel fraction, and (D) diffusivity. Sand is shown as light stipple, gravel as "stony" ornament. Direction of transport is left to right The thin dashed and solid lines are isochrons drawn every 104 years. The graph at the left shows the form of the variation in the forcing parameter; the intervals represented by dashed lines indicate maxima in this parameter. Reproduced from Paola *et al.* (1992). "The large-scale dynamics of grain-size variation in alluvial basins. 1. Theory", *Basin Research*, by permission of Blackwell Science Ltd

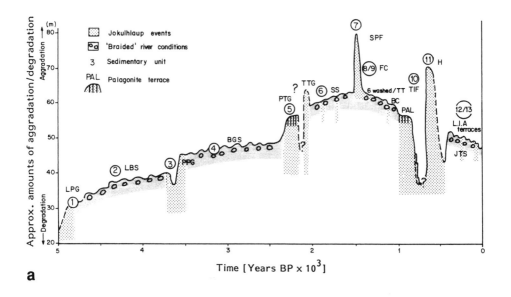

a

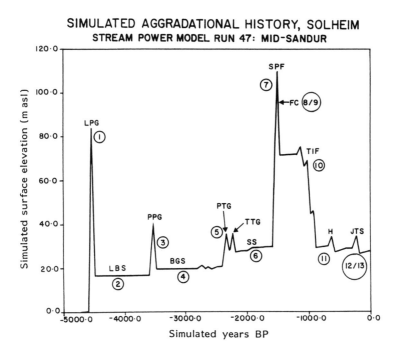

b

Figure 12.28 Example of the use of Maizels (1993a) regime model for simulating alluvial stratigraphy. (A) Reconstructed over the past 5 ka of the geomorphic history, and patterns of aggradation and incision on Solheimasandur, a glacial outwash braid plain. This is derived from extensive field work. Numbers and initials refer to the main sedimentary units identified. A "jokulhaup" is a large flood event associated with rapid ice-sheet melting. (B) The geomorphic history of the same sandur simulated by the SANDSIM regime model. based solely on setting the regime parameters. The match with the real stratigraphy in (A) is most encouraging. Reproduced from Maizels (1993a) by permission of the Geological Society

model has been tested with some success by comparing predicted with actual fluvial stratigraphies on Icelandic outwash braidplains (Figure 12.28). It has the advantage over many models in that the flow regime can change during a model run. Both Maizels' model and that of Paola *et al.* have clear value for improving our understanding of the long-term controls on fluvial stratigraphy They are particularly attractive because they are process-based and deterministic. But, because they require as input the primary factors such as runoff, eustatic and tectonic variation, it is not a simple matter to use them to predict subsurface stratigraphy in any given basin. Being forward models, conditioning is also a major difficulty.

There are many other forward simulation models that try to recreate basin-filling sequences, taking into account eustatic and tectonic variations, but virtually none make specific predictions of the details of fluvial stratigraphy, focusing instead on the large-scale sediment architecture, produced in long timespans ($>10^6$ years), such as is detectable on seismic sections. Many incorporate and investigate sequence stratigraphic concepts – Thorne & Swift (1991) have produced one to test the different approaches to sequence stratigraphy. For an overview, see the introduction in Kendall *et al.* (1991); Lawrence *et al.* (1990) discuss some fundamental concepts. One package of passing interest though is the SEDPAK software (Helland-Hansen *et al.*, 1988; Strobel *et al.*, 1989), which has been applied to the delta-plain sequences of the Middle Jurassic Ness Formation in the northern North Sea (Helland-Hansen *et al.*, 1989). SEDPAK, a deterministic technique, does not attempt analytical simulation of the physical processes of sedimentation, but instead mimics them with a geometric method of depositional triangles that represent processes such as infilling of topographic depressions, and development of fans and wedges. Comparison of model runs with actual Ness Formation stratigraphy has been used to try and assess the relative role of eustasy and tectonic subsidence.

INDIRECT NUMERICAL MODELLING AND USE OF ANALOGUES

Stochastic (probabilistic) methods

Intuitively, natural systems are deterministic. Yet we will probably never understand fully enough the processes of sedimentation to be able to replicate them with total certainty. Even if we did understand the processes, because of the immense number of interactions and feedbacks, and the inherent apparent chaos this brings about, it is unlikely we could ascertain well enough all the starting values for all the parameters so as to reproduce exactly the same results. It is the character of complex non-linear systems that extremely small changes to the starting conditions can lead to very large and unpredictable changes in outcome, as the creators of the SEDSIM package have discovered.

In the absence of complete information about processes and the products they produce, one solution is to rely on statistical analysis of available data and make predictions using stochastic simulations. The general principles and mathematical procedures for stochastic simulation follow those in standard texts such as Ripley

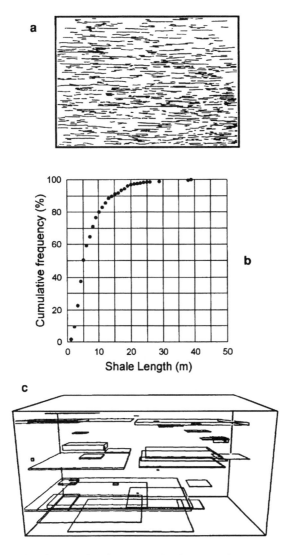

Figure 12.29   Example of the object-based stochastic modelling approach of Haldorsen & Lake (1984) to distribute shales in unsampled volumes. Using photographs of outcrops, maps of shale distribution are produced such as that in (A). Shale length is measured on such maps to produce, via a frequency distribution plot, the cumulative probability distribution function of shale length (B). Random sampling of the frequency distribution function is used to choose the centre point, width and length of each new shale, to create a stochastic three-dimensional representation (C) of shale distribution in the subsurface. Reproduced from Haldorsen & Lake (1984) by permission of SPE

(1987) and Stoyan *et al.* (1987). A variety of stochastic simulations have been developed and applied to fluvial sequences, mainly to aid improved description of hydrocarbon reservoirs. The impetus behind this came particularly from reservoir engineers, who needed to make allowance in their fluid-flow simulations for

a

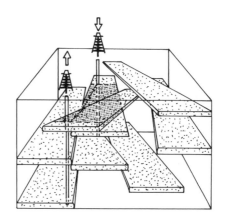

b

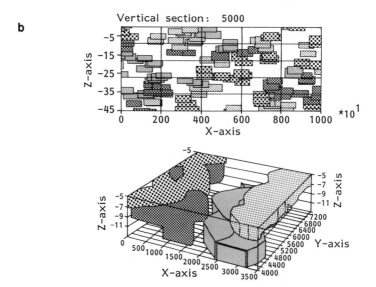

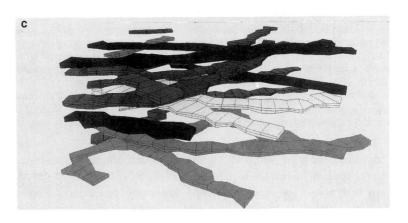

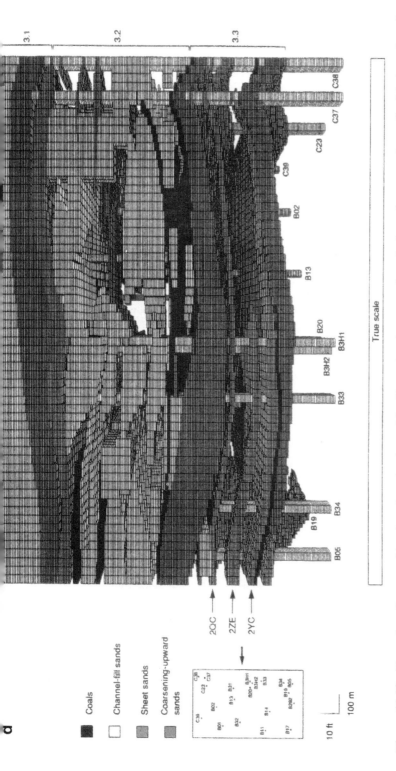

Figure 12.30 Examples of newer object-based stochastic models, illustrating the range of methods used to approximate fluvial channel sand geometry. (A) The SISABOSA package (Stanley *et al.*, 1990) uses simple parallelepipeds to build a model from which it is possible to estimate the amount of sand that would be swept by a water-flood to recover oil between any pair of boreholes. Reproduced by permission of the Norwegian Institute of Technology. (B) Cross-section (upper diagram) and 3D perspective view (lower) from the FLUREMO package (Clemetsen *et al.*, 1990) Reproduced by permission of the Norwegian Institute of Technology (C) Oblique 3D perspective view of a model from the SIRCH package (Hirst *et al.* 1993). Different colours indicate connected channel belts. Reproduced from Hirst *et al.* (1993), "Stochastic modelling of fluvial sandstone bodies" in *The geological modelling of hydrocarbon reservoirs and outcrop analogues*, by permission of Blackwell Science Ltd (D) 3D perspective view, from the east, of a reconstruction using the MONARCH software of the fluvio-deltaic Lower Ness Formation in the Brent Field, northern North Sea. The transparent voxels represent shales. The columns projecting from the base of the model are the boreholes. From Bryant *et al.* (1991); reprinted by permission of AAPG

permeability barriers that were inadequately sampled by boreholes and had dimensions less than the inter-well spacing. These models are generally of two types, namely discrete models or continuous models, or they may be a hybrid of these types (Haldorsen & Damsleth, 1993). Nearly all of them depend to a greater or lesser degree on information other than the direct measurements from the subsurface – they draw on information from analogue outcrops, comparable subsurface examples, modern-day analogue environments, or from geologists' conceptual models, in order to add "geologic knowledge" into the modelling. This is why they are described here as "indirect" methods.

Discrete models (or Boolean models) have particularly been designed for geologic problems requiring the prediction of discrete variables. Each point in space can belong to just one of a limited number of classes of geological object (hence another name for this approach, "object-based", which must not be confused with a similar name in computing science which has many more connotations). These objects may be lithofacies types (gravel, sand, mud), or may be genetic units (channel, crevasse splay, floodplain). The simulation (Figure 12.29C) distributes these objects in space by randomly sampling probability distribution functions (PDFs) that describe the frequency of occurrence and dimensions of each object (Figure 12.29B). The PDFs must be supplied in advance, and are derived from analysis of available borehole data, analogue outcrops (Figure 12.29A), or both. Process-based simulations such as described above are another useful way of deriving PDFs (e.g. Bridge & Mackey, 1993a).

The packages that have been developed for discrete modelling differ in the sophistication with which they handle the dimensions and shape of each object type, the interactions between object types, and the mix they allow between deterministic and stochastic elements. The earliest models produced by reservoir engineers were concerned with a binary system of permeable and non-permeable rock objects (the latter loosely referred to as "shales"). The objects were considered as square or rectangular in plan, with a negligible or constant thickness (Haldorsen & Lake, 1984; Haldorsen & Chang, 1986) (Figure 12.29C). Despite the simplicity of the approach, it was found that reservoir descriptions built in this way better explain observed oil production histories than non-stochastic models had done; for example, Geehan *et al.* (1986) applied it to the braided fluvial deposits in the Prudhoe Bay field of Alaska.

Geologists demurred at the geological naivete in the early attempts and have driven the development of more complex models. This has gone hand-in-hand with the addition of sophisticated three-dimensional computer graphics. Most now explicitly model reservoir rock (sand) first, instead of shale first, and try to approximate the natural shape of fluvial channels, for instance by the use of parallelepipeds (Figure 12.30A). There are several examples of note in connection with simulation of fluvial sequences. Dranfield *et al.* (1987) use a piecewise application of the Haldorsen & Lake (1984) shale modelling technique on separate parts of the Triassic Sherwood Sandstone of southern England. The permeability of rock units observed in core is averaged into units appropriate to larger-scale modelling to derive the permeability to assign to the reservoir rock predicted by the stochastic model. Moraes & Surdam (1993) found this simple approach useful in testing the importance of shale barriers in their geological models. Chessa & Martinius (1992) describe another object-based model for simulating fluvial reservoirs, and Chessa (1992) discusses refinements to the procedures that enable the simulations to generate the correct number of sand bodies at well locations. The SISABOSA package (Augedal et al., 1986; Stanley et al., 1990) is a straightforward stochastic simulator, but its successor FLUREMO (Clemetsen et al. 1990; Henriquez et al., 1990) allows individual sand bodies to be clustered within channel belts (Figure 12.30B). The MOHERES package can use Markov fields instead of PDFs to distribute the geological objects spatially (Falt et al., 1991). Yet further increased sophistication is included in the model described by Tyler et al. (1992, 1993, 1994). Godi & Cosentino (1992) describe a geostatistical approach based on sequential indicator simulation, a variety of non-parametric statistics.

The SESIMERA package (Gundesø & Egeland, 1990) and the MONARCH package (Keijzer & Kortekaas, 1990; Bryant et al., 1991; Budding et al., 1992; van Vark et al., 1992) both allow the initial geological framework to be built deterministically by the initial specification of units which are known to correlate between wells. They then fill in the unsampled volume stochastically, stopping when specified criteria are met, such as the sand–shale ratio reaching that observed at the nearby wells. MONARCH allows more than two classes of object, and incorporates features to honour expected spatial relationships, such as between crevasse splays and channels, and it stochastically models palaeoflow direction to ensure that preferred orientation of sand bodies can occur (Figure 12.30D). The SIRCH software (Hirst et al., 1993) is aimed very much at fluvial deposits, as it creates sophisticated channel geometries (Figure 12.30C), and models avulsions and crevasse splays in considerable detail. The HERESIM package differs from the above in that, instead of using PDFs to specify the distribution and dimensions of objects, it uses values from semi-variogram analysis and applies a kriging methodology (Rudkiewicz et al., 1990). This is a technique where indicator kriging is proving of great value (Journel & Alabert, 1990).

Continuous stochastic models are appropriate for continuous variables such as porosity and permeability. Though not used to predict fluvial stratigraphy directly, they are commonly used to predict the fluid-flow behaviour in reservoirs formed in alluvial sequences. A common approach is first to use a discrete model of the above type to create a lithology framework that represents the large-scale heterogeneity, then to use a continuous model to include variation in porosity and permeability within each discrete object (Tjølsen & Damsleth, 1991; Damsleth et al., 1992; Tyler et al., 1992, 1993); see Figure 12.31 for an explanation of the main stages of this procedure. A number of statistical methods can be used for this, including Markov and Gaussian random variables, or geostatistical methods such as kriging (Rudkiewicz et al., 1990).

By their nature, stochastic models do not produce a single outcome but produce multiple realizations, no single one of which is necessarily correct. This is one reason why some workers dislike this approach – what do you do with all the possible answers? It is also the reason that caution needs to be exercised when considering examples in the literature, as papers commonly present just a single realization. Perhaps the most useful thing to do is to analyse statistically the models as a way to

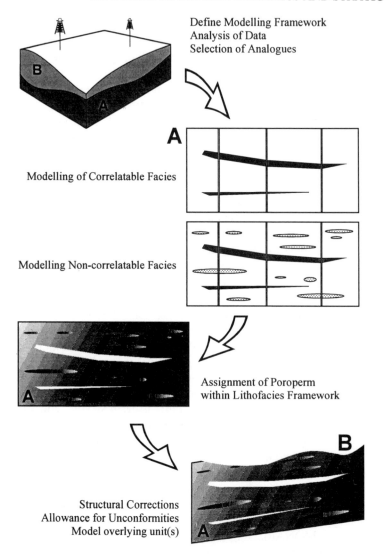

Define Modelling Framework
Analysis of Data
Selection of Analogues

A

Modelling of Correlatable Facies

Modelling Non-correlatable Facies

Assignment of Poroperm
within Lithofacies Framework

B

Structural Corrections
Allowance for Unconformities
Model overlying unit(s)

Figure 12.31 Stages commonly followed in hybrid stochastic modelling. Initial modelling may be deterministic, followed by phases of discrete and continuous stochastic modelling

quantify the uncertainty in geological knowledge (Haldorsen & Damsleth, 1990; Aasen *et al.*, 1990; Meling *et al.*, 1990). For example, the volume of hydrocarbons in place can be calculated for each realization to produce a probability distribution function of hydrocarbon volume. Srivastava (1994) has developed a method to animate multiple realizations smoothly, in a way that highlights the regions of greatest change between models. Stochastic models can be used for sensitivity analysis, such as to understand sand-body connectivity (e.g. King, 1990).

   There is no agreement on whether stochastic models are worth the effort or not. Martin *et al.* (1988) reported great difficulty in getting a match between stochastic models of low-sinuosity channel sands and actual hydrocarbon production, which they attributed to uncertainty in the input data. They developed a deterministic approach which they found better matched well test data than did any of their stochastic models. Clemetsen *et al.* (1990) argued for stochastic models, and against process-based models, on three grounds: conditioning to observations is simpler; they are not so sensitive to specification of geological events; and, they require much simpler parameter set-up to run them. In contrast, Henriquez *et al.* (1990), working with the same package (FLUREMO), argue against stochastic models on four grounds: difficulties in obtaining unique input parameters; absence of sedimentological understanding in the model (the model can contravene known geological processes); there is commonly an overestimation of sand-body connectedness; and they produce large numbers of realizations between which it is difficult to distinguish. Henriquez *et al.* advocate that greater emphasis be placed on physical process models. Van Vark *et al.* (1992), working with the MONARCH system, also noted the tendency of the stochastic approach towards overoptimism with regard to channel connectivity. Budding *et al.* (1992) note that, despite the large number of special features incorporated in the MONARCH package to ensure geological reasonableness, it is still difficult to generate three-dimensional models with the same vertical facies relationships as found in the wells. They suggest that the incorporation of Markov transition statistics may help, but are aware of the above-noted difficulties with this route.

Stochastic models were produced for the Triassic Sherwood Sandstone in the Wytch Farm oilfield of southern England (Dranfield *et al.*, 1987; Begg *et al.*, 1989) in order to improve the reservoir description. It is interesting to note that a few years later, when many infill wells have been drilled, it is clear that the stochastic model underestimated the lateral extent of many units in the reservoir, and fluid flow is better modelled with a distinctly layered description (Bowman *et al.*, 1993). This is in part because the original models were oversimplistic, and did not take account of pedogenic carbonate horizons in the floodplain, and layers of reworked pedogenic carbonate in channel sands.

Keijzer & Kortekaas (1990) carried out a comparison, on the fluvial Statfjord Formation in the Brent Field, between a conventional deterministic model (based on deterministic mapping supplemented by conceptual models) and a stochastic model produced with the MONARCH software. They tested the geological models by comparing fluid-flow simulations on these models with fluid production data. The stochastic model appeared to produce a substantially improved reservoir description, because it had a better estimation of the shale distribution – deterministic models are known to underestimate heterogeneity in the inter-well volumes. But they also comment that these results do not necessarily imply the need for stochastic modelling techniques. They might only reflect the necessity of a more detailed geological description. Perhaps the greatest value of carrying out stochastic modelling is that it forces one to look much closer than normal at the available geological data.

Conditioning to observations

An important consideration with stochastic models is whether the resultant prediction produces the same values at sample points as have actually been measured. Much effort can be expended to ensure the model honours actual data, which is referred to as conditioning the model. Note conditional simulation is not restricted to geostatistical approaches but is an aspect relevant to all stochastic models. The general solution is for a deterministic step to precede the stochastic step: the measured values are forced into the realization, and the stochastic modelling is only allowed to fill in volumes not affected by this (Figure 12.31). However, even this deterministic step usually has a stochastic element. For example, in packages such as MONARCH, geological units such as channel sands that have been observed in boreholes are extended out into the inter-well volume using a random number to select lateral dimension and orientation from user-supplied PDFs of these values (Budding *et al.*, 1992) An alternative conditioning route is to ensure consistency with pressure test data (Alabert, 1989; Lorenz *et al.*, 1991).

Stochastic methods depend heavily on statistical information derived from analogues of some sort: comparable subsurface examples such as adjacent oil fields, outcrops, modern depositional environments, process-based simulations. For example, discrete stochastic simulations of fluvial sequences typically use data on channel sand-body width and thickness such as in Fielding & Crane (1987) (Figure 12.18) and Dreyer (1990). It is interesting that experience with the MONARCH package led Budding *et al.* (1992) to comment that sensitivity analysis shows the modelling results are influenced less by the PDFs for geometric parameters than by the accuracy of information on the stratigraphic position of the objects. The use of analogues is fraught with many difficulties (e.g. see review by Alexander (1993)). There are limitations within the collected dataset, and there are limitations with applying the dataset to other systems.

When it comes to collecting the necessary data, the size of available outcrop or modern environment, coupled with physical accessibility, means there is usually a bias towards small-scale phenomena, and large features are inadequately sampled. Geehan & Underwood (1993) have suggested a procedure that partially compensates for such undersampling. Kossack (1989) notes that many published outcrop studies attempting to derive lateral extent from observed thickness are statistically invalid due to insufficient data. Then it is essential that the observed features are correctly interpreted. When it comes to applying the data, it is essential that the analogue is relevant. If the target under study has been misinterpreted, because of paucity of data, then inevitably the analogue will mislead. But the analogue must also be compatible in fundamental features such as scale, climatic and tectonic setting, and diagenetic overprint.

Despite the inherent difficulties, much effort is still being put into collecting data at outcrop (e.g. Cuevas-Gozalo & Martinius, 1993; Dreyer, 1993a,b; Dreyer *et al.*, 1993; Falkner & Fielding, 1993). Attempts are being made to add into the database information gleaned from mature, densely drilled oil and gas fields (e.g. Mijnssen *et al.*, 1993). A great deal of information is also being collected from modern environments (e.g. Jordan & Pryor, 1992) and outcrop (e.g. Bromley, 1991; Cowan, 1991; Hirst, 1991; Dreyer, 1993a) about the detailed character and architecture of fluvial deposits. Some of these results are being submitted to fluid-flow simulations, in an attempt to establish which aspects of the heterogeneity matter most (e.g. Høimyr *et al.*, 1993). At present, however, much is being published in the format of multiple vertical profiles, photomontages and line drawings because we still do not really know how to handle all the available facts – but at least it is being published!

CLOSING REMARKS

Dependence on understanding of modern fluvial systems

An issue repeatedly arising during this review of subsurface fluvial modelling methods is that deficiencies and limitations arise largely because of our lack of knowledge about present-day river systems. More data on modern rivers are desperately needed. In addition, data are needed on more types of rivers, from a wider range of tectonic and climatic settings. Furthermore, studies of modern rivers must be integrated, embracing hydrology, sedimentology, and regional setting and tectonics. A lot can be achieved by careful observation of modern river systems without the need for sophisticated computer packages or fancy statistical treatment – for example, Leeder (1993) and Leeder & Jackson (1993) have established some of the fine detail on the way drainage patterns are modified in regions of normal faulting.

Hybrid modelling techniques – the way forward?

It is human nature to want a single method to solve all problems. In reality, as we cannot yet model physical processes precisely, because our understanding of those processes is incomplete, all modelling approaches are partly or wholly empirical and depend on datasets which are known to be inadequate. The best results come from parallel use of several methods – conceptual modelling will always be one, then deterministic analysis such as contouring (or its 3D equivalent), in combination with stochastic models. Such integration of approaches will limit more tightly the estimate of subsurface geology.

The combination of discrete and continuous modelling, to create a full description of hydrocarbon reservoirs, has already been mentioned (Figure 12.31). What is also needed is the combination of multiple discrete methods, or, better still, the conjunction of stochastic and process-based techniques. In such combinations, one would perhaps choose a deterministic technique to build the larger-scale framework, and a stochastic technique (or a process-based simulation?) to infill the small scale. Conceptual models such as the sequence stratigraphy paradigm have an important role to play here. Improvements have already been noted from elementary examples of such hybrid modelling (Tetzlaff, 1991; MacDonald & Halland, 1993; Tyler *et al.*, 1994).

Validity of the models

If the lack of data about the subsurface means that most of our efforts have to go into prediction rather than description, how are we to know if the result is right? Of course, the usual test is when the prediction is put to some practical use, such as to decide where to drill a well or drive a mine shaft. But it would be better (and potentially much less expensive) if we could validate the model earlier. Two aspects should be considered: geological consistency and statistical consistency (Pang, 1993). In general, a model of the subsurface would be considered valid (geologically consistent) as long as it does not contravene accepted geological knowledge. In addition, it is usual to check that the model has similar statistical characteristics (statistical consistency) to the available data. The statistical test has to be used sparingly: the many ways in which limited profiles through sequences can mislead have already been noted above. Note the example of the Sherwood Sandstone in the Wytch Farm field, where the earlier model (Dranfield *et al.*, 1987) was consistent with the data available at that time, but not with subsequently acquired data (Bowman *et al.*, 1993). There is a danger of circular reasoning here. What is accepted geological knowledge evolves with time, partly through experiences gained with attempts to model the subsurface.

At the end, the most important thing one can do to validate a model is to ensure that as much time and effort as possible has gone into data analysis, understanding the available information, and checking its quality. It is crucial to keep an open mind, and to be aware of the philosophy and weaknesses of the chosen methods. If approached in this way, application of the modelling methods described usually brings useful results, even if this is only a better understanding of the geological data one has collected. Finally, one has to beware the false sense of security that is imparted by a complex numerical model that may have taken many months of effort to produce – the model is only as good as the data and assumptions that went into it.

ACKNOWLEDGEMENTS

This chapter has greatly benefited from the helpful comments on an earlier draft from Jan Alexander, Frank Ethridge, Peter Friend and Nigel Trewin, and to these persons I extend grateful thanks. I wish also to thank the editors for giving me the opportunity to write this review. The author takes sole responsibility for the views expressed.

REFERENCES

Aasen, J.O., Silseth, J.K., Holden, L., Omre, H., Halvorsen, K.B. & Hoiberg, J. (1990) A stochastic reservoir model and its use in evaluations of uncertainties in the results of recovery processes. In: Buller, A.T., Berg, E., Hjelmeland, O., Kleppe, J., Torsaeter, O. & Aasen, J.O. (eds) *North Sea oil and gas reservoirs – II*. Graham & Trotman, London, 425–436.

Aasum, Y., Kelkar, M.G. & Gupta, S.P. (1991) An application of geostatistics and fractal geometry for reservoir characterization. *SPE Formation Evaluation* **6**, 11–19.

Alabert, F.G. (1989) Constraining description of randomly heterogeneous reservoirs to pressure test data: A Monte Carlo study. SPE Technical Paper **19600**, presented at *SPE annual technical conference*, San Antonio, Texas, 8–11 October.

Alexander, J. (1986) Idealised flow models to predict alluvial sandstone body distribution in the Middle Jurassic Yorkshire Basin. *Marine and Petroleum Geology* **3**, 298–306.

Alexander, J. (1993) A discussion on the use of analogues for reservoir geology. In: Ashton, M. (ed.) *Advances in reservoir geology*, Geological Society Special Publication **69**, London, 175–194.

Alexander, J. & Leeder, M.R. (1987) Active tectonic control on alluvial architecture. In: Ethridge, F.G., Flores, R.M. & Harvey, M.D (eds) *Recent developments in fluvial sedimentology*. SEPM Special Publication **39**, Tulsa, 243–252.

Allen, J.R.L. (1963) The classification of cross-stratified units with notes on their origin. *Sedimentology* **2**, 93–114.

Allen, J.R.L. (1965) A review of the origin and characteristics of Recent alluvial sediments. *Sedimentology* **5**, 89–191.

Allen, J.R.L. (1966) On bedforms and palaeocurrents. *Sedimentology* **6**, 153–190.

Allen, J.R.L. (1970) Studies in fluviatile sedimentation: a comparison of fining-upwards cyclothems with special reference to coarse member composition and interpretation. *Journal of Sedimentary Petrology* **40**, 298–323.

Allen, J.R.L. (1974) Studies in fluviatile sedimentation: implications of pedogenic carbonate units, Lower Old Red Sandstone, Anglo-Welsh outcrop. *Geological Journal* **9**, 181–208.

Allen, J.R.L. (1978) Studies in fluviatile sedimentation: an exploratory quantitative model for the architecture of avulsion-controlled alluvial sites. *Sedimentary Geology* **21**, 129–147.

Allen, J.R.L. (1983) Studies in fluviatile sedimentation: bars, bar-complexes and sandstone sheets (low sinuosity braided streams) in the Brownstones (L. Devonian), Welsh Borders. *Sedimentary Geology* **33**, 237–293.

Allen, J.R.L. & Williams, B.P.J. (1982) The architecture of an alluvial suite: rocks between the Townsend Tuffand Pickard Bay Tuff Beds (Early Devonian), southwest Wales. *Philosophical Transactions of the Royal Society of London* **B297**, 51–89.

Ambrose, W.A., Tyler, N. & Parsley, M.J. (1991) Facies heterogeneity, pay continuity, and infill potential in barrier-island, fluvial, and submarine-fan reservoirs: examples from the Texas Gulf Coast and Midlands Basin. In: Miall, A.D. & Tyler, N. (eds) *The three-dimensional facies architecture of terrigenous clastic sediments and its implications for hydrocarbon discovery and recovery*. SEPM Concepts in Sedimentology and Palaeontology Vol. **3**, Tulsa, 13–21.

Anderton, R. (1985) Clastic facies models and facies analysis. In: Brenchley, P.J. & Williams, B.P.J. (eds) *Sedimentology: recent developments and applied aspects*. Geological Society Special Publication **18**, Blackwell, Oxford, 31–47.

Ashworth, P.J., Best, J.L., Leddy, J.O. & Geehan, G.W. (1993) The geometry and internal architecture of braided alluvium: scaled physical models and reservoir heterogeneities (Abstract). *Proceedings of the 5th International Conference on Fluvial Sedimentology*, Brisbane, 5–9 July.

Atkinson, C.D. (1986) Tectonic control on alluvial sedimentation as revealed by an ancient catena in the Capella Formation (Eocene) of northern Spain. In: Wright, V.P. (ed.) *Palaeosols: their recognition and interpretation*. Blackwell Scientific Publications, Oxford, 139–179.

Atkinson, C.D., McGowen, J.H., Bloch, S., Lundell, L.L. & Trumbly, P.N. (1990) Braidplain and deltaic reservoir, Prudhoe Bay Field, Alaska. In: Barwis, J.H., McPherson, J.G. & Studlick, J.R.J. (eds) *Sandstone Petroleum Reservoirs*. Springer Verlag, New York, 7–29.

Augedal, H.O., Stanley, K.O. & Omre, H. (1986) SISABOSA, a program for stochastic modelling and evaluation of reservoir geology. In: Nitteberg, J. (ed.) *Proceedings from Conference on Reservoir Description and Simulation with Emphasis on EOR*, Oslo, September 1986. Institute for Energy Technology.

Autin, W.J., Burns, S.F., Miller, R.T., Saucier, R.T. & Snead, J.I. (1991) Quaternary geology of the lower Mississippi valley. In: Morrison, R.B. (ed.) *Quaternary nonglacial geology: Conterminious U.S.* The Geology of North America Volume **K–2**, Geological Society of America, 547–582.

Baldwin, B. & Butler, C.O. (1985) Compaction curves. *The American Association of Petroleum Geologists Bulletin* **69**, 622–626.

Banks, R. (1991) Contouring algorithms. *Geobyte* October, 15–23.

Banks, R.B. & Sukkar, J.K. (1992) Computer processing of multiple 3-D fault blocks containing multiple surfaces. *Geobyte* August, 58–62.

Begg, S.H., Carter, R.R. & Dranfield, P. (1989) Assigning effective values to simulator grid-block parameters for heterogeneous reservoirs. *SPE Reservoir Engineering* **4**, 455–463.

Belcher, R.C. & Hoffman, K.S. (1992) Geologic structural models: improvements in 3-D modeling. *Geobyte* August, 50–53.

Bentham, P.A., Talling, P.J. & Burbank, D.W. (1993) Braided stream and flood-plain deposition in a rapidly aggrading basin: the Escanilla formation, Spanish Pyrenees. In: Best, J.L. & Bristow, C.S. (eds) *Braided rivers*. Geological Society special publication **75**, London, 177–194.

Berg, R.R. (1968) Point-bar origin of Fall River sandstone reservoirs, northeastern Wyoming. *The American Association of Petroleum Geologists Bulletin* **52**, 2116–2122.

Blair, T.C. & McPherson, J.G. (1992) The Trollheim alluvial fan and facies model revisited. *Geological Society of America Bulletin* **104**, 762–769.

Blatt, H., Middleton, G.V. & Murray, R.C. (1972) *Origin of sedimentary rocks*. Prentice-Hall, New Jersey.

Blum, M.D. (1992) *Modern depositional environments and recent alluvial history of the lower Colorado River, Gulf coastal plain, Texas*. PhD Thesis, University of Texas at Austin, Texas.

Blum, M.D., Toomey, R.S.I. & Valastro, S.J. (1994) Fluvial response to Late Quaternary climatic and environmental change, Edwards Plateau, Texas. *Palaeogeography, Palaeoclimatology, Palaeoecology* **108**, 1–21.

Bouma, A.H. (1962) *Sedimentology of some flysch deposits*. Elsevier, Amsterdam.

Bourgine, B. (1992) Advanced interpolation: kriging with external drift and conditional simulations. *Geobyte* October, 42–45.

Bourke, L.T. (1992) Sedimentological borehole image analysis in clastic rocks: a systematic approach to interpretation. In: Hurst, A., Griffiths, C.M. & Worthington, P.F. (eds) *Geological applications of wireline logs II*. Geological Society special publication **65**, London, 31–42.

Bowdon, R.D. (1992) Computer contouring in the real world. *Geobyte* June, 12–13.

Bowman, M., McClure, N.M. & Wilkinson, D.W. (1993) Wytch Farm oilfield: deterministic reservoir description of the Triassic Sherwood Sandstone. In: Parker, J.R. (ed.) *Petroleum Geology of NW Europe. Proceedings of the 4th Conference*. Geological Society, London, 1513–1518.

Bown, T.M. & Kraus, M.J. (1987) Integration of channel and floodplain suites, I. Developmental sequence and lateral relations of alluvial paleosols. *Journal of Sedimentary Petrology* **57**, 587–601.

Bridge, J. S. (1975) Computer simulation of sedimentation in meandering streams. *Sedimentology* **22**, 3–44.

Bridge, J.S. (1977) Flow, bed topography and sedimentary structures in open channel bends: a three-dimensional model. *Earth Surface Processes and Landforms* **2**, 281–294.

Bridge, J.S. (1979) A Fortran IV program to simulate alluvial stratigraphy. *Computers and Geosciences* **5**, 335–348.

Bridge, J.S. (1982) A revised mathematical model and Fortran IV program to predict flow, bed topography and grainsize in open-channel bends. *Computers and Geosciences* **8**, 91–95.

Bridge, J.S. (1984) Large-scale facies sequences in alluvial overbank environments. *Journal of Sedimentary Petrology* **54**, 583–588.

Bridge, J.S. (1985) Paleochannel patterns inferred from alluvial deposits: a critical evaluation. *Journal of Sedimentary Petrology* **55**, 579–589.

Bridge, J.S. (1993a) Description and interpretation of fluvial deposits: a critical perspective. *Sedimentology* **40**, 801–810.

Bridge, J.S. (1993b). Keynote address: Recent developments in understanding water flow, sediment transport, erosion and deposition in rivers: a personal perspective. In: Yu, B. & Fielding, C.R. (eds) *Keynote addresses and abstracts, 5th International Conference on*

Fluvial Sedimentology, Brisbane, Australia, 5–9 July, Geological Society of Australia, Canberra, K1–K13.

Bridge, J.S. & Leeder, M.R. (1979) A simulation model of alluvial stratigraphy. *Sedimentology* **26**, 617–644.

Bridge, J.S. & Mackey, S.D. (1993a) A theoretical study of fluvial sandstone body dimensions. In: Flint, S.S. & Bryant, I.D. (eds) *The geological modelling of hydrocarbon reservoirs and outcrop analogues*. International Association of Sedimentologists special publication no. **15**, Blackwell, Oxford, 213–236.

Bridge, J.S. & Mackey, S.D. (1993b) A revised alluvial stratigraphy model. In: Marzo, M. & Puidefabregas, C. (eds) *Alluvial sedimentation*. International Association of Sedimentologists special publication **17**, Blackwell Scientific Publishers, Oxford, 319–336.

Brierley, G.J. (1991) Bar sedimentology of the Squamish River, British Columbia: definition and application of morphostratigraphic units. *Journal of Sedimentary Petrology* **61**, 211–225.

Bromley, M.H. (1991) Variations in fluvial style as revealed by architectural elements, Kayenta Formation, Mesa Creek, Colorado, USA: Evidence for both ephemeral and perennial fluvial processes. In: Miall, A.D. & Tyler, N. (eds) *The three-dimensional facies architecture of terrigenous clastic sediments and its implications for hydrocarbon discovery and recovery*. SEPM Concepts in Sedimentology and Palaeontolgy Vol. **3**, Tulsa, 94–103.

Brooker, P.I. (1991) *A geostatistical primer*. World Scientific Publishing Co., Singapore.

Bryant, I.D. & Flint, S.S. (1993) Quantitative clastic reservoir geological modelling: problems and perspectives. In: Flint, S. & Bryant, I.D. (eds) *The geological modelling of hydrocarbon reservoirs and outcrop analogues*. International Association of Sedimentologists special publication **15**, Blackwell Scientific Publishers, Oxford, 3–20.

Bryant, I.D., Paardekam, A.H.M., Davies, P. & Budding, M.C. (1991) Integrated reservoir characterisation of Cycle III, Brent Group, Brent Field, UK North Sea for reservoir management. In: Sneider, R., Massell, W., Mathis, R., Loren, D. & Wichmann, P. (eds) *The integration of geology, geophysics, petrophysics and petroleum engineering in reservoir delineation, description and management. Proceedings of the first Archie Conference*. American Association of Petroleum Geologists, Tulsa, 405–422.

Budding, M.C., Paardekam, A.H.M. & Van Rossem, S.J. (1992) 3D connectivity and architecture in sandstone reservoirs. SPE Technical Paper **22342**, presented at *SPE International Meeting on Petroleum Engineering*, Beijing, China, 24–27 March.

Burrough, P.A. (1983) Multiscale sources of spatial variation in soil. II. A non-Brownian fractal model and its application in soil survey. *Journal of Soil Science* **34**, 599–620.

Cameron, G.I.F. (1992) Analysis of dipmeter data for sedimentary orientation. In: Hurst, A., Griffiths, C.M. & Worthington, P.F. (eds) *Geological applications of wireline logs II*. Geological Society special publication **65**, London, 141–154.

Cameron, G.I.F., Collinson, J.D., Rider, M.H. & Xu, L. (1993) Analogue dipmeter logs through a prograding deltaic sandbody. In: Ashton, M. (ed.) *Advances in reservoir geology* Geological Society special publication **69**, London, 195–217.

Cant, D.J. (1992) Subsurface facies analysis. In: Walker, R.G. & James, N.P. (eds) *Facies models: response to sea level change*. Geological Association of Canada, St. Johns, Newfoundland, 27–45.

Cant, D.J. & Walker, R.G. (1976) Development of a braided-fluvial facies model for the Devonian Battery Point Sandstone, Quebec, Canada. *Canadian Journal of Earth Sciences* **13**, 102–119.

Carlston, C.W. (1965) The relation of free meander geometry to stream discharge and its geomorphic implications. *American Journal of Science* **263**, 864–885.

Charles, H.H. (1941) Bush City field, Anderson County, Kansas. In: *Stratigraphic type oil fields*. The American Association of Petroleum Geologists, Tulsa, 43–56.

Chessa, A.G. (1992) On the object-based method for simulating sandstone deposits. In: Christie, M.A., Da Silva, F.V., Farmer, C.L., *et al. ECMOR III: Proceedings of the Third*

European Conference on the Mathematics of Oil Recovery. Delft University Press, Delft, Netherlands, 67–77.

Chessa, A.G. & Martinius, A.W. (1992) Object-based modelling of the spatial distribution of fluvial sandstone deposits. In: Christie, M.A., Da Silva, F.V., Farmer, C.L., *et al. ECMOR III: Proceedings of the Third European Conference on the Mathematics of Oil Recovery.* Delft University Press, Delft, Netherlands, 5–14.

Clemetsen, R., Hurst, A.R., Knarud, R. & Omre, H. (1990) A computer program for evaluation of fluvial reservoirs. In: Buller, A.T., Berg, E., Hjelmeland, O., Kleppe, J., Torsaeter, O. & Aasen, J.O. (eds) *North Sea oil and gas reservoirs – II.* Graham & Trotman, London, 373–385.

Cloetingh, S. (1988) Intraplate stresses: a new element in basin analysis. In: Kleinspehn, K.L. & Paola, C. (eds) *New perspectives in basin analysis.* Springer-Verlag, New York, 205–230.

Cloetingh, S. (1991) Tectonics and sea-level changes: a controversy? In: Müller, D.W., McKenzie, J.A. & Weissart, H. (eds) *Controversies in modern geology.* Academic Press, London, 249–277.

Cloetingh, S. & Kooi, H. (1992) Tectonics and global change – inferences from Late Cenozoic subsidence and uplift patterns in the Atlantic/Mediterranean region. *Terra Nova* **4**, 340–350.

Coleman, J.H. (1966) Ecologic changes in a massive fresh-water clay sequence. *Transactions of the Gulf Coast Association of Geological Societies* **16**, 159–174.

Collinson, J.D. (1978) Vertical sequence and sand body shape in alluvial sequences. In: Miall, A.D. (ed.) *Fluvial sedimentology,* Canadian Society for Petroleum Geology Memoir **5**, 577–586.

Collinson, J.D. (1986) Alluvial sediments. In: Reading, H.G. (ed.) *Sedimentary environments and facies.* 2nd edition. Blackwell Scientific Publications, Oxford, 20–62.

Corbett, P.W.M. & Jensen, J.L. (1992) Variation of reservoir statistics according to sample spacing and measurement type for some intervals in the Lower Brent Group. *The Log Analyst* Jan-Feb, 22–41.

Cornish, F.G. (1984) Fluvial environments and palaeohydrology of the Upper Morrow "A" (Pennsylvanian) meander belt sandstone, Beaver County, Oklahoma. *Shale Shaker* **34**(6), 70–80.

Cowan, E.J. (1991) The large-scale architecture of the Westwater Canyon Member, Morrison Formation (Upper Jurassic), San Juan Basin, New Mexico. In: Miall, A.D. & Tyler, N. (eds) *The three-dimensional facies architecture of terrigenous clastic sediments and its implications for hydrocarbon discovery and recovery.* SEPM Concepts in Sedimentology and Palaeontology Vol. **3**, Tulsa, 80–93.

Crane, S.D. & Tubman, K.M. (1990) Reservoir variability and modeling with fractals. SPE Technical Papers **20606**, presented at *65th SPE Annual Technical Conference,* New Orleans, 23–26 September.

Cuevas-Gozalo, M.C. & Martinius, A.W. (1993) Outcrop data-base for the geological characterization of fluvial reservoirs: an example from distal fluvial fan deposits in the Lorance Basin, Spain. In: North, C.P. & Prosser, D.J. (eds) *Characterization of fluvial and aeolian reservoirs.* Geological Society special publication **73**, London, 79–94.

Dahlberg, E.C. (1975) Relative effectiveness of Geologists and computers in mapping potential hydrocarbon exploration targets. *Mathematical Geology* **7**, 373–394.

Damsleth, E., Tjølsen, C.B., Omre, K.H. & Haldorsen, H.H. (1992) A two-stage stochastic model applied to a North Sea reservoir. *Journal of Petroleum Technology* **44**, 402–408.

Davies, D.K., Williams, B.P.J. & Vessell, R.K. (1993) Dimensions and quality of reservoirs originating in low and high sinuosity channel systems, Lower Cretaceous Travis Peak formation, East Texas, USA. In: North, C.P. & Prosser, D.J. (eds) *Characterization of fluvial and aeolian reservoirs.* Geological Society special publication **73**, London, 95–121.

Davis, J.C. (1986) *Statistics and data analysis in geology.* 2nd edition. John Wiley, New York.

Davis, J.C. & Sampson, R.J. (1992) Trend-surface analysis: early computer technique remains popular, useful. *Geobyte* August, 38–43.

Dawson, M.R. & Bryant, I.D. (1987) Three-dimensional facies geometry in Pleistocene outwash sediments, Worcestershire, U.K. In: Ethridge, F.G., Flores, R.M. & Harvey, M.D. (eds) *Recent developments in fluvial sedimentology.* SEPM special publication **39**, Tulsa, 189–196.

DeCelles, P.G., Gray, M.B., Ridgway, K.D. *et al.* (1991) Controls on synorogenic alluvial fan architecture, Beartooth Conglomerate (Palaeocene), Wyoming and Montana. *Sedimentology* **38**, 567–590.

Denver, L.E. & Phillips, D.C. (1990) Stratigraphic geocellular modeling. *Geobyte* **5**, 45–47.

Deutsch, C.V. & Journel, A.G. (1992) *GSLIB geostatistical software library and user's guide.* Oxford University Press, New York.

Di Julio, S.S. (1993) North Vickers East waterflood performance prediction with fractal geostatistics. *SPE Formation Evaluation* **8** (June), 89–95.

Doyen, P.M. (1988) Porosity from seismic data: a geostatistical approach. *Geophysics* **53**, 1263–1275.

Doyen, P.M., Guidish, T.M. & de Buyl, M.H. (1989) Monte Carlo simulation of lithology from seismic data in a channel-sand reservoir. SPE Technical Paper **19588**, presented at *SPE Annual Technical Conference*, San Antonio, 8–11 October.

Dranfield, P., Begg, S.H. & Carter, R.R. (1987) Wytch Farm Oilfield: reservoir characterization of the Triassic Sherwood Sandstone for input to reservoir simulation studies. In: Brooks, J. & Glennie, K.W. (eds) *Petroleum Geology of North West Europe.* Graham & Trotman, London, 149–160.

Dreyer, T. (1990) Sand body dimensions and infill sequences of stable, humid-climate delta plain channels. In: Buller, A.T., Berg, E., Hjelmeland, O., Kleppe, J., Torsaeter, O. & Aasen, J.O. (eds) *North Sea oil and gas reservoirs – II.* Graham & Trotman, London, 337–351.

Dreyer, T. (1993a) Geometry and facies of large-scale flow units in fluvial-dominated fan–delta-front sequences. In: Ashton, M. (ed.) *Advances in reservoir geology.* Geological Society special publication **69**, London, 135–174.

Dreyer, T. (1993b) Quantified fluvial architecture in ephemeral stream deposits of the Esplugafreda Formation (Palaeocene), Temp-Graus Basin, northern Spain. In: Marzo, M. & Puidefabregas, C. (eds) *Alluvial sedimentation.* International Association of Sedimentologists special publication **17**, Blackwell Scientific Publications, Oxford, 337–362.

Dreyer, T., Scheie, A. & Walderhaug, O. (1990) Minipermeameter-based study of permeability trends in channel sand bodies. *The American Association of Petroleum Geologists Bulletin* **74**, 359–374.

Dreyer, T., Falt, L.-M., Hoy, T., Knarud, R., Steel, R. & Cuevas, J.-L. (1993) Sedimentary architecture of field analogues for reservoir information (SAFARI): a case study of the fluvial Escanilla Formation, Spanish Pyrenees. In: Flint, S. & Bryant, I.D. (eds) *The geological modelling of hydrocarbon reservoirs and outcrop analogues.* International Association of Sedimentologists special publication **15**, Blackwell Scientific Publishers, Oxford, 57–80.

Dubiel, R.F. (1991) Architectural-facies analysis of nonmarine depositional systems in the Upper Triassic Chinle Formation, Southeastern Utah. In: Miall A.D. & Tyler N. (eds) *The three-dimensional facies architecture of terrigenous clastic sediments and its implications for hydrocarbon discovery and recovery.* SEPM Concepts in Sedimentology and Palaeontology Vol. 3, Tulsa, 103–110.

Dubiel, R.F. (1992) Sedimentology and depositional history of the Upper Triassic Chinle Formation in the Uinta, Piceance, and Eagle Basins, northeastern Colorado and southeastern Utah. *US. Geological Survey Bulletin* **1787-W**.

Dubrule, O. (1989) A review of stochastic models for petroleum reservoirs. In: Armstrong, M. (ed.) *Geostatistics, Vol. 2.* Kluwer Academic Publishers, Dordrecht, Netherlands, 493–506.

Elliot, T. (1976) The morphology, magnitude and regime of a Carboniferous fluvial-distributary channel. *Journal of Sedimentary Petrology* **46**, 70–76.

Emanuel, A.S., Alameda, G.K., Behrens, R.A. & Hewett, T.A. (1989) Reservoir performance prediction methods based on fractal geostatistics. *SPE Reservoir Engineering* **4**, 311–318.

Englund, E. & Sparks, A. (1988) *Geostatistical environmental assessment software. User's guide*. US Environmental Protection Agency, Las Vegas, Nevada, USA.

Eschard, R., Ravenne, C., Houel, P. & Knox, R. (1991) Three-dimensional reservoir architecture of a valley-fill sequence and a deltaic aggradational sequence: Influences of minor relative sea-level variations (Scalby formation, England). In: Miall, A.D. & Tyler, N. (eds) *The three-dimensional facies architecture of terrigenous clastic sediments and its implications for hydrocarbon discovery and recovery*. SEPM Concepts in Sedimentology and Palaeontology Vol. 3, Tulsa, 133–147.

Ethridge, F.G., Jackson, T.J. & Youngberg, A.D. (1981) Floodbasin sequence of a fine-grained meander belt system: the coal-bearing Lower Wasatch and Upper Fort Union Formations, southern Powder River Basin, Wyoming. In: Ethridge. F.G. & Flores, R.M. (eds) *Recent and ancient nonmarine depositional environments: models for exploration*. SEPM special publication **31**, Tulsa, 191 –209.

Ethridge, R.G. & Schumm, S.A. (1978) Reconstructing paleochannel morphologic and flow characteristics: Methodology, limitations, and assessment. In: Miall, A.D. (ed.) *Fluvial sedimentology*. Canadian Society of Petroleum Geologists Memoir **5**, Calgary, 703–721.

Falkner, A. & Fielding, C. (1993) Quantitative facies analysis of coal-bearing sequences in the Bowen Basin, Australia: applications to reservoir description. In: Flint, S. & Bryant, I.D. (eds) *The geological modelling of hydrocarbon reservoirs and outcrop analogues*. International Association of Sedimentologists special publication **15**, Blackwell Scientific Publishers, Oxford, 81–98.

Fält, L.M., Henriquez, A., Holden, L. & Tjelmeland, H. (1991) Moheres, a program system for simulation of reservoir architecture and properties. In: *Proceedings of the Sixth European Symposium on Improved Oil Recovery*, Stavanger, Norway, 21–23 May 1991.

Farmer, C.L. 1989. Numerical rocks: the mathematical generation of reservoir geology. In: *Proceedings of the Joint Institute of Mathematics and Society of Petroleum Engineers European Conference on the Mathematics of Oil Recovery*, Cambridge, England, July 25–27, 437–447.

Farrell, K.M. (1987) Sedimentology and facies architecture of overbank deposits of the Mississippi River, False River region, Louisiana. In: Ethridge, F.G., Flores, R.M. & Harvey, M.D. (eds) *Recent developments in fluvial sedimentology*. SEPM special publications **39**, Tulsa, 111–120.

Fielding, C.R. (1986) Fluvial channel and overbank deposits from the Westphalian of the Durham coalfield, NE England. *Sedimentology* **33**, 119–140.

Fielding, C.R. & Crane, R.C. (1987) An application of statistical modelling to the prediction of hydrocarbon recovery factors in fluvial reservoir sequences. In: Ethridge, F.G., Flores, R.M. & Harvey, M.D. (eds) *Recent developments in fluvial sedimentology*. SEPM special publication **39**, Tulsa, 321–327.

Fielding, C.R., Falkner, A.J. & Scott, S.G. (1993) Fluvial response to foreland basin overfilling; the Late Permian Rangal Coal Measures in the Bowen Basin, Queensland, Australia. *Sedimentary Geology* **85**, 475–497.

Fisher, T.R. & Wales, R.Q. (1990) 3-D Solid modelling of sandstone reservoirs using NURBS: A case study of Noonen Ranch Field, Denver Basin, Colorado. *Geobyte* **5**, 39–41.

Flores, R.M., Ethridge, F.G., Miall, A.D., Galloway, W.E. & Fouch, T.D. (1985) *Recognition of fluvial depositional systems and their resource potential*. SEPM short course no. **19**, Tulsa.

Flynn, J.J. (1990) 3-D computing geosciences update. Hardware advances set the pace for software developers. *Geobyte* **5**, 33–36.

Frakes, L.A. (1979) *Climates through geologic time*. Elsevier, Amsterdam.

Fried, C.C. & Leonard, J.E. (1990) Petroleum 3-D models come in many flavors. *Geobyte* **5**, 27–30.

Friend, P.F. (1983) Towards the field classification of alluvial architecture or sequence. In:

Collinson, J.D. & Lewin, J. (eds) *Modern and ancient fluvial systems*. International Association of Sedimentologists special publication **6**, 345–354.

Galay, V.J., Kellerhals, R. & Bray, D.I. (1973) Diversity of river types in Canada. In: *Fluvial Process and Sedimentation: Proceedings of the Hydrology Symposium*. National Research Council of Canada, 217–250.

Galli, A., Guerillot:, D. & Ravenne, C. (1990) Integration of geology, geostatistic and multiphasic flow for 3D reservoir studies. In: Guerillot, D. & Guillon, O. (eds) *Second European Conference on the Mathematics of Oil Recovery*. Editions Technip, Paris, 11–19.

Galloway, W.E. (1981) Depositional architecture of Cenozoic Gulf Coast plain fluvial systems. In: Ethridge, F.G. & Flores, R.M. (eds) *Recent and ancient nonmarine depositional environments: models for exploration*. SEPM special publication **31**, Tulsa, 127–155.

Galloway, W.E. (1989a) Genetic stratigraphic sequences in basin analysis I: architecture and genesis of flooding-surface bounded depositional units. *The American Association of Petroleum Geologists Bulletin* **73**, 125–142.

Galloway, W.E. (1989b) Genetic stratigraphic sequences in basin analysis II: application to northwest Gulf of Mexico Cenozoic basin. *The American Association of Petroleum Geologists Bulletin* **73**, 143–154.

Galloway, W.E. & Hobday, D.K. (1983) *Terrigenous clastic depositional systems*. Springer-Verlag, New York.

Gardiner, S., Thomas, D.V., Bowering, E.D. & McMinn, L.S. (1990) A braided fluvial reservoir, Peco Field, Alberta, Canada. In: Barwis, J.H., McPherson, J.G. & Studlick, J.R.J. (eds) *Sandstone Petroleum Reservoirs*. Springer Verlag, New York, 31–56.

Geehan, G. & Underwood, J. (1993) The use of shale length distributions in geological modelling. In: Flint, S. & Bryant, I.D. (eds) *The geological modelling of hydrocarbon reservoirs and outcrop analogues*. International Association of Sedimentologists special publication **15**, Blackwell Scientific Publishers, Oxford, 205–212.

Geehan, G.W., Lawton, T.F., Sakurai, S. *et al.* (1986) Geologic prediction of shale continuity: Prudhoe Bay field. In: Lake, L.W. & Carroll, H.B.J. (eds) *Reservoir characterization*. Academic Press, Orlando, 63–82.

Gingerich, P.D. (1969) Markov analysis of cyclic alluvial sediments. *Journal of Sedimentary Petrology* **39**, 330–332.

Godi, A. & Cosentino, L. (1992) Geostatistical simulation and flow modeling of a fluvial reservoir: a case study. In: Christie, M.A., Da Silva, F.V., Farmer, C.L. *et al. ECMOR III: Proceedings of the Third European Conference on the Mathematics of Oil Recovery*. Delft University Press, Delft, Netherlands, 99–108.

Goggin, D.J., Chandler, M.A., Kocurek, G. & Lake, L.W. (1992) Permeability transects of eolian sands and their use in generating random permeability fields. *SPE Formation Evaluation* **7**, 7–16.

Goodchild, M.F. & Mark, D.M. (1987) The fractal nature of geographic phenomena. *Annals of the Association of American Geographers* **77**, 265–278.

Gregory, K.J. (1983) Introduction. In: Gregory, K.J. (ed.) *Background to palaeohydrology*. Wiley, Chichester, 3–23.

Guccione, M.J. (1993) Grain-size distribution of overbank sediment and its use to locate channel position. In: Marzo, M. & Puidefabregas, C. (eds) *Alluvial sedimentation*. International Association of Sedimentologists special publication **17**, Blackwell Scientific Publications, Oxford, 185–194.

Gundesø, R. & Egeland, O. (1990) SESIMIRA – a new geological tool for 3D modelling of heterogeneous reservoirs. In: Buller, A.T., Berg, E., Hjelmeland, O., Kleppe, J., Torsaeter, O. & Aasen, J.O. (eds) *North Sea oil and gas reservoirs – II*. Graham & Trotman, London, 363–371.

Haecker, M.A. (1992) Convergent gridding: a new approach to surface reconstruction. *Geobyte* June, 48–53.

Haldorsen, H.H. & Chang, D.M. (1986) Notes on stochastic shales: from outcrop to simulation model. In: Lake, L.W. & Carroll, H.B.J. (eds) *Reservoir characterization.* Academic Press. Orlando, 445–485.

Haldorsen, H.H. & Damsleth, E. (1990) Stochastic modeling. *Journal of Petroleum Technology* **42**, 404–412.

Haldorsen, H.H. & Damsleth, E. (1993) Challenges in reservoir characterization. *The American Association of Petroleum Geologists Bulletin* **77**, 541–551.

Haldorsen, H.H. & Lake, L.W. (1984) A new approach to shale management in field-scale models. *Society of Petroleum Engineers Journal* **24**, 447–457.

Hall, S.A. (1990) Channel trenching and climatic change in the southern U.S. Great Plains. *Geology* **18**, 342–345.

Hamilton, D.E. & Jones, T.A. (Eds). (1992) *Computer modeling of geologic surfaces and volumes.* AAPG Computer Applications in Geology no. **1**, Tulsa.

Handley, E.J. (1954) Contouring is important. *World Oil* **138**, 106–107.

Haq, B.U., Hardenbol, J. & Vail, P.R. (1987) Chronology of fluctuating sea levels since the Triassic (250 million years ago to present). *Science* **235**, 1156–1167.

Haq, B.U., Hardenbol, J. & Vail, P.R. (1988) Mesozoic and Cenozoic chronostratigraphy and cycles of sea-level change. In: Wilgus, C.J., Hastings, B.S., Kendall, C.G.S.C., Posamentier, H.W., Ross, C.A. & Van Wagoner, J.C. (eds) *Sea-level changes: an integrated approach.* SEPM special publication **42**, Tulsa, 71–108.

Harbaugh, J.W. & Bonham-Carter, G. (1970) *Computer simulation in geology.* Wiley Interscience, New York.

Harden, D.R. (1990) Controlling factors in the distribution and development of incised meanders in the central Colorado Plateau. *Geological Society of America Bulletin* **102**, 233–242.

Harker, S.D. McGann, G.J., Bourke, L.T. & Adams, J.T. (1990) Methodology of Formation MicroScanner image interpretation in Claymore and Scapa fields (North Sea). In: Hurst, A., Lovell, M.A. & Morton, A.C. (eds) *Geological applications of wireline logs.* Geological Society of London special publication **48**, 11–25.

Harms, J.C. (1966) Stratigraphic traps in a valley fill, western Nebraska. *The American Association of Petroleum Geologists Bulletin* **50**, 2119–2149.

Hattori, I. (1976) Entropy in Markov chains and discrimination of cyclic patterns in lithologic successions. *Mathematical Geology* **8**, 477–497.

Helland-Hansen, W., Kendall, C.G.S.C., Lerche, I. & Nakayama, K. (1988) A simulation of continental basin margin sedimentation in response to crustal movements, eustatic sea level change, and sediment accumulation rates. *Mathematical Geology* **20**, 777–802.

Helland-Hansen, W., Steel, R., Nakayama, K. & Kendall, C.G.S.C. (1989) Review and computer modelling of the Brent Group stratigraphy. In: Whateley, M.K.G. & Pickering, K.T. (eds) *Deltas: sites and traps for fossil fuels.* Geological Society special publication **41**, London, 237–252.

Heller, P.L. & Paola, C. (1992) The large-scale dynamics of grain-size variation in alluvial basins, 2. Application to syntectonic conglomerate. *Basin Research* **4**, 91–102.

Henriquez, A., Tyler, K.J. & Hurst, A. (1990) Characterization of fluvial sedimentology for reservoir simulation modeling. *SPE Formation Evaluation* **5**, 211–216.

Herweijer, J.C., Hocker, C.F.W., Williams, H. & Eastwood, K.M. (1990) The relevance of dip profiles from outcrops as reference for the interpretation of SHDT logs. In: Hurst, A., Lovell, M.A. & Morton, A.C. (eds) *Geological applications of wireline logs.* Geological Society of London special publication **48**, London, 39–43.

Hewett, T.A. (1986) Fractal distributions of reservoir heterogeneity and their influence on fluid transport. *SPE Technical Paper* **15386**, presented at the *61st SPE Annual Technical Conference,* New Orleans, 5–8 October.

Hewett, T.A. & Behrens, R.A. (1990) Conditional simulation of reservoir heterogeneity with fractals. *SPE Formation Evaluation* **5** (Sept), 217–225.

Hillier, A.P. & Williams, B.P.J. (1991) The Leman Field, blocks 49/26, 49/27, 49/28, 53/1, 53/2, UK North Sea. In: Abbotts, I.L. (ed.) *United Kingdom oil and gas fields, 25 years commemorative volume.* Geological Society Memoir No. **14**, 451–458.

Hirst, J.P.P. (1991) Variations in alluvial architecture across the Oligo-Miocene Huesca Fluvial system, Ebro Basin, Spain. In: Miall, A.D. & Tyler, N. (eds) *The three-dimensional facies architecture of terrigenous clastic sediments and its implications for hydrocarbon discovery and recovery*. SEPM Concepts in Sedimentology and Palaeontology Vol. 3, Tulsa, 111–121.

Hirst, P., Blackstock, C. & Tyson, S. (1993) Stochastic modelling of fluvial sandstone bodies. In: Flint, S.S. & Bryant, I.D. (eds) *The geological modelling of hydrocarbon reservoirs and outcrop analogues*. International Association of Sedimentologists special publication **15**, Blackwell, Oxford, 237–251.

Hohn, M.E. (1988) *Geostatistics and petroleum geology*. Van Nostrand Reinhold, New York.

Høimyr, O., Kleppe, A. & Nystuen, J.P. (1993) Effects of heterogeneities in a braided stream channel sandbody on the simulation of oil recovery: a case study from the Lower Jurassic Statfjord Formation, Snorre Field, North Sea. In: Ashton, M. (ed.) *Advances in reservoir geology*. Geological Society special publication **69**, London, 105–134.

Horne, J.C., Ferm, J.C., Caruccio, F.T. & Baganz, B.P. (1978) Depositional models in coal exploration and mine planning in Appalachian region. *The American Association of Petroleum Geologists Bulletin* **62**, 2379–2411.

Hurst, H.E., Black:, R.P. & Simaika, Y.M. (1965) *Long-term storage: an experimental study*. Constable, London.

Iglehart, C.F. (1992) Computer contouring gains acceptance: from jeers to cheers in 40 years. *Geobyte* October, 7–11.

Isaaks, E.H. & Srivastava, R.M. (1989) *An introduction to applied geostatistics*. Oxford University Press, Oxford.

Jackson, R.G.2. (1978) Preliminary evaluation of lithofacies models for meandering alluvial streams. In: Miall, A.D. (ed.) *Fluvial sedimentology*. Canadian Society of Petroleum Geologists memoir **5**, Calgary, 543–576.

Jervey, M.T. (1988) Quantitative geological modeling of siliciclastic rock sequences and their seismic expression. In: Wilgus, C.J., Hastings, B.S., Kendall, C.G.S.C., Posamentier, H.W., Ross, C.A. & Van Wagoner, J.C. (eds) *Sea-level changes: an integrated approach*. SEPM special publication **42**, Tulsa, 47–69.

Johnson, C.R. & Jones, T.A. (1988) Putting geology into reservoir simulations: a three-dimensional modeling approach. SPE Technical Paper **18321**, presented at *63rd SPE Annual Technical Conference*, Houston, 2–5 October.

Johnson, H.D. & Stewart, D.J. (1985) Role of clastic sedimentology in the exploration and production of oil and gas in the North Sea. In: Brenchley, P.J. & Williams, B.P.J. (eds) *Sedimentology: recent developments and applied aspects*. Geological Society of London special publication **18**, London, 249–310.

Jones, C.B. (1989) Data structures for three-dimensional spatial information systems in geology. *International Journal of Geographical Information Systems* **3**, 15–31.

Jones, N.L. & Nelson, J. (1992) Geoscientific modeling with TINs. *Geobyte* August, 44–49.

Jones, T.A. (1988a) Geostatistical models with stratigraphic control. *Computers and Geosciences* **14**, 135–138.

Jones, T.A. (1988b) Modeling geology in three dimensions. *Geobyte* **3**, 14–20.

Jones, T.A. (1989) The three faces of geological computer contouring. *Mathematical Geology* **21**, 271–283.

Jones, T.A. & Krum, G.L. (1992) Pitfalls in computer contouring – part 2. *Geobyte* August, 31–37.

Jones, T.A. & Leonard, J.E. (1990) Why 3-D modeling? *Geobyte* **5**, 25–26.

Jones, T.A., Hamilton, D.E. & Johnson, C.R. (1986) *Contouring geologic surfaces with the computer*. Van Nostrand Reinhold, New York.

Jordan, D.W. & Pryor, W.A. (1992) Hierarchical levels of heterogeneity in a Mississippi River Meander Belt and application to reservoir systems. *The American Association of Petroleum Geologists Bulletin* **76**, 1601—1624.

Journel, A.G. (1983) Non-parametric estimation of spatial distributions. *Mathematical Geology* **15**, 445–468.

Journel, A.G. (1986) Geostatistics: models and tools for the earth sciences. *Mathematical Geology* **18**, 119–139.

Journel, A.G. & Alabert, F.G. (1988) Focusing on spatial connectivity of extreme-valued attributes: stochastic indicator models of reservoir heterogeneities. SPE Technical Paper **18324**, presented at the *63rd SPE Annual Technical Conference*, Houston, Texas, 2–5 October.

Journel, A.G. & Alabert, F.G. (1990) New method for reservoir mapping. *Journal of Petroleum Technology* **42** (Feb.), 212–218.

Journel, A.G. & Huijbregts, C.J. (1978) *Mining geostatistics.* Academic Press, New York.

Kearey, P. & Brooks, M. (1991) *Introduction to Geophysical Exploration.* 2nd edition. Blackwell Scientific Publishers, Oxford.

Keijzer, J.H. & Kortekaas, T.F.M. (1990) Comparison of deterministic and probabilistic simulation models of channel sands in the Statfjord Reservoir, Brent Field. SPE Technical Paper **20947**, presented at *SPE Europec 90 Conference*, The Hague, Netherlands, 22–24 October.

Kendall, C.G.S.C., Moore, P., Strobel, J., Cannon, R., Bezdek, J. & Biswas, G. (1991) Simulation of sedimentary fill of basins. In: Watney, L., Franseen, E., Kendall, C.G.S.C. & Ross, W.C. (eds) *Sedimentary modeling: computer simulation and methods for improving parameter definition.* Subsurface geology series **12**, bulletin 233, Kansas Geological Survey, Lawrence, Kansas, 9–30.

Kendall, M.G. & Buckland, W.R. (1982) *A dictionary of statistical terms.* 4th edition. Longmans, London.

King, P.R. (1990) The connectivity and conductivity of overlapping sand bodies. In: Buller, A.T., Berg, E., Hjelmeland, O., Kleppe, J., Torsaeter, O. & Aasen, J.O. (eds) *North Sea oil and gas reservoirs – II.* Graham & Trotman, London.

Knighton, A.D. & Nanson, G.C. (1993) Anastomosis and the continuum of channel pattern. *Earth Surface Processes and Landforms* **18**, 613–625.

Kolla, V. & Perlmutter, M.A. (1993) Timing of turbidite sedimentation on the Mississippi Fan. *The American Association of Petroleum Geologists Bulletin* **77**, 1129–1141.

Koss, J.E., Ethridge, F.G. & Schumm, S.A. (1994) An experimental study of the effects of base-level change on fluvial, coastal plain and shelf systems. *Journal of Sedimentary Research* **B64**, 90–98.

Kossack, C.A. (1989) Prediction of layer lengths from layer heights for reservoir simulation: a statistical analysis of outcrop data. *Journal of Petroleum Technology* **41**, 867–871.

Koster, E.H. (1987) Vertebrate taphonomy applied to the analysis of ancient fluvial systems. In: Ethridge, F.G., Flores, R.M. & Harvey, M.D. (eds) *Recent developments in fluvial sedimentology.* SEPM special publication **39**, Tulsa, 159–168.

Kraus, M.J. (1987) Integration of channel and floodplain suites, II. Vertical relations of alluvial paleosols. *Journal of Sedimentary Petrology* **57**, 602–612.

Kraus, M.J. & Bown, T.M. (1988) Pedofacies analysis; a new approach to reconstructing ancient fluvial sequences. In: Reinhardt, J. & Sigleo, W.R. (eds) *Paleosols and weathering through geologic time: principles and applications.* Geological Society of America special paper **216**, Boulder, Colorado, 143–152.

Kraus, M.J. & Bown, T.M. (1993) Palaeosols and sandbody prediction in alluvial sequences. In: North, C.P. & Prosser, D.J. (eds) *Characterization of fluvial and aeolian reservoirs.* Geological Society special publication **73**, London, 23–31.

Krinitzsky, E.L. & Smith, F.L. (1969) *Geology of backswamp deposits in the Atchafalaya Basin, Louisiana.* US Army Waterways Experiment Station, Corp of Engineers, Vicksburg, Mississippi, report no. **S–69–8**.

Kruit, C. (1987) Sedimentologic reservoir study of a stream-drive project in deltaic river sands, East Tia Juna field, Venezuela. In: Tillman, R.W. & Weber, K.J. (eds) *Reservoir sedimentology.* SEPM special publication **40**, Tulsa, 293–310.

Krum, G.L. & Johnson, C.R. (1993) A 3-D modelling approach for providing a complex reservoir description for reservoir simulation. In: Flint, S. & Bryant, I.D. (eds) *The geological modelling of hydrocarbon reservoirs and outcrop analogues.* International

Association of Sedimentologists special publication **15**, Blackwell Scientific Publishers, Oxford, 253–258.

Krum, G.L. & Jones, T.A. (1992) Pitfalls in computer contouring – part 1. *Geobyte* June, 30–35.

Kushnir, G. (1992) Incorporating accumulated geological knowledge in computer mapping. *Geobyte* June, 54–56.

Lasseter, T.J., Waggoner, J.R. & Lake, L.W. (1986) Reservoir heterogeneities and their influence on ultimate recovery. In: Lake, L.W. & Carroll, H.B.J. (eds) *Reservoir characterization*. Academic Press, Orlando, 545–559.

Lawrence, D.T., Doyle, M. & Aigner, T. (1990) Stratigraphic simulation of sedimentary basins: concepts and calibration. *The American Association of Petroleum Geologists Bulletin* **74**, 273–295.

Le Roux, J.P. (1992) Determining the channel sinuosity of ancient fluvial systems from paleocurrent data. *Journal of Sedimentary Petrology* **62**, 283–291.

Lee, Y., Martinez, P.A. & Harbaugh, J.W. (1990) Dynamic 3-D graphics critical element in Stanford's SEDSIM project. *Geobyte* **5** (Feb.), 37–38.

Leeder, M.R. (1973) Fluviatile fining-upward cycles and the magnitude of paleochannels. *Geological Magazine* **110**, 265–276.

Leeder, M.R. (1975) Pedogenic carbonates and flood sediment accretion rates: a quantitative model for alluvial arid-zone lithofacies. *Geological Magazine* **112**, 257–270.

Leeder, M.R. (1978) A quantitative stratigraphic model for alluvium, with special reference to channel deposit density and interconnectedness. In: Miall, A.D. (ed.) *Fluvial sedimentology*. Canadian Society of Petroleum Geologists Memoir **5**, Calgary, 587–596.

Leeder, M.R. (1993) Tectonic controls upon drainage basin development, river channel migration and alluvial architecture: implications for hydrocarbon reservoir development and characterization. In: North, C.P. & Prosser, D.J. (eds) *Characterization of fluvial and aeolian reservoirs*. Geological Society special publication **73**, London, 7–22.

Leeder, M.R. & Alexander, J. (1987) The origin and tectonic significance of asymmetric meander-belts. *Sedimentology* **34**, 217–226.

Leeder, M.R. & Jackson, J.A. (1993) The interaction between normal faulting and drainage in active extensional basins, with examples from the western United States and central Greece. *Basin Research* **5**, 79–102.

Leopold, L.B. & Wolman, M.G. (1960) River meanders. *Geological Society of America Bulletin* **71**, 769–794.

Lin, C. & Harbaugh, J.W. (1984) *Graphic display of two and three dimensional Markov computer models in geology*. Van Nostrand Reinhold, New York.

Livera, S.E. (1989) Facies associations and sand-body geometries in the Ness Formation of the Brent Group, Brent Field. In: Whateley, M.K.G. & Pickering, K.T. (eds) *Deltas: sites and traps for fossil fuels*. Geological Society of London special publication **41**, London, 269–289.

Lorenz, E.N. (1969) The predictability of a flow which possesses many scales of motion. *Tellus* **21**, 289–307.

Lorenz, J.C., Heinze, D.M., Clark, J.A. & Searls, C.A. (1985) Determination of widths of meander-belt sandstone reservoirs from vertical downhole data, Mesaverde Group, Piceance Creek Basin, Colorado. *The American Association of Petroleum Geologists Bulletin* **69**, 710–721.

Lorenz, J.C., Warpinski, N.R. & Branagan, P.T. (1991) Subsurface characterization of Mesaverde reservoirs in Colorado: Geophysical and reservoir-engineering checks on predictive sedimentology. In: Miall, A.D. & Tyler, N. (eds) *The three-dimensional facies architecture of terrigenous clastic sediments and its implications for hydrocarbon discovery and recovery*. SEPM Concepts in Sedimentology and Palaeontology Vol. 3, Tulsa, 57–79.

Love, S.E. (1993) *Floodbasin deposits as indicators of sandbody geometry and reservoir architecture*. Unpublished PhD Thesis, University of Aberdeen, Scotland.

Luthi, S.M. (1990) Sedimentary structures of clastic rocks identified from electrical borehole images. In: Hurst, A., Lovell, M.A. & Morton, A.C. (eds) *Geological applications of wireline logs*. Geological Society of London special publication **48**, London, 3–10.

MacDonald, A.C. & Halland, E.K. (1993) Sedimentology and shale modeling of a sandstone-rich fluvial reservoir: Upper Statfjord Formation, Statfjord Field, northern North Sea. *The American Association of Petroleum Geologists Bulletin* **77**, 1016–1040.

McEachran, D.B. (1992) Mathematical and logical manipulation of grid models. *Geobyte* October, 31–35.

Mackey, S.D. & Bridge, J.S. (1992) A revised Fortran program to simulate alluvial stratigraphy. *Computers and Geosciences* **18**, 119–181.

Maizels, J.K. (1988) Palaeochannels: Plio-Pleistocene raised channel systems of the western Sharqiyah. *Journal of Oman Studies Special Report* **3**, 95–112.

Maizels, J. (1990) Raised channel systems as indicators of palaeohydrologic change: a case study from Oman. *Palaeogeography, Palaeoclimatology, Palaeoecology* **76**, 241–277.

Maizels, J.K. (1993a) Quantitative regime modelling of fluvial depositional sequences: application to Holocene stratigraphy of humid-glacial braid-plains (Icelandic sandurs). In: North, C.P. & Prosser, D.J. (eds) *Characterization of fluvial and aeolian reservoirs*. Geological Society special publication **73**, London, 53–78.

Maizels, J.K. (1993b) Lithofacies variations within sandur deposits: the role of runoff regime, flow dynamics and sediment supply characteristics. *Sedimentary Geology* **85**, 299–325.

Mandelbrot, B.B. (1967) How long is the coast of Britain? Statistical self-similarity and fractal dimension. *Science* **156**, 636–638.

Mandelbrot, B.B. (1977) *Fractals: form, chance, and dimension*. W.H. Freeman, San Francisco.

Mandelbrot, B.B. & Wallis, J.R. (1969) Robustness of the rescaled range R/S in the measurement of noncyclic long run statistical dependence. *Water Resources Research* **5**, 967–988.

Marriott, S.B. & Wright, V.P. (1993) Palaeosols as indicators of geomorphic stability in two Old Red Sandstone alluvial suites, South Wales. *Journal of the Geological Society of London* **150**, 1109–1120.

Martin, J.H., Evans, A.J. & Raper, J.K. (1988) Reservoir modelling of low-sinuosity channel sands: a network approach. SPE Technical Paper **18364**, presented at the *SPE European Petroleum Conference*, London, 17–19 October.

Martinsen, O.J., Martinsen, R.S. & Steidtmann, J.R. (1993) Mesaverde Group (Upper Cretaceous), southeastern Wyoming: Allostratigraphy versus sequence stratigraphy in a tectonically active area. *The American Association of Petroleum Geologists Bulletin* **77**, 1351–1373.

Matheron, G. (1963) Principles of geostatistics. *Economic Geology* **58**, 1246–1266.

Matheron, G. (1967) Kriging, or polynomial interpolation procedures. *The Canadian Mining and Metallurgical Bulletin* **60**, 1041–1045.

Matheron, G., Beucher, H., de Fouquet, C., Galli, A., Guerillot, D. & Ravenne, C. (1987) Conditional simulation of the geometry of fluvio-deltaic reservoirs. SPE Technical Paper **16753**, presented at *SPE Annual Technical Conference*, Dallas, Texas, 27–30 September.

May, R.M. (1976) Simple mathematical models with very complicated dynamics. *Nature* **261**, 459–467.

Mayoraz, R., Mann, C.E. & Parriaux, A. (1992) Three-dimensional modeling of complex geological structures: new development tools for creating 3-D volumes. In: Hamilton, D.E. & Jones, T.A. (eds) *Computer modeling of geologic surfaces and volumes*. AAPG Computer Applications in Geology No. **1**, Tulsa, 261–271.

Meling, L.M., Morkeseth, P.O. & Langeland, T. (1990) Production forecasting for gas fields with multiple reservoirs. *Journal of Petroleum Technology* **42**, 1580–1587.

Mesa, O.J. & Poveda, G. (1993) The Hurst effect: the scale of fluctuation approach. *Water Resources Research* **29**, 3995–4002.

Miall, A.D. (1973) Markov chain analysis applied to an ancient alluvial plain. *Sedimentology* **20**, 347–364.

Miall, A.D. (1977) A review of the braided-river depositional environment. *Earth Science Reviews* 13, 1–62.

Miall, A.D. (1978) Lithofacies types and vertical profile models in braided river deposits: a summary. In: Miall, A.D. (ed.) *Fluvial sedimentology*. Canadian Society of Petroleum Geologists memoir 5, Calgary, Canada, 597–604.

Miall, A.D. (1985) Architectural-element analysis: a new method of facies analysis applied to fluvial deposits. *Earth Science Reviews* 22, 261–308.

Miall, A.D. (1986) Eustatic sea level changes interpreted from seismic stratigraphy: a critique of the methodology with particular reference to the North Sea Jurassic record. *The American Association of Petroleum Geologists Bulletin* 70, 131–137.

Miall, A.D. (1988a) Architectural elements and bounding surfaces in fluvial deposits: anatomy of the Kayenta Formation (Lower Jurassic), southwest Colorado. *Sedimentary Geology* 55, 233–262.

Miall, A.D. (1988b) Reservoir heterogeneities in fluvial sandstones: lessons from outcrop studies. *The American Association of Petroleum Geologists Bulletin* 72, 682–697.

Miall, A.D. (1991a) Hierarchies of architectural units terrigenous clastic rocks, and their relationship to sedimentation rate. In: Miall, A.D. & Tyler, N. (eds) *The three-dimensional facies architecture of terrigenous clastic sediments and its implications for hydrocarbon discovery and recovery*. SEPM Concepts in Sedimentology and Palaeontology Vol. 3, Tulsa, 6–12.

Miall, A.D. (1991b) Stratigraphic sequences and their chronostratigraphic correlation. *Journal of Sedimentary Petrology* 61, 497–505.

Miall, A.D. (1992) Alluvial deposits. In: Walker, R.G. & James, N.P. (eds) *Facies models: response to sea level change*. Geological Association of Canada, St. John's Newfoundland, 119–142.

Miall, A.D. (1993) The architecture of fluvial-deltaic sequences in the Upper Mesaverde Group (Upper Cretaceous), Book Cliffs, Utah. In: Best, J.L. & Bristow, C.S. (eds) *Braided rivers*. Geological Society special publication 75, London, 305–332.

Miall, A.D. (1994a) Reconstructing fluvial macroform architecture from two-dimensional outcrops: examples from the Castlegate Sandstone, Book Cliffs, Utah. *Journal of Sedimentary Research* B64, 146–158.

Miall, A.D. (1995) Description and interpretation of fluvial deposits: a critical perspective: Discussion. *Sedimentology* 42, 379–384.

Miall, A.D. & Tyler, N. (1991) *The three-dimensional facies architecture of terrigenous clastic sediments and its implication for hydrocarbon discovery and recovery*. SEPM Concepts in Sedimentology and Palaeontology Vol. 3, Tulsa.

Mijnssen, F.C.J., Tyler, N. & Weber, K.J. (1993) Knowledge base development for the estimation of reservoir rock properties in the interwell area: examples from the Texas Gulf coast. In: Flint, S.S. & Bryant, I.D. (eds) *The geological modelling of hydrocarbon reservoirs and outcrop analogues*. International Association of Sedimentologists special publication 15, Blackwell Scientific Publications, Oxford, 169–180.

Miller, D.M., McPherson, J.G. & Covington, T.E. (1990) Fluviodeltaic reservoir, South Belridge Field, San Joaquin Valley, California. In: Barwis, J.H., McPherson, J.G. & Studlick, J.R.J. (eds) *Sandstone petroleum reservoirs*. Springer Verlag, New York, 109–130.

Mitchum, R.M.J., Vail, P.R. & Thompson, S.I. (1977) Seismic stratigraphy and global changes of sea level, part 2: the depositional sequence as a basic unit for stratigraphic analysis. In: Payton, C.E. (ed.) *Seismic stratigraphy – applications to hydrocarbon exploration*. AAPG memoir 26, Tulsa, 53–62.

Mjøs, R., Walderhaug, O. & Prestholm, E. (1993) Crevasse splay sandstone geometries in the Middle Jurassic Ravenscar Group of Yorkshire, UK. In: Marzo, M. & Puidefabregas, C. (eds) *Alluvial sedimentation*. International Association of Sedimentologists special publication 17, Blackwell Scientific Publications, Oxford, 167–184.

Mollard, J.D. (1973) Airphoto interpretation of fluvial features. In: *Fluvial Process and Sedimentation: Proceedings of the Hydrology Symposium*. National Research Council of Canada, 341–380.

Moody-Stuart, M. (1966) High and low-sinuosity stream deposits, with examples from the Devonian of Spitsbergen. *Journal of Sedimentary Petrology* **36**, 1102–1117.

Moraes, M.A.S. & Surdam, R.C. (1993) Diagenetic heterogeneity and reservoir quality: fluvial, deltaic, and turbiditic sandstone reservoirs, Potiguar and Recôncavo rift basins, Brazil. *The American Association of Petroleum Geologists Bulletin* **77**, 1142–1158.

Myers, K.J. & Bristow, C.S. (1989) Detailed sedimentology and gamma-ray log characteristics of a Namurian deltaic succession. II: gamma-ray logging. In: Whateley, M.K.G. & Pickering, K.T. (eds) *Deltas: sites and traps for fossil fuels*. Geological Society of London special publication **41**, London, 81–88.

NACSN – North American Commission on Stratigraphic Nomenclature. (1983) North American Stratigraphic Code. *The American Association of Petroleum Geologists Bulletin* **67**, 841–875.

Nami, M. & Ashworth, S.G.E. (1993) Principles of a sediment sorting model and its application for predicting economic values in placer deposits. In: Marzo, M. & Puidefabregas, C. (eds) *Alluvial sedimentation*. International Association of Sedimentologists special publication **17**, Blackell, Oxford.

Nanz, R.H.J. (1954) Genesis of Oligocene sandstone reservoir, Seeligson field, Jim Wells and Kleberg counties, Texas. *The American Association of Petroleum Geologists Bulletin* **38**, 96–117.

Nathanail, C.P. & Rosenbaum, M.S. (1992) The use of low cost geostatistical software in reserve estimation. In: Annels, A.E. (ed.) *Case histories and methods in mineral resource evaluation*. Geological Society special publication **63**, London, 169–177.

Nobre, M.M. & Sykes, J.F. (1992) An application of geostatistics in subsurface characterization using hard data and soft information. In: Christie, M.A., Da Silva, F.V., Farmer, C.L., et al. *ECMOR III: Proceedings of the Third European Conference on the Mathematics of Oil Recovery*. Delft University Press, Delft, Netherlands, 109–118.

North, C.P. & Halliwell, D.I. (1994) Bias in estimating fractal dimension with the rescaled-range (R/S) technique. *Mathematical Geology* **26**, 531–555.

Olea, R.A. (1992) Kriging: understanding allays intimidation. *Geobyte* October, 12–17.

Olsen, T. (1993) Determining the channel sinuosity of ancient fluvial systems from palaeocurrent data – discussion. *Journal of Sedimentary Petrology* **63**, 306–307.

Pang, J. (1993) *Geological reservoir modelling of fluvial channel sands*. Unpublished PhD Thesis, University of Aberdeen, Scotland.

Paola, C. & Borgman, L. (1991) Reconstructing random topography from preserved stratification. *Sedimentology* **38**, 553–565.

Paola, C., Heller, P.L. & Angevine, C.L. (1992) The large-scale dynamics of grain-size variation in alluvial basins, 1. Theory. *Basin Research* **4**, 73–90.

Paradis, A. & Bellcher, B. (1990) Interactive volume modelling: A new product for 3-D mapping. *Geobyte* **5**, 42–44.

Patton, P.C. & Schumm, S.A. (1981) Ephemeral stream processes: implications for studies of Quaternary valley fills. *Quaternary Research* **15**, 24–43.

Payton, C.E. (Ed.). (1977) *Seismic stratigraphy – applications to hydrocarbon exploration*. AAPG memoir **26**, Tulsa.

Perrier, R. & Quiblier, J. (1974) Thickness changes in sedimentary layers during compaction history: methods for quantitative evaluation. *The American Association of Petroleum Geologists Bulletin* **58**, 507–520.

Pflug, R. & Harbaugh, J.W. (1992) *Computer graphics in geology: three-dimensional computer graphics in modeling geologic structures and simulating geologic processes*. Lectures in Earth Sciences No. **41**, Springer-Verlag, Berlin.

Platt, N.H. & Keller, B. (1992) Distal alluvial deposits in a foreland basin setting – the Lower Freshwater Molasse (Lower Miocene), Switzerland: sedimentology, architecture and palaeosols. *Sedimentology* **39**, 545–565.

Posamentier, H.W. & Vail, P.R. (1988) Eustatic controls on clastic deposition II—sequence and systems tract models. In: Wilgus, C.J., Hastings, B.S., Kendall, C.G.S.C., Posamentier,

H.W., Ross, C. A. & Van Wagoner, J.C. (eds) *Sea-level changes: an integrated approach.* SEPM special publication **42**, Tulsa, 125–154.

Posamentier, H.W. & Weimar, P. (1993) Siliciclastic sequence stratigraphy and petroleum geology – where to from here? *The American Association of Petroleum Geologists Bulletin* **7**, 731–742.

Posamentier, H.W., Jervey, M.T. & Vail, P.R. (1988) Eustatic controls on clastic deposition I – conceptual framework. In: Wilgus, C.J., Hastings, B.S., Kendall, C.G.S.C., Posamentier, H.W., Ross, C..A. & Van Wagoner, J.C. (eds) *Sea-level changes: an integrated approach.* SEPM special publication **42**, Tulsa, 109–124.

Potter, P.E. & Blakely, R.F. (1967) Generation of a synthetic vertical profile of a fluvial sandstone body. *Journal of the Society of Petroleum Engineers* **6**, 243–251.

Potter, P.E. & Blakely, R.F. (1968) Random processes and lithologic transitions. *Journal of Geology* **76**, 154–170.

Putnam, P.E. (1983) Fluvial deposits and hydrocarbon accumulations: examples from the Lloydminster area, Canada. In: Collinson, J.D. & Lewin, J. (eds) *Modern and ancient fluvial systems.* International Association of Sedimentologists special publication **6**, Oxford, 517–532.

Qiu, Y., Xue, P. & Xiao, J. (1987) Fluvial sandstone bodies as hydrocarbon reservoirs in lake basins. In: Ethridge, F.G., Flores, R.M. & Harvey, M.D. (eds) *Recent developments in fluvial sedimentology.* SEPM special publication **39**, Tulsa, 323–342.

Ramos, A. & Sopeña, A. (1983) Gravel bars in low-sinuosity streams (Permian and Triassic, central Spain). In: Collinson, J.D. & Lewin, J. (eds) *Modern and ancient fluvial systems.* International Association of Sedimentologists special publication **6**, Blackell Scientific Publishers, Oxford, 301–312.

Ravenne, C., Eschard, R., Galli, A., Mathieu, Y., Montadert, L. & Rudkiewicz, J.L. (1989) Heterogeneities and geometry of sedimentary bodies in a fluvio-deltaic reservoir. *SPE Formation Evaluation* June, 239–246.

Reading, H.G. (1986) Facies. In: Reading, H.G. (ed.) *Sedimentary environments and facies.* 2nd edition. Blackwell Scientific Publications, Oxford, 4–19.

Reading, H.G. & Orton, G.J. (1991) Sediment calibre: a control on facies models with special reference to deep-sea depositional systems. In: Müller, D.W., McKenzie, J.A. & Weissart, H. (eds) *Controversies in modern geology.* Academic Press, London, 85–111.

Retallack, G.J. (1986) Fossil soils as grounds for interpreting long-term controls on ancient rivers. *Journal of Sedimentary Petrology* **56**, 1–18.

Retallack, G.J. (1990) *Soils of the past.* Unwin-Hyman, London.

Rettger, R.E. (1929) On specifying the type of structural contouring. *The American Association of Petroleum Geologists Bulletin* **13**, 1559–1560.

Rider, M.H. (1990) Gamma-ray log shape used as a facies indicator: critical analysis of an oversimplified methodology. In: Hurst, A., Lovell, M.A. & Morton, A.C. (eds) *Geological applications of wireline logs.* Geological Society of London Special Publication **48**, London, 27–37.

Rider, M.H. (1991) *The geological interpretation of well logs.* 2nd edition. Blackie (John Wiley), London.

Ripley, B.D. (1987) *Stochastic simulation.* John Wiley & Sons, New York.

Rongey, T. (1992) Mapping with fractal geometry. *Geobyte* October, 18–23.

Royle, A.G. (1992) A personal overview of geostatistics. In: Annels A.E. (ed.) *Case histories and methods in mineral resource evaluation.* Geological Society special publication **63**, London, 233–241.

Rubin, D.M. (1987) *Cross-bedding, bedforms and paleocurrents.* SEPM Concepts in Sedimentology and Palaeontology No. **1**, SEPM, Tulsa.

Rubin, D.M. & Hunter, R.E. (1982) Bedform climbing in theory and nature. *Sedimentology* **29**, 121–138.

Rudkiewicz, J.L., Guerillot, D., Galli, A. & Heresi, G. (1990) An integrated software for stochastic modelling of reservoir lithology and property with an example from the Yorkshire Middle Jurassic. In: Buller, A.T., Berg, E., Hjelmeland, O., Kleppe, J.,

Torsaeter, O. & Aasen, J.O. (eds) *North Sea oil & gas reservoirs – II*. Graham & Trotman, London, 399–406.

Ruijtenberg, P.A., Buchanan, R. & Marke, P. (1990) Three-dimensional data improve reservoir mapping. *Journal of Petroleum Technology* **42**, 22–61.

Rust, B.R. (1978) Depositional models for braided alluvium. In: Miall, A.D. (ed.) *Fluvial sedimentology*. Canadian Society of Petroleum Geologists memoir **5**, Calgary, 605–625.

Rust, B.R. & Gibling, D.A. (1990) Three-dimensional antidunes as HCS mimics in a fluvial sandstone: the Pennsylvanian South Bar Formation near Sydney, Nova Scotia. *Journal of Sedimentary Petrology* **60**, 540–548.

Salter, T. (1993) Fluvial scour and incision: models for their influence on the development of realistic reservoir geometries. In: North, C.P. & Prosser, D.J. (eds) *Characterization of fluvial and aeolian reservoirs*. Geological Society special publication **73**, London, 33–51.

Schumm, S.A. (1965) Quaternary paleohydrology. In: Wright, H.E. & Frey, D.G. (eds) *Quaternary of the United States*. Princeton University Press, Princeton, New Jersey, 783–794.

Schumm, S.A. (1968) Speculations concerning palaeohydrologic controls of terrestrial sedimentation. *Geological Society of America Bulletin* **79**, 1573–1588.

Schumm, S.A. (1977) *The fluvial system*. Wiley, New York.

Schumm, S.A. (1981) Evolution and response of the fluvial system, sedimentologic implications. In: Ethridge, F.G. & Flores, R.M. (eds) *Recent and ancient nonmarine depositional environments: models for exploration*. SEPM special publication **31**, Tulsa, 19–29.

Schumm, S.A. (1991) *To interpret the Earth: ten ways to be wrong*. Cambridge University Press, Cambridge.

Schumm, S.A. (1993) River response to baselevel change: implications for sequence stratigraphy. *Journal of Geology* **101**, 279–294.

Schwarzacher, W. (1964) An application of statistical time-series analysis of a limestone–shale sequence. *Journal of Geology* **72**, 195–213.

Schwarzacher, W. (1969) The use of Markov chains in the study of sedimentary cycles. *Mathematical Geology* **1**, 17–39.

Sclater, J.G. & Christie, P.A.F. (1980) Continental stretching: an explanation of the post-mid-Cretaceous subsidence of the Central North Sea Basin. *Journal of Geophysical Research* **B7**, 3711–3739.

Scott, N. (1986) *Modern vs. ancient braided stream deposits: a comparison between simulated sedimentary deposits and the Ivishak Formation of the Prudhoe Bay Field, Alaska*. Unpublished MS Thesis, Stanford University, USA.

Shanley, K.W. & McCabe, P.J. (1991) Predicting facies architecture through sequence stratigraphy – an example from the Kaiparowits Plateau, Utah. *Geology* **19**, 742–745.

Shanley, K.W. & McCabe, P.J. (1993) Alluvial architecture in a sequence stratigraphic framework: a case history from the Upper Cretaceous of southern Utah, USA. In: Flint, S.S. & Bryant, I.D. (eds) *The geological modelling of hydrocarbon reservoirs and outcrop analogues*. International Association of Sedimentologists special publication **15**, Blackwell, Oxford, 21–56.

Shanley, K.W. & McCabe, P.J. (1994) Perspectives on the sequence stratigraphy of continental strata. *The American Association of Petroleum Geologists Bulletin* **78**, 544–568.

Sheriff, R.E. (1985) Aspects of seismic resolution. In: Berg, O.R. & Woolverton, D.G. (eds) *Seismic stratigraphy II: an integrated approach*. AAPG Memoir **39**, Tulsa, 1–12.

Slatt, R.M., Jordan, D.W., D'Agostino, A.E. & Gillespie, R.H. (1992) Outcrop gamma-ray logging to improve understanding of subsurface well log correlations. In: Hurst, A., Griffiths, C.M. & Worthington, P.F. (eds) *Geological explications of wireline logs II*. Geological Society special publication **65**, London, 3–19.

Sloss, L.L. (1963) Sequences in the cratonic interior of North America. *Geological Society of America Bulletin* **74**, 93–114.

Sloss, L.L. (1979) Global sea level change: a view from the craton. In: Watkins, J.S.,

Montadert, L. & Dickerson, P.W. (eds) *Geological and geographical investigations of continental margins*. AAPG memoir **29**, Tulsa, 461–467.

Sloss, L.L. (1988) 40 years of sequence stratigraphy. *Geological Society of America Bulletin* **100**, 1661.

Sloss, L.L. (1991) The tectonic factor in sea level change: a countervailing view. *Journal of Geophysical Research* **96**, 6609–6617.

Smith, N.D., Cross, T.A., Dufficy, J.P. & Clough, S.R. (1989) Anatomy of an avulsion. *Sedimentology* **36**, 1–23.

Snow, R.S. (1989) Fractal sinuosity of stream channels. *Pageoph* **131**, 99–109.

Srivastava, R.M. (1994) The visualization of spatial uncertainty. *Abstract presented at the 79th AAPG Annual Convention*, Denver, Colorado, 12–15 June.

Stanley, K.O., Forde, K., Raestad, N. & Stockbridge, C.P. (1990) Stochastic modelling of reservoir sand bodies for input to reservoir simulation, Snorre Field, Northern North Sea, Norway. In: Buller, A.T., Berg, E., Hjelmeland, O., Kleppe, J., Torsaeter, O. & Aasen, J.O. (eds) *North Sea oil and gas reservoirs – II*. Graham & Trotman, London, 91–101.

Stokes, W.L. (1961) Fluvial and eolian sandstone bodies in Colorado Plateau. In: Peterson, J.A. & Osmond, J.C. (eds) *Geometry of sandstone bodies*. American Association of Petroleum Geologists, Tulsa, 151–178.

Stoyan, D., Kendall, W.S. & Mecke, J. (1987) *Stochastic geometry and its applications*. John Wiley & Sons, Chichester.

Strobel, J., Cannon, R., Kendall, C.G.S.C., Biswas, G. & Bezdek, J. (1989) Interactive (SEDPAK) simulation of clastic and carbonate sediments in shelf to basin settings. *Computers and Geosciences* **15**, 1279–1290.

Struijk, A.P. & Green, R.T. (1991) The Brent Field, block 211/29, UK North Sea. In: Abbotts, I.L. (ed.) *United Kingdom oil and gas fields, 25 years commemorative volume*. Geological Society Memoir **14**, 63–72.

Swanson, D.C. (1993) The importance of fluvial processes and related reservoir deposits. *Journal of Petroleum Technology* **45**, 368–377.

Tearpock, D.J. (1992) Contouring: art or science? *Geobyte* June, 40–43.

Tearpock, D.J. & Bischke, R.E. (1991) *Applied subsurface geological mapping*. Prentice-Hall, Englewood Cliffs, New Jersey.

Tetzlaff, D.M. (1991) The combined use of sedimentary process modeling and statistical simulation in reservoir characterization. SPE Technical Paper **22759**, presented at *66th SPE Annual Technical Conference*, Dallas, Texas, 6–9 October.

Tetzlaff, D.M. & Harbaugh, J.W. (1989) *Simulating clastic sedimentation*. Van Nostrand Reinhold, New York.

Thorne, J.A. (1992) An analysis of the implicit assumptions of the methodology of seismic sequence stratigraphy. In: Watkins, J.S. (ed.) *Geology and Geophysics of Continental Margins: Proceedings of the M.T. Harbouty Continental Margins Conference*. AAPG Memoir **53**, Tulsa, 375–394.

Thorne, J.A. & Swift, D.J.P. (1991) Sedimentation on continental margins, VI: a regime model for depositional sequences, their component sequence systems tracts, and bounding surfaces. In: Swift, D.J.P., Oertel, G.F., Tillman, R.W. & Thorne, J.A. (eds) *Shelf sand and sandstone bodies: geometry, facies, and sequence stratigraphy*. International Association of Sedimentologists special publication **14**, Blackwell Scientific Publishers, Oxford, 189–255.

Tillman, R.W. & Jordan, D.W. (1987) Sedimentology and subsurface geology of deltaic facies, Admire 650' Sandstone, El Dorado field, Kansas. In: Tillman, R.W. & Weber, K.J. (eds) *Reservoir Sedimentology*. SEPM special publication **40**, Tulsa, 221–291.

Tjølsen, C.B. & Damsleth, E. (1991) A model for the simultaneous generation of core-controlled stochastic absolute and relative permeability fields. SPE Technical Paper **22691**, presented at *66th SPE Annual Technical Conference*, Dallas, Texas, 6–9 October.

Trewin, N.H. & Bramwell, M.G. (1991) The Auk Field, Block 30/16, UK North Sea. In: Abbotts, I.L. (ed.) *United Kingdom oil and gas fields, 25 years commemorative volume*. Geological Society Memoir **14**, 227–236.

Tye, R.S. (1991) Fluvial-sandstone reservoirs of the Travis Peak Formation, East Texas Basin. In: Miall, A.D. & Tyler, N. (eds) *The three-dimensional facies architecture of terrigenous clastic sediments and its implications for hydrocarbon discovery and recovery*. SEPM Concepts in Sedimentology and Palaeontology Vol. 3, Tulsa, 172–188.

Tyler, K., Henriquez, A., Georgsen, F., Holden, L. & Tjelmeland, H. (1992) A program for 3-D modelling of heterogeneities in a fluvial reservoir. In: Christie, M. A., Da Silva, F.V., Farmer, C.L., et al. *ECMOR III: Proceedings of the third European conference on the mathematics of oil recovery*. Delft University Press, Delft, Netherlands, 31–40.

Tyler, K.J., Svanes, T. & Omdal, S. (1993) Faster history matching and uncertainty in predicted production profiles with stochastic modeling. SPE Technical Paper **26420**, presented at *68th SPE Annual Technical Conference*, Houston, Texas, 3–6 October.

Tyler, K.J., Svanes, T. & Henriquez, A. (1994) Heterogeneity modeling used for production simulation of a fluvial reservoir. *SPE Formation Evaluation* **9**, 85–92.

Tyler, N. (1988) New oil from old fields. *Geotimes* **33**, 8–10.

Tyler, N. & Finley, R.J. (1991) Architectural controls on the recovery of hydrocarbons from sandstone reservoirs. In: Miall, A.D. & Tyler, N. (eds) *The three-dimensional facies architecture of terrigenous clastic sediments and its implications for hydrocarbon discovery and recovery*. SEPM Concepts in Sedimentology and Palaeontology Vol. 3, Tulsa, 1–5.

Underhill, J.R. & Partington, M.A. (1993) Use of genetic sequence stratigraphy in defining and determining a regional tectonic control on the "Mid-Cimmerian Unconformity" – implications for North Sea basin development and the global sea level chart. In: Weimer, P. & Posamentier, H.W. (eds) *Siliciclastic sequence stratigraphy: recent developments and applications*. AAPG memoir **58**, Tulsa, 449–484.

Vail, P.R., Mitchum, R.M., Thompson, S. *et al.* (1977) Seismic stratigraphy and global changes of sea level. In: Payton, C.E. (ed.) *Seismic stratigraphy – applications to hydrocarbon exploration*. AAPG Memoir **26**, Tulsa, 49–211.

van de Graaff, W.J.E. & Ealey, P.J. (1989) Geological modeling for simulation studies. *The American Association of Petroleum Geologists Bulletin* **73**, 1436–1444.

van Vark, W., Paardekam, A.H.M., Brint, J.F., van Lieshout, J.B. & George, P.M. (1992) The construction and validation of a numerical model of a reservoir consisting of meandering channels. SPE Technical Paper **25057**, presented at *SPE European Petroleum Conference*, Cannes, France, 16–18 November.

Van Wagoner, J.C.., Mitchum, R.M., Campion, K.M. & Rahmanian, V.D. (1990) *Siliciclastic sequence stratigraphy in well logs, cores and outcrops: concepts for high-resolution correlation of time and facies*. AAPG Methods in Exploration Series no. **7**, Tulsa.

Visher, G.S. (1965) Use of vertical profile in environmental reconstruction. *The American Association of Petroleum Geologists Bulletin* **49**, 41–61.

Visher, G.S. (1972) Physical characteristics of fluvial deposits. In: Rigby, J.K. & Hamblin, W.K. (eds) *Recognition of ancient sedimentary environments*. SEPM special publication **16**, Tulsa, 84–97.

Voss, R.F. (1985) Random fractal forgeries. In: Earnshaw, R.A. (ed.) *Fundamental algorithms for computer graphics*. NATO ASI Series, Volume F.17, Springer-Verlag, Berlin, 805–835.

Wagner, F.J. & Busbey, A.B. (1992) International directory mapping/contouring software – June 1992. *Geobyte* June, 14–21.

Walker, R.G. (1979) *Facies models*. Geological Association of Canada, Geoscience Reprint Series **1**, Toronto.

Walker, R.G. (1984a) *Facies models*. 2nd edition. Geological Association of Canada, Geoscience Canada Reprint Series **1**, Toronto.

Walker, R.G. (1984b) General introduction: facies, facies sequences and facies models. In: Walker, R.G. (ed.) *Facies models*. 2nd edition. Geological Association of Canada, Geoscience Canada Reprint Series **1**, 1–10.

Walker, R.G. (1990) Facies modeling and sequence stratigraphy. *Journal of Sedimentary Petrology* **60**, 777–786.

Walker, R.G. (1992) Facies, facies models and modern stratigraphic concepts. In: Walker,

R.G. & James, N.P. (eds) *Facies models: response to sea level change*. Geological Association of Canada, St. John's, Newfoundland, 1–14.

Walker, R.G. & Cant, D.J. (1984) Sandy fluvial systems. In: Walker, R.G. (ed.) *Facies models*. 2nd edition. Geological Association of Canada, Geoscience Canada Reprint Series **1**, 71–90.

Watts, A.B. & Thorne, J. (1984) Tectonics, global changes in sea level and their relationship to stratigraphical sequences at the U.S. Atlantic continental margin. *Marine and Petroleum Geology* **1**, 319–339.

Webb, E.K. (1994) Simulating the three-dimensional distribution of sediment units in braided-stream deposits. *Journal of Sedimentary Research* **B64**, 219–231.

Weber, K.J. (1993) The use of 3-D seismic in reservoir geological modelling. In: Flint S. & Bryant, I.D. (eds) *The geological modelling of hydrocarbon reservoirs and outcrop analogues*. International Association of Sedimentologists special publication **15**, Blackwell Scientific Publishers, Oxford, 181–188.

Weber, K.J. & van Geuns, L.C. (1990) Framework for constructing clastic reservoir simulation models. *Journal of Petroleum Technology* **42** (Oct.), 1248–1297.

Weimar, R.J. & Tillman, R.W. (1980) *Tectonic influence on deltaic shoreline facies, Fox Hills Sandstone, west-central Denver Basin*. Colorado School of Mines Professional Contributions **10**. Denver, Colorado.

Werren, E.G., Shew, R.D., Adams, E.R. & Stancliffe, R.J. (1990) Meander-belt reservoir geology, Mid-Dip Tuscaloosa, Little Creek Field, Mississippi. In: Barwis, J.H., McPherson, J.G. & Studlick, J.R.J. (eds) *Sandstone petroleum reservoirs*. Springer Verlag, New York, 85–107.

Wescott, W.A. (1993) Geomorphic thresholds and complex response of fluvial systems – some implications for sequence stratigraphy. *The American Association of Petroleum Geologists Bulletin* **77**, 1208–1218.

Weston, P.J. & Alexander, J. (1993) Computer modelling of flow lines over deformed surfaces: the implications for prediction of alluvial facies distribution. In: Marzo, M. & Puidefabregas, C. (eds) *Alluvial sedimentation*. International Association of Sedimentologists special publication **17**, Blackwell Scientific Publications, Oxford, 211–217.

Wilgus, C.K., Hastings, B.S., Kendall, C.G.S.C., Posamentier, H.W., Ross, C.A., & Van Wagoner, J.C. (Eds). (1988) *Sea-level changes: an integrated approach*. SEPM special publication **42**, Tulsa.

Willetts, B.B., Maizels, J.K. & Florence, J. (1987) The simulation of stream bed armouring and its consequences. *Proceedings of the Institution of Civil Engineers* **82**, 799–814.

Williams, B.P.J., Love, S.E. & Davies, D.K. (1993) Fluvial channel reservoirs: prediction and location using floodbasin paleosols. *Abstract presented at the 78th AAPG Annual Convention*, New Orleans, 25–28 April.

Williams, G.P. (1984) Paleohydrologic equations for rivers. In: Costa, J.E. & Fleisher, P.J. (eds) *Development and application of geomorphology*. Springer-Verlag, Berlin, 343–367.

Williams, H. & Soek, H.F. (1993) Predicting reservoir sandbody orientation from dipmeter data: the use of sedimentary dip profiles from outcrop studies. In: Flint, S.S. & Bryant, I.D. (eds) *The geological modelling of hydrocarbon reservoirs and outcrop analogues*. International Association of Sedimentologists special publication **15**, Blackwell, Oxford, 143–156.

Willis, B.J. (1989) Palaeochannel reconstruction from point bar deposits: a 3-D perspective. *Sedimentology* **36**, 757–766.

Willis, B.J. (1993) Interpretation of bedding geometry within ancient river deposits. In: Marzo, M. & Puidefabregas, C. (eds) *Alluvial sedimentation*. International Association of Sedimentologists special publication **17**, Blackwell Scientific Publishers, Oxford, 101–114.

Wills, J.M. (1991) The Forties Field, block 21/10, 22/6a, UK North Sea. In: Abbotts, I.L. (ed.) *United Kingdom oil and gas fields, 25 years commemorative volume*. Geological Society Memoir **14**, 301–308.

Wilson, R.C.L. (1992) Sequence stratigraphy: an introduction. In: Brown, G.C., Hawkesworth, C.J. & Wilson, R.C.L. (eds) *Understanding the Earth.* 2nd edition. Cambridge University Press, Cambridge, 388–414.

Winter, D.A. & King, B. (1991) The West Sole Field, block 48/6, UK North Sea. In: Abbotts, I.L. (ed.) *United Kingdom oil and gas fields, 25 years commemorative volume.* Geological Society Memoir **14**, 517–523.

Wong, P.Z. & Lin, J.S. (1988) Studying fractal geometry on sub-micron length scales by small-angle scattering. *Mathematical Geology* **20**, 655–665.

Wright, V.P. (1989) Geomorphic and stratigraphic relationships of alluvial soils: a guide for interpreting ancient palaeosol-bearing sequences. In: Allen, J.R.L,. & Wright, V.P. (eds) *Palaeosols in siliciclastic sequences.* Postgraduate Research Institute for Sedimentology, University of Reading, UK, 26–48.

Wright, V.P. (1990) Estimating rates of calcrete formation and sediment accretion in ancient alluvial deposits. *Geological Magazine* **127**, 273–276.

Wright, V.P. & Alonso Zarza, A.M. (1990) Pedostratigraphic models for alluvial fan deposits: a tool for interpreting ancient sequences. *Journal of the Geological Society of London* **147**, 8–10.

Yarus, J.M. (1992) Computer mapping, biased or unbiased? That is the question. *Geobyte* October, 63–64.

Zeitlin, M. (1992) Visualization brings a new dimension to oil exploration and production. *Geobyte* June, 36–39.

Zeller, E.J. (1964) Cycles and psychology. *Kansas Geological Survey Bulletin* **169**, 631–636.

13 Perspectives for the Future

P.A. CARLING[1] AND M.R. DAWSON[2]
[1]*Department of Geography, University of Lancaster, UK*
[2]*PGS Tigress Ltd, Aberdeen, UK*

INTRODUCTION

Fluvial sedimentology as a discipline has had a distinct existence for perhaps only the last 30 years. As a discipline its basis has been the interpretation of sedimentary sections and features by reference to measurements and observations made on modern analogues. Advances in the science, particularly in the interpretation, modelling and prediction of features in the ancient record and the subsurface, thus depend on a better understanding of modern processes and forms.

Modelling the facies of sedimentary sequences by appealing to the first principles of hydraulics can be traced back to Henry Clifton Sorby amongst others (Allen, 1993). The approach is fundamental to reconstructing environments of deposition, as without this simplification and manipulation of data we would not be able to relate observed sections to genetic controls. In the 1950s the emphasis was firmly on the hydraulic interpretation that could be derived from grain size and the statistical parameters obtained from size distributions (Reading & Orton, 1991). Numerous researchers sought to identify stable assemblages of suite statistics pertaining to known modern environments and so apply this tool to identify ancient environments (e.g. Tanner, 1991). Throughout the 1960s geologists increasingly made use of the burgeoning data-base initially constructed by engineers, but later driven by experimental sedimentologists, relating bedform characteristics to hydraulic parameters usually under steady flow conditions (e.g. Guy *et al.*, 1966; Harms, 1969; Southard, 1971; Costello, 1974; Allen, 1984). By linking flume studies with the theoretical mechanics of uniform or gradually varying flow, water depth and the velocity of palaeoflows were predicted, although an emphasis on the control of grain size remained a key element in most analyses. In fluvial systems, for example, the fining-upward sequence of Allen (1963) epitomizes this grain-size focus especially contributing to stratigraphic models of meander evolution. Although these models are now being challenged, the fining-upward sequence still represents a bench-mark in the development of linkages between fluvial process and sedimentary sequence (Reading & Orton, 1991).

The fining-up sequence model presented by Allen (1963) also represents an early, but lasting, example of the development and use of qualitative analogue models in

Advances in Fluvial Dynamics and Stratigraphy. Edited by P.A. Carling and M.R. Dawson.
© 1996 John Wiley & Sons Ltd.

the interpretation of fluvial sequences. As described in greater detail by Brierley (Chapter 8) and North (Chapter 12) facies models have become a key method used to synthesize the inherent variability of fluvial deposits as a basis for comparative interpretation. The attractions of facies models are that they provide an easily assimilated reference point in understanding a fluvial sequence, which, although primarily qualitative tools, have been used as the basis for quantitative interpretation. Early models, such as those proposed by Miall (1978), concentrated on the prediction of vertical sequences, an approach perhaps best illustrated by the volume of Walker (1984). The objective of facies models has been to provide descriptive and predictive models of the fluvial environment based on a set of typical, distinctive characteristics. Usually, the major aim of these models has been to relate internal characteristics to a river planform. In part, the drive to develop facies models which associate characteristics to planform has been because it has been seen necessary in interpretative studies of subsurface fluvial sequences to identify planform as an aid to the determination of the geometry within a sequence.

The limitations of facies models as an aid to interpretation, either in determining planform or the geometry of fluvial deposits, highlighted by Bridge (1985, 1993a, 1995), has led to the development of systematic schemas which directly describe the three dimensional geometry of fluvial deposits (Miall, 1985, 1991a,b, 1995). Differing approaches are proposed within this volume by Brierley (Chapter 8) and Bristow (Chapter 10), whilst their application is discussed by Todd (Chapter 9) and North (Chapter 12). A major driving force behind the development of three-dimensional approaches has been the demands of the petroleum industry for better description of the geometry of fluvial sandbodies and permeability barriers, together with a need to improve the understanding of the internal heterogeneities of these sandbodies (Brayshaw *et al.*, Chapter 11). Although techniques for describing the three-dimensional geometry of sequences have now been well tested and practical classification schemes have been developed, the problem remains of how to apply the results of studies which have used these methods in a meaningful quantitative way. Here, as described by North (Chapter 12), the use of computer-based modelling of morphological development may represent a long-term solution.

A recent trend in fluvial sedimentology has been to attempt to predict geometry not only at the localized, section scale, where channel processes control the detailed morphology, but also to consider the allogenic controls on the development of a sequence. Here, the need has been to develop a framework which assists in the understanding of large-scale sequence boundaries and tracts that can be identified in the subsurface from seismic surveys as an aid to hydrocarbon trap prediction. The approach has developed from the study of marine sequences and essentially considers the effect of cycles of eustatic sea-level rise and fall on the succession of depositional sequences within a fluvial succession (Posamentier & Vail, 1988). Although, such models have been shown to be of value in marine sequences (Posamentier & Weimar, 1993), the applicability of simplified models to fluvial sequences has been questioned (Miall, 1991, 1992; Schumm, 1993; Westcott, 1993), because of the possibility of both a complex response in a fluvial system to a single allogenic trigger and the occurrence of coexistent intrinsic and external causes of change. Despite the possible limitations of the method, it has led to a re-emergence of a consideration of basin-scale processes and their influence on fluvial stratigraphy. Such an approach is not, however, new since geological studies as far back as those of Playfair (1802) have concentrated on the effect of external change on fluvial sedimentation and it is perhaps significant that it has been necessary for workers such as Westcott (1993) to highlight well understood geomorphic concepts and their application to subsurface analysis. Despite the limitations of present sequence stratigraphic models for fluvial environments, the approach provides a potentially powerful framework for understanding the controls on alluvial architecture (Todd, Chapter 9).

FLOW DYNAMICS

Williams (Chapter 1) has shown that further advances in the study of turbulence are needed to resolve the observation that scaling of bursts correlates with outer-layer hydraulic variables although bursting is known to be generated by uplift of streaks from the bed. It is the poor understanding of the mechanisms leading to the growth of turbulence structures which is a principal weakness in all models of turbulence. In the immediate term no new theoretical advances seem imminent: rather the focus is on developing instrumentation which better characterizes turbulence structure and so, through refined statistical analysis of time series, lead to intuitive insight into the nature of the problem (cf. Clifford et al., 1993). Consequently in many situations it is the inability to measure at fine temporal and spatial resolution which will retard collection of data needed to validate non-linear turbulence closure terms. Laboratory studies of turbulence in sediment-laden waters will provide consistent scaling for mathematical modelling but these tightly controlled experimental scenarios are not readily transferred to applications within natural rivers. However, increasing precision in measuring turbulence is leading to some interesting observations with direct relevance to how we view natural systems. McLean et al. (1994) for example have argued, from detailed observation of turbulent flow structure over duneforms, that the statistical distributions of turbulence components do not scale well with the bed shear stress because of the spatial evolution of the turbulence field; they conclude that it is probably inappropriate to use the shear stress to characterize sediment flux over dunes.

Hey and Rainbird (Chapter 2) note that the increasing ability to sample at a higher rate and finer spatial resolution in accord with flow model resolution will lead to improvements in numerical models. However, effective interfacing of flow and sediment transport in models remains a challenge.

Recent developments in the understanding of bedform dynamics are highlighted by Best (Chapter 3). Future research into bedform development will need to address the detail of turbulent flow structure above the bedforms, especially at the transitions from one recognized bedform to another, in order to better appreciate the genetic controls on bedforms. Further, consideration will need to be given to the concept and definition of equilibrium bedforms as Baas (1994), for example, has argued that straight and sinuous ripples are in fact non-equilibrium bedforms which inevitably develop to linguoid form given due time-lag. Finally, despite a burgeoning body of data concerning bedforms developed in gravel, it is not clear if many of these are equilibrium forms and indeed how and if these coarse structures relate to phase diagrams of recognized sandy bedforms.

Komar (Chapter 4) has shown that relatively sophisticated semi-empirical relationships are now developed defining the critical stress imposed by a flow to entrain given grain sizes for mixtures of sediment grains. In part these have some theoretical justification if not full exposition. However, geologists more frequently require an estimation of bulk velocity, discharge or water depth and considerable further work will be required to develop these latter relationships from grain-competence evaluations. This will require not only further field and experimental data representing entrainment from differing sedimentary environments but also an increased understanding of the coupling of near-bed laminar and turbulent flow structure with flow in the outer layer. In many environments the common assumption that some form of logarithmic function describes the flow throughout the depth is untenable. More complex coupled models including wake functions may be appropriate over beds of mixed roughness scales. It follows that there is an urgent need to quantitatively express the detailed nature of bed roughness and the effect of particle shape, sorting and attitude on both initial motion conditions and particle interaction within bedload transport.

That unsteadiness in bedload transport is characteristic across a wide size range of bed sediments including both sands and gravels is described by Kuhnle (Chapter 5). For low-frequency events this may be related to controls within the catchment. At one extreme, debris flow may feed sediment into channels periodically, but such controls are not restricted to headwater streams. In lowland alluvial streams, unsteadiness in the form of large-scale bed waves may range over scales including several decades and the annual cycle, perhaps mediated by a glacial legacy, for example, or the seasonal release of sediments to water courses (Meade, 1985; Roberts & Church, 1986). As hydrographs wax and wane remobilization or storage of slugs of sediments from within pools also produces unsteady downstream transport (Lisle & Hilton, 1992). The challenge at this low-frequency end of the spectrum is to determine the catchment controls and the mechanisms coupling sediment transfer from the slopes to the channel. Unsteadiness of transport at higher frequencies ranging between seconds and hours has also been widely reported from both field and laboratory experiments even though the flow appears steady and uniform (Gomez et al., 1989). The classic explanation for some scales of this phenomenon in sand-bed rivers is related to the periodic passage of ripples or dunes past the point of observation. However, some unsteadiness may be related to quasi-periodic fluctuations in turbulence structure such as the "bursting" phenomena discussed by Williams (Chapter 1), the break-up of such bedforms as clusters of pebbles, or systematic variation in bed roughness induced by downstream sorting processes. Our understanding of these later controls is limited and chiefly determined by narrow-flume observations conducted within prescribed experimental constraints. Just how these observations relate to transport in broad natural channels is a major challenge for the future. One obvious problem is sampling of bedload in nature. Variations in bed roughness and topography may induce bedload to move in discrete longitudinal ribbons which meander across the channel so that spot measurements at a given point might indicate unsteady transport as the ribbon repeatedly crosses the sampling point. This highlights the need to both improve instrumentation for measuring and observing bedload motion and to determine more precisely the statistical robustness of sampling in natural rivers (eg. Gaweesh & van Rijn, 1992).

Although quality field data will continue to be required for scaling and validation exercises, it is now increasingly possible to model initial motion, deposition and the associated bedload transport processes and sorting processes at the scale of individual interacting grains in a turbulent flow rather than considering mean quantities of sediment driven by bulk flow parameters. Inevitably such models will be largely scaled from precise laboratory data rather than field data, but manipulation within realistic scenarios will provide new insights into natural sediment transport and sorting processes which have important implications for bedform evolution and ultimately for the large-scale genetic modelling of suites of channel sediments. It would appear (Kelsey, Chapter 7) that such developments presently are largely restricted by the availability of computing power and an inadequate understanding of the particle interaction processes which give rise to spatial segregation of grains by size and density (Whiting, Chapter 6).

FIELD AND PROCESS STUDIES

In the field, parallel advances can be anticipated by improved characterization of bedform and channel morphological changes. For many years morphological modelling has been hampered by the difficulty and tediousness of surveying channel topography and sedimentary sequences accurately at sufficient spatial and temporal resolution, usually using plane survey and/or echo-sounding for topography and trenching and bore-hole logging for the sedimentary sequences. Advances in photogrammetry (Lane *et al.*, 1993), remote sensing (including GPS), and developments in the presentation and manipulation of spatial data (GIS especially) raise the expectation that morphological adjustment will soon be computer-mapped more precisely and readily at a variety of scales (see also Chaper 2). Additionally, the development of ground-penetrating radar (Huggenberger, 1993) or seismic methods (Steeples & Miller, 1990) to log sections in shallow fluvial sediments poses the prospect of more rapid and accurate controls on forward modelling of channel geometry (e.g. Howard, 1992). In the latter procedure, a variety of scenarios are generated and presently controlled by a rule-base rather than strictly deterministic constraints. In the specific case of the evolution of stratigraphic complexes owing to bedform migration, forward modelling has provoked new hypotheses concerning process control (e.g. Rubin, 1987). The evolving application of stochastic and boolean techniques to predict morphology and geometry is highlighted by North (Chapter 12) who argues that ultimately the greatest advances are likely to made through a combination of both process-based and stochastic techniques in a hybrid fashion (Tetslaff, 1991; MacDonald & Halland, 1993; Tyler *et al.*, 1994).

STRATIGRAPHIC STUDIES AND SUBSURFACE MODELLING

A major driving force behind the recent development of fluvial sedimentology as a discipline has been the demand of extractive industries which depend on accurate geological descriptions to enable them to both locate and produce resources. As pointed out by North (Chapter 12), similar demands exist both within the coal and petroleum industries, but it is the latter which has encouraged the growth of research to the greatest extent. Here the demand has been for better predictive models of the location, distribution and properties of fluvial deposits which form a significant source of hydrocarbons (Miall, 1991). The requirements of the industry have been two-fold. Firstly, there has been a need for large-scale models of sedimentary sequences and the location of geometry of fluvial deposits within them to predict the likely occurrence of hydrocarbon trap sequences given relatively sparse and often remotely sensed data. Heavy use has been made of conceptual facies models, often developed from observations of modern fluvial environments. Secondly, the exigencies of hydrocarbon production have encouraged detailed studies of the geometry of fluvial sandbodies, which potentially form reservoirs, and their internal heterogeneities, a major control on production potential. This has led to a greater concentration on the description of the three-dimensional form of fluvial sandbodies and to the development of quantitative studies of the variation of internal properties (Brayshaw *et al.*, Chapter 11). A common issue, highlighted by reviews both in this volume (Brierley, Chapter 8; Bristow, Chapter 10; North, Chapter 12) and elsewhere (Bridge, 1993b), is the present absence of detailed quantitative information about modern fluvial environments that can be applied effectively as analogues in subsurface studies.

Early approaches in interpreting fluvial environments attempted to identify facies and facies associations diagnostic of a given channel planform (e.g. Cant & Walker, 1976; Walker, 1984; Miall, 1985). The limitations of such a methodology are discussed by Brierley (Chapter 8). He argues that it is not possible to unequivocally determine planform from sedimentary characteristics and that the development of distinct, planform-based facies models artificially contrains the interpretative process. In particular he highlights the difficulties of applying such facies models beyond the scope of the environment for which they were derived and the impossibility of associating specific geomorphic processes to a given channel planform. In order to overcome these difficulties Brierley proposes the adoption of a technique which has as its basis the identification of geomorphic elements. These are intended to be scale-independent geomorphic units such as basal channel lags or unit bars which can be differentiated according to their geometry, texture and depositional process. He argues that process–form relationships are best determined at this element scale. These elements can be grouped into three components representing within-channel deposits, floodplain elements and channel-margin deposits and it is the stacking arrangement of the components and their respective elements within an alluvial suite that is diagnostic of alluvial style.

The value of this approach is in the refinement of the process–form interpretative process by introducing an intermediate diagnostic step when attempting to relate small-scale sedimentary to a supposed alluvial style. This should result in more rigorous descriptions of modern fluvial environments and better interpretations of subsurface sequences. In proposing the use of geomorphic elements, Brierley recognizes a number of current limitations and argues for further research in generating more detailed data on element-scale assemblages for modern fluvial depositional systems and in determining the controls on the preservation potential of element-scale depositional units. He, like other workers (Bridge, 1993a), also identifies the need for additional work in relating three-dimensional element inventories in the modern record to their likely representation in one and two dimensions in the subsurface if the approach is to be of use.

Bristow (Chapter 10), from a perspective of reconstructing channel morphology from sedimentary sequences, draws a very similar conclusion and argues that more quantitative studies of the area, dimensions, volumes, and geometries of recent fluvial sediments, are required as a foundation for reconstructions of ancient fluvial sequences. However, in reviewing techniques for reconstructing geometry he identifies that a three-dimensional approach, applied to the subsurface and outcrops, is an ideal goal which may never be attained and highlights the advantages, and difficulties of applying existing one- (vertical profile) and two-dimensional (cross-profile) methodologies. Like Brierley, he recognizes the value of classification schemes that provide a descriptive terminology covering both internal facies structure and bed geometry and highlights the importance of the analysis of bounding surfaces. However, in contrast to other workers (Allen, 1983; Miall, 1985, 1988, 1991) he suggests that the emphasis should be upon the geometric relationships between bounding surfaces, and the type of such surfaces, rather than on attempting to analyse hierarchies of bounding surfaces using a numerical classification approach supported by Bridge (1993a, 1995).

Todd (Chapter 9) evaluates recent advances in the understanding of relationships between the structures observed in both conglomerate and sandstone fluvial sequences and formative process, emphasizing not only the requirement to consider internal lithofacies structures but also external geometries. He argues that the developing techniques for analysing two and three dimensions has represented one of the major recent advances in fluvial sedimentology. Todd also considers the importance and implications of mud deposits and palaeosols to the interpretation of fluvial systems. Both he and North (Chapter 12) highlight the diagnostic power of the analysis of palaeosols in the understanding of channel-belt geometries and the chronostratigraphic development of floodplain sequences. To fully understand the evolution of alluvial sequences Todd believes that a consideration of the spatial and temporal variability is required. The implications of both longitudinal and lateral facies variability are considered and Todd argues that the techniques of sequence stratigraphy provide a valuable framework for understanding the controls on sequence development at this scale.

The close relationship between fluvial sedimentology and the exploration and production demands of the petroleum industry is illustrated by the subjects covered in the chapters by Brayshaw *et al.* (Chapter 11) and North (Chapter 12). Brayshaw *et al.* consider the primary controls on the variability of permeability and porosity in fluvial sandstones. Successful prediction and modelling of such heterogeneity away from a limited number of points of data control represents a major objective of the petroleum geologist in attempting to refine and improve reservoir models. They describe the effects of texture and fabric on permeability and porosity and consider trends in these properties within fluvial sandstones. Recent advances in describing the variability of porosity and permeability have developed from the use of portable measuring apparatus such as probe permeameters, together with laboratory flooding experiments in slabbed sandstones. These approaches have enabled a better understanding of the effects of small-scale heterogeneity. The data collected by such methods, together with a better understanding of fluvial processes, when applied within a suitable reservoir modelling framework will lead to improved understanding and prediction of the performance of fluvial petroleum reservoirs.

The current "state of the art" of techniques of fluvial stratigraphic analysis and modelling in this economic context is described in a comprehensive review by North (Chapter 12). He considers the requirements of subsurface analysis and the difficulties presented by fluvial deposits in this context, before examining the philosophy behind modelling and the approaches that can be taken. North argues that models are idealized representations of reality established to assist the understanding of complex natural phenomena and processes designed to have descriptive and interpretative value. In addition, of particular value in the exploitation of economic resources, models also have predictive power in that they should enable deductions to be made about the properties at locations for which measurements have not been made.

North argues that subsurface prediction typically incorporates a combination of modelling philosophies in hybrid methodologies such as a combination of forward models of process based on parameters obtained from preliminary, inverse, interpretation of data. He identifies a number of alternative approaches to modelling in current use. These include conceptual modelling techniques like facies modelling and sequence stratigraphic analysis; direct numerical modelling from observations including methods such as palaeohydrology, surface modelling, geostatistics and fractals; simulation studies using Markov methods and deterministic process studies; and indirect numerical methods such as probabilistic simulation. In considering future advances he notes the limitations of the techniques and proposes that advances in modelling will arise not only from the application of hybrid philosophies but also from the application of hybrid modelling techniques such as the combination of both stochastic and process-based methods. He concludes, however, by stating that the validity of subsurface models depends directly on the quality of the data and the assumptions made, and identifies that current models are fundamentally limited because of our lack of knowledge about present-day river systems.

CONCLUSION

This volume has highlighted the very great recent advances that have occurred in the understanding of modern fluvial processes, in the techniques of interpreting ancient sequences, and in the modelling and representation of subsurface geometries and heterogeneities. Notwithstanding these advances much remains to be achieved and it is significant that, amongst the authors considering the morphology and architecture and geometry of fluvial deposits, a common conclusion is that there needs to be a greater understanding about present-day river systems. Much work thus remains to be done, particularly in developing predictive models of the morphology, geometry and preservation of depositional features. In addition there needs to be further refinement in the methods of describing, analysing and modelling subsurface geology. As a consequence it is apparent that fluvial sedimentology is not only a dynamic field of current research, but is likely to remain so for some considerable time.

REFERENCES

Allen, J.R.L. (1963) Henry Clifton Sorby and the sedimentary structures of sands and sandstones in relation to flow conditions. *Geologie Mijnb.*, **42**, 223–228.

Allen, J.R.L. (1983) Studies in fluviatile sedimentation: bars, bar-complexes and sandstone sheets (low sinuosity braided streams) in the Brownstones (L. Devonian), Welsh Borders. *Sedimentary Geology*, **33**, 237–293.

Allen, J.R.L. (1984) *Sedimentary Structures: their Character and Physical Basis.* Elsevier, Amsterdam.

Allen, J.R.L. (1993) Sedimentary structures: Sorby and the last decade. *Journal of the Geological Society*, **150**, 417–425.

Baas, J.H. (1994) A flume study on the development and equilibrium morphology of current ripples in very fine sand. *Sedimentology*, **41**, 185–209.

Bridge, J.S. (1985) Paleochannel patterns inferred form alluvial deposits: a critical evaluation. *Journal of Sedimentary Petrology*, **55**, 579–589.

Bridge, J.S. (1993a) Description and interpretation of fluvial deposits: a critical perspective. *Sedimentology*, **40**, 801–810.

Bridge, J.S. (1993b) The interaction between channel geometry, water flow, sediment transport and deposition in braided rivers. In: Best, J.L. & Bristow, C.S. (eds) *Braided Rivers. Geological Society, London, Special Publication*, **75**, 13–71.

Bridge, J.S. (1995) Description and interpretation of fluvial deposits: A reply. *Sedimentology*, **42** 384–389.

Cant, D.J. & Walker, R.G. (1976) Development of a braided-fluvial facies model for the Devonian Battery Point Sandstone, Quebec, Canada. *Canadian Journal of Earth Science*, **13**, 102–119.

Clifford, N.J., French, J.R. & Hardisty, J. (1993) *Turbulence: Perspectives on Flow and Sediment Transport.* Wiley, Chichester, 360 pp.

Costello, W.R. (1974) *Development of Bed Configurations in Coarse Sands.* Massachusetts Institute of Technology, Dept. of Earth and Planetary Sciences. Report **74–1**, 120 p.

Gaweesh, M.T.K. & van Rijn, L.C. (1992) Laboratory and field investigation of a new bedload sampler for rivers. In: *Hydraulic and Environmental Modelling: Estuarine and River Waters. Proceedings 2nd International Conference on Hydraulic and Environmental Modelling of Coastal and River Waters. Vol.* **2**, 479–488.

Gomez, B., Naff, R.L. & Hubbell, D.W. (1989) Temporal variations in bedload transport rates associated with the migration of bedforms. *Earth Surface Processes and Landforms*, **14**, 135–156.

Guy, H.P., Simons, D.B. & Richardson, E.V. (1966) *Summary of Alluvial Channel Data from Flume Experiments, 1956–61.* US Geological Survey Professional Paper **462-I**.

Harms, J.C. (1969) Hydraulic significance of some sand ripples. *Geological Society of America Bulletin*, **80**, 363–396.

Howard, A.D. (1992) Modelling channel migration and floodplain sedimentation in meandering streams. In: Carling, P.A. & and Petts, G.E. (eds) *Lowland Floodplain Rivers: Geomorphological Perspectives*, Wiley, Chichester, 1–41.

Huggenberger, P. (1993) Radar facies: recognition of facies patterns and heterogeneities within Pleistocene Rhine gravels, NE Switzerland. In: Best, J.L. & Bristow, C.S. (eds) *Braided Rivers. Geological Society, London, Special Publication* **75**, 163–176.

Lane, S.N., Richards, K.S. & Chandler, J.H. (1993) Developments in photogrammetry; the geomorphological potential. *Progress in Physical Geography*, **17**(3), 306–328.

Lisle, T.E. & Hilton, S. (1992) The volume of fine sediment in pools: an index of sediment supply in gravel-bed streams. *Water Resources Bulletin*, **28**, 371–383.

Meade, R.M. (1985) Wave-like movement of bedload sediment, East Fork River, Wyoming. *Environmental Geology and Water Science*, **7**, 215–225.

MacDonald, A.C. & Halland, E.K. (1993) Sedimentology and shale modelling of a sandstone rich fluvial reservoir: Upper Statfjord Formation, Statfjord Field, northern North Sea. *AAPG Bulletin*, **77**, 1016–1040.

McLean, S.R., Nelson, J.M. & Wolfe, S.R. (1994) Turbulence structure over two-dimensional bedforms: implications for sediment transport. *Journal of Geophysical Research*, **99**(66), 12 729–12 747.

Miall, A.D. (1978) Lithofacies types and vertical profile models in braided river deposits: a summary. In: Miall A.D. (ed.) *Fluvial Sedimentology*. Canadian Society of Petroleum Geologists, Memoir **5**, 597–604.

Miall, A.D. (1985) Architectural-element analysis: a new method of facies analysis applied to fluvial deposits. *Earth Science Reviews*, **22**, 261–308.

Miall, A.D. (1988) Architectural elements and bounding surfaces in fluvial deposits: anatomy of the Kayenta Formation (Lower Jurassic), southwest Colorado. *Sedimentary Geology*, **55**, 233–262.

Miall, A.D. (1991a) Hierarchies of architectural units in terrigenous clastic rocks and their relationship to sedimentation rate. In: Miall, A.D. & Tyler, N. (eds) *The Three-dimensional Facies Architecture of Terrigenous Clastic Sediments and its Implications for Hydrocarbon Discovery and Recovery*. SEPM Concepts in Sedimentology and Palaeontology, **3**, 6–12.

Miall, A.D. (1991b) Stratigraphic sequences and their chronostratigraphic correlation. *Journal of Sedimentary Petrology*, **61**, 497–505.

Miall, A.D. (1992) Alluvial deposits. In: Walker, R.G. & James N.P. (eds) *Facies Models: Response to Sea Level Change*. Geological Association of Canada, St John's, Newfoundland.

Miall, A.D. (1995) Description and interpretation of fluvial deposits: a critical perspective: Discussion. *Sedimentology*, **41**, 379–383.

Playfair, J. (1802) *Illustration of the Huttonian Theory of the Earth*. Edinburgh.

Posamentier, H.W. & Vail, P.R. (1988) Eustatic controls on clastic deposition II – sequence and systems tract models. In: Wilgus, C.J., Hastings, B.S., Kendall, C.G.S.C., Posamentier. H.W., Ross, C.A. & Van Wagoner, J.C. (eds) *Sea-level Changes: an Integrated Approach*. SEPM Special Publication, **42**, 125–154.

Posamentier, H.W. & Weimar, P. (1993) Siliclastic sequence stratigraphy and petroleum geology – where to from here? *AAPG Bulletin*, **7**, 731–742.

Reading, H.G. & Orton, G.J. (1991) Sediment calibre: a control on facies models with special reference to deep-sea depositional systems. In: Müller, D.W., McKenzie, J.A. & Weissert, H. (eds) *Controversies in Modern Geology*. Academic Press, London, Ch. 6.

Roberts, R.G. & Church, M. (1986) The sediment budget of severely disturbed watersheds, Queen Charlotte Ranges, British Columbia. *Canadian Journal of Forest Research*, **1**, 1092–1096.

Rubin, D.M. (1987) Formation of scalloped cross-bedding without unsteady flows. *Journal of Sedimentary Petrology*, **57**, 39–45.

Schumm, S.A. (1993) River response to baselevel change: implications, for sequence stratigraphy. *Journal of Geology*, **72**, 195–213.

Southard, J.B. (1971) Representation of bed configurations in depth-velocity-size diagrams. *Journal of Sedimentary Petrology*, **41**, 903–915.

Steeples, D.W. & Miller, R.D. (1990) Seismic reflection methods applied to engineering, environment and groundwater problems. In: Ward, S.H. (ed.) *Geotechnical and Environmental Geophysics*. Investigations in Geophysics, **5**, Vol. 1, SEG, Tulsa, 1–30.

Tanner, W.F. (1991) Suite statistics: The hydrodynamic evolution of the sediment pool. In:

Syvitski, J.P.M. (ed.) *Principles, Methods and Applications of Particle Size Analysis.* University of Cambridge Press, Cambridge, 225–236.

Tetslaff, D.M. (1991) The combined use of sedimentary process modeling and statistical simulation in reservoir characterisation. SPE Technical Paper **22759**, *66th SPE Annual Technical Conference*, Dallas, Texas.

Tyler, K.J., Svanes, T. & Henriquez, A. (1994) Heterogeneity modeling used for production simulation of a fluvial reservoir. *SPE Formation Evaluation*, **9**, 85–92.

Walker, R.G. (1984) *Facies Models.* 2nd edn, Geoscience Canada Reprint Series **1**, Geological Association of Canada, St John's, Newfoundland, 1–10.

Westcott, W.A. (1993) Geomorphic thresholds and complex response of fluvial systems – some implications for sequence stratigraphy. *AAPG Bulletin*, **77**, 1208–1218.

Index